Rescue Breathing

Adults and Children
(5 years and older)

1. While someone else calls or goes for help, such as dialing 9-1-1 or phoning for an ambulance, lay victim face up.

2. Remove any obstructions from mouth or airway.

3. Tilt the head back slightly to open airway **(see sketch (a))**.

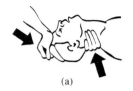

(a)

4. Pinch the victim's nostrils shut. Placing your mouth over the victim's open mouth, breathe slowly into the victim, two full breaths, watching for the chest to rise **(see sketch (b))**. Remove your mouth between breaths, allowing the chest to fall.

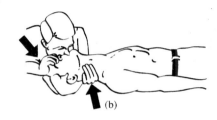

(b)

5. Continue breathing into the victim's mouth, one breath every five seconds, until the victim begins to breathe without help or until medical help arrives.

Infants and Small Children
(through age 4)

1. Place child on back; lift neck and tilt head back—do not force **(see sketch (c))**.

2. Tightly seal child's mouth and nose with your mouth and breathe slowly into the victim, using only 2 small puffs of air for infants.

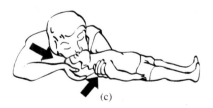

(c)

3. If child's chest does not expand, invert child and strike between shoulders **(see sketch (d))**; resume breathing.

4. Continue breathing into child's mouth, every four seconds for a child and every three seconds for an infant, until child begins to breathe without help or until medical help arrives.

(d)

Operating, Testing, and Preventive Maintenance of Electrical Power Apparatus

Charles I. Hubert, P. E.

United States Merchant Marine Academy

Prentice Hall

Upper Saddle River, New Jersey
Columbus, Ohio

Editor in Chief: Stephen Helba
Assistant Vice President and Publisher: Charles E. Stewart, Jr.
Production Editor: Alexandrina Benedicto Wolf
Production Supervision: Clarinda Publication Services
Design Coordinator: Diane Ernsberger
Cover Designer: Jeff Vanik
Cover art: Fluke Instruments
Production Manager: Matthew Ottenweller
Marketing Manager: Ben Leonard

Pearson Education Ltd.
Pearson Education Australia Pty. Limited
Pearson Education Singapore Pte. Ltd.
Pearson Education North Asia Ltd.
Pearson Education Canada, Ltd.
Pearson Educación de Mexico, S.A. de C.V.
Pearson Education—Japan
Pearson Education Malaysia Pte. Ltd.
Pearson Education, *Upper Saddle River, New Jersey*

ISBN: 0-13-041774-2

Preface

This book is an outgrowth of the author's previously published *Preventive Maintenance of Electrical Equipment,* long acclaimed as the "electrical bible" for operation, maintenance, testing, troubleshooting, and emergency repair of electrical power apparatus, and a recommended reference by the U.S. Coast Guard for those preparing for the marine engineering license exam.

Many of the changes that have taken place in the electrical industry are addressed in this book, including the development of solid-state test equipment; the use of thermography to locate overheating connections, overheating bearings, etc.; new emphasis on safety brought about by OSHA, including grounding and bonding; and the proper handling and disposal of PCBs in transformers and capacitors. New insulation systems are in use, and a more comprehensive understanding of the factors involved in insulation aging has been developed. New equipment such as ground-fault protective devices and circuit breakers with solid-state tripping circuits require new maintenance and testing techniques. Aluminum conductors are now used extensively and they must meet special installation and maintenance requirements to prevent overheating and electrical fires. Consideration must be given to the damaging effects of harmonic currents and harmonic voltages generated by the expanded use of electronic controllers and computers. Harmonics are often the underlying reason for many electrical failures.

This book is based on the author's teaching experience and continuing review of *Transactions* and other communications from the IEEE Electrical Insulation Society, the IEEE Power Engineering Society, the IEEE Industry and Applications Society, the Underwriters Laboratory, and his own consulting practice. All topics in the author's previous versions of this book have been retained, and updated where required, to reflect new techniques, improved testing equipment, and the latest IEEE, NEMA, USAS, NEC, and OSHA standards. A chapter on the cost–benefit relationship of preventive maintenance and a chapter on scheduling, conducting, and evaluating preventive maintenance programs provide trainees with an awareness of the economic aspects of the maintenance problem.

The many additional topics included in this book also make it ideal for a variety of industrial training programs, for motor and industrial control inspectors, and as a general reference for operating engineers and maintenance electricians. This text also serves the needs of community colleges and special programs for the Navy, Coast Guard and Maritime Academies, and union schools. Vocational schools and technical institutes, with programs in electric power technology, special-purpose programs in 4-year engineering schools, continuing education courses, industry-sponsored courses for electric motor mechanics and power station engineers and technical representatives for insurance underwriters will find this text a welcome addition to their instructional programs. It is also effective as a survey course in circuits and machines for students in nonelectrical programs.

The minimum prerequisite for the effective use of this text is a working knowledge of elementary algebra and trigonometry. Complex equations needed to understand transformer and machine behavior are presented without derivation. Pertinent equations are stated and then applied to real-world situations. References are provided for those who desire further study. This low-level approach enhances self-study and eliminates the need for separate texts on electrical circuits and machines.

This book sets forth up-to-date methods for preventive maintenance, provides logical methods by which the more common problems can be identified and localized, recommends emergency repairs that will keep the equipment in operation until it can be scheduled out of service, suggests correct operating procedures, and outlines inspection programs to ensure safe, efficient, economical, and dependable operation. Case studies of electrical and electromechanical failures caused by improper operating procedures or inadequate maintenance are introduced at appropriate points in the text. Real-life cases, from insurance company files, dramatize some of the disastrous consequences resulting from poor maintenance, inadequate plant protection, and incorrect operating procedures. Included are examples that illustrate and explain how switchgear may be wrecked by magnetic forces; how migrating arcs are created, causing complete burndown of distribution panels; and so on.

Supporting principles pertinent to the operation of specific equipment are woven into the text at appropriate places. The book is extensively cross-referenced to provide the reader with quick access to related or dependent material in other chapters, and supporting references are provided at the end of each chapter.

This book provides considerable flexibility in course planning. The selection and arrangement of chapters may be easily changed to suit a desired emphasis on either AC or DC equipment. Questions and problems are included at the end of each chapter, and answers to all problems are provided at the back of the book. Problem numbers are keyed to chapter sections by a triple-number system. For example, Problem 8–13/5 indicates that Chapter 8, Problem 13, requires knowledge of the material in Section 8.5 to solve. This makes it easier for faculty to assign homework problems, and easier for a student to pick additional problems for review.

An instructor's manual includes complete solutions to all homework problems and recommended course outlines that will assist faculty in preparing specialized courses for electrical machinery inspectors, power plant operators, and maintenance technicians, and a very practical survey course in the principles of circuits and machines for students in nonelectrical power-oriented programs.

Overview of the Text

The first nine chapters in this 32-chapter book provide the essential background in AC circuits, DC circuits, and magnetics necessary for an understanding of the behavior and testing of *electrical power apparatus,* and thus may obviate the need for a separate circuits text.

Students who have satisfactorily completed a course in electric circuits covering direct current, single-phase, and three-phase circuits may start with Chapter 10, which deals with transformers. However, a review of the first nine chapters will be beneficial because they emphasize high-current applications as well as dangerous situations that can occur in inductive, capacitive, and magnetic circuits. Basic circuit faults such as short circuits, open circuits, flashovers, ground faults, and migrating arcs that can occur in power distribution systems are also defined and illustrated.

Chapter 1 introduces electric charge, current, driving voltage, and insulation; differentiates between conductors for energy transmission and conductors for current limiting and conversion to heat energy; calculations of conductor temperature from its resistance; the application of ampacity tables for the selection of conductor size; and the use of ohmmeters for the direct measurement of resistance. Chapter 2 introduces series and parallel resistive circuits, Ohm's law, Kirchhoff's laws, power and energy loss in conductors, and circuit faults (short circuits, open circuits, ground faults, and flashovers).

Chapter 3 introduces the magnetic field and magnetic forces in electric power applications. Included are mechanical forces between parallel conductors, between adjacent turns in a coil, and between a conductor and a magnet. Motor torque, arcing faults, and the magnetic forces that cause are migration are also included. Chapter 4 introduces inductance, Lenz's law, Faraday's law, time-constant, energy storage, and the burning and arcing that can occur when a highly inductive electric power circuit is opened. The discussion of capacitors, introduced in Chapter 5, includes energy storage, voltage buildup, time-constant, capacitor testing, and the correct procedure for discharging power capacitors.

Chapter 6 is devoted to single-phase AC power applications. Current, voltage, impedance, phase angles, and phasor diagrams are introduced. All calculations use simple algebra or trigonometry; complex algebra is not necessary for these applications and is not used in examples, nor is it required for homework problems. Resonance and harmonics, and their adverse effects on electric power systems, are introduced in Chapter 7. Power and its measurement in the single-phase system are introduced in Chapter 8, which includes discussions of active power, reactive power, apparent power, power factor, motor efficiency, and power factor improvement. Chapter 9 introduces the three-phase system for balanced wye and delta loads. Included are three-phase power, power factor correction, power measurement, phase sequence, and cautions in the use of capacitors.

Chapter 10 presents the principles of operation and maintenance of transformers in power applications, and includes single- and three-phase connections, autotransformers, instrument transformers, harmonic heating, and the transformer k factor.

Chapter 11 discusses three-phase induction motors, both squirrel-cage and wound-rotor types. It includes the principles of operation, factors affecting motor speed, behavior during acceleration and loading, characteristics of NEMA design motors, multispeed motors, in-rush current, and how to operate three-phase motors from single-phase lines. Synchronous motors, presented in Chapter 12, tie in nicely with the rotating field theory discussed in Chapter 11. Their starting, stopping, and reversing problems are covered.

Chapter 13 discusses the problems that may be experienced by three-phase motors due to improper operating procedures or poor power quality. These include operating above or below the rated nameplate voltage, the effect of unbalanced line voltages and unbalanced motor line currents, the effect of the number of starts on motor life, the perils of reclosing out of phase, operating 60-Hz motors from a 50-Hz system, and the damaging effects of an open phase.

Chapter 14 presents the principles, operation, and paralleling of synchronous generators, transfer of power between machines, and division of oncoming load. Also included are operational problems such as motorization, loss of excitation, loss of synchronism, and operating outside the nameplate rating. Chapter 15 presents troubleshooting and emergency repair of AC machines. The principles of single-phase induction motors, speed control methods, and common faults are covered in Chapter 16.

Chapters 17, 18, 19, and 20 are devoted to DC machines: principles, operation, maintenance, troubleshooting, and emergency repairs.

Chapter 21 provides a discussion on the disassembly and reassembly of electric machines, maintenance of bearings, mechanical vibration, shaft currents, and bearing insulation.

Operation, maintenance, and troubleshooting of motor controls are covered in Chapter 22. It discusses reading and interpreting control diagrams for AC and DC motors, basic starting circuits, and reversing and hoist controllers.

The most important factor in the life of electrical apparatus, whether motors, generators, transformers, cable, switchgear, etc., is its insulation. Chapters 23, 24, 25, and 26 are devoted to this critical area of insulation maintenance. These chapters cover the classification, characteristics, aging, and failure mechanisms of electrical insulation; the measurement and interpretation of insulation resistance; high-potential maintenance testing; and the cleaning and drying of electrical insulation in motors, generators, and transformers.

Chapter 27 presents the operation and maintenance of lead-acid and nickel-cadmium cells, both of which make up the preponderance of battery systems used in industrial and utility operations. Maintenance, testing, charging, safety considerations, and the requirements for an uninterruptible power supply are discussed.

Chapter 28 covers bonding, grounding, earthing, and ground-fault protection of distribution systems, causes of overvoltage, and the detection and location of ground faults. Overcurrent protection by means of fuses and circuit breakers and the coordination of overcurrent protective devices are presented in Chapter 29. The maintenance of switchgear and other miscellaneous electrical apparatus is discussed in Chapter 30. Included are the effects of overcurrent, arcing faults, and bolted faults on switchgear; maintenance and testing of power capacitors; and infrared and visual inspection of buses, circuit breakers, switches, and cable connections.

Chapter 31 presents the economics of preventive maintenance, including economic factors, failure rate, specifying wear and operating limits, assigning service reliability levels, spare parts, safety, and the need for proper staffing and training. Chapter 32 provides a guide for scheduling, conducting, and evaluating a preventive maintenance program for electrical power apparatus. Included topics are scheduling strategy, the preventive maintenance routine, record taking and interpretation, procedures to follow when troubleshooting intermittent faults, and investigating failures.

ACKNOWLEDGMENTS

I acknowledge with gratitude the many contributions made by my colleagues, friends, and students in the preparation of this book.

A special thanks to my son Thomas Hubert, senior marine engineer with AMSEC LLC, for his very detailed review of the manuscript and many helpful suggestions from the standpoint of a "man in the field." During his 20 years in the marine industry he has engaged in a wide variety of design, assessment and maintenance tasks on numerous classes of commercial and navy ships. Thomas is currently performing reliability centered maintenance studies for the U.S. Navy.

Thanks also to Dr. George J. Billy (chief librarian), Mr. Donald Gill, Ms. Marilyn Stern, Ms. Laura Cody, and Ms. Barbara Adesso, of the United States Merchant Marine Academy Library, for their kind assistance in obtaining needed references.

I would also like to thank the following reviewers for their invaluable feedback: Robert Borns, Purdue University; Ilya Greenberg, Buffalo State College; William Hessmiller; Costas Vassiliadis, Ohio University; and Richard Windley, ECPI College of Technology.

Many thanks to Holly Henjum and The Clarinda Company for their outstanding dedication and performance in preparing this book for publication.

An affectionate thanks to my wonderful wife Josephine for her encouragement, faith, counsel, patience, and companionship during the many years of preparing this and other manuscripts. Her early reviews of the manuscript, while in its formative stages, assisted in clarity of expression and avoidance of ambiguity. Her apparently endless years of pounding typewriters and word processors, for this and other texts, were truly a work of love.

Charles I. Hubert

Contents

10 TRANSFORMERS IN POWER APPLICATIONS: PRINCIPLES, OPERATION, AND MAINTENANCE   181

14 SYNCHRONOUS GENERATORS: PRINCIPLES AND OPERATIONAL PROBLEMS   273

15   TROUBLESHOOTING AND EMERGENCY REPAIR OF THREE-PHASE MOTORS   307

16   SINGLE-PHASE INDUCTION MOTORS   325

17 DIRECT-CURRENT GENERATORS: PRINCIPLES AND OPERATIONAL PROBLEMS   337

18 DIRECT-CURRENT MOTORS: PRINCIPLES AND OPERATIONAL PROBLEMS   371

19 COMMUTATOR, SLIP-RING, AND BRUSH MAINTENANCE   395

20   TROUBLESHOOTING AND EMERGENCY REPAIR OF DC MACHINES   427

24 INSULATION RESISTANCE: ITS MEASUREMENT AND INTERPRETATION   503

27 OPERATION AND MAINTENANCE OF BATTERY SYSTEMS FOR INDUSTRIAL, MARINE, AND UTILITY OPERATIONS   579

28 BONDING, GROUNDING, EARTHING, AND GROUND-FAULT PROTECTION OF DISTRIBUTION SYSTEMS   603

29   PROTECTION AGAINST SUSTAINED OVERLOADS AND SHORT CIRCUITS   623

30   MAINTENANCE OF SWITCHGEAR AND OTHER MISCELLANEOUS ELECTRICAL APPARATUS   647

31 COST–BENEFIT RELATIONSHIP OF PREVENTIVE MAINTENANCE   665

32 SCHEDULING, CONDUCTING, AND EVALUATING A PREVENTIVE MAINTENANCE PROGRAM   677

Reference Tables
in Chapters

1

Current, Voltage, Resistance, Insulation, and Conductors in Electric Power Systems

1.0 INTRODUCTION

A prerequisite to the study of electrical maintenance is an understanding of the effects produced by the movement of extremely small negatively charged particles called *electrons*. By controlling the number and movement of these electrons, and directing them along prescribed paths, the designer of electrical equipment has provided mankind with electric lights, motors, generators, heaters, and the great variety of things that depend on an abundance of electric power. It is the function of the electrical maintenance technician to ensure that these electrons follow the prescribed path, with no shortcuts or unplanned outages.

1.1 ELECTRIC CHARGE, CURRENT, CONDUCTORS, AND DRIVING VOLTAGE

An electric current is defined as the rate of movement of electric charges. The current may be the result of movement of positive charges, movement of negative charges, or a combination of positive and negative charges moving in opposite directions. In metals, the current is a movement of small negatively charged particles called *electrons*. In gases the current is a movement of negatively charged electrons in one direction and a drift of positively charged ions in the opposite direction.[1] In salt solutions the current is a movement of positive ions and negative ions in opposite directions. In solid-state electronic components (diodes, transistors, etc.), the current is a movement of electrons in one direction and a movement of positive charges in the opposite direction.

---

[1]An ion is an atom or group of atoms that has acquired an electric charge by gaining or losing one or more electrons. Atoms that gain electrons are negatively charged, and atoms that lose electrons are positively charged.

1

Conductors

Any material that allows the essentially free passage of current, when connected to a battery or other source of electric energy, is called a *conductor*. In metallic conductors, illustrated in Figure 1–1a, "free electrons" wander aimlessly about the crystal structure of the material, resulting in an average current equal to zero.

However, if some means is provided to drive these "free electrons" in one direction, as shown in Figure 1–1b, the average current will be nonzero. The driving voltage in this case is supplied by a battery. Connecting a driver (battery or other electric generator) to the two ends of the wire causes a good number of free electrons to be driven in the direction of the driving voltage. The direction of this electron movement is from the negative (−) terminal of the driver, through the conductor, to the positive (+) terminal of the driver, and constitutes an electric current. The closed loop formed by the battery and the conductor is called an *electric circuit*.

The driver does not supply the electrons. The electrons are always present in the conductor as free electrons, and are caused to move by the application of a driving voltage. The negative terminal of the driving source repels the free electrons, and the positive terminal attracts them. As long as the circuit is closed, the electrons continue to circle around and around the loop. This unidirectional movement of electrons is called *direct current*. The total number of free electrons in the conductor remains the same; each time an electron enters the positive terminal of the driver, another electron leaves the negative terminal for its journey around the loop.

Although, as shown in Figure 1–1b, the actual direction of current is from the negative terminal of the driver to the positive terminal, to conform with conventional practice and to avoid confusion with most other literature, the direction of current in all circuits throughout this text has been standardized as going from positive to negative. Using this conventional direction of current, instead of the actual direction of electron flow, will in no way affect the solution of electric circuit problems. The conventional direction was established before the electron was discovered, and to attempt to change it now is to beat one's head needlessly against a stone wall. Figure 1–1c contrasts the conventional direction of current (+ to −) with the actual direction of electron movement.

The application of an alternating voltage to a conductor causes the electrons to flow first in one direction and then in the opposite direction, oscillating continuously about their central position. This repetitive back-and-forth movement of electric charges constitutes an alternating current. Thus, any apparatus that generates a repetitive alternating driving voltage is called an *alternating-current (AC) generator*. Figure 1–1d illustrates the alternations of current supplied by an AC generator, and Figure 1–1e is a graph showing how the magnitude and direction of a sinusoidally varying alternating current change with time. The frequency of an alternating current is the number of cycles completed per second. One cycle per second is called one hertz (Hz), 60 cycles per second is 60 Hz, etc.

The unit of electric charge is the coulomb, and the rate of flow of these electric charges, in coulombs per second, is the current in amperes. Expressed mathematically,

$$I = \left[\frac{dq}{dt} \right] \tag{1–1}$$

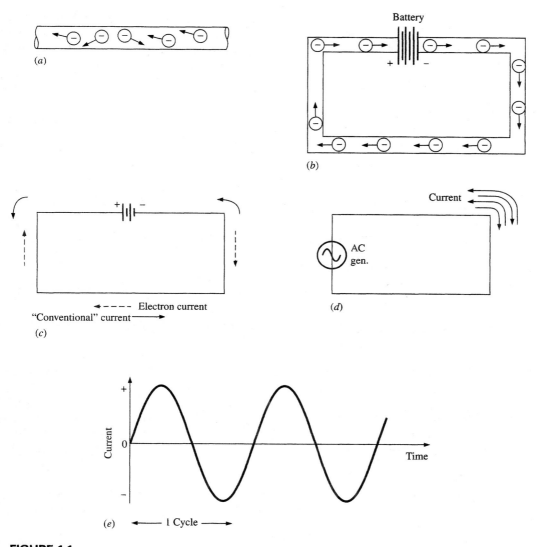

FIGURE 1.1
Electron motion in a conductor: (a) random motion of electrons when no driver is applied; (b) unidirectional motion of electrons caused by the application of a driver; (c) comparison of actual direction of electron flow with the conventional direction; (d) application of an alternating voltage to a conductor; (e) sinusoidally varying current.

where: I = current in amperes (A)
dt = increment change in time in seconds (s)
dq = increment change in charge that occurs in time dt, expressed in coulombs (C).

One coulomb of electric charge is equivalent to the charge produced by 6.25×10^{18} (6.25 million million million) electrons. A current of one ampere is equivalent to one coulomb of charge (6.25×10^{18} electrons) passing a specified point in one second.

1.2 ELECTRICAL INSULATION

Materials that are extremely poor conductors are classified as nonconductors and are better known as insulators or dielectrics. A vacuum is the only known perfect dielectric.[2] All other insulating materials, such as paper, cloth, wood, rubber, plastic, glass, mica, ceramic, shellac, varnish, air, and special-purpose oils, are imperfect dielectrics.

The function of electrical insulation is to offer high opposition to the passage of current between two or more conductors. To do this, the conductors may be covered with cotton, plastic or rubber tape, mica tubing, ceramic tubing, etc. The choice of insulation depends on the electrical, thermal, and mechanical stresses involved in the application, and whether the conductors will be located in wet or dry locations. The twin conductors in ordinary 120-V lamp cord are separated by a rubber or plastic coating, but the conductors on long-distance high-voltage transmission lines are insulated from one another by air. The desired spacing of the transmission-line conductors is obtained with porcelain insulators.

The electrons in an insulating material are tightly bound to the parent atoms and are not free to wander within the materials, as they do in a conductor. Hence, at normal room temperatures, the application of rated to moderately high driving voltages to the insulating material causes only a very few electrons to break away. In good insulation this very small "leakage" current is generally less than one-millionth of an ampere (microampere, μA). However, if a driver of sufficiently high voltage is applied, such as lightning or some other high-voltage surge, the electrons will be torn away from their molecular bonds, destroying the insulating properties of the material. In the case of organic materials, such as wood and rubber, this electron avalanche through the material may generate sufficient heat to cause carbonization or fire. The voltage per millimeter of insulation thickness that the insulation can withstand without breaking down is called the *dielectric strength* of the material.

Electrical insulation surrounding the conductors in generators, motors, transformers, cables, etc., gradually deteriorates with usage and age.[3] Thus, for critical apparatus, periodic condition monitoring of electrical insulation is essential, so that remedial action may be taken before significant deterioration occurs.

---

[2]In vacuum tubes, such as diodes, triodes, and other multielement vacuum tubes, a filament heated to incandescence emits electrons that provide a conducting path.

[3]For a detailed explanation of the mechanism of insulation failure, see Chapter 23.

A word of warning about one type of thermoplastic insulation called polyvinylchloride (PVC): When heated to about 230°C (446°F), PVC insulation gives off hazardous hydrogen chloride gas (HCl). The gas appears as a white mist that has an irritating and corrosive effect on the respiratory system. At much higher temperatures, a dark sooty smoke is released.

1.3 CONDUCTORS FOR ENERGY TRANSMISSION

Conductors selected for the efficient transmission of electric energy, and for the windings of motors, generators, and transformers, must have good electrical conductivity; that is, they must offer little opposition to current. The relatively high electrical conductivity, good heat-conduction capability, and relatively low cost of copper result in its extensive use in motors, generators, transformers, controls, switches, cables, buses, etc. Buses are solid or hollow conductors, generally of large cross-sectional area, from which taps are made to supply three or more circuits. Aluminum, because of its light weight, good casting properties, and high conductivity (although not as good as copper), is used in induction-motor rotors, cables, bus bars, long-distance transmission lines, and some transformers.

Cadmium-plated and silver-plated contacts are used in motor-control equipment, and gold plating is sometimes used to provide the excellent mating surfaces of bayonet-type contacts in electronic equipment. Although silver has a higher conductivity than copper, the relatively high cost makes its general use prohibitive.[4]

Carbon, in very short lengths, has applications as sliding contacts for rheostats, commutators, and slip rings.

For safety reasons, all water must be treated as having good conductivity. Although distilled water is not a conductor, any water around electrical equipment, tap water, rainwater, seawater, etc., is contaminated with salts and conducts electricity.

1.4 CONDUCTORS FOR LIMITING CURRENT AND FOR CONVERTING ELECTRICAL ENERGY TO HEAT ENERGY

Conductors of low conductivity are used in those applications where the magnitude of the current must be limited or where it is desired to convert electrical energy to heat energy (toaster, electric range, etc.). Examples of relatively low conductivity materials are iron, tungsten, nickel alloys, and compositions of carbon and inert materials. Figure 1–2a illustrates some different arrangements of current-limiting conductors, called *resistors*, and Figure 1–2b illustrates an adjustable resistor called a *rheostat*. The standard color code for small composition resistors is provided in Appendix 4.

---

[4]During World War II, due to a shortage of copper, the coil windings of some power transformers were wound with silver.

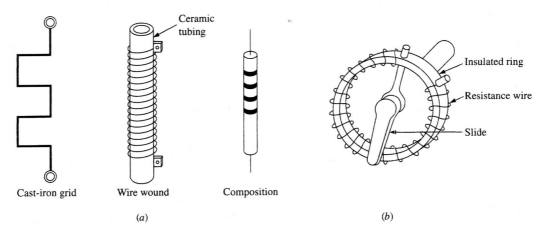

Cast-iron grid   Wire wound   Composition

(a)   (b)

FIGURE 1.2
Different arrangements of current-limiting conductors.

1.5 RESISTANCE, RESISTIVITY, AND CONDUCTANCE

When electrons are driven through a metallic conductor, they collide with some of the atoms that make up the material. Such collisions interfere with the free movement of the electrons and generate heat. This property of a material that limits the current and converts electric energy to heat energy is called *resistance*.

For a given length and cross-sectional area, materials of low conductivity have a higher resistance to current than materials of high conductivity; thus, to cause the same current in both, the material of low conductivity (higher resistance) requires a higher driving voltage.

Resistance is generally measured with an ohmmeter and is commonly expressed in ohms (Ω), micro-ohms ($\mu\Omega$), milliohms (mΩ), kilohms (kΩ), or megohms (MΩ), where

$$1\ \mu\Omega \text{ (one micro-ohm)} = 10^{-6}\ \Omega \text{ (one millionth of an ohm)}$$
$$1\ \text{m}\Omega \text{ (one milliohm)} = 10^{-3}\ \Omega \text{ (one thousandth of an ohm)}$$
$$1\ \text{k}\Omega \text{ (one kilohm)} = 10^{3}\ \Omega \text{ (one thousand ohms)}$$
$$1\ \text{M}\Omega \text{ (one megohm)} = 10^{6}\ \Omega \text{ (one million ohms)}$$

Conductance

Conductance (G) is a measure of the ease with which a material conducts an electric current. It is numerically equal to the reciprocal of the resistance. That is,

$$G = \frac{1}{R} \tag{1--2}$$

where: G = conductance, whose unit is the Siemen[5] (S)
         R = resistance in ohms (Ω).

EXAMPLE 1.1   Determine the conductance of each of the following resistors: 25 Ω and 60 kΩ.

Solution

$$G = \frac{1}{R} = \frac{1}{25} = 0.040 \text{ S}$$

$$G = \frac{1}{R} = \frac{1}{60,000} = 16.667 \times 10^{-6} = 16.7 \ \mu\text{S}$$

Determining Conductor Resistance from Its Dimensions and Its Resistivity

The resistivity of a material is the resistance of a specified length and cross section of that material. Thus, if the resistivity of a material is known, the resistance of a conductor of any length and any cross section may be determined by substituting into the following formula:

$$R = \rho \frac{\ell}{A} \tag{1–3}$$

where: R = resistance (Ω)
         ℓ = length of conductor (ft or m)
         ρ = resistivity (see Table 1–1)
         A = cross-sectional area (cmil or m^2).

The resistivity of common metallic conductors in both English units and in SI units (the international system of measurements) is given in Table 1–1. The reciprocal of resistivity ($1/\rho$) is called *conductivity*.

EXAMPLE 1.2   Given a 2.0-m length of 10.0-mm-diameter round aluminum conductor at 20°C. Determine (a) its cross-sectional area; (b) its resistance.

Solution

$$\pi \approx 3.1416$$

a. $A = \dfrac{\pi D^2}{4} = \dfrac{\pi \times (0.01)^2}{4} = 78.54 \times 10^{-6} \text{ m}^2$

b. $R = \rho \dfrac{\ell}{A} = 2.83 \times 10^{-8} \times \dfrac{2.0}{78.54 \times 10^{-6}} = 721 \times 10^{-6} \ \Omega$ or 721 $\mu\Omega$

---

[5]The unit of conductance was previously called the *mho* (ohm spelled backwards).

TABLE 1.1
Resistivity (ρ) of Common Metallic Conductors[a]

Material	(SI)[b] ohm-meter at 20°C (68°F)	(English)[c] ohm-cmil/ft at 20°C (68°F)
Aluminum	2.83×10^{-8}	17
Antimony	41.7×10^{-8}	251
Bismuth	120×10^{-8}	722
Brass	7×10^{-8}	42
Cadmium	7.5×10^{-8}	45
Cesium, solid	22×10^{-8}	132
Chromium	2.7×10^{-8}	16
Cobalt	9.7×10^{-8}	58
Constantan (40% Ni, 60% Cu)	49×10^{-8}	295
Copper, annealed	1.724×10^{-8}	10.37
Copper, hard-drawn	1.77×10^{-8}	10.65
German silver (18% Ni)	33×10^{-8}	199
Gold, pure drawn	2.4×10^{-8}	14
Iron, cast	9×10^{-8}	54
Lead	22×10^{-8}	132
Magnesium	4.6×10^{-8}	28
Manganin (8% Cu, 12% Mn, 4% Ni)	44×10^{-8}	265
Mercury	96×10^{-8}	577
Molybdenum	5.7×10^{-8}	34
Monel metal	42×10^{-8}	253
Nichrome	100×10^{-8}	602
Nickel	7.8×10^{-8}	47
Platinum	10×10^{-8}	60
Silver (99.98% pure)	1.64×10^{-8}	9.9
Steel, annealed sheet	11×10^{-8} to 50×10^{-8}	66–300
Tin	11.5×10^{-8}	69
Tungsten	5.5×10^{-8}	33
Zinc	6×10^{-8}	36

[a] Values taken from Smithsonian Physical Tables.
[b] If using SI units in Eq. (1–3), the length must be in meters, and the cross-sectional area must be in square meters. If the area is given in square millimeters, multiply it by 10^{-6} to obtain square meters before substituting into Eq. (1–3).
[c] If using English units in Eq. (1–3), the length must be in feet, and the area must be in circular mils. See Section 1–6.

1.6 CIRCULAR-MIL AREA

The unit of cross-sectional area commonly used for wire and cable sizes in the United States is the circular mil (cmil). One circular mil is defined as the area of a circle whose diameter is 1 mil (0.001 inch). Thus the area of a circle whose diameter is 1 mil is

$$A_{\text{of 1-mil-diam. conductor}} = \frac{\pi D^2}{4} = \frac{\pi (0.001)^2}{4} = 7.854 \times 10^{-7} \text{ in.}^2 \qquad (1\text{--}4)$$

To obtain the area in circular mils for any cross section regardless of shape, divide the square-inch area of the cross section by the square-inch area of 1 cmil. Thus, for any cross section,

$$A_{\text{cmil}} = \frac{A_{\text{in.}^2}}{\dfrac{\pi (0.001)^2}{4}} = A_{\text{in.}^2} \times 1.273 \times 10^6 \qquad (1\text{--}5)$$

For the special case of circular cross-section conductors, the area in circular mils may be obtained by dividing the area of a circular cross section by the area of a circle whose diameter is 1 mil and simplifying. Thus, for a circular cross section,

$$A_{\text{cmil}} = \frac{\dfrac{\pi D^2}{4}}{\dfrac{\pi (0.001)^2}{4}} = D_{\text{in.}}^2 \times 10^6 \qquad (1\text{--}6)$$

where: D_{in} = diameter in inches.

EXAMPLE 1.3

Determine the circular mil area of (a) round copper conductor whose diameter is 0.5 in; (b) aluminum conductor of rectangular cross section, 6 in. × 0.75 in.

Solution

a. $A_{\text{cmil}} = D_{\text{in.}}^2 \times 10^6 = 0.5^2 \times 10^6 = 250{,}000 \text{ cmil}$

b. $A_{\text{cmil}} = A_{\text{in.}^2} \times 1.273 \times 10^6 = 6 \times 0.75 \times 1.273 \times 10^6 = 5.728 \times 10^6$

EXAMPLE 1.4

Given a 6-ft section of rectangular annealed copper bus bar whose cross section is 0.5 in. × 6 in. Determine (a) circular mil area; (b) resistance of the conductor at a temperature of 20°C.

Solution

a. From Table 1–1, resistivity (ρ) = 10.37 ohm-cmil/ft.

$$A = 0.5 \times 6 = 3 \text{ in}^2$$

$$A_{\text{cmil}} = A_{\text{in}^2} \times 1.273 \times 10^6 = 3 \times 1.273 \times 10^6 = 3.819 \times 10^6 \Rightarrow 3.82 \times 10^6 \text{ cmil}$$

b. $R = \rho\dfrac{\ell}{A} = 10.37 \times \dfrac{6}{3.819 \times 10^6} = 16.29 \times 10^{-6}\,\Omega \quad$ or $16.29\,\mu\Omega$

(a)   (b)

FIGURE 1.3
(a) Analog meter (courtesy Simpson Electric Company); (b) digital meter (courtesy Fluke Corporation).

1.7 DIRECT MEASUREMENT OF RESISTANCE

Most resistance measurements of conductors can be made with an all-purpose measuring instrument like the analog-type multimeter shown in Figure 1–3a or the digital-type multimeter shown in Figure 1–3b.[6]

Although analog instruments put a load on the circuit being measured, causing a slightly different reading than if a digital meter is used, the difference is generally insignificant when making measurements in electric power systems. Both instruments use internal batteries to supply power for resistance measurements. Although a digital instrument can provide accuracy to more decimal places than can an analog instrument, the analog instrument is better for observing rapidly changing conditions.

1.8 TEMPERATURE-RESISTANCE CHARACTERISTIC OF METALLIC CONDUCTORS

The resistance of all conductors is temperature dependent to some extent. Hence, valid measurements call for the determination of conductor temperature at the time of resistance measurement. Most conducting materials, such as copper, aluminum, iron, nickel, and tungsten, increase in resistance with increasing temperature. However, carbon, which decreases in resistance with increasing temperature, is used for sliding contacts in direct-current machines, in some alternating-current machines, and in some rheostats.

---

[6]A comparison of the many different methods commonly used to measure electrical resistance in electrical power apparatus is given in Reference [1].

FIGURE 1.4

(a) Temperature-resistance characteristic for annealed copper; (b) similar triangles formed by the linear section and the temperature axis.

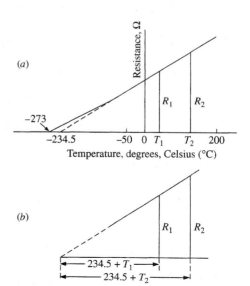

The temperature-resistance characteristic for annealed copper[7] is shown in Figure 1–4a. The curve is essentially a straight line (linear) from about $-50°C$ to approximately $+200°C$. Below $-50°C$ the curve becomes nonlinear, and the resistance approaches $0\ \Omega$ at about $-273°C$. Note that $-273°$ on the Celsius temperature scale is the absolute zero temperature on the Kelvin temperature scale.

If the linear section of the curve in Figure 1–4a is extended, as shown by the broken line, it will touch the temperature axis at $-234.5°C$; this temperature is called the *inferred absolute zero temperature* of the material and is used only for calculation purposes. The true absolute zero is $-273°C$, as shown in Figure 1–4a.

The similar triangles formed by the linear section and the temperature axis provide a convenient method for determining the resistance of a conductor at different temperatures, providing the resistance at one temperature is known. Similar triangles are redrawn in Figure 1–4b, and the corresponding sides identified. From the geometry of similar triangles, the ratio of the lengths of the corresponding sides is

$$\frac{R_2}{R_1} = \frac{234.5 + T_2}{234.5 + T_1} \tag{1–7}$$

where: R_1 = resistance at temperature T_1 (Ω)
 R_2 = resistance at temperature T_2 (Ω)
 T_1 and T_2 are in $°C$.

Fahrenheit–Celsius Conversion

If the temperature is given in Fahrenheit, it must be converted to Celsius before substituting into Eqs. (1–7) through (1–11). The following easy-to-remember formulas are useful when converting from Fahrenheit ($°F$) to Celsius ($°C$), and vice versa:

$$°C = (°F + 40) \times \frac{5}{9} - 40 \qquad °F = (°C + 40) \times \frac{9}{5} - 40 \tag{1–8}$$

EXAMPLE 1.5 The ambient temperature of a certain boiler room is $120°F$. Determine the temperature in $°C$.

Solution

$$°C = (°F + 40) \times \frac{5}{9} - 40 = (120 + 40) \times \frac{5}{9} - 40 = 48.9°C$$

---

[7]Annealed copper (heat treated) is easy to bend and wind into coils and is thus used in most electrical power apparatus, such as motors, generators, transformers, and cables. Hard-drawn copper is used in some telephone lines to reduce the sag between poles.

FIGURE 1.5
Geometry for the general case of any metallic conductor
that is linear over a given range.

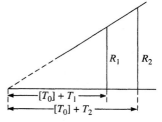

General Case for Any Metallic Conductor

Using Figure 1–4b as a guide, the general case of any metallic conductor whose
temperature-resistance characteristic is linear over a given range of temperature is
shown in Figure 1–5. A list of different conductor materials and their inferred absolute
zeros is given in Table 1–2. Note that the inferred absolute zero (T_0) is a negative number.
From the geometry of similar triangles in Figure 1–5,

$$\frac{R_2}{R_1} = \frac{|T_0| + T_2}{|T_0| + T_1} \tag{1–9}$$

where: $|T_0|$ = the *magnitude* of the inferred absolute zero temperature
for the conductor material. The minus sign is omitted.

**EXAMPLE
1.6**

A resistor constructed from Monel wire has a resistance of 14.6 Ω at 10°C. What is its
resistance at 90°C?

Solution
From Table 1–2,

$$T_0 = -480°C$$

From Eq. (1–8),

$$R_2 = R_1 \times \frac{|T_0| + T_2}{|T_0| + T_1} = 14.6 \times \frac{480 + 90}{480 + 10} = 16.98 \ \Omega$$

Temperature Determination by the Change in Resistance Method

Equation (1–8) can also be used to determine the temperature of coils of wire (windings) in electric machines and transformers.

TABLE 1.2
Inferred absolute zero and temperature coefficients of conductor materials[a]

Material	α_{20} Temperature coefficent (Ω per °C per Ω at 20°C)	T_0 Inferred absolute zero (°C)
Aluminum	0.0039	−236
Antimony	0.0036	−258
Bismuth	0.004	−230
Brass	0.002	−480
Constantan (60% Cu, 40% Ni)	0.000008	−125,000
Copper, annealed	0.00393	−234.5
Copper, hard-drawn	0.00382	−242
German silver	0.0004	−2480
Iron	0.005	−180
Lead	0.0041	−224
Magnesium	0.004	−230
Manganin (84% Cu, 12% Mn, 4% Ni)	0.000006	−167,000
Mercury	0.00089	−1100
Molybdenum	0.0034	−274
Monel metal	0.002	−480
Nichrome	0.0004	−2480
Nickel	0.006	−147
Platinum	0.003	−310
Silver (99.98% pure)	0.0038	−243
Steel, soft	0.0042	−218
Tin	0.0042	−218
Tungsten	0.0045	−200
Zinc	0.0037	−250

[a] Values taken from Smithsonian Physical Tables.

The resistance of the windings of electric machinery and transformers is available from the manufacturer at some specified temperature, usually 25°C. Thus, to determine the winding temperature, after the machine had been in operation for some time, it is only necessary to substitute the manufacturer's data and the measured resistance when hot into Eq. (1–9) and solve.

EXAMPLE 1.7 Data supplied by a motor manufacturer list the resistance of the motor coils to be 15.6 Ω at 25°C. The coils are wound with annealed copper. After 12 hours of operation, the winding resistance is measured and found to be 19.5 Ω. Determine the temperature of the winding.

Solution

From Table 1–2, the inferred absolute temperature of annealed copper is $-234.5°C$. Using Eq. (1–9),

$$\frac{R_2}{R_1} = \frac{|T_0| + T_2}{|T_0| + T_1} \Rightarrow \frac{19.5}{15.6} = \frac{234.5 + T_2}{234.5 + 25}$$

$$T_2 = \frac{19.5}{15.6} \times (234.5 + 25) - 234.5 = 89.87°C$$

1.9 TEMPERATURE COEFFICIENT OF RESISTANCE

The *temperature coefficient of resistance* (α) is defined as the change in resistance of a given conductor due to a 1°C change from the initial temperature divided by the resistance at the initial temperature. Because the temperature coefficient of resistance is different for different temperatures, a subscript is used to indicate the temperature at which the coefficient was calculated; for example, α_0, α_{20}, α_{50}, etc.

Table 1–2 lists the temperature coefficients of resistance at 20°C for some metallic conductors. The temperature coefficient of resistance given in Table 1–2 provide a quick comparison of the effect of temperature on the resistance of some of the more commonly used metallic conductors; lower α numbers indicate that changes in the conductor temperature will have less effect on the resistance. Note that Manganin is least affected by changes in temperature and is therefore used extensively in ammeter shunts and voltmeter multiplier-resistors, as a temperature-resistance stabilizer within the instrument.

The mathematical relationship between the temperature coefficient of resistance of a conductor and its inferred absolute zero is

$$T_0 = t - \frac{1}{\alpha_t} \tag{1–10}$$

where: T_0 = inferred absolute zero (°C)
            t = temperature (°C)
            α_t = temperature coefficient at temperature t.

EXAMPLE 1.8   The temperature coefficient of resistance for a certain alloy at 20°C is 0.0018. The alloy, wound in the form of a coil, has a resistance of 142.6 Ω at 70°C. Determine (a) the inferred absolute zero; (b) the resistance at 130°C.

Solution
From Eq. (1–10),

a. $T_0 = 20 - \dfrac{1}{\alpha_{20}} = 20 - \dfrac{1}{0.0018} = -535.6°C$

b. $\dfrac{R_2}{R_1} = \dfrac{|T_0| + T_2}{|T_0| + T_1} \Rightarrow R_2 = R_1 \times \dfrac{|T_0| + T_2}{|T_0| + T_1}$

$R_2 = 142.6 \times \dfrac{535.6 + 130}{535.6 + 70} = 156.7 \ \Omega$

1.10 WIRE TABLES AND CABLE AMPACITY

Table 1–3 provides dimensions and resistance data for the most commonly used sizes of copper and aluminum conductors, in AWG (American wire gauge) and in kcmil (thousand circular mils).

A concentric-lay stranded conductor consists of a central wire surrounded by one or more layers of spirally laid wires. All wires have the same diameter, and each layer has six more wires than the preceding layer. This is shown in Figure 1–6 for a 37-strand conductor.

FIGURE 1.6
Concentric-lay stranded conductor.

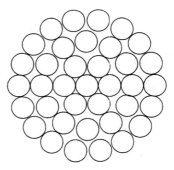

EXAMPLE 1.9 A magnet coil is wound with 200 ft of AWG-16 tinned copper wire. Determine the resistance of the coil.

Solution
From Table 1–3, the resistance per 1000 ft of AWG-16 tinned copper wire is 4.26 Ω at 25°C.

$$\text{Resistance per foot} = 4.26 \div 1000 = 0.00426 \ \Omega$$
$$\text{Resistance of 200 ft} = 200 \times 0.00426 = 0.852 \ \Omega$$

TABLE 1.3

Resistance of Copper and Aluminum Conductors

Size AWG MCM	Area (cmils)	Concentric-lay stranded conductors		Bare conductors		DC resistance [Ω/k ft at 25°C (77°F)]		
		No. Wires	Diam. each wire (in.)	Diam. (in.)	Area[a] in.^{2}	Copper[b] Bare cond.	Tin'd cond.	Aluminum[b]
AWG 18	1620	Solid	0.0403	0.0403	0.0013	6.51	6.79	10.7
16	2580	Solid	0.0508	0.0508	0.0020	4.10	4.26	6.72
14	4110	Solid	0.0641	0.0641	0.0032	2.57	2.68	4.22
12	6530	Solid	0.0808	0.0808	0.0051	1.62	1.68	2.66
10	10380	Solid	0.1019	0.1019	0.0081	1.018	1.06	1.67
8	16510	Solid	0.1285	0.1285	0.0130	0.6404	0.659	1.05
6	26240	7	0.0612	0.184	0.027	0.410	0.427	0.674
4	41740	7	0.0772	0.232	0.042	0.259	0.269	0.424
3	52620	7	0.0867	0.260	0.053	0.205	0.213	0.336
2	66360	7	0.0974	0.292	0.067	0.162	0.169	0.266
1	83690	19	0.0664	0.332	0.087	0.129	0.134	0.211
0	105600	19	0.0745	0.372	0.109	0.102	0.106	0.168
00	133100	19	0.0837	0.418	0.137	0.0811	0.0843	0.133
000	167800	19	0.0940	0.470	0.173	0.0642	0.0668	0.105
0000	211600	19	0.1055	0.528	0.219	0.0509	0.0525	0.0836
kcmil 250	250000	37	0.0822	0.575	0.260	0.0431	0.0449	0.0708
(MCM) 300	300000	37	0.0900	0.630	0.312	0.0360	0.0374	0.0590
350	350000	37	0.0973	0.681	0.364	0.0308	0.0320	0.0505
400	400000	37	0.1040	0.728	0.416	0.0270	0.0278	0.0442
500	500000	37	0.1162	0.813	0.519	0.0216	0.0222	0.0354
600	600000	61	0.0992	0.893	0.626	0.0180	0.0187	0.0295
700	700000	61	0.1071	0.964	0.730	0.0154	0.0159	0.0253
750	750000	61	0.1109	0.998	0.782	0.0144	0.0148	0.0236
800	800000	61	0.1145	1.030	0.833	0.0135	0.0139	0.0221
900	900000	61	0.1215	1.090	0.933	0.0120	0.0123	0.0197
1000	1000000	61	0.1280	1.150	1.039	0.0108	0.0111	0.0177
1250	1250000	91	0.1172	1.289	1.305	0.00863	0.00888	0.0142
1500	1500000	91	0.1284	1.410	1.561	0.00719	0.00740	0.0118
1750	1750000	127	0.1174	1.526	1.829	0.00616	0.00634	0.0101
2000	2000000	127	0.1255	1.630	2.087	0.00539	0.00555	0.00885

[a] Area given is that of a circle having a diameter equal to the overall diameter of a stranded conductor.

[b] The resistance values given in the last three columns are applicable only to direct current. When conductors larger than No. 4/0 are used with alternating current the multiplying factors in Appendix 10 should be used to compensate for skin effect.

SOURCE: The values given in the table are those given in Handbook 100 of the National Bureau of Standards except that those shown in the eighth column are those given in Specification B33 of the American Society for Testing and Materials, and those shown in the ninth column are those given in Standard No. S-19-81 of the Insulated Power Cable Engineers Association and Standard No. WC3-1964 of the National Electrical Manufacturers Association.

Skin Effect

A phenomenon called *skin effect* causes a conductor to offer a higher resistance to an alternating current than to a direct current. In a DC circuit, the current is distributed uniformly over the entire cross section of the conductor. However, with alternating current, the density of the current is greatest at the wall of the conductor (skin), and has its lowest value at the center of the conductor. The skin effect, caused by the magnetic effect of the alternating current, is more pronounced at higher frequencies and is appreciably greater if the conductor is enclosed in a steel conduit. The resistance offered by a conductor to direct current is called *DC resistance* or *ohmic resistance*, and the resistance offered by the same conductor to an alternating current is called *AC resistance* or *effective resistance*.

A table of multiplying factors for converting DC resistance to 60-Hz AC resistance is given in Appendix 10 for conductors with different cross-sectional areas, with and without steel conduit.

EXAMPLE 1.10

The ohmic resistance of a certain length of 1750-kcmil copper conductor is 0.012 Ω. Determine the effective resistance at 60 Hz if the conductor is enclosed in a steel pipe. From Appendix 10, the multiplying factor is 1.67.

Solution:

$$R_{60\text{Hz}} = 0.012 \times 1.67 = 0.020 \ \Omega$$

Cable Ampacity

The *ampacity* of a conductor is the current in amperes that a conductor can carry continuously without exceeding the temperature rating of its insulation. The factors considered are the thermal stability of the insulation that is covering the conductor, the heat dissipation capability to the surrounding media, the number of conductors in the cable or pipe, and the particular combination of work and rest periods (duty cycle). Operating a conductor at temperatures higher than the temperature rating of its insulation will degrade or damage the insulation. Ampacity tables for conductor sizes with specific types of insulation, in an ambient temperature of 30°C (86°F), are given in Appendix 1 and Appendix 2. Correction factors for ambient temperatures other than 30°C are given in Appendix 3.

EXAMPLE 1.11

A cable containing three size AWG-2 insulated copper conductors, each insulated with type TW insulation, is in an ambient temperature of 50°C. Determine the ampacity of the cable.

Solution

From Appendix 1, in an ambient temperature of 30°C, the three-conductor cable with size AWG-2 copper conductors and TW (60°C) insulation has an ampacity of 95 A. However, in accordance with Appendix 3, if 60°C insulation is operating in an ambient of 50°C, instead of 30°C, its ampacity must be derated by a factor 0.58. Thus, the ampacity of the cable operating in a 50°C ambient is 95 × 0.58 = 55.1 A.

1.11 SEMICONDUCTORS

A semiconductor material, such as silicon or germanium, has a conductivity about midway between good conductors and good insulators. Useful semiconductor devices are grown as crystals from a melt of semiconductor material to which extremely small amounts of impurities have been added. Semiconductor devices grown in this manner include diodes, silicon-controlled rectifiers (SCRs; also called thyristors), and transistors. A diode has the characteristics of a conductor when the applied voltage is in one direction and those of an insulator when the applied voltage is reversed. This is illustrated in Figure 1–7 for forward and reverse directions of applied voltage (also called forward and reverse bias).

The circuit in Figure 1–7 shows four diodes, connected in a bridge-type full-wave rectifier circuit. The rectifier supplies unidirectional current to the load, even though the current from the generator is alternating. The solid arrows indicate the direction of current when generator terminal L_2 is positive, and the broken arrows indicate the direction of current when generator terminal L_1 is positive. Note that the direction of current in the load is the same in both cases. Other applications of semiconductor devices are discussed in connection with specific apparatus.

FIGURE 1.7
Four diodes connected in a bridge-type full-wave rectifier.

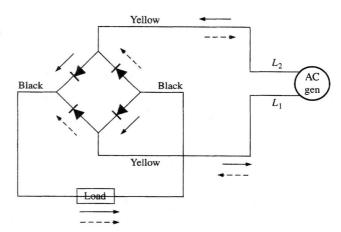

SUMMARY OF EQUATIONS FOR PROBLEM SOLVING

$$I = \frac{dq}{dt} \tag{1-1}$$

$$G = \frac{1}{R} \tag{1-2}$$

$$R = \rho \frac{\ell}{A} \tag{1-3}$$

$$A_{cmil} = A_{in.^2} \times 1.273 \times 10^6 \tag{1-5}$$

$$A_{cmil} = D_{in.}^2 \times 10^6 \tag{1-6}$$

$$^\circ C = (^\circ F + 40) \times \frac{5}{9} - 40^\circ \qquad ^\circ F = (^\circ C + 40) \times \frac{9}{5} - 40 \tag{1-8}$$

$$\frac{R_2}{R_1} = \frac{|T_0| + T_2}{|T_0| + T_1} \tag{1-9}$$

$$T_0 = t - \frac{1}{\alpha_t} \tag{1-10}$$

SPECIFIC REFERENCE KEYED TO THE TEXT

[1] Standard Test Code for Resistance Measurement. IEEE Std. 118–1992.

REVIEW QUESTIONS

1. What is the relationship between electric charge and electric current?
2. A flow of electrons occurs when a battery and conductors form a closed loop. Where do these electrons come from?
3. What is the essential difference between conductors and insulators?
4. What is the function of electrical insulation?
5. How do conductors used for energy transmission differ from those used for limiting current? Give examples of each.
6. Explain why the passage of current through a conductor generates heat.
7. What effect does increasing temperature have on (a) copper; (b) Manganin; (c) carbon?
8. Define *ampacity* and state the factors that affect it.
9. Define *resistivity, resistance*, and *conductance*.
10. What is the characteristic behavior of a semiconductor diode? Sketch a circuit that uses diodes and explain its function.
11. What factors must be considered when determining the ampacity of a cable?
12. Sketch a full-wave rectifier circuit and explain how DC is obtained from an AC source.

PROBLEMS

1–1/5. Determine the conductance of each of the following resistors: (a) 50 kΩ; (b) 465 Ω; (c) 0.018 Ω; (d) 0.00040 Ω; (e) 52 MΩ.

1–2/5. Determine the resistance of each of the following conductances: (a) 0.0041 S; (b) 3.25 S; (c) 180.55 S; (d) 1000 S; (e) 400 mS.

1–3/5. Determine the resistance of a 30-m length of aluminum bus whose rectangular cross section is 20 mm × 200 mm.

1–4/5. Determine the resistance of a 5.64-m length of annealed copper conductor whose rectangular cross section is 4.0 mm × 2.0 mm.

1–5/6. Determine (a) the circular-mil area of a 200-ft length of annealed copper bar 1/2 in. × 1/4 in.; (b) its resistance at 20°C.

1–6/6. Determine (a) the circular-mil area of an annealed copper bus whose cross section is 0.5 in. × 8.0 in.; (b) the resistance of a 300-ft length of the same bus at 20°C.

1–7/6. Determine (a) the circular-mil area of a round aluminum conductor whose diameter is 4.0 in.; (b) the resistance of 1000 ft of the same conductor at 20°C.

1–8/6. A 60-ft length of circular annealed hollow copper bus has inside and outside diameters of 4.0 in and 4.5 in, respectively. Determine its resistance at 20°C.

1–9/8. Determine (a) the resistance of a 6-ft length of 2-in.-diameter annealed copper at 20°C; (b) its resistance at 90°C.

1–10/8. The resistance of a length of aluminum conductor is 80 Ω at 40°C. If the conductor is allowed to cool to 20°C, determine its resistance.

1–11/8. A coil wound with annealed copper wire has a resistance of 25 Ω at a temperature of 20°C. After several hours of operation, the temperature of the coil rises to 50°C. Determine the resistance of the coil at the higher temperature.

1–12/8. The copper field winding of a generator has a resistance of 138 Ω at 104°F. After 24 hr of continuous operation, its resistance rises to 163.1 Ω. Determine the temperature of the field winding for this hot condition.

1–13/8. The resistance of a length of iron wire is 126 Ω at 25°C. What is its resistance at 96°C?

1–14/8. A resistor wound with nichrome wire has a resistance of 256 Ω at 60°C. What is its resistance at 195°C?

1–15/8. A resistor wound with Manganin wire has a resistance of 142 Ω at 10°C. What is its resistance at 130°C?

1–16/9. A certain motor is operating at 162°F. What is its temperature in °C?

1–17/9. The temperature of a certain transformer is 28°C. What is its temperature in °F?

1–18/9. The copper field windings of a generator have a resistance of 162 Ω at 60°C. After 8 hours of cooling, the resistance was determined to be 140 Ω. Determine the winding temperature after cooling.

1–19/9. A coil wound with aluminum wire has a resistance of 25 Ω at a temperature of 20°C. After several hours of operation, the temperature of the coil rises to 50°C. Determine the resistance of the coil at this temperature.

1–20/10. A certain alloy has a temperature coefficient of resistance of 0.0040 at 20°C. An electrical apparatus utilizing this alloy has a resistance of 185 Ω when operating at 70°C. Determine (a) the inferred absolute zero of the material; (b) its resistance at 140°C.

1–21/11. Determine the resistance of an 800-ft length of size 600-kcmil tinned copper conductor at 25°C.

1–22/11. Determine the resistance of a 654-ft length of size AWG-00 aluminum conductor at 25°C.

1–23/11. A single copper conductor, size AWG-0000 in free air, is insulated with Type RHW insulation. The ambient temperature surrounding the cable is 60°C. Determine the ampacity of the conductor.

1–24/11. A cable containing three copper conductors, size AWG-8, each insulated with Type FEP insulation, is in an ambient temperature of 60°C. Determine the ampacity of the conductors.

1–25/11. Three 250-kcmil copper conductors, each insulated with Type UF insulation are in a raceway. The ambient temperature surrounding the raceway is 54°C. Determine the ampacity of the conductors.

2

DC Resistive Circuits and Circuit Faults

2.0 INTRODUCTION

This chapter incorporates those aspects of DC circuit theory that are necessary for an understanding of the operation, maintenance testing, and troubleshooting of electrical power apparatus. Included are discussions on the fundamental laws of Ohm and Kirchhoff; series, parallel, and series–parallel circuits; power and energy loss in conductors; and circuit faults such as shorts, opens, grounds, and flashovers.

2.1 THE ELECTRIC CIRCUIT AND OHM'S LAW

An electric circuit is any arrangement of conducting parts that form a closed loop. A few examples of simple electric circuits are shown in Figure 2–1. Note that a closed loop constitutes a circuit whether or not it includes a useful load or driving voltage. In

FIGURE 2.1
Examples of simple electric circuits:
(a) closed loop without a useful load or driving voltage; (b) closed loop with a battery; (c) closed loop with a battery and a lamp.

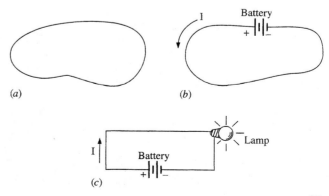

23

the case of Figure 2–1a there is no current in the loop unless it is induced by a changing magnetic field, such as that caused by moving magnets, a lightning discharge, or a changing current in a nearby circuit. The current in the circuits shown in Figures 2–1b and 2–1c is the result of the application of a battery.

Ohm's Law

The current in an electric circuit is directly proportional to the driving voltage and inversely proportional to the opposition that the circuit offers to the current. Expressed mathematically for a resistive circuit,

$$I = \frac{V}{R} \tag{2–1}$$

where V = driving voltage, volts (V)
R = resistance, ohms (Ω)
I = current, amperes (A).

Note: The driving voltage is also called the driver or source voltage, and the opposition to the current is the resistance of the circuit.

EXAMPLE 2.1   A 12-V source is connected to a circuit whose resistance is 48 Ω. Determine the current.

Solution
The circuit is similar to that shown in Figure 2–1c, where the resistance of the circuit is equal to the resistance of the lamp plus the resistance of the connecting wires. The battery is assumed to have negligible resistance.

$$I = \frac{V}{R} = \frac{12}{48} = 0.25 \text{ A}$$

EXAMPLE 2.2   A coil wound with 200 ft of AWG-16 solid copper wire is connected to a 12-V battery. Determine the current.

Solution
From Table 1–3 in Chapter 1, the resistance per 1000 ft of AWG-16 solid copper wire at 25°C is 4.10 Ω. Thus, the ohms per foot is calculated as follows:

$$\frac{4.10}{1000} = 0.0041 \ \Omega$$

The resistance of 200 ft is 200 $\times$ 0.0041 = 0.82 Ω. From Ohm's law,

$$I = \frac{V}{R} = \frac{12}{0.82} = 14.63 \text{ A}$$

2.2 VOLTMETER–AMMETER METHOD FOR DETERMINING RESISTANCE

The voltmeter–ammeter method is often used to determine the resistance of a motor, coil of wire, cable, resistor, or other electrical apparatus. The measuring circuit requires an ammeter, voltmeter, and a direct-current source. This is shown in Figure 2–2, where the resistance of a coil of wire is to be determined.

Note: The ammeter must always be connected in series with the apparatus or component whose current it is to measure, and the voltmeter must always be connected in parallel with the apparatus or component whose voltage it is to measure. To prevent a reversed deflection of the instrument pointer, which may bend it, the positive terminal of each instrument should be connected to the wire leading from the positive terminal of the driving source. The fuse is used to prevent damage to the circuit in the event of excessive current.[1]

EXAMPLE 2.3

For the circuit in Figure 2–2, assume the ammeter reads 10 A, the voltmeter reads 36 V, and the resistance of the connecting conductors is negligible. Determine the resistance of the coil.

Solution
From Ohm's law

$$R_{coil} = \frac{V}{I} = \frac{36}{10} = 3.6 \ \Omega$$

Unless otherwise specified, when making voltage and current measurements on electric power apparatus, such as motors, generators, and transformers, ammeters are assumed to have zero resistance and voltmeters are assumed to have essentially infinite resistance. Furthermore, because resistance is temperature dependent, to obtain

---

[1]For information on fuses and their application, see Chapter 29.

FIGURE 2.2
Voltmeter-ammeter method for
determining resistance.

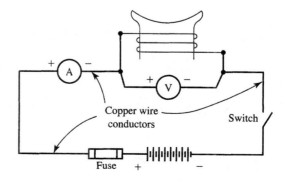

Copper wire conductors

Switch

Fuse

the "hot resistance," the ammeter and voltmeter readings should be taken when the apparatus has reached its rated operating temperature. If "cold resistance" is desired, the readings should be taken immediately after the switch is closed.

Ammeters for Measuring High Values of Direct Current

An ammeter for measuring high values of direct current (hundreds and thousands of amperes) is shown in Figure 2–3a. It consists of a millivoltmeter in parallel with a low-resistance calibrated resistor, called an ammeter-shunt. Because of its very low resistance, the ammeter shunt has an insignificant effect on the current. The resistance strips in the shunt are generally made of Manganin for resistance stability over a wide range of temperatures, and are terminated in large blocks of copper that act as heat sinks. Manganin is a trademark for an alloy of copper, manganese, and nickel.

The shunt is connected in series with the conductor whose current it is to measure, and the millivoltmeter is connected to the voltage terminals (called the *IR*-drop terminals) with a pair of calibrated leads. The millivoltmeter together with its shunt and leads constitute an ammeter. The current and millivolt ratings of the shunt are stamped on the heat sink. The ampere rating of the shunt indicates the maximum allowable current, and the millivolt rating of the shunt indicates the rating of the required millivoltmeter. Standard millivolt ratings for ammeter shunts are 50 mV and 100 mV, although sometimes 200-mV shunts are used. With rated current in the shunt, the meter will deflect to full scale. *Note:* The millivoltmeter, its shunt, and the calibrated leads must be a matching set! An elementary circuit with an external ammeter-shunt is shown in Figure 2–3b.

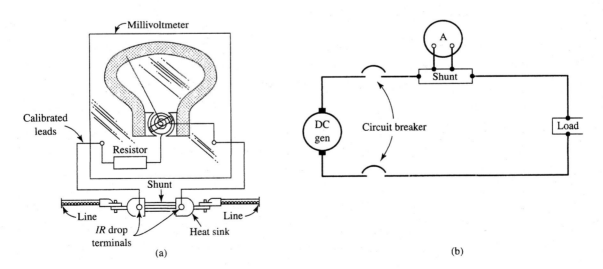

FIGURE 2.3
(a) Ammeter for measuring direct current; (b) elementary diagram showing proper ammeter connection in a DC circuit.

2.3 RESISTORS IN SERIES

A representative series circuit, shown in Figure 2–4a, has four resistors connected in series with a battery. The graphical symbol for a resistor is either the zigzag symbol shown in Figure 2–4, or a rectangle with the value of resistance or the letter R inserted as shown in Figure 2–5. Both are American National Standards for graphic symbols, and both will be used in this text.

Assuming the battery and connecting wires in Figure 2–4a have negligible resistance, the equivalent resistance of the series circuit ($R_{eq.S}$) is the sum of all resistors connected in series.

$$R_{eq.S} = R_1 + R_2 + R_3 + R_4 + \cdots + R_n \qquad (2\text{–}2)$$

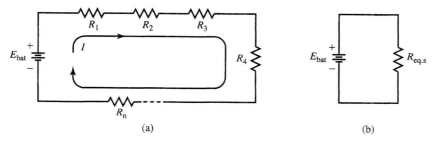

(a) (b)

FIGURE 2.4
(a) Representative series circuit; (b) equivalent series circuit.

FIGURE 2.5
(a) Series circuit for Example 2–4;
(b) polarity of the voltage drops.

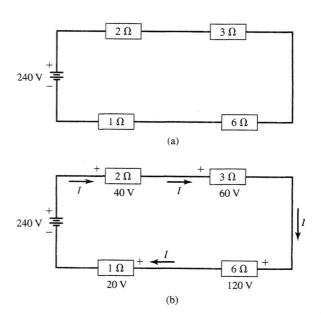

(a)

(b)

Note that *the current in any element of a series circuit is the same as in any other element*. Hence, if the current in any one element is measured, the current in the other elements will be known. Thus the current in R_1, R_2, R_3, R_4, R_n, and E_{bat} will be the same.

The equivalent series circuit of Figure 2–4a is shown in Figure 2–4b. Applying Ohm's law to the equivalent circuit,

$$I = \frac{E_{bat}}{R_{eq.S}}$$

where I = current (A)
E_{bat} = battery voltage (V)
$R_{eq.S}$ = resistance of equivalent series circuit (Ω).

Voltage Drop Across a Resistor

Current in a resistor causes a voltage to appear across the resistor terminals. This voltage, called a *voltage drop*, is equal to the resistance of the resistor multiplied by the current in it. Expressed mathematically,

$$V_{drop} = IR \tag{2–3}$$

where I = current in resistor (A)
R = resistance of resistor (Ω)
V_{drop} = voltage across resistor (V).

Note: Equation (2–3) is an application of Ohm's law.

EXAMPLE 2.4

For the series circuit shown in Figure 2–5a determine (a) the equivalent series-circuit resistance; (b) the circuit current; (c) the voltage drop across each resistor.

Solution
a. $R_{eq.S} = 2 + 3 + 6 + 1 = 12 \; \Omega$
b. $I = \dfrac{E_{bat}}{R_{eq.S}} = \dfrac{240}{12} = 20 \; A$
c. Using Ohm's law:

$V_2 = 20 \times 2 = 40 \; V$ $V_3 = 20 \times 3 = 60 \; V$

$V_6 = 20 \times 6 = 120 \; V$ $V_1 = 20 \times 1 = 20 \; V$

Note: The sum of the voltage drops is equal to the driving voltage: $40 + 60 + 120 + 20 = 240$ V.

Polarity of a Voltage Drop

The terminal at which the current enters a resistor is the positive terminal of that resistor. This is illustrated in Figure 2–5b, along with the direction of current and corresponding voltage drop.

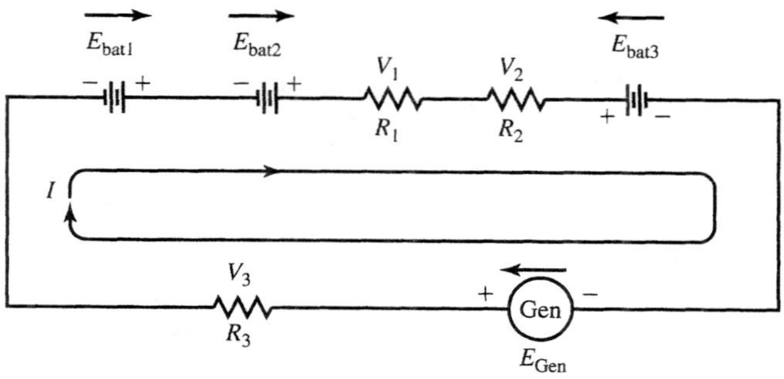

FIGURE 2.6
Series circuit and Kirchhoff's voltage law.

2.4 KIRCHHOFF'S VOLTAGE LAW

For any closed loop, such as the series circuit shown in Figure 2–6, containing resistors and voltage sources (batteries, generators, etc.), the algebraic sum (Σ) of the source voltages is equal to the algebraic sum (Σ) of the voltage drops. Thus, for the circuit in Figure 2–6, traveling the loop in a clockwise direction, the algebraic sum of the source voltages is

$$\Sigma_{\text{source voltages}} = E_{\text{bat1}} + E_{\text{bat2}} - E_{\text{bat3}} + E_{\text{gen}}$$

Note that E_{bat3} was assigned a minus sign because it is opposite in direction to the assigned direction of travel around the loop.

The algebraic sum of the voltage drops around the loop is

$$\Sigma_{\text{voltage drops}} = V_1 + V_2 + V_3 = IR_1 + IR_2 + IR_3$$

In accordance with Kirchhoff's law,

$$\Sigma_{\text{source voltages}} = \Sigma_{\text{voltage drops}}$$

Thus,

$$E_{\text{bat1}} + E_{\text{bat2}} - E_{\text{bat3}} + E_{\text{gen}} = IR_1 + IR_2 + IR_3 = I(R_1 + R_2 + R_3)$$

Kirchhoff's voltage law has many applications in analyzing and troubleshooting electric systems.

EXAMPLE 2.5 For the circuit in Figure 2–6, assume $R_1 = 7.0 \, \Omega$, $R_2 = 12 \, \Omega$, $R_3 = 8 \, \Omega$, $E_{\text{bat1}} = 6.0 \, \text{V}$, $E_{\text{bat2}} = 24 \, \text{V}$, $E_{\text{bat3}} = 12 \, \text{V}$, and $E_{\text{gen}} = 80 \, \text{V}$. Determine (a) current; (b) voltage drop across R_3.

Solution
a. Applying Kirchhoff's voltage law,

$$E_{bat1} + E_{bat2} - E_{bat3} + E_{gen} = I(R_1 + R_2 + R_3)$$
$$6 + 24 - 12 + 80 = I(7 + 12 + 8)$$
$$I = 3.63 \text{ A}$$

b. $IR_3 = 3.63 \times 8 = 29.04 \text{ V}$

Line Drop

When a voltage source is connected to a lamp, motor, or other load, the resistance of the connecting lines (wires or cables) causes the voltage at the load to be less than that at the source. The difference between the source voltage and the voltage across the load is called the voltage drop in the connecting lines, or simply the *line drop*.

EXAMPLE 2.6   A toaster plugged into a wall socket draws 12.5 A. The voltage measured at the wall socket is 121.5 V, and the voltage measured at the toaster is 118.6 V. The circuit is shown in Figure 2–7. Determine (a) the line drop in the connecting cord; (b) the total resistance of the connecting lines.

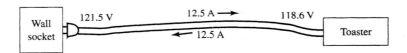

FIGURE 2.7
Illustration of line drop in the connecting lines between a wall outlet and a toaster.

Solution
a. From Kirchhoff's voltage law,

$$V_{source} = V_{cord} + V_{toaster} \Rightarrow 121.5 = V_{cord} + 118.6$$
$$V_{cord} = 2.9 \text{ V}$$

b. From Ohm's law,

$$R_{cord} = \frac{V_{cord}}{I_{cord}} = \frac{2.9}{12.5} = 0.232 \ \Omega$$

Note: There are two conductors in the connecting cord. Thus, the resistance of the connecting cord includes the resistance of both conductors in series.

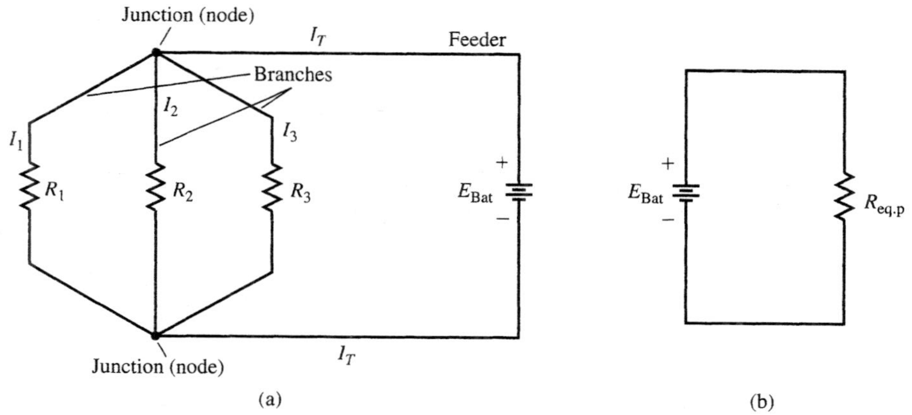

FIGURE 2.8
(a) Representative parallel circuit with three resistors and one battery; (b) equivalent parallel circuit.

2.5 RESISTORS IN PARALLEL

Figure 2–8a shows a representative parallel circuit with three resistors and one battery.[2] The voltage is the same across each element of a parallel circuit. The connection point where the resistors join is called the *junction* or *node*, and the line leads that connect the source to the junction are called the *feeders*.

The equivalent parallel resistance ($R_{eq.P}$) of any group of parallel resistors, shown in Figure 2–8b, may be determined from either of the following equations:

$$R_{eq.P} = \frac{1}{\dfrac{1}{R_1} + \dfrac{1}{R_2} + \dfrac{1}{R_3} + \cdots + \dfrac{1}{R_n}} \quad \text{or} \quad \frac{1}{R_{eq.P}} = \frac{1}{R_1} + \frac{1}{R_2} + \frac{1}{R_3} + \cdots + \frac{1}{R_n} \quad (2\text{–}4)$$

[2]Parallel and series–parallel connections of batteries and uninterruptible power supplies containing batteries are presented in Chapter 27.

EXAMPLE 2.7

Assume the resistors in Figure 2–8a have the following values: $R_1 = 2\ \Omega$, $R_2 = 4\ \Omega$, and $R_3 = 5\ \Omega$. Determine (a) the equivalent parallel resistance of the circuit; (b) the feeder current if the battery voltage is 120 V.

Solution

a.

$$\frac{1}{R_{eq.P}} = \frac{1}{2} + \frac{1}{4} + \frac{1}{5} = 0.95$$

$$R_{eq.P} = \frac{1}{0.95} = 1.0526 \Rightarrow 1.05\ \Omega$$

Note: The equivalent resistance of parallel resistors is less than the resistance of the smallest resistor.

b. $I_T = \dfrac{E_{bat}}{R_{eq.P}} = \dfrac{120}{1.0526} = 114$ A

Special Case Formula for Any Two Resistors in Parallel

A simple special case formula that *applies to only two resistors in parallel* is

$$R_{eq.P2} = \frac{R_1 R_2}{R_1 + R_2} \tag{2-5}$$

where $R_{eq.P2}$ = equivalent parallel resistance of two resistors (Ω).

Special Case Formula for Identical Resistors in Parallel

A special case formula that *applies only to identical resistors in parallel* is

$$R_{eq.Pn} = \frac{R}{n} \tag{2-6}$$

where $R_{eq.Pn}$ = equivalent parallel resistance of n identical resistors (Ω)
R = resistance of one resistor (Ω)
n = number of identical resistors.

EXAMPLE 2.8 Using the special case formula for two parallel resistors, determine the equivalent resistance of a 2400-Ω resistor in parallel with a 9100-Ω resistor.

Solution

$$R_{eq.P2} = \frac{R_1 R_2}{R_1 + R_2} = \frac{2400 \times 9100}{2400 + 9100} = 1899 \ \Omega$$

EXAMPLE 2.9 Determine the equivalent resistance of twenty-five 2-Ω resistors connected in parallel.

Solution

$$R_{eq.Pn} = \frac{R}{n} = \frac{2}{25} = 0.08 \ \Omega$$

2.6 KIRCHHOFF'S CURRENT LAW

The algebraic sum of all currents going to a junction is equal to the algebraic sum of all currents going away from the junction. This law, illustrated in Figure 2–9, is very useful for determining an unknown current in a branch when all other currents at the

FIGURE 2.9
Parallel circuit showing feeder and branches.

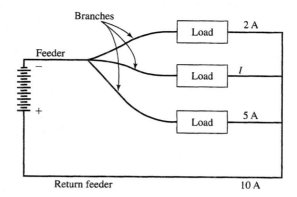

junction are known. Since the feeder current and two of the branch currents in Figure 2–8 are known, the unknown current may be determined using Kirchhoff's current law. Thus, for the circuit in Figure 2–9,

$$2 + I + 5 = 10$$
$$I = 3 \text{ A}$$

2.7 SERIES–PARALLEL RESISTIVE CIRCUITS

An elementary series–parallel resistive circuit is shown in Figure 2–10a. Analysis of such circuits involves the application of Kirchhoff's voltage law, Kirchhoff's current law, and the equations for both equivalent series resistance and equivalent parallel resistance. The following example ties these concepts together.

FIGURE 2.10
(a) Elementary series–parallel resistive circuit; (b) combining series elements; (c) combining parallel elements.

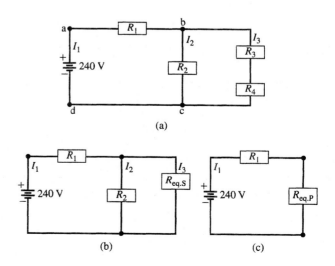

EXAMPLE 2.10

For the circuit shown in Figure 2–10a, $R_1 = 45\ \Omega$, $R_2 = 12\ \Omega$, $R_3 = 18\ \Omega$, and $R_4 = 12.4\ \Omega$. Determine (a) equivalent series–parallel resistance ($R_{eq.SP}$); (b) current supplied by the battery; (c) voltage drop across resistor R_1; (d) voltage between nodes b and c; (e) current in resistor R_2; (f) current in resistor R_4.

Solution

a. The equivalent series–parallel resistance ($R_{eq.SP}$) of the circuit, shown in Figure 2–10a, may be obtained by starting at the section farthest from the voltage source, combining elements as you progress toward the input terminals.

Thus, starting with the series section composed of R_3 and R_4, as shown in Figure 2–10a, the equivalent series-circuit combination is

$$R_{eq.S} = R_3 + R_4 = 18 + 12.4 = 30.4\ \Omega$$

Combining the parallel combination of $R_{eq.S}$ with R_2, as shown in Figure 2–10b,

$$R_{eq.P} = \frac{R_{eq.S} \times R_2}{R_{eq.S} + R_2} = \frac{30.4 \times 12}{30.4 + 12} = 8.60\ \Omega$$

Combining the series combination of $R_{eq.P}$ with R_1, as shown in Figure 2–10c,

$$R_{eq.SP} = R_{eq.P} + R_1 = 8.60 + 45 = 53.6\ \Omega$$

b. Applying Ohm's law to the reduced circuit in Figure 2–10c,

$$I_1 = \frac{E_{bat}}{R_{eq.SP}} = \frac{240}{53.6} = 4.477\ A \Rightarrow 4.48\ A$$

c. The voltage drop across R_1 is

$$V_{R1} = I_1 \times R_1 = 4.477 \times 45 = 201.48\ V$$

d. Applying Kirchhoff's voltage law to loop abcd in Figure 2–10a,

$$E_{bat} = V_{ab} + V_{bc} \Rightarrow 240 = 201.48 + V_{bc}$$
$$V_{bc} = 38.52\ V$$

e. $I_2 = \dfrac{V_{bc}}{R_2} = \dfrac{38.52}{12} = 3.21\ A$

f. Applying Kirchhoff's current law to node b,

$$I_1 = I_2 + I_3 \Rightarrow 4.48 = 3.21 + I_3$$
$$I_3 = 1.27\ A$$

Or, applying Ohm's law to the I_3 branch,

$$I_3 = \frac{V_{bc}}{R_3 + R_4} = \frac{38.51}{18 + 12.4} = 1.27\ A$$

2.8 POWER LOSS IN CONDUCTORS

The forced movement of electrons through a conductor, by the application of a driving voltage, results in the generation of heat. The heating is caused by electron collisions with atomic particles of the material. The *rate* at which electric energy is converted to heat energy is called *electric power*, and its basic unit is the watt. The electric power expended in the resistance of a conductor is equal to the resistance of the conductor times the square of the current. Expressed mathematically,

$$P = I^2 R \qquad (2-7)$$

where P = rate of conversion of electric energy to heat energy in watts (W)
I = current in conductor (A)
R = resistance of conductor (Ω).

Equation (2–7) provides the most convenient method for determining the heat-power loss caused by electron collision in conductors. Two other useful formulas for calculating heat-power loss in resistors are as follows:

$$P = \frac{V^2}{R} \qquad (2-8)$$

$$P = VI \qquad (2-9)$$

where V = voltage drop measured across the resistor (V)
I = current in the resistor (A).

Equations (2–8) and (2–9) were derived by substituting Ohm's law equation into power equation (2–7). In electric power systems involving thousands or millions of watts, the power is generally expressed in kilowatts (kW) or megawatts (MW) respectively, where

$$1 \text{ kW} = 1000 \text{ W}$$
$$1 \text{ MW} = 1,000,000 \text{ W}$$

EXAMPLE 2.11   A 240-V DC source is connected to an 8-Ω resistor. Determine the heat power expended in the resistor.

Solution

$$P = \frac{V^2}{R} = \frac{(240)^2}{8} = 7200 \text{ W} = 7.2 \text{ kW}$$

EXAMPLE 2.12   A 240-V DC generator is supplying 45 amperes to a DC motor through a two-conductor cable as shown in Figure 2–11a. The resistance of each conductor is 0.04 Ω. Determine (a) the total heat-power loss in the cable; (b) the total voltage drop in the cable. The equivalent series circuit is shown in Figure 2–11b.

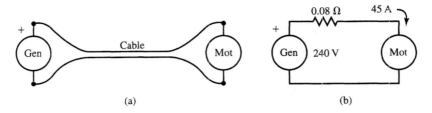

FIGURE 2.11
Circuit for Example 2–12.

Solution
Since the two conductors that make up the cable are connected in series, as shown in Figure 2–11a, the equivalent series resistance of the cable is

$$R_{cable} = 2 \times 0.04 = 0.08 \ \Omega$$

a. $P_{cable} = I^2 R_{cable} = 45^2 \times 0.08 = 162 \ W$
b. $V_{cable} = I \times R_{cable} = 45 \times 0.08 = 3.6 \ V$

2.9 ENERGY LOSS IN CONDUCTORS

Equations (2–7), (2–8), and (2–9) are very useful for calculating heat power, which is the *rate* at which electric energy is converted to heat energy. The total energy expended, however, is equal to the power multiplied by the total time that the circuit was energized. Expressed mathematically,

$$W = Pt \qquad \qquad \textbf{(2–10)}$$

where W = energy in watt-seconds (W·s)
P = power (W)
t = time (s).

Note: Watt-seconds are also called joules (J), where 1 joule = 1 watt-second.
Other useful equations for energy, derived from the power relationships in Section 2–8 and Ohm's law, are as follows:

$$W = I^2 Rt \qquad \qquad \textbf{(2–11)}$$

$$W = \frac{V^2}{R} t \qquad \qquad \textbf{(2–12)}$$

$$W = VIt \qquad\qquad (2\text{–}13)$$

If the power is in kilowatts (kW) and the time is in hours (h), the energy expended may be expressed in watt-hours (Wh) or in kilowatt-hours (kWh), where one kilowatt is equal to one thousand watts.

Depending on the I^2R heat losses in the conductor, the total operating time, and the environmental conditions, the temperature of the conductor can range from warm to the melting point of the material. A well-known example of heat loss in conductors is the heating of extension cords and cables when overloaded. Almost everyone has observed the heating of a plug where it mates with a wall receptacle for such household appliances as toasters, grills, and coffee pots; if the contact resistance between the two mating surfaces is excessively high, the heat produced can start a fire. Mating surfaces for plug-in devices and contactors depend on spring pressure and clean surfaces for low contact resistance. Corroded or dirty surfaces, worn contacts, or weak spring pressure due to loss of temper results in a higher than normal contact resistance.

EXAMPLE 2.13   If the motor in Example 2–12 were operating continuously at 45 A for 60 h, determine the total heat energy expended in the two-conductor cable.

Solution

$$W_{cable} = P_{cable} \times t$$
$$W_{cable} = (I^2R) \times t = (45^2 \times 0.08) \times 60 = 9720 \text{ Wh or } 9.72 \text{ kWh}$$

EXAMPLE 2.14   Given a two-wire cable whose total resistance is 1.0 ohm. Determine which of the following combinations of current and time results in the greatest energy loss, and which combination results in the least energy loss: 100 A for 1 second, 50 A for 3 seconds, or 25 A for 20 seconds.

Solution

$$W = I^2R \cdot t$$
$$W_1 = 100^2 \times 1.0 \times 1 = 10{,}000 \text{ W·s or } 10{,}000 \text{ J}$$
$$W_2 = 50^2 \times 1.0 \times 3 = 7500 \text{ W·s or } 7500 \text{ J}$$
$$W_3 = 25^2 \times 1.0 \times 20 = 12{,}500 \text{ W·s or } 12{,}500 \text{ J}$$

The 25-A current for 20 seconds produces the most heat loss, and the 50-A current for 3 seconds produces the least heat loss.

EXAMPLE 2.15   A plot of voltage across a 10-Ω resistor vs. time is shown in Figure 2–12. Determine the total heat energy expended in 8 seconds.

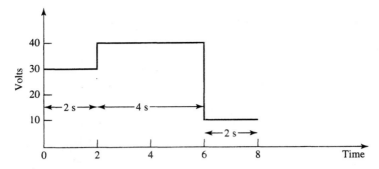

FIGURE 2.12
Plot of voltage applied to a 10-Ω resistor vs. time.

Solution

$$W = \sum \left[\frac{V^2}{R} \times t \right] = \left[\frac{30^2}{10} \times 2 \right] + \left[\frac{40^2}{10} \times 4 \right] + \left[\frac{10^2}{10} \times 2 \right]$$

$$= 180 + 640 + 20 = 840 \text{ J}$$

2.10 CIRCUIT FAULTS

Common circuit faults that occur in electrical power apparatus include short circuits, flashovers, grounds, and opens. Damage from such faults must be repaired, the cause of the fault determined, and action taken to prevent recurrence.

Short Circuit

A short circuit, commonly called a short, is an electrical connection that bypasses part or all of an electric circuit, thus providing an additional route for the current. Figure 2–13 illustrates some examples of undesirable shorts in electric circuits. Figure 2–13a shows a *solid short*, where a solid metallic connection across the incoming wires to the lamp may result in a blown fuse or tripped circuit breaker.

Figure 2–13b shows a *partial short* between two commutator bars, where an accumulated mixture of carbon dust and oil vapors formed a conducting path; it is called a partial short, or *high-resistance short*, because the resistance of the conducting path is higher than that for a metal-to-metal contact.

Figure 2–13c shows the shorted coil in a field winding; the insulation, worn away by either vibration, wind erosion, or excess heat, caused adjacent wires to make contact. Short circuits can also be caused by loose connections, mechanical faults, rodents gnawing on insulation, and wrong connections.

If electric conduction between adjacent wires of the coil is due to dirt, it is called a high-resistance or partial short; if due to metal-to-metal contact, it is called a dead

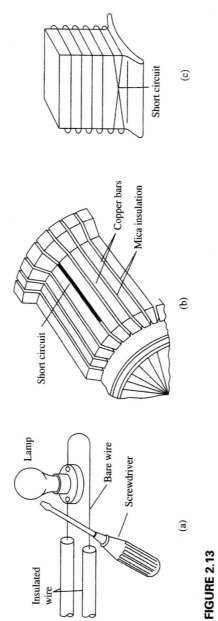

FIGURE 2.13

Examples of short circuits: (a) solid short; (b) dirt short-circuiting copper commutator bars; (c) short between two turns of a field coil.

short. Short circuits involving high currents can cause arcing, burning, melting of conductors, severe mechanical stresses, or an explosion.[3]

---

[3]Case studies of explosions that actually occurred in motor terminal boxes due to short circuits within the motor or within the terminal box are discussed in Reference [1].

EXAMPLE 2.16

A 240-V DC generator supplies power through a 900-ft two-conductor cable to a heating load whose resistance is 5.76 Ω. The circuit is shown in Figure 2–14a, and an equivalent series-circuit model representing Figure 2–14a is shown in Figure 2–14b. The two conductors that make up the cable are each size AWG-6 tinned copper wire. Determine (a) current in the cable; (b) voltage drop in the cable; (c) voltage at the load; (d) power loss in the cable; (e) power drawn by the load; (f) current in the cable if an accidental short occurs across the heater terminals.

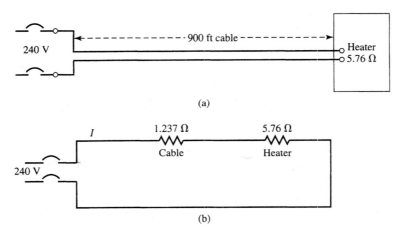

(a)

(b)

FIGURE 2.14
(a) Circuit for Example 2–16; (b) equivalent series-circuit model.

Solution
a. From Table 1–3 in Chapter 1, the resistance of AWG-6 tinned copper wire is 0.427 Ω/kft at 25°C. Thus,

$$R_{cable} = 2 \times 900 \times \frac{0.427}{1000} = 0.7686 \ \Omega$$

The total resistance of the series circuit is

$$R_{eq.S} = R_{cable} + R_{heater} = 0.7686 + 5.76 = 6.5286 \Rightarrow 6.5 \ \Omega$$

Applying Ohm's law,

$$I = \frac{V}{R_{eq.S}} = \frac{240}{6.5286} = 36.76 \Rightarrow 36.8 \ A$$

b. $V_{cable} = IR_{cable} = 36.76 \times 0.7686 = 28.25$ V
c. $V_{load} = V - IR_{cable} = 240 - 28.25 = 211.74$ V
d. $P_{cable} = I^2R_{cable} = 36.76^2 \times 0.7686 = 1038.7$ W
e. $P_{load} = I^2R_{load} = 36.76^2 \times 5.76 = 7784$ W $\Rightarrow$ 7.8 kW
f. A solid short across the heater terminals bypasses the heater, leaving only the cable resistance to limit the current.

$$I_{short\ circuit} = \frac{V}{R_{cable}} = \frac{240}{0.7686} = 312.3 \text{ A}$$

Flashover

A flashover is a violent disruptive discharge around or across the surface of solid or liquid insulation. Flashovers occur suddenly, involve heavy currents, and generally cause considerable damage. A flashover is always preceded by ionization, a process by which the surrounding atmosphere is made into a conductor. Flashovers may be caused by "electron creepage" across dirty and moist insulators; very high temperatures, such as that caused by a nearby arc; switching overvoltages; or lightning.

Creepage Distance

The creepage distance is the shortest distance between two bare conductors of opposite polarity or between a conductor and ground, measured along the surface of an insulating material. Although the current through creepage paths may not be appreciable, the arcing and sparking produced can ionize the surrounding atmosphere and cause a flashover. Electric arcs are highly mobile and can travel away from the initial striking point, causing extensive damage to nearby equipment.[4] Furthermore, when arcing occurs in metal-enclosed switchgear, the arc can be expected to spread to the equipment ground.[5]

Ground Fault

A ground fault is an accidental electrical connection between the wiring of an apparatus and its metal framework or enclosure, the wiring and a water pipe, the wiring and the armored sheath of a cable, the wiring and the chassis of a car, the wiring and the hull of a ship, etc. Figure 2–15 illustrates an accidental ground fault caused by the chafing of the insulation against a sharp edge in the opening of a connection box.

---

[4] See arcing faults and arc migration, Section 3–6, Chapter 3.

[5] The equipment ground is the metallic framework of switchboards, machines, cable armour, etc. Equipment grounds are connected to earth through a common-equipment-grounding conductor. See Chapter 28 for more detailed information.

FIGURE 2.15
Conductor grounded on a connection box.

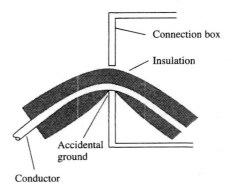

Ground faults can also occur in electrical machinery. The effects of age, heat, and vibration can damage the insulation, permitting the entry of conducting dust and moisture. If the dust forms a bridge between the exposed conductor and the frame, the wiring is grounded. Apparatus exposed to atmospheric conditions, even though totally enclosed and "weatherproofed," sometimes develop ground faults due to breathing. During the heat of the day the air inside the apparatus expands, forcing some internal air through the seals to the outside. In the evening, when the apparatus cools, a partial vacuum is formed inside the apparatus, and cool moist air is sucked in. The moisture condenses inside the apparatus, and after many heating and cooling cycles, a small pool of water begins to form. Eventually enough water may collect to make an electrical connection between the wiring and the frame, and the equipment suffers a ground fault.

Open

An open, or open circuit, is a break in the continuity of the circuit. This break can be deliberate or accidental. Deliberate opens are made by the manual opening of a switch or circuit breaker. Accidental opens may be caused by blowing of a fuse, melting of a soldered connection, blowing out of a lamp, a loose connection, a broken wire, a broken resistor due to vibration, etc.

SUMMARY OF EQUATIONS FOR PROBLEM SOLVING

$$I = \frac{V}{R} \tag{2-1}$$

$$R_{\text{eq.S}} = R_1 + R_2 + R_3 + R_4 + \cdots + R_n \tag{2-2}$$

$$R_{\text{eq.P}} = \cfrac{1}{\cfrac{1}{R_1} + \cfrac{1}{R_2} + \cfrac{1}{R_3} + \cdots + \cfrac{1}{R_n}} \quad \text{or} \quad \frac{1}{R_{\text{eq.P}}} = \frac{1}{R_1} + \frac{1}{R_2} + \frac{1}{R_3} + \cdots + \frac{1}{R_n} \tag{2-4}$$

$$R_{\text{eq.P2}} = \frac{R_1 R_2}{R_1 + R_2} \tag{2-5}$$

$$R_{\text{eq.P}n} = \frac{R}{n} \tag{2-6}$$

$$P = I^2 R \tag{2-7}$$

$$P = \frac{V^2}{R} \tag{2-8}$$

$$P = VI \tag{2-9}$$

$$W = Pt \tag{2-10}$$

$$W = I^2 Rt \tag{2-11}$$

$$W = \frac{V^2}{R}t \tag{2-12}$$

$$W = VIt \tag{2-13}$$

SPECIFIC REFERENCE KEYED TO TEXT

[1] Crawford, K. S., Motor Terminal Box Explosions Due to Faults, *IEEE Trans. Industry Applications*, Vol. IA-29, No. 1, January/February 1993.

REVIEW QUESTIONS

1. How can Ohm's law be used to determine resistance?
2. State the formula for resistors in series.
3. What is a voltage drop and what causes it?
4. State Kirchhoff's voltage law and illustrate it with your own example.
5. What is meant by a line drop?
6. State the general formula for resistors in parallel.
7. State the special case formulas for two different resistors in parallel.
8. State the special case formulas for two or more identical resistors in parallel.
9. State Kirchhoff's current law and illustrate it with your own example.
10. Differentiate between power loss and energy loss in conductors.
11. Describe a short circuit and illustrate it with an example not given in the text.
12. What is a flashover? How can flashovers be prevented?
13. Make a three-dimensional sketch showing an accidental ground that should be corrected.
14. Define "open" as it pertains to an electric circuit.
15. Describe three accidental opens that can occur in electrical apparatus or wiring systems, other than those described in the text.

PROBLEMS

2–1/1. Determine the voltage required to cause 15 A in a coil wound with 600 ft of size 18 AWG copper wire.

2–2/1. Repeat Problem 2–1/1 for size 18 AWG aluminum cable.

2–3/1. Determine the current required to obtain 420 V across a 16-MΩ resistor.

2–4/2. If the voltage measured across a resistor is 156 V and the measured current to the resistor is 4.0 mA, determine the ohmic value of the resistor.

2–5/3. Determine the equivalent resistance of the following series-connected resistors: 6 Ω, 10 Ω, 5 Ω, and 30 Ω.

2–6/3. The equivalent resistance of six series-connected resistors is 10 kΩ. If the values of five of the resistors are 1000 Ω, 2400 Ω, 1800 Ω, 4000 Ω, and 500 Ω, respectively, determine the resistance of the sixth resistor.

2–7/3. A 120-V source of negligible resistance is connected in series with the following resistors: 20 Ω, 200 Ω, 2000 Ω, and 20 kΩ. Sketch the circuit and determine (a) the current; (b) the voltage drop across each resistor.

2–8/3. Four field coils, each with a resistance of 30 ohms, are connected in series and supplied by a 240-V DC generator. Sketch the circuit and calculate (a) the current in each coil; (b) the voltage across each coil.

2–9/3. A hair dryer with a hot resistance of 13.71 Ω is connected to a 118.5-V source. The cord connecting the hair dryer to the electrical outlet has a total resistance of 0.18 Ω. Sketch the circuit and determine (a) the current in the hair dryer; (b) the voltage at the hair dryer; (c) the line drop in the cord.

2–10/3. A motor field circuit consisting of four coils of wire, each wrapped around a separate iron core, is connected in series and supplied by a 126-V source. The resistance of the coils are 56 Ω, 54 Ω, 53 Ω, and 57 Ω, respectively. Sketch the circuit and determine (a) the resistance of the field circuit; (b) circuit current; (c) voltage drop across the 54-Ω coil.

2–11/4. A 36-volt battery and a 12-V battery are in series with a 2-Ω, a 4-Ω, and a 6-Ω resistor. Determine the current if (a) the batteries are additive; (b) the batteries are subtractive.

2–12/4. A 120-V battery, a 72-V battery, and a 58-Ω resistor are connected in series. Determine the current if (a) the battery voltages are additive; (b) the battery voltages are subtractive. (c) Sketch the two circuits.

2–13/5. Determine the equivalent resistance of the following parallel-connected resistors: 6 Ω, 10 Ω, 5 Ω, 30 Ω.

2–14/5. The equivalent resistance of six parallel resistors is 0.29 Ω. If five of the resistors are 3.2 Ω, 4.5 Ω, 2.5 Ω, 5.2 Ω, and 6.1 Ω, respectively, determine the resistance of the sixth resistor.

2–15/5. Using the special case formula, determine the equivalent resistance of each of the following sets of paralleled resistors: (a) 100 Ω, 80 Ω; (b) 5 Ω, 500 Ω; (c) 2 kΩ, 2 kΩ; (d) 0.01 Ω, 100 Ω.

2–16/5. Twenty 12-Ω resistors are connected in parallel, and the combination is connected to a 40-V source. Determine (a) the equivalent parallel resistance; (b) the total current supplied by the source.

2–17/6. A 12-V battery is connected to a feeder that supplies current to three branch circuits composed of a 3-ohm resistor, a 120-ohm resistor, and a 60-ohm resistor, respectively. Sketch the circuit and, using Kirchhoff's current law, determine the battery current.

2–18/6. A 3-V battery is connected to a parallel circuit consisting of the following three resistors: 0.50 Ω, 0.333 Ω, and 0.20 Ω. Sketch the circuit and determine (a) the equivalent parallel resistance; (b) the overall circuit current.

2–19/6. Sketch the circuit and determine (a) the component currents and (b) the total circuit current for a system composed of a 12-V battery connected to a parallel arrangement of the following three resistors: 3 Ω, 2 Ω, and 6 Ω.

2–20/6. A 120-V DC generator supplies a total of 6 A to two resistors connected in parallel. The current drawn by one resistor is 2 A. Sketch the circuit and determine (a) the current drawn by the other resistor; (b) the resistance of the other resistor.

2–21/7. For the series–parallel circuit shown in Figure 2–16, $R_1 = 14\ \Omega$, $R_2 = 20\ \Omega$, $R_3 = 16\ \Omega$, $R_4 = 12\ \Omega$. Determine the equivalent resistance.

2–22/7. For the series–parallel circuit shown in Figure 2–17, $R_1 = 100\ \Omega$, $R_2 = 300\ \Omega$, $R_3 = 500\ \Omega$, $R_4 = 400\ \Omega$, $E_{gen} = 240$ V. Determine (a) the equivalent resistance; (b) the current supplied by the generator.

FIGURE 2.16
Series–parallel circuit for Problem 2–21/7.

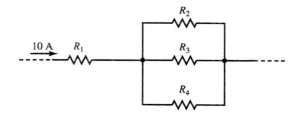

FIGURE 2.17
Series–parallel circuit for Problem 2–22/7.

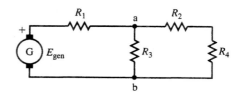

FIGURE 2.18
Series–parallel circuit for Problem 2–23/7.

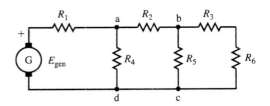

2–23/7. For the series–parallel circuit in Figure 2–18, $R_1 = 150\ \Omega$, $R_2 = 500\ \Omega$, $R_3 = 1000\ \Omega$, $R_4 = 100\ \Omega$, $R_5 = 2000\ \Omega$, $R_6 = 4000\ \Omega$, $E_{gen} = 600$ V. Determine (a) the equivalent resistance; (b) the current supplied by the generator.

2–24/8. A 240-V generator supplies 50 A to a feeder that supplies four branch circuits. The currents in three of the branches are 12 A, 6 A, and 25 A, respectively. Sketch the circuit and determine (a) current in the fourth branch; (b) total power supplied by the generator.

2–25/8. A 120-V generator supplies power to the following parallel connected lamps: 60 W, 100 W, and 150 W. Each lamp has a voltage of 120 V. Sketch the circuit and determine (a) the current drawn by each lamp; (b) the total current; (c) the total current if an open occurs in the 150-W lamp.

2–26/8. The total heat-power loss in a two-conductor extension cord is 10 watts when it carries a current of 30 A. Sketch the circuit and determine the resistance of each of the two wires.

2–27/8. What is the heat-power loss in 150 ft of size 2 AWG two-conductor untinned copper cable that supplies 90 A to a 25-hp, 230-V DC motor?

2–28/8. A 600-ft-long cable containing two rubber-insulated size AWG-6 copper conductors is used to connect an electric range to a 240-V DC generator. Sketch the circuit and determine (a) the resistance of the cable; (b) the total heat-power losses in the cable when the range draws 60 A; (c) the voltage drop in the cable; (d) the voltage drop across the range.

2–29/8. A 240-V DC generator supplies power to an electromagnet through a 450-ft cable. The two conductors that make up the cable are each size AWG-8 tinned copper wire. The resistance of the electromagnet is 6.2 Ω. Sketch the circuit and determine (a) the current in the magnet; (b) the voltage drop in the cable; (c) the voltage across the magnet; (d) the power loss in the cable; (e) the current in the cable if an accidental solid short occurs across the magnet terminals.

2–30/8. A 240-V driver is connected to a series circuit consisting of a lamp, a 4-Ω resistor, a 6-Ω resistor, and a 10-Ω resistor. An ammeter clipped to the circuit indicates 2 A. Sketch the circuit and, using Kirchhoff's voltage law, determine (a) the voltage across the lamp; (b) the heat power dissipated by each of the three resistors and by the lamp.

2–31/8. Four heating elements each rated at 600 W and 120 V are required for use in an oven. The only voltage available is 240 V. Make a sketch showing the proper way to connect the elements to obtain 2400 W in the oven. (a) What is the resistance of each element? (b) What is the overall circuit current? (c) What is the overall circuit resistance?

2–32/9. An electric heater draws 2000 W from a 100-V DC system. Sketch the circuit and determine (a) the current drawn by the heater; (b) the resistance of the heater; (c) the total heat energy expended in 8 h.

2–33/9. An electric soldering iron uses 6 kWh of energy in 12 h when connected to a 120-V source. Determine (a) the power rating of the soldering iron; (b) the current; (c) the resistance of the soldering iron.

2–34/9. Determine which combination of current and time will result in the greatest expenditure of heat energy in a 20-Ω resistor: 100 A for 1 s, 60 A for 5 s, or 25 A for 10 s?

2–35/9. Which of the following combinations of current and time will result in the greatest expenditure of heat energy in a 10-Ω resistor: 50 A for 1 s, 80 A for 0.5 s, 10 A for 10 s, or 1 A for 1000 s?

2–36/9. A motor field circuit, consisting of four coils of wire connected in series, is connected to a 240-V DC driver. Each coil has a resistance of 26 Ω. Sketch the circuit and determine (a) the resistance of the field circuit; (b) the circuit current; (c) the total heat-power loss; (d) the voltage drop across each coil; (e) the total heat energy expended in 8 hours of operation.

2–37/9. Referring to the circuit in Figure 2–19, $R_1 = 2\ \Omega$, $R_2 = 6\ \Omega$, $R_3 = 8\ \Omega$, $R_4 = 4\Omega$, $E_{bat} = 12$ V, $E_{gen1} = 24$ V, $E_{gen2} = 36$ V. Determine (a) the current; (b) the voltage drop across the 6-Ω resistor; (c) the power drawn by the 6-Ω resistor; (d) the energy expended in the 6-Ω resistor if the circuit was energized for 20 hours.

2–38/9. Repeat Problem 2–37/9 assuming the polarity of generator E_{gen1} is reversed.

2–39/10. A circuit consisting of four 50-Ω resistors connected in series draws 1.6 A from a DC generator. Sketch the circuit and, assuming that an accidental

FIGURE 2.19
Circuit for Problem 2–37/9.

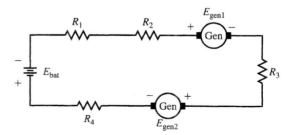

partial short occurs across one resistor, reducing its value to 4.2 Ω, determine the current for this fault condition.

2–40/10. A generator supplies 5000 W at 246 V to an electric heater via a 50-ft two-conductor cable. The conductors are AWG-10 copper. Sketch the circuit and determine (a) current in cable; (b) resistance of the cable; (c) generator voltage; (d) current in cable if an accidental solid short occurs at the heater terminals.

3

Magnetic Fields and Magnetic Forces in Electric Power Systems

3.0 INTRODUCTION

Electromagnetic forces are present whenever electric circuits are energized. The magnitude of these forces, however, is exceptionally high in electrical power apparatus when large currents are present. If these forces are not adequately controlled, damage to insulation, warped switchboards, severely damaged machines, fire, and explosions can occur. Thus, for safety to personnel, and for the protection of electrical power apparatus, maintenance and operating personnel must have a thorough understanding of the mechanical forces produced by magnetic action. This chapter provides the student with a basic understanding of the interaction of magnetic fields and illustrates some of the damaging effects of uncontrolled current.

3.1 THE MAGNETIC FIELD

A magnetic field is present whenever electric charges are in motion. For example, a flash of lightning between clouds, or between a cloud and earth, generates a magnetic field that can be detected miles away from the actual stroke. A similar but much lesser magnetic field will be produced if an object is given an electric charge by rubbing, and the object is then hurled across the room; the magnetic field will be present only while the electric charge is in motion. Figure 3–1a illustrates the magnetic field surrounding a lightning stroke, and Figure 3–1b illustrates the magnetic field around a charged object moving through dry air or a vacuum. The concentric circles around the moving charges represent the magnetic field and are called *magnetic lines of flux*. The unit of magnetic flux is the weber (Wb); one weber is equal to 10^8 lines of flux.

In a similar manner, as shown in Figure 3–2a, the movement of electric charges in a conductor, caused by the application of a driving voltage, will generate a magnetic field around the conductor. The magnetic field, or flux as it is called, completely

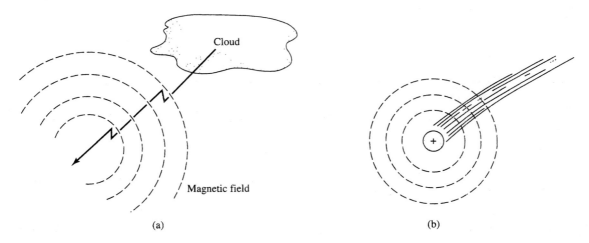

FIGURE 3.1
(a) Magnetic field surrounding a lightning stroke; (b) magnetic field surrounding a charged body moving through dry air or through a vacuum.

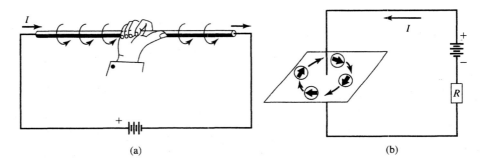

FIGURE 3.2
(a) Determining the direction of the magnetic field around a conductor, using the right-hand rule; (b) detecting a magnetic field with a magnetic compass.

surrounds the current that causes it and is perpendicular to it. Note that the *battery voltage is the driver or electromotive force (emf) that causes the current, and the current causes the magnetomotive force (mmf) that causes the magnetic field.*

The direction of the magnetic field around a current is dependent on the direction of the current and may be determined by the right-hand rule: Grasp the conductor with the right hand with the thumb pointing in the direction of the current, as shown in Figure 3–2a, and the fingers will point in the direction of the magnetic field around the conductor. The presence of this magnetic field and its perpendicularity relative to the current can easily be detected with a magnetic compass, as shown in Figure 3–2b.

If a long insulated conductor carrying an electric current is wound in the form of a coil, the flux around the wire will be concentrated in a relatively small area, as shown

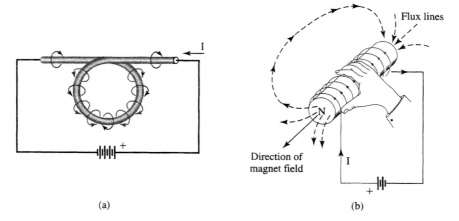

(a)   (b)

FIGURE 3.3

(a) Coiling a conductor into one or more turns concentrates the flux. (b) Using the right-hand rule to determine the direction of the magnetic flux in a coil.

in Figure 3–3a. Coiling a conductor into two or more turns has the effect of using the same current more than once. For example, a four-turn coil carrying a current of 10 amperes produces the same magnetic flux as a one-turn coil carrying 40 amperes.

The magnetic polarity of a coil may be determined by using the right-hand rule: Grasp the coil with the right hand, as shown in Figure 3–3b, so that the fingers point or curl in the direction the current takes around the loops, and the thumb will then point in the direction of the north pole. The direction of the flux will be from north to south outside the coil, and from south to north inside the coil.

3.2 MECHANICAL FORCES PRODUCED BY THE INTERACTION OF MAGNETIC FIELDS

When two or more sources of magnetic fields are arranged so that their fluxes, or a component of their fluxes, are parallel within a common region, a mechanical force will be produced that tends to either push the sources of flux together (attract) or push them apart (repel).

Through experiment, and as shown in Figure 3–4, it was learned that like poles of magnets repel each other, and unlike poles attract. If a repelling or separating force occurs, as shown in Figures 3–4a and 3–4b, it is observed that the component fluxes in the common region are parallel and in the same direction, resulting in a net increase in flux, hereafter called *flux bunching*. However, if a force of attraction occurs, as shown in Figures 3–4c and 3–4d, it is observed that the component fluxes in the common region are parallel but in opposite directions, resulting in a net decrease in flux.

Thus, the direction of the mechanical force, produced by the interaction of any two magnetic fields, may be determined from the relative directions of the parallel

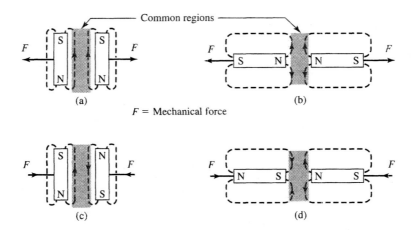

FIGURE 3.4
Mechanical forces produced by magnetic action: (a, b) force of repulsion; (c, d) force of attraction.

components of the respective fluxes in a common region. A net increase in flux results in a force of repulsion, while a net decrease in flux results in a force of attraction.

3.3 MECHANICAL FORCES BETWEEN ADJACENT PARALLEL CONDUCTORS

Figure 3–5 shows the direction of the mechanical forces (F) produced by the magnetic fields of adjacent current-carrying conductors. If the currents in adjacent conductors are in opposite directions, as shown in Figures 3–5a and 3–5b, the respective fluxes in the common region will be parallel and in the same direction (flux bunching), causing a separating force to be produced. However, if the respective currents in adjacent conductors are in the same direction, as shown in Figures 3–5c and 3–5d, the fluxes in the common region will be parallel and in opposite directions, causing a force of attraction.

The forces developed between adjacent conductors, under conditions of very high currents such as may occur during fault conditions, can literally bend the conductors out of shape and even tear them from their supporting structures. Damage to a switchboard caused by a very severe short circuit is illustrated in Figure 29–11 in Chapter 29. Mechanical forces on switchboard buses, caused by short-circuit current can reach several tons per foot.

In those applications where the available short-circuit current is of such a magnitude as to cause destruction of equipment if a fault occurs, special fuses (called current limiters) are installed. These fuses can open the circuit in less than 4 ms, preventing the current from attaining damaging values.[1]

---

[1]See Section 29–3, Chapter 29.

FIGURE 3.5

Mechanical force (*F*) produced by magnetic fields around current-carrying conductors: (a, b) forces of repulsion; (c, d) forces of attraction.

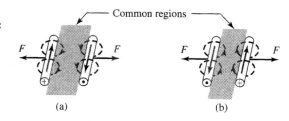

(a)   (b)

F = Mechanical force

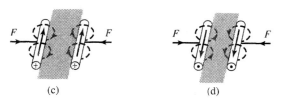

(c)   (d)

The magnitude of the force of attraction or repulsion produced by direct current or alternating current on each of two long, straight parallel conductors in air, such as the conductors in Figure 3–5, may be determined by substituting into the following formula:

$$F = 5.4 I_1 I_2 \frac{\ell}{D} \times 10^{-7} \qquad (3\text{–}1)$$

where D = distance between centers of conductors (in.)
          ℓ = length of conductors (ft)
          F = force on each conductor (lb)
          I_1 = current in conductor 1 (A)
          I_2 = current in conductor 2 (A).

If the current in each conductor is the same, Eq. (3–1) reduces to

$$F = 5.41 \, I^2 \frac{\ell}{D} \times 10^{-7} \qquad (3\text{–}2)$$

Note: To express the force in newtons (N), multiply the force in pounds by 4.448.

As indicated in Eq. (3–2), for a given length, and given distance between conductors carrying the same current, the mechanical force is proportional to the square of the current. That is,

$$F \propto I^2 \qquad (3\text{–}3)$$

EXAMPLE 3.1 Two parallel cables, 3 inches between centers, are supported at 10-foot intervals and each carries 780 amperes of direct current in opposite directions. An accidental dead short causes a fault current of 50,000 amperes through each conductor. Determine (a) the mechanical force developed per 10-foot length of cable for rated current; (b) repeat (a) for the fault current.

Solution

a. $F = 5.4I^2 \dfrac{\ell}{D} \times 10^{-7} = 5.4 \times 780^2 \times \dfrac{10}{3} \times 10^{-7} = 1.095 \Rightarrow 1.1$ lb

b. $F = 5.4I^2 \dfrac{\ell}{D} \times 10^{-7} = 5.4 \times 50,000^2 \times \dfrac{10}{3} \times 10^{-7} = 4500$ lb

The separating force caused by the fault current would probably rip the cables off the supports.

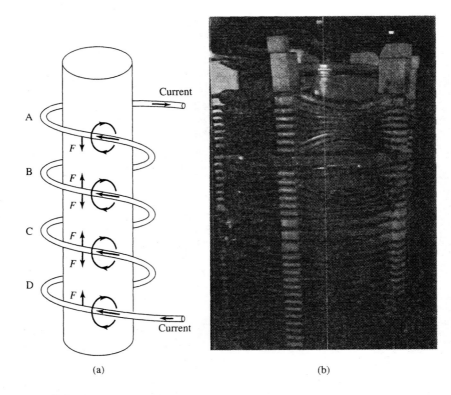

(a) (b)

FIGURE 3.6
(a) Direction of mechanical forces between adjacent turns in a coil; (b) transformer windings squashed by forces due to excessively high currents (courtesy Hartford Steam Boiler Inspection and Insurance Co.).

3.4 MECHANICAL FORCES BETWEEN ADJACENT TURNS OF A COIL

The direction of the forces exerted on the adjacent turns of a coil may be determined from the directions of the respective fluxes in the common regions. Thus, referring to Figure 3–6a, turn B is attracted in opposite directions by turn A and turn C, respectively. Thus, the net force on turn B is zero. Turn C is attracted in opposite directions by turn B and turn D, respectively, and thus it too has a net force of zero.

However, end turn A is attracted only by turn B, and end turn D is attracted only by turn C. The net result is a distortion of the end turns. An extremely high current, such as that caused by severe overload or a short circuit, could result in forces high enough to distort the windings and crush the coil insulation. The transformer in Figure 3–6b shows the damaging effect that excessively high current can have on transformer windings. Similar effects can occur in motor and generator windings.

3.5 MECHANICAL FORCES BETWEEN A CONDUCTOR AND A MAGNET

Figure 3–7a illustrates the interaction of the magnetic field produced by the current in a conductor, and the magnetic field produced by a magnet, when the current is perpendicular to the magnetic lines of the magnet. Figure 3–7b is an end view of the conductor showing the flux due to the magnet (ϕ_{magnet}) and the flux due to the current in the conductor ($\phi_{conductor}$). Note that the flux from the magnet is upward on both sides of the conductor, whereas the flux produced by the current in the conductor is downward

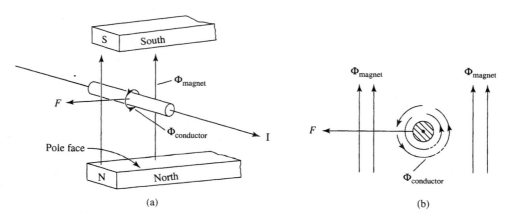

(a) (b)

FIGURE 3.7

(a) Mechanical force produced by the interaction of the magnetic field of a current and the magnetic field of a magnet; (b) end view of the conductor.

on the left side of the conductor and upward on the right side. In the common region to the left of the conductor the respective fluxes are in opposite directions, and in the common region to the right of the conductor the respective fluxes are in the same direction (flux bunching).

Thus, the direction of the force on the conductor will be toward the left, and the direction of the force on the magnet will be toward the right. If the magnet is locked in position, and the mechanical force is great enough, the conductor will be thrown out of the field.

3.6 ARCING FAULTS AND ARC MIGRATION

Normal Arcing

Arcing occurs in distribution and power systems whenever a circuit carrying an electric current is opened. Such arcing occurs normally in switches, circuit breakers, and contactors; the relatively large contact area of switch blades and circuit breaker contacts (when in the closed position) is reduced to a point contact just before the blades or contacts separate. At the moment of separation, the concentration of current (current density) at the point of contact raises the contact temperature high enough to cause incandescence. The evaporation of metal combined with electron emission that normally occurs at incandescence causes an arc to be drawn as the blades separate. The arc, consisting of ionized gas, evaporated metal, and electrons, can be rapidly extinguished by interrupter switches, circuit breakers, and contactors with magnetic blowouts.

Arcing Faults

Arcing faults caused by line-to-line or line-to-ground shorts are not easily extinguished. Such arcs, once started, travel in a direction away from the power source.

The theory behind arc migration is illustrated in Figure 3–8. The two conductors are connected to a power source, and an arcing fault is assumed to occur between the two conductors. The direction of the magnetic field surrounding the arc and the direction of the magnetic field surrounding the conductors were obtained by applying the right-hand rule to the known directions of current. Examination of the shaded area in Figure 3–8 shows that the direction of the magnetic flux due to the arc is in the same direction (upward) as that due to the current in the two conductors. As indicated by the bunching of flux in the shaded region, the resultant mechanical force on the arc causes it to move to the left, away from the power source. As the arc migrates downstream, it burns everything in its path. In some cases, additional arcs called *secondary arcs* emanate from the initial fault area, and follow the first arc downstream [1].

The enormous heat generated by the arc prevents self-extinguishment. This is true for both AC and DC. The incandescent material does not have time to cool in the short time that the AC voltage goes through zero. Hence, the arc restrikes every half

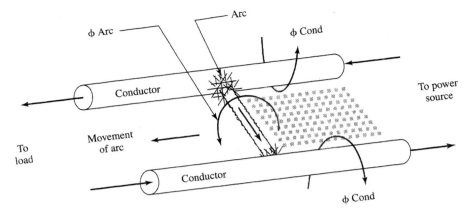

FIGURE 3.8
Arc migration caused by the interaction of the magnetic flux surrounding the conductors, with the magnetic flux surrounding the arc, when an arcing fault is initiated.

cycle. The heat is so intense that it easily vaporizes copper and aluminum conductors as well as steel enclosures. If the arcing fault occurs in a confined space, the violently expanding gases generated by the arc can cause intense pressure waves and an explosion.

The only way to prevent an arcing fault from causing serious damage or complete burndown of the distribution system is to have upstream breakers and/or fuses properly sized and coordinated. Since the intense heat and ionization associated with line-to-line arcing generally initiate line-to-ground arcing, the presence of correctly sized ground-fault protection devices will trip the upstream breaker, clearing the fault.

Arcing faults can be caused by the accumulation of dirt and moisture, rodents, loose connections, swinging wires, accidents, and voltage surges. Thus, good housekeeping practice can be very helpful in reducing the probability of arcing faults.

SUMMARY OF EQUATIONS FOR PROBLEM SOLVING

$$F = 5.4\, I_1\, I_2 \frac{\ell}{D} \times 10^{-7} \qquad\qquad (3\text{--}1)$$

SPECIFIC REFERENCE KEYED TO THE TEXT

[1] Dunki-Jacobs, J. R., The Escalating Arcing Ground-Fault Phenomenon, *IEEE Trans. Industry Applications*, Vol. IA-22, No. 6. November/December 1986.

REVIEW QUESTIONS

1. With appropriate sketches to show the direction of current and flux, determine the direction of the mechanical force developed on each of two parallel wires connecting a lamp to a battery.
2. What bad effect can excessively high currents have on conductors and supporting structures?
3. Using an appropriate sketch, explain how abnormally high current can squash the coils in a transformer.
4. Using an appropriate sketch, explain why an arcing fault migrates away from the source of power.

PROBLEMS

3–1/2. In Figure 3–9a, copper conductor A is resting on top of copper conductor B. Both conductors are insulated. Using the interaction of magnetic fluxes, determine the mechanical motion if any that may occur when the switch is closed.

3–2/2. Copy the sketch in Figure 3–9b and indicate on the sketch (a) direction of flux from the magnet; (b) direction of flux around the conductor; (c) direction of mechanical force on the conductor.

3–3/2. Figure 3–9c represents an end view of the field winding and the rotor winding of a motor. Using the interaction of magnetic fluxes, determine the direction of rotor rotation (clockwise or counterclockwise).

3–4/3. A short circuit at an electrical switchboard causes a fault current of 28,000 amperes in opposite directions in two parallel cables. The cables are 1.2 inches between centers. (a) Sketch the cables and show the directions of current, flux, and mechanical force developed. (b) Determine the force per foot of cable.

3–5/3. A current of 5000 A through two 6-foot-long parallel conductors causes a total separating force of 45 lb. Determine the force if the current is raised to 20,000 amperes.

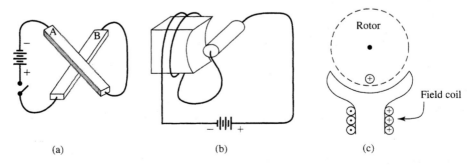

(a)   (b)   (c)

FIGURE 3.9
Sketches for (a) Problem 3–1/2; (b) Problem 3–2/2; (c) Problem 3–3/2.

4

Magnetic Circuits and Inductance in Electric Power Systems

4.0 INTRODUCTION

To safeguard electrical equipment and ensure personnel safety, it is imperative that operating engineers and safety personnel have a thorough understanding of magnetic-energy storage, magnetic-energy discharge, and induced voltages produced by magnetic action. Induced voltages, caused by a collapsing magnetic field, can result in severely damaged machines, damaged insulation, burned switches, and contactors that can lead to fire or an explosion.

4.1 MAGNETIC CIRCUIT

Two types of magnetic circuits are illustrated in Figure 4–1. Each magnetic circuit is composed of an arrangement of ferromagnetic materials, called a core, which form a path to contain and guide the magnetic flux in a specific direction. The core shape shown in Figure 4–1a is used in transformers. Figure 4–1b shows the magnetic circuit for a simple two-pole motor; it includes a stationary core, called the *stator*, a rotor core, and two air gaps. Note that the flux (ϕ) always takes the shortest path across an air gap.

Ferromagnetic materials used in motors, generators, and transformers are basically alloys of iron containing 0.5 percent to 3.5 percent silicon. Materials used for permanent magnets are alloys of two or more elements such as iron and chromium; aluminum, nickel, and cobalt; copper, nickel, and iron; etc.

Magnetomotive Force

The driving force that causes magnetic flux in the core is called the magnetomotive force (mmf), $\mathscr{F}$. It is equal to the number of series-connected turns in the coil times the magnitude of the current in the coil:

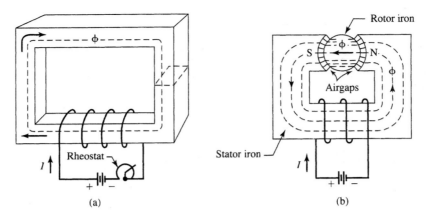

FIGURE 4.1
Magnetic circuits: (a) for a transformer; (b) for a two-pole motor.

$$\mathscr{F} = NI \tag{4-1}$$

where $\mathscr{F}$ = magnetomotive force in ampere-turns (A-t)
         I = current (A)
         N = number of series-connected turns in the coil.

The mmf of a coil is dependent solely on the ampere-turns of the coil, and is independent of its physical dimensions. A one-turn coil of large diameter has the same ampere-turns as a one-turn coil of small diameter, carrying the same current.

Reluctance of a Magnetic Circuit

The reluctance, $\mathscr{R}$, of a magnetic circuit is a measure of the opposition that the magnetic circuit offers to the flux; it is analogous to resistance in the electric circuit. The reluctance of a magnetic circuit depends on its geometry, dimensions, and the material of which it is constructed.

Ferromagnetic core materials, such as iron and steel, offer considerably less opposition to the flux than do nonmagnetic materials such as air, copper, and aluminum. Due to this phenomenon, iron and steel are used extensively for magnetic core materials in electrical apparatus that require large magnetic fields.

Magnetic-Circuit Law

The flux in a magnetic circuit is directly proportional to the applied mmf, and inversely proportional to the reluctance of the magnetic circuit. Expressed mathematically,

$$\Phi = \frac{\mathcal{F}}{\mathcal{R}} = \frac{NI}{\mathcal{R}} \tag{4–2}$$

where Φ = magnetic flux in webers (Wb)
   $\mathcal{R}$ = reluctance of the magnetic circuit in ampere-turns per weber (A-t/Wb).

EXAMPLE 4.1   Assume the coil in Figure 4–1a has 284 turns and a resistance of 5.72 Ω. With a 48-V battery and the rheostat set for 3.2 Ω, the flux in the core is 0.86 Wb. Determine (a) current; (b) mmf; (c) reluctance of core.

Solution

a. $R_{eq.S} = R_{coil} + R_{rheo} = 5.72 + 3.2 = 8.92 \; \Omega$

$$I = \frac{E_{bat}}{R_{eq.S}} = \frac{48}{8.92} = 5.38 \; A$$

b. $\mathcal{F} = NI = 284 \times 5.38 = 1528 \; A\text{-}t$

c. $\Phi = \dfrac{\mathcal{F}}{\mathcal{R}} \Rightarrow \mathcal{R} = \dfrac{\mathcal{F}}{\Phi} = \dfrac{1528}{0.86} = 1777 \; A\text{-}t/Wb$

4.2   BEHAVIOR OF FERROMAGNETIC MATERIALS

A magnetic field produced within a ferromagnetic material such as iron or steel is caused primarily by the spinning of electrons about their own axis within the atomic structure of the material. Groups of adjacent atoms with parallel electron spins are called *magnetic domains*. Each domain is a region of strong magnetization. Simulated arrangements of magnetic domains for unmagnetized and magnetized materials are shown in Figures 4–2a and 4–2b, respectively. In unmagnetized materials (Figure 4–2a) the random orientation of the domains results in no net external magnetic field. However, if the magnetic material is placed in a region possessing a magnetic field, for example, the earth's magnetic field, or the magnetic field around a current, the domains will tend to align themselves in the direction of the field as shown in Figure 4–2b.

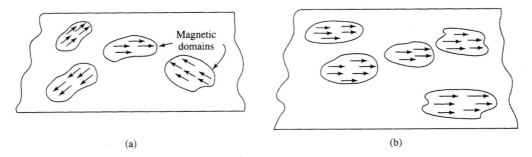

(a)   (b)

FIGURE 4.2
Simulated magnetic domains in ferromagnetic materials: (a) unmagnetized material; (b) magnetized material.

When all the domains that are available for alignment are shifted in the direction of the magnetizing force, the core is saturated, and further increases in magnetizing force will not produce a significant increase in flux.

When the magnetic material is removed from the field, the domains tend to go back to a random pattern. In materials used for permanent magnets, the magnetic domains are more difficult to align; but once aligned, and then removed from the external magnetizing field, a good amount of alignment is retained.

Magnetization Curve

The behavior of ferromagnetic materials in response to a magnetizing force (mmf) may be visualized by using the circuit shown in Figure 4–3. The curve shown in Figure 4–3 is a plot of magnetic flux vs. current as the current in the coil (and hence the mmf) is increased from zero.

The curve has three principal regions: the linear region, the knee region, and the "flattened" section called the saturation region. In the linear region, the flux is proportional to the current (straight-line characteristic). Although the magnetization curve has a slight reverse curvature for low values of current, the effect on electrical power apparatus is insignificant. As magnetization proceeds into and beyond the knee region,

FIGURE 4.3
Representative magnetization curve and associated circuit.

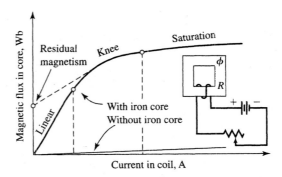

magnetic saturation begins to set in, resulting in less flux for a given increment increase in current. If increases in current produce no "useful" increase in magnetic flux, the material is said to be saturated. The curve shown in Figure 4–3 is representative of ferromagnetic materials used in electrical machines and transformers.

A plot of magnetic flux vs. current for the same coil, but without a ferromagnetic core, would be a straight line with a shallow slope, as shown in Figure 4–3.

Depending on the specific application, the magnetic core of electric power apparatus may be operated in the linear region, the knee region, or the saturation region. For example, at rated or below-rated conditions, transformers and AC machines operate in the linear region and in the lower end of the knee; self-excited DC generators and DC motors operate in the upper end of the knee region, extending into the saturation region; separately excited DC generators operate in the linear and lower end of the knee region.

Magnetic Hysteresis

When the circuit in Figure 4–3 is opened, the current decays to zero. However, a phenomenon known as *magnetic hysteresis* prevents the magnetic flux in the core from going to zero. This is shown by the broken line in Figure 4–3. The flux remaining in the core when the coil current is reduced to zero is called *residual magnetism*. Materials used for permanent magnets retain a considerable amount of residual magnetism. Direct current motors using permanent magnets are available in sizes ranging from 0.5 hp to 200 hp.

4.3 LENZ'S LAW AND INDUCTANCE

Whenever an electric current is made to increase or decrease, for example, when starting or stopping the current in a circuit, the magnetic field that surrounds the conductors increases or decreases, respectively. The changing magnetic field induces a voltage in the conductors that delays the change in current. When the current is increasing, the increasing flux generates a voltage within the conductors in opposition to the driving voltage; this delays the buildup of current. When the current is decreasing, the decreasing flux generates a voltage within the conductors that is in the same direction as the driving voltage; this tends to keep the electrons in motion and thus delays the decay in current. Simply stated, the voltage generated by a change in flux is always in a direction that opposes the change; this phenomenon is called *Lenz's law*.

The inductive effect in an electrical system is somewhat analogous to the flywheel effect in a mechanical system. A flywheel has a fixed value of inertia, which is a function of its mass and physical dimensions. As long as its velocity does not change, the opposition to motion offered by the flywheel is zero. However, if any attempt is made to change its velocity, the flywheel would develop a torque in a direction that opposes the change. The opposing torque serves only to delay the change; it cannot prevent it. Similarly, the inductive property, called *inductance*, which is present to some extent in all circuits, does not prevent the current from changing; it merely delays the change.

Lumped Inductance

Coiling an insulated wire around an iron core, as shown in Figure 4–4, concentrates the inductive effect in a relatively small package, called a *lumped inductance* or an *inductor*.

The inductance of the connecting wires, switches, cables, etc., in a circuit is negligible when compared with the inductance of electrical apparatus containing coils of insulated wire wound around an iron core such as in motors, generators, transformers, and other iron-core apparatus. The unit of inductance is the henry (H), and its magnitude (for a coil) is dependent on the number of turns of wire in the coil and the reluctance of the magnetic circuit. Expressed mathematically,

$$L = \frac{N^2}{\mathcal{R}} \qquad \qquad (4\text{--}3)$$

where $\mathcal{R}$ = reluctance of the core material (A-t/Wb)
   L = inductance in henrys (H)
   N = number of series-connected turns of wire in the coil.

If the coil has an air core, or the core is ferromagnetic and is operating in the linear region, the reluctance will be constant, and the inductance in Eq. (4–3) may be written as a proportionality. That is,

$$L \propto N^2 \qquad \qquad (4\text{--}4)$$

The inductance of a coil may also be expressed in terms of the ratio of the change in flux to the change in current that caused it, multiplied by the number of series-connected turns in the coil:

$$L = N\frac{\Delta\phi}{\Delta I} \qquad \qquad (4\text{--}5)$$

where ΔI = increment change in current (A)
   $\Delta\phi$ = change in flux due to the change in current (Wb).

FIGURE 4.4
Coiling the insulated wire around an iron core concentrates the inductive effect.

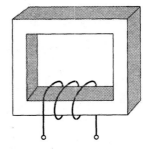

EXAMPLE 4.2 A coil of 300 turns wound around an iron core, as shown in Figure 4–4, has an inductance of 0.54 H. Determine the inductance if 100 turns are removed. Assume operation is in the linear region.

Solution
Using proportionality (4–4),

$$\frac{L_2}{L_1} = \left[\frac{N_2}{N_1}\right]^2 \Rightarrow L_2 = L_1 \left[\frac{N_2}{N_1}\right]^2 = 0.54 \times \left[\frac{200}{300}\right]^2 = 0.24 \text{ H}$$

Steady-State Current

An inductive circuit, consisting of an inductor, a battery, and a switch, is shown in Figure 4–5a. The equivalent circuit in Figure 4–5b shows the resistance and the inductance of the inductor as separate series-connected lumped values. When the switch is closed, the inductance delays, but cannot prevent, the buildup of current. The current is zero at the instant the switch is closed, building up in time to its final value. The final value of current, called the *steady-state current*, depends solely on the magnitude of the driving voltage and the resistance of the circuit. The resistance of the circuit includes the internal resistance of the battery, the resistance of the coil, and the resistance of the connecting wires. The steady-state current for the circuit in Figure 4–5 may be determined using Ohm's law:

$$I_{ss} = \frac{V}{R_{cir}} \qquad (4\text{–}6)$$

where I_{ss} = steady-state current (A)
V = voltage applied to the circuit (V)
R_{cir} = resistance of the circuit (Ω).

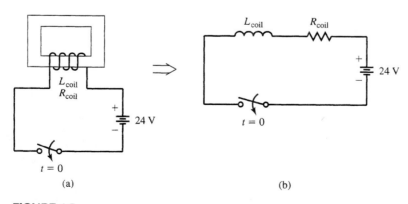

FIGURE 4.5
(a) Inductive circuit; (b) equivalent circuit.

EXAMPLE 4.3 For the circuit shown in Figure 4–5, assume that the inductance and resistance of the coil are 3.6 H and 4 Ω, respectively. Determine (a) the current at the instant the switch is closed; (b) the final (steady-state) current.

Solution

a. Zero.

b. $I_{ss} = \dfrac{V}{R_{cir}} = \dfrac{24}{4} = 6$ A

4.4 VOLTAGE OF SELF-INDUCTION

The magnitude of the voltage (V_L) induced in a coil by a changing current may be expressed in terms of the coil inductance and the rate of change of current, or in terms of the number of turns in the coil and the rate of change of flux (called *Faraday's law*). Expressed mathematically,

$$V_L = L\left(\frac{di}{dt}\right) \tag{4–7}$$

$$V_L = N\left(\frac{d\phi}{dt}\right) \tag{4–8}$$

where $\left(\dfrac{di}{dt}\right)$ = rate of change of current in amperes per second (A/s)

$\left(\dfrac{d\phi}{dt}\right)$ = rate of change of flux in webers/second (Wb/s).

EXAMPLE 4.4 The current in a 4-Ω, 8-H inductor drops from 10 amperes to 3 amperes in 0.05 second. Determine the induced voltage.

Solution

$$V_L = L\left(\frac{di}{dt}\right) = 8 \times \frac{10 - 3}{0.05} = 1120 \text{ V}$$

4.5 INDUCTORS IN SERIES AND IN PARALLEL

The equivalent inductance of two or more inductors in series, as illustrated in Figure 4–6a, is equal to the sum of the individual inductances:

$$L_{eq.S} = L_1 + L_2 + L_3 + \cdots + L_n \tag{4–9}$$

FIGURE 4.6
(a) Inductors in series; (b) inductors in parallel.

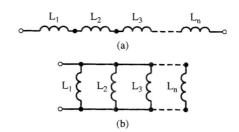

The equivalent inductance of two or more inductors in parallel, as illustrated in Figure 4–6b, is the reciprocal of the sum of the reciprocals:

$$L_{eq.P} = \cfrac{1}{\cfrac{1}{L_1} + \cfrac{1}{L_2} + \cfrac{1}{L_3} + \cdots + \cfrac{1}{L_n}} \quad \text{or} \quad \frac{1}{L_{eq.P}} = \frac{1}{L_1} + \frac{1}{L_2} + \frac{1}{L_3} + \cdots + \frac{1}{L_n} \quad \textbf{(4–10)}$$

Note: Equations (4–9) and (4–10) are valid only if no magnetic coupling exists between coils. Magnetic coupling, also called transformer action, is present whenever the flux caused by the current in one coil passes through another coil.

EXAMPLE 4.5

Given the following ideal inductors (zero resistance): 2.5 H, 6.3 H, and 5.2 H. Determine: (a) equivalent inductance if connected in series; (b) equivalent inductance if connected in parallel.

Solution

a. $L_{eq.S} = L_1 + L_2 + L_3 = 2.5 + 6.3 + 5.2 = 14$ H

b. $L_{eq.P} = \cfrac{1}{\cfrac{1}{L_1} + \cfrac{1}{L_2} + \cfrac{1}{L_3} + \cdots + \cfrac{1}{L_n}} = \cfrac{1}{\cfrac{1}{2.5} + \cfrac{1}{6.3} + \cfrac{1}{5.2}} = \cfrac{1}{0.751} = 1.33$ H

Note 1: The equivalent inductance of parallel-connected inductances is always less than the equivalent inductance of series-connected inductances.
Note 2: The equivalent inductance of parallel inductances is always less than the smallest of the parallel inductances.

EXAMPLE 4.6

The resistance and inductance of each coil of a certain four-pole, 15-hp DC motor are 44.3 Ω and 17 H, respectively. The coils, shown in Figure 4–7a, are connected in series to a 240-V source. Determine: (a) steady-state current; (b) induced voltage if, at the instant the switch is opened, the current decays at a rate equal to 100 amperes per second.

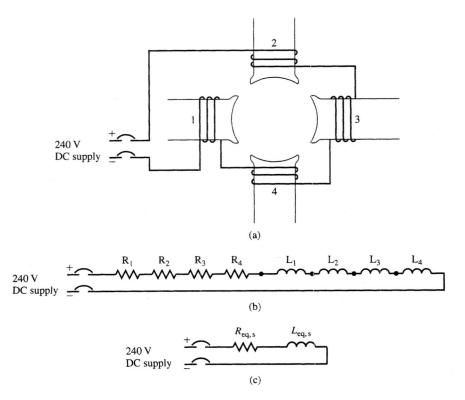

FIGURE 4.7
Field winding for a four-pole DC motor: (a) connection diagram; (b) elementary diagram; (c) simplified equivalent series circuit.

Solution

a. The equivalent circuit for the four-pole field winding is shown in Figure 4–7b. Note that the resistances of all four coils are shown connected in series, and the inductances of all four coils are shown connected in series. A further simplification is shown in Figure 4–7c, where:

$$R_{eq.S} = R_1 + R_2 + R_3 + R_4 = 4 \times 44.3 = 177.2 \ \Omega$$

$$L_{eq.S} = L_1 + L_2 + L_3 + L_4 = 4 \times 17 = 68 \ \text{H}$$

Using Ohm's law,

$$I_{ss} = \frac{V}{R_{eq.S}} = \frac{240}{177.2} = 1.35 \ \text{A}$$

b. $V_L = L_{eq.S} \left[\dfrac{di}{dt} \right] = 68 \times [100] = 6800 \ \text{V}$

4.6 ENERGY STORAGE IN A MAGNETIC FIELD

Energy is accumulated in the magnetic field during the buildup of current, and it is released when the current decays.

The energy stored in a magnetic field is proportional to the self-inductance of the coil and the square of the current. Expressed mathematically,

$$W_\phi = \frac{1}{2} L I^2 \qquad \textbf{(4–11)}$$

where: L = inductance (H)
           I = current (A)
         W_ϕ = magnetic energy (J).

Opening an inductive circuit releases the energy stored in the magnetic field. The released energy manifests itself by arcing and burning at the opening contacts, as well as inducing high voltage stresses that are detrimental to the electrical insulation. Unless adequately protected, such arcing and burning of switches and contacts can cause serious maintenance problems. Figure 4–8 shows an electrical contactor with burned and pitted contacts, caused by arcing when it opened a highly inductive circuit.

Before opening a circuit containing appreciable inductance, some provision should be made to dissipate the energy stored in the magnetic field. If a resistor is placed in parallel with the coil before the switch is opened, it will dissipate the stored magnetic energy as heat when the circuit is opened. A field-discharge switch is sometimes used for this purpose. It has an extra blade that connects a resistor across the coil immediately before the coil is disconnected from the line. The resistance of the

FIGURE 4.8
Electrical contactor with burned and pitted contacts, caused by arcing.

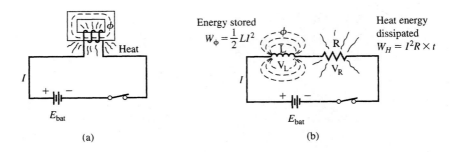

FIGURE 4.9
(a) Connection diagram and (b) equivalent energy diagram of an actual inductor.

field-discharge resistor is generally designed to be equal to the resistance of the coil it protects. Semiconductors, such as diodes or varistors, are often permanently connected across the coil to dissipate the energy released during fluctuations of coil current.

The energy stored in the magnetic field is distinct from the I^2Rt energy loss in the conductors, which is expended as heat energy. Thus, when analyzing the energy relationships in an inductor, it is convenient to make an equivalent-circuit model, showing the inductance and the resistance as separate lumped values. Figure 4–9a shows an actual inductor, and Figure 4–9b shows the equivalent-circuit model of the inductor, where an ideal (pure) inductor and an ideal (pure) resistor take the place of the actual inductor.

EXAMPLE 4.7 The shunt-field circuit of a certain 900-hp cement-pump motor used on a drilling rig has an inductance of 4.1 H, a resistance of 1.85 Ω, and is connected to a 106-V DC source. The equivalent series-circuit diagram is shown in Figure 4–10. Determine: (a) circuit current; (b) magnetic energy stored in the magnetic field; (c) heat power expended by the coil; (d) kilowatt-hours of heat energy expended in 24 hours.

FIGURE 4.10
Equivalent series-circuit diagram for Example 4–7.

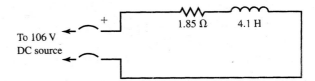

Solution

a. $I = \dfrac{V}{R} = \dfrac{106}{1.85} = 57.3 \text{ A}$

b. $W_\phi = \dfrac{1}{2}LI^2 = \dfrac{1}{2} \times 4.1 \times 57.3^2 = 6730.7 \text{ J}$

c. $P_H = I^2R = 57.3^2 \times 1.85 = 6073.51 \text{ W} \Rightarrow 6.07 \text{ kW}$

d. $W_H = I^2Rt = 6073.51 \times 24 = 145,764 \text{ Wh} \Rightarrow 145.8 \text{ kWh}$

4.7 VOLTAGE AND CURRENT TRANSIENTS IN INDUCTIVE CIRCUITS

Figure 4–11a shows a coil of insulated wire wrapped around an iron core and connected in series with a battery and a switch. The resistance and inductance of the coil are 1.2 Ω and 0.6 H, respectively. Figure 4–11b is an equivalent series circuit that shows the resistance and inductance as separate lumped values. The behavior of the current and induced voltage, from the instant the switch is closed to the final steady-state condition, is shown in Figure 4–11c. The part of each curve that extends over a period of time beginning with the closing of the switch (called time-zero) and ending when the change is complete is called a *transient*.

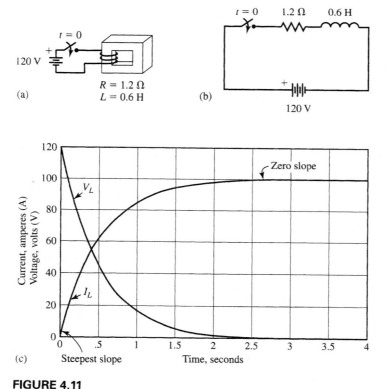

FIGURE 4.11

(a) Connection diagram; (b) elementary diagram; (c) current and voltage transients.

The equations for the current and voltage transients in Figure 4–11c are

$$I_L = \frac{E_{bat}}{R} \left(1 - \varepsilon^{-(R/L)t}\right) \qquad V_L = E_{bat}\left(\varepsilon^{-(R/L)t}\right) \qquad \textbf{(4–12)}$$

where R = resistance (Ω)
L = inductance (H)
t = time (s)
ε = 2.718.

The points on the induced voltage curve are related to the slope or "steepness" of the current curve. Note that the induced voltage has its greatest value when the current curve has its steepest slope, and has zero value when the current curve has zero slope. This should be expected, since the slope of the current curve is equal to the rate of change of current (di/dt), and $V_L = L\,(di/dt)$.

As previously mentioned, the inductance does not prevent the current from changing, nor does it limit the current; it merely delays the change. The only factors limiting the current in an inductive circuit are the resistance of the coil and the applied voltage. If the coil in Figure 4–11a had no inductance (which is not possible) there would be no inductive time delay; and at the instant the switch is closed, the current would "jump" to its steady-state Ohm's law value of

$$I_{ss} = \frac{120}{1.2} = 100 \text{ A}$$

Time Constant of an Inductive Circuit

The actual time delay experienced in the buildup of current in an inductive circuit is related to the ratio of inductance to resistance (L/R) in the circuit. The time delay may be on the order of microseconds, seconds, minutes, or even hours.

The time that it takes for the voltage and current to make a 63.2 percent change in their respective values, starting from the instant the switch is closed, is called the *time constant*. It is, in effect, a measure of the speed of response of the circuit. The time constant for a series circuit, or equivalent series circuit, containing resistance and inductance is

$$\tau_L = \frac{L}{R} \qquad \textbf{(4–13)}$$

where τ_L = inductive time constant (s)
L = inductance (H)
R = resistance (Ω).

The inductive time constant provides a convenient means for predicting how rapidly (or slowly) the current will build up or decay in an inductive circuit. For example, referring to the circuit in Figure 4–11, the time constant is

$$\tau_L = \frac{L}{R} = \frac{0.6}{1.2} = 0.5 \text{ s}$$

Time to Steady State

After a time interval equal to five time constants ($5\tau_L$), the change in current will be 99.3 percent complete and, for all practical purposes, the current is considered to be at steady state. Thus, for the circuit in Figure 4–11a, the current is considered to have reached its steady-state value in five time constants:

$$5\tau_L = 5 \times 0.5 = 2.5 \text{ s}$$

Generalized Time-Constant Curves

Curves[1] showing the buildup of current with time and the decay of induced voltage with time, for circuits containing resistance and inductance, expressed in terms of time constants are shown in Figure 4–12.

---

[1]Generalized time-constant curves are easy to sketch, and may also be applied to transients in other physical systems, such as mechanical, hydraulic, and pneumatic.

EXAMPLE 4.8 The shunt-field circuit of a certain generator has a resistance of 41.5 Ω and an inductance of 78.8 H. The field circuit is connected to a 120-V source through a switch. Determine (a) time constant; (b) steady-state current; (c) current and induced voltage at one time constant after the switch is closed; (d) time it will take the current to reach steady state.

Solution
The circuit is similar to that shown in Figure 4–12.

a. $\tau_L = \dfrac{L}{R} = \dfrac{78.8}{41.5} = 1.898 \text{ s}$

b. $I_{ss} = \dfrac{V}{R} = \dfrac{120}{41.5} = 2.89 \text{ A}$

c. Referring to the generalized curves in Figure 4–12, in one time constant the current will have increased to 63.2 percent of its final 100 percent value. Thus,

$$I_L = 0.632 \times 2.89 = 1.83 \text{ A}$$

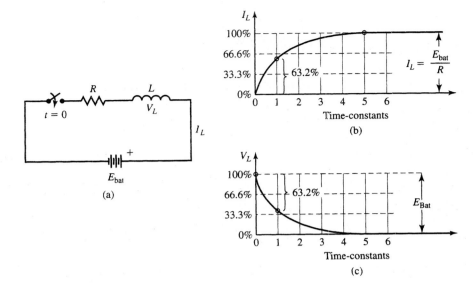

FIGURE 4.12
Generalized time-constant curves for inductive circuits: (a) equivalent circuit; (b) current transient; (c) voltage transient.

In one time constant, the induced voltage will have decreased 63.2 percent from its initial 100 percent value. Thus,

$$100\% - 63.2\% = 36.8\%$$

$$V_L = 0.368 \times 120 = 44.16 \text{ V}$$

d. $5\tau_L = 5 \times 1.898 = 9.49 \text{ s}$

Comparing Time Constants When Closing and Opening Inductive Circuits

Figure 4–13 shows the direction of the induced voltage relative to the battery voltage when the switch is closed and when the switch is opened.

At the instant the switch is closed, the current is zero, and the induced voltage will be equal in magnitude but opposite in direction to the battery voltage. The current rises with time, approaching its steady-state value in five time constants $(5\tau_L)$, at which time the opposing voltage of self-induction approaches zero. This is shown on the left side of Figure 4–13b for the switch-closed position.

When the circuit is opened, the rapid collapse of magnetic flux induces a voltage in a direction that opposes the decrease in current, causing the voltage of self-induction to be in the same direction as the driving voltage. This is shown on the right side of Figure 4–13b for the switch-open position.

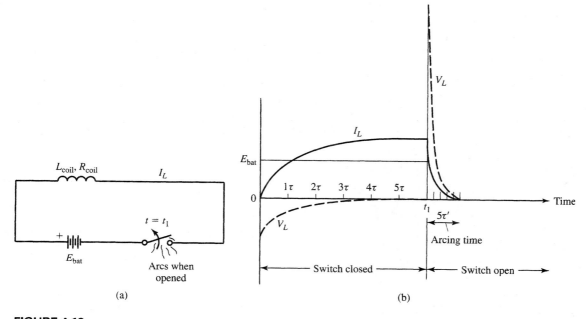

FIGURE 4.13

a) Inductive circuit; b) induced voltage caused by closing and opening an inductive circuit.

Opening the switch adds the resistance of the electric arc in series with the resistance of the coil. The time constant when the switch is opened is much shorter than the time constant when the switch is closed. The respective time constants are:

Switch closed　　　**Switch opened**

$$\tau_L = \frac{L_{coil}}{R_{coil}} \qquad \tau'_L = \frac{L_{coil}}{R_{coil} + R_{arc}}$$

Since the time constant is smaller when the switch is opened than when it is closed, a more rapid drop in current will occur when the switch is opened, resulting in a higher induced voltage. Although the resistance of the arc is not constant (nonlinear), the curves shown in Figure 4–13b are representative of the behavior of the induced voltage when closing and opening the switch in an inductive circuit.

4.8　FIELD-DISCHARGE CIRCUITS

To prevent burning and arcing at the contacts of a switch or a circuit breaker, a diode or varistor is sometimes permanently connected across the terminals of coils with high inductance. This is shown in Figure 4–14. The solid lines indicate the path of current through the circuit when the switch is closed, and the broken lines indicate the path of current when the switch is opened. Since the inductance of the coil prevents an

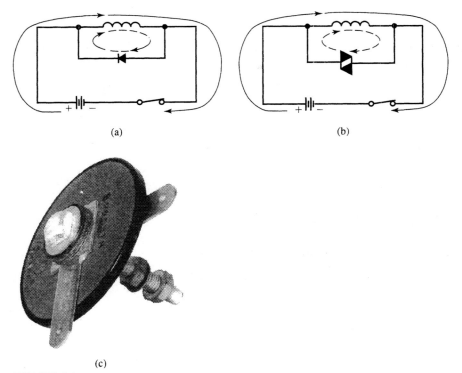

(a)   (b)

(c)

FIGURE 4.14
Discharging the energy stored in a magnetic field: (a) with a diode; (b) with a varistor;
(c) with a Thyrite® thyristor (courtesy GE Industrial Systems).

instantaneous change in the direction of the current, the direction of the current
through the coil at the instant the switch is opened will be the same as when the switch
was closed.

When the switch is opened, the magnetic energy will be dissipated as heat in the
resistance of the closed loop formed by the coil and the discharge element. If a diode
is used as the discharge element (Figure 4–14a), almost all of the magnetic energy will
be dissipated in the resistance of the coil; because of its low forward resistance, the I^2R
dissipation in the diode will be insignificant. If a varistor is used as the discharge element (Figure 4–14b), some of the energy will be dissipated in the varistor and some in
the resistance of the coil.

The time-constant for the diode loop is

$$\tau_{\text{diode loop}} = \frac{L_{\text{coil}}}{R_{\text{coil}}}$$

The time-constant for the varistor loop is

$$\tau_{\text{varistor loop}} = \frac{L_{\text{coil}}}{R_{\text{coil}} + R_{\text{varistor}}}$$

Note that the time constant for the varistor loop is less than the time constant for the diode loop. A varistor discharge loop will discharge the energy stored in the magnetic field more quickly than will a diode discharge loop. One type of varistor used in electric power applications is shown in Figure 4–14c.

4.9 EFFECT OF INDUCTANCE IN AN ALTERNATING CURRENT SYSTEM

The inductive effect is present whether the driving voltage is alternating or direct (AC or DC). However, with an alternating-voltage generator, the driving voltage is continually reversing, causing the current to be in a continuous state of delay. Figures 4–15a

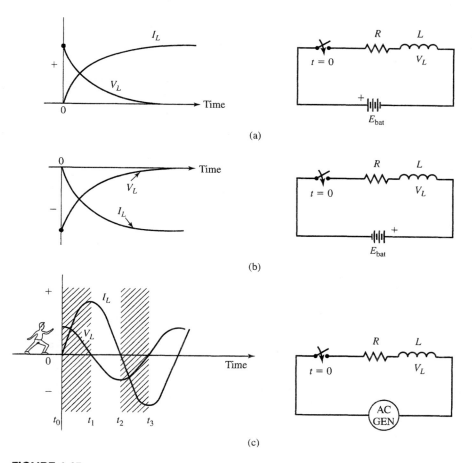

FIGURE 4.15
Effect of an alternating driving voltage on an inductor current and voltage: (a, b) current curves for two different directions of battery driving voltage; (c) current and voltage curves for a sinusoidal driving voltage.

and 4–15b illustrate the current curves for two directions of battery driving voltage (DC). Figure 4–15c shows the current and driving voltage when an alternating driving voltage is applied. The lag or lead of current and voltage waves may be determined by an observer "walking" along the time axis from left to right, starting at time zero, and observing the positive peaks of the waves. For the waves shown in Figure 4–15c, the observer passes the positive peak of the voltage wave first, and later the positive peak of the current wave. Hence, the current wave is said to lag the voltage wave.

From time t_0 to t_1, the behavior of the alternating current is similar to that shown in Figure 4–15a, and from t_2 to t_3 it is similar to that shown in Figure 4–15b. The difference in curvature between the curves shown in Figure 4–15c and their counterparts in Figures 4–15a and 4–15b is caused by different driving voltages; the driving voltage of the battery is constant in each direction, whereas the driving voltage of the AC generator alternates in a sinusoidal manner.

4.10 EFFECT OF MAGNETIC SATURATION ON INDUCTANCE

The inductance of a coil wound around a ferromagnetic core is related to the slope (steepness) of its magnetization curve; the steeper the slope, the greater its inductance. This relationship is shown in Figure 4–16 for a representative magnetization curve.

The slope of the magnetization curve, at any point on the curve, may be determined from a tangent line at that point. For example, the slope and corresponding inductance at point p in Figure 4–16 are

$$\text{slope} = \frac{\Delta \phi}{\Delta I}$$

$$L = N\frac{\Delta \phi}{\Delta I} \tag{4–5}$$

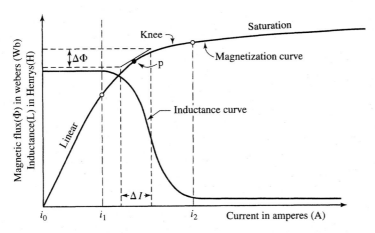
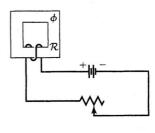

FIGURE 4.16
Representative magnetization curve and associated curve of inductance.

where ΔI = increment change in current (A)
$\Delta \phi$ = increment change in flux (due to ΔI) (Wb).

Linear Region

The slope of the magnetization curve and, hence, the coil inductance will have their greatest values for current below the knee of the magnetization curve; this is called the *linear region*. The inductance for that range of currents is essentially constant for that inductor, and is the value generally stamped on the inductor nameplate. Unless otherwise specified, the inductors referred to in discussions and problems throughout this text are assumed to be operating in the linear region.

Knee Region

For values of increasing current in the *knee region*, the slope of the magnetization curve and, hence, the coil inductance will be decreasing in value.

Saturation Region

For values of current in the *saturation region*, the slope of the magnetization curve is very small. Hence, the coil inductance is very small, and very little opposition is offered to a changing current in that region.

SUMMARY OF EQUATIONS FOR PROBLEM SOLVING

$$\mathscr{F} = NI \tag{4-1}$$

$$\Phi = \frac{\mathscr{F}}{\mathscr{R}} = \frac{NI}{\mathscr{R}} \tag{4-2}$$

$$L = \frac{N^2}{\mathscr{R}} \tag{4-3}$$

$$L = N\frac{\Delta \phi}{\Delta I} \tag{4-5}$$

$$I_{ss} = \frac{V}{R_{cir}} \tag{4-6}$$

$$V_L = L\left[\frac{di}{dt}\right] \tag{4-7}$$

$$V_L = N\left(\frac{d\phi}{dt}\right) \tag{4-8}$$

$$L_{eq.S} = L_1 + L_2 + L_3 + \cdots + L_n \tag{4-9}$$

$$L_{eq.P} = \cfrac{1}{\cfrac{1}{L_1} + \cfrac{1}{L_2} + \cfrac{1}{L_3} + \cdots + \cfrac{1}{L_n}} \quad \text{or} \quad \frac{1}{L_{eq.P}} = \frac{1}{L_1} + \frac{1}{L_2} + \frac{1}{L_3} + \cdots + \frac{1}{L_n} \tag{4-10}$$

$$W_\phi = \frac{1}{2}LI^2 \tag{4-11}$$

$$I_L = \frac{E_{bat}}{R}\left(1 - \varepsilon^{-(R/L)t}\right) \qquad V_L = E_{bat}\left(\varepsilon^{-(R/L)t}\right) \tag{4-12}$$

$$\tau_L = \frac{L}{R} \tag{4-13}$$

REVIEW QUESTIONS

1. What is a magnetic circuit?
2. What is the relationship between reluctance, magnetic flux, and magnetomotive force?
3. Sketch a magnetization curve, label the axis, and indicate the linear region, knee region, and saturation region.
4. What is magnetic hysteresis?
5. What is the condition called magnetic saturation?
6. What are the defining characteristics of inductance?
7. State and explain Lenz's law.
8. What effect does the inductive property of a coil have on the buildup of current? Explain.
9. Assuming a ferromagnetic core of constant reluctance, what effect will tripling the number of turns of wire have on the inductance? Explain.
10. Given two inductors, which connection (series or parallel) will result in the greater total inductance? Explain.
11. What is the relationship between the current in a coil and the energy stored in its magnetic field.
12. When the circuit to an inductance is opened, what happens to the energy stored in the magnetic field?
13. Explain why an inductor gets warm when it is energized.
14. State and explain Faraday's law.
15. What is an electrical transient?
16. When the switch to a generator field circuit is opened, the current arcs across the switch, burning the blades. Explain why this happens.
17. If two inductive circuits have the same current and the same resistance values, but different time constants, which circuit will produce the greatest amount of arcing when its switch is opened? Explain.

18. Sketch the generalized time-constant curve for current in an inductive circuit, and mark off the time axis in time constants.

PROBLEMS

4–1/1. A coil of 132 turns wound around an iron core has a resistance of 8.42 Ω. When connected to a 120-V DC source, it causes a flux of 1.36 Wb in the core. Determine (a) current; (b) magnetomotive force; (c) reluctance of the core.

4–2/1. A doughnut-shaped magnetic circuit has a reluctance of 1500 A-t/Wb. A 200-turn coil wound through the doughnut draws 3 A when connected to a 6-V battery. Determine (a) flux in the magnetic circuit; (b) resistance of coil.

4–3/1. A DC field coil of a certain generator has 400 turns and a resistance of 200 Ω. The reluctance of the magnetic circuit is 1000 A-t/Wb. Determine the flux in the magnetic circuit if the voltage across the coil is 40 V.

4–4/3. A 1000-turn coil has an inductance of 1.2 H. Assuming the reluctance is constant, determine the new inductance if 400 turns are added to the coil.

4–5/3. A certain 40-turn coil has an inductance of 6.0 H. Assuming the reluctance is constant, determine the new inductance if 10 turns are added to the coil.

4–6/3. A coil of 50 turns has an inductance of 10 H and a resistance of 2.0 Ω. Determine the reluctance of the magnetic circuit.

4–7/3. A 300-turn coil wound around an iron core has a resistance of 71.11 Ω, an inductance of 16 H, and is connected to a 240-V direct-current source. Sketch the circuit and determine (a) current; (b) reluctance of the core; (c) heat-power loss in coil.

4–8/4. If the current in a 16-Ω, 0.34-H coil is changing at the rate of 2698 A/s, determine the induced emf.

4–9/4. An 84.4-Ω, 114-H inductance is connected in series with a switch and energized from a 240-V DC source. (a) Sketch the circuit. (b) Determine the steady-state current. (c) If quick-opening the switch caused 4695 V to be induced in the coil, determine the associated rate of change of current.

4–10/5. Determine the equivalent inductance of the following series-connected inductors: 5 H, 1 H, 100 H, and 12 H.

4–11/5. Using the inductors from Problem 4–10/5, determine the equivalent inductance if connected in parallel.

4–12/5. Determine the equivalent inductance of the following series-connected inductors: 2 H, 4 H, 6 H, and 8 H.

4–13/5. Using the inductors from Problem 4–12/5, determine the equivalent inductance if connected in parallel.

4–14/6. The shunt-field circuit of a certain 10-hp, 230-V, 3500-rpm DC motor has a resistance of 208.3 Ω and an inductance of 63 H. Assuming the motor is operating at rated voltage; determine (a) shunt-field current; (b) energy stored in the magnetic field.

4–15/6. A 16-H coil energized at 240 volts stores 3200 J of magnetic energy. Determine (a) current in coil; (b) resistance of coil.

4–16/6. The field winding of a 230-V, 125-hp, 850-rpm motor has a resistance of 22.1 Ω and an inductance of 42 H. Assuming the motor is operating at rated voltage, determine (a) field current; (b) energy stored in the magnetic field; (c) heat power expended in the coil.

4–17/6. The field winding of a certain DC generator have resistance and inductance values of 20 Ω and 5.4 H, respectively. The field winding is supplied from a 240-V DC source in series with a rheostat set for 3.4 Ω. Determine (a) field current; (b) energy stored in the magnetic field; (c) heat-power loss.

4–18/7. The electrical wiring of a certain machine has an inductance of 0.1 H and a resistance of 0.5 Ω. Determine (a) time constant; (b) time it will take for the current to reach its steady-state value when connected to a 120-V battery; (c) steady-state current.

4–19/7. A 16-H coil has a resistance of 7 Ω and is connected in series with a 24-V battery and a switch. Assume that the total resistance of the connecting wires, switch, and battery is 1 Ω. Sketch the circuit and the curve of current buildup and determine (a) final value of current; (b) time constant; (c) current at one time constant.

4–20/7. A 10-H, 16-Ω coil is connected in series with a 4.0-Ω resistor, a 120-V battery, and a switch. Sketch the circuit and determine (a) the time constant of the coil; (b) the time constant of the circuit; (c) the steady-state current. (d) If the 4.0-Ω resistor is to be replaced by another that would cause the time constant of the circuit to be 0.34 s, determine the resistance value of the new resistor.

4–21/7. An 8-H, 4-Ω coil is connected in series with a 2-Ω resistor, a 6-V battery, and a switch. Sketch the equivalent circuit and determine (a) initial current when the switch is closed; (b) steady-state current; (c) current in the 2-Ω resistor at $t = 1$ time constant; (d) voltage across the 2-Ω resistor at $t = 1$ time constant. (e) If the 2-Ω resistor is to be replaced by another that would cause the time constant to be one-third of a second, determine the ohmic value of the new resistor.

4–22/7. A 10-H, 1-Ω coil is connected in series with a 4-Ω resistor, a 12-V battery, and a switch. Sketch the equivalent circuit and determine (a) time constant of the coil; (b) time constant of the circuit; (c) steady-state current; (d) current at $t = 1$ time constant after the switch is closed; (e) induced emf in the coil at $t = 1$ time constant.

5

Capacitors, Their Applications, and Effects in Electric Power Systems

5.0 INTRODUCTION

Capacitors are the simplest of all electrical apparatus. They have no moving parts and draw essentially no power from a battery or generator. Yet they are essential to the operation of a large variety of electrical apparatus. Capacitors can be found in numerous electric power applications. They are used for motor starting, surge protection, energy absorption (reduce sparking), energy storage (e.g., laser power supplies, flash photography, and hot-shot wind tunnels), harmonic filtering, power-factor improvement, and reducing current drain on large AC motor distribution systems.

Because of the dangers inherent in charged capacitors, maintenance personnel must know how to safely discharge them before working on the associated apparatus.

5.1 CAPACITIVE PROPERTY OF A CIRCUIT

Capacitance is that property of an electric circuit or circuit element that delays a change in the voltage across it. The delay is accompanied by the absorption or release of energy and is associated with a change in the electric charge.

Figure 5–1 shows two conductors separated by a nonconducting material, such as air, paper, or glass, and connected to a DC generator or battery. The two conductors and the associated insulation represent a simple capacitor. The battery, acting as an "electron pump," transfers some of the "free electrons" from conductor A to conductor B. The transfer of electrons causes conductor B to become increasingly negative, and conductor A increasingly positive. The material that lost electrons is said to be *positively charged*, and that which gained electrons is said to be *negatively charged*. As the charging process continues, conductor B eventually becomes sufficiently negative to prevent any additional transfer of electrons. When this occurs, the voltage across the conductors, called *countervoltage*, is equal and opposite to the driving

FIGURE 5.1
Elementary capacitor.

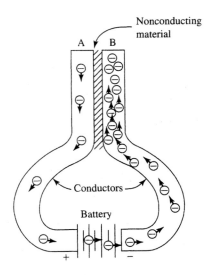

voltage. Because the rate of movement of these electrons is affected by the resistance of the conducting materials, the charging process will take longer if higher resistance conductors are used.

Lengthening the conductors will increase their capacity for electron storage. This will permit more of the "free electrons" to be shifted from A to B before the accumulated charge is sufficiently concentrated to build up an equal countervoltage. However, although a greater charge can be accumulated, it will take a longer time to attain the same countervoltage. Shorter conductors will have less capacity for electron storage. Hence, the same density of electrons, and same countervoltage, will occur with less electron transfer and in less time.

The unit of capacitance is the farad, which is the ratio of accumulated electric charge divided by the voltage across the capacitor. Expressed mathematically,

$$C = \frac{q}{V_C} \tag{5–1}$$

where C = capacitance, in farads (F)
q = electric charge, in coulombs (C)
(1 coulomb = 6.28×10^{18} electrons)
V_C = voltage across capacitor (V).

The farad is a rather large unit, and in most applications the capacitance is on the order of microfarads (μF). One microfarad is one-millionth of a farad, 1×10^{-6} farad.

In those applications where significant energy storage capability is required or where it is desirable to increase the time delay in the buildup of voltage, "blocks" of capacitance, called *capacitors*, are added to the circuit. Some of the many varieties of capacitors are shown in Figure 5–2a.

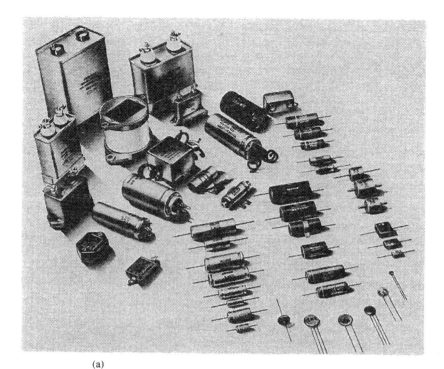

(a)

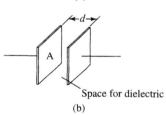

Space for dielectric

(b)

FIGURE 5.2

(a) Assorted capacitors (courtesy Cornell Dubilier); (b) simple two-plate capacitor.

Capacitors are composed of two parallel conductors called *plates*, generally made of metal foil. The plates are separated by insulating material called the *dielectric*. The various types of capacitors differ primarily in the type of dielectric used. Air, paper, mica, ceramic, and electrochemical (electrolytic) dielectrics are some of the more common types of dielectrics used in the construction of capacitors. Large capacitors are constructed of two very long strips of aluminum foil separated by insulating paper and rolled into compact units for enclosure in a protective can. The can is then filled with a nonconducting liquid and sealed.

Figure 5–2b shows the construction details of a simple parallel plate capacitor. The capacitance of a parallel plate capacitor is a function of the cross-sectional area of the plates, the distance between the plates, and the dielectric constant of the insulating material.[1] The dielectric constant is the ratio of the capacitance of a capacitor containing a particular dielectric material to the capacitance of the same capacitor with air as the dielectric medium. The capacitance of a capacitor expressed in terms of its dimensions is

$$C = \frac{KA(8.854 \times 10^{-12})}{d} \qquad (5\text{-}2)$$

where C = capacitance (F)
     A = surface area common to both plates, in square meters (m^2)
     d = thickness of dielectric, in meters (m)
     K = dielectric constant of the insulating material.

EXAMPLE 5.1   For the circuit in Figure 5–3, determine the electric charge in the capacitor.

FIGURE 5.3
Capacitor connected to a DC generator.

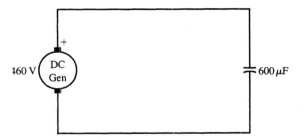

Solution
From Eq. (5–1),

$$q = CV_C = (600 \times 10^{-6}) \times 460 = 0.276 \ C$$

5.2 ENERGY STORAGE IN A CAPACITIVE CIRCUIT

The process of transferring electric charges from one capacitor plate to the other results in the storage of energy. This energy, in the form of displaced electric charges (static electricity), remains stored for some time after the applied voltage is disconnected. The duration of the charge, whether for minutes, hours, or days, depends on such factors as the resistance of the dielectric, leakage between terminals (due to surface contaminants between terminals), humidity, and radioactivity of the environment, all of which serve to gradually dissipate the stored energy.

---

[1] The dielectric constant is also called the *relative permittivity*.

The amount of energy stored in the capacitor depends on the capacitance times the square of the voltage developed across it. Expressed mathematically,

$$W_C = \frac{1}{2}CV_C^2 \qquad (5\text{–}3)$$

where C = capacitance (F)
V_C = voltage across capacitor (V)
W_C = energy stored in joules (J).

EXAMPLE 5.2   For the circuit in Figure 5–3, determine the energy stored in the capacitor.

Solution

$$W_C = \frac{1}{2}CV_C^2 = \frac{1}{2}(600 \times 10^{-6}) \times 460^2 = 63.48 \text{ J}$$

5.3 CAPACITORS IN SERIES AND IN PARALLEL

The equivalent capacitance of two or more capacitors in series, as illustrated in Figure 5–4a, is determined by

$$C_{eq.S} = \cfrac{1}{\cfrac{1}{C_1} + \cfrac{1}{C_2} + \cfrac{1}{C_3} + \cdots + \cfrac{1}{C_n}} \quad \text{or} \quad \frac{1}{C_{eq.S}} = \frac{1}{C_1} + \frac{1}{C_2} + \frac{1}{C_3} + \cdots + \frac{1}{C_n} \quad (5\text{–}4)$$

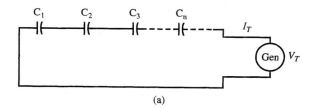

(a)

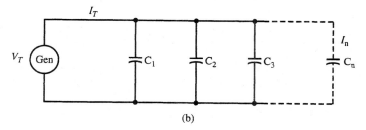

(b)

FIGURE 5.4
(a) Capacitors in series; (b) capacitors in parallel.

The equivalent capacitance of two or more capacitors in parallel, as illustrated in Figure 5–4b, is determined by:

$$C_{eq.P} = C_1 + C_2 + C_3 + \cdots + C_n \tag{5–5}$$

EXAMPLE 5.3

Given the following capacitors: 125 μF, 65.0 μF, and 425 μF. Determine (a) the equivalent capacitance if connected in series; (b) the equivalent capacitance if connected in parallel.

Solution

a. $C_{eq.S} = \dfrac{1}{\dfrac{1}{C_1} + \dfrac{1}{C_2} + \dfrac{1}{C_3}} = \dfrac{1}{\dfrac{1}{125 \times 10^{-6}} + \dfrac{1}{65.0 \times 10^{-6}} + \dfrac{1}{425 \times 10^{-6}}}$

$C_{eq.S} = \dfrac{1}{8000 + 15{,}384.61 + 2352.94} = 38.9 \ \mu F$

Note that the equivalent capacitance of series-connected capacitors is always less than the smallest of the individual capacitors.

b. $C_{eq.P} = C_1 + C_2 + C_3 = 125 \times 10^{-6} + 65.0 \times 10^{-6} + 425 \times 10^{-6} = 615 \ \mu F$

5.4 VOLTAGE AND CURRENT TRANSIENTS IN CAPACITIVE CIRCUITS

The circuit in Figure 5–5 shows an *uncharged* capacitor in series with a battery, a resistor, and a switch. The equivalent series resistance of the capacitor itself, which includes the internal lead resistance and internal terminal connections, is small and assumed to be negligible.[2]

At the instant the switch is closed (called time zero), the current to an uncharged capacitor will be limited by the driving voltage, and any external resistance in series with the capacitor including resistance of the conductors. This is called *inrush current*.

---

[2] The equivalent internal series resistance, called ESR, has significance in applications at frequencies equal to and greater than 120 Hz.

FIGURE 5.5
Uncharged capacitor in series with a battery, resistor, and switch.

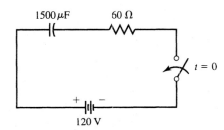

$$I_0 = \frac{E_{bat}}{R} \qquad \qquad \text{(5-6)}$$

where I_0 = current at time zero (A)
 R = resistance of circuit including the connecting wires and any external resistors (Ω).

Thus, for the circuit in Figure 5–5, at the instant the switch is closed,

$$I_0 = \frac{E_{bat}}{R} = \frac{120}{60} = 2 \text{ A}$$

Although an uncharged capacitor has no effect on the initial value of the current, as charging progresses, the buildup of an opposing voltage across the capacitor plates reduces the current to zero. At full charge, the capacitor voltage (countervoltage) is equal to the applied voltage, causing the capacitor current to be zero. The capacitor, in fact, is a discontinuity (break, or open) in an otherwise closed loop and is further evidence that any current must be transitory, decaying to zero as the charging progresses.

Figure 5–6 illustrates voltage and current transients for a representative capacitive circuit. The curves show the buildup of capacitor voltage with time and the decay of capacitor current with time. The part of each curve that extends over the period of time beginning with the closing of the switch (time zero) and ending when the charge

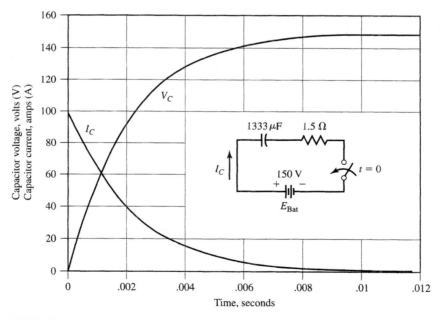

FIGURE 5.6
Voltage and current transients for the given capacitive circuit.

is complete is called a *transient*. The equations for the voltage and current transients shown in Figure 5–6 are as follows:

$$V_C = E_{bat} \times \left[1 - \varepsilon^{-t/RC}\right] \qquad I_C = \frac{E_{bat}}{R} \times \varepsilon^{-t/RC} \qquad (5\text{–}7)$$

where V = capacitor voltage (V)
C = capacitance (F)
R = resistance of battery, connecting wires, switch, and any external resistors (Ω)
t = elapsed time, starting at time zero (s)
ε = 2.718.

Time Constant of a Capacitive Circuit

The time delays experienced in the buildup of voltage and the decay of current are related to the amount of resistance and capacitance in the circuit, and may be of the order of microseconds, minutes, or even hours. A circuit parameter, called the time constant, is very useful for approximating the time delay caused by the capacitive effect.

The time constant is the time that it takes for the voltage and current to make a 63.2 percent change in value, starting from time zero; *it is a measure of the speed of response of the circuit*. After an interval equal to five time constants, the charging process will be 99.3 percent complete, and the circuit may be assumed to be at *steady state*. Expressed mathematically, the time constant for a capacitor in series with resistance is

$$\tau_C = RC \qquad (5\text{–}8)$$

where τ_C = capacitive time constant (s)
R = resistance (Ω)
C = capacitance (F).

Generalized Time-Constant Curves

Curves showing the buildup of capacitor voltage with time and the decay of capacitor current with time, expressed in terms of time constants, are shown in Figure 5–7 for a representative capacitive circuit. These generalized curves are easily sketched, and may be applied to all circuits, or component parts of circuits, containing resistance and capacitance in series.[3]

---

[3] The generalized curves in Figure 5–7 may also be applied to many types of transient phenomena that occur in electrical, mechanical, hydraulic, and other systems.

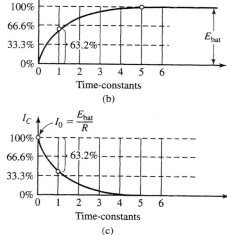

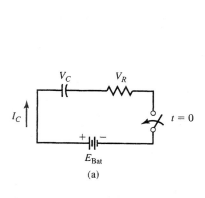

FIGURE 5.7

Voltage and current transients in a capacitive circuit: (a) circuit diagram; (b, c) generalized time-constant curves for a capacitor.

EXAMPLE 5.4

For the circuit in Figure 5–7, assume $R = 1500 \; \Omega$, $C = 3000 \; \mu F$ uncharged capacitor, and $E_{bat} = 120$ V. Determine (a) time constant; (b) current at time zero; (c) current at one time constant; (d) time that it takes to reach steady state.

Solution

a. $\tau_C = RC = 1500 \times 3000 \times 10^{-6} = 4.5$ s

b. $I_0 = \dfrac{E_{bat}}{R} = \dfrac{120}{1500} = 0.08$ A

c. Referring to the time-constant curves in Figure 5–7, in one time constant the current will have decreased to 63.2 percent of its initial 100 percent value.

$$100\% - 63.2\% = 36.8\%$$

$$I_{\text{one time constant}} = 0.368 \times I_0 = 0.368 \times 0.08 = 0.0294 \text{ A}$$

d. $5\tau_C = 5 \times 0.45 = 2.25$ s

EXAMPLE 5.5

A series circuit consisting of a 40-Ω resistor and a 0.003-F uncharged capacitor is connected to a 272-V DC source through a switch. Determine (a) capacitor voltage and current at time zero; (b) time constant; (c) capacitor voltage, capacitor current, and energy stored at one time constant after the switch is closed; (d) voltage across the resistor at one time constant; (e) time it takes to acquire essentially full charge (99.3 percent); (f) capacitor voltage, current, and energy stored at steady-state; (g) voltage across the resistor at steady state.

FIGURE 5.8
Circuit for Example 5–5.

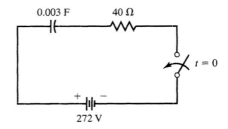

Solution
The circuit is shown in Figure 5–8.

a. At time zero, $I_C = \dfrac{E_{bat}}{R} = \dfrac{272}{40} = 6.8$ A $V_C = 0$

b. $\tau_C = RC = 40 \times 0.003 = 0.12$ s

c. Referring to the time-constant curves in Figure 5–7, at one time constant,

$$V_C = 0.632 \times 272 = 172 \text{ V}$$

$$I_C = 0.368 \times I_0 = 0.368 \times 6.8 = 2.5 \text{ A}$$

$$W_C = \frac{1}{2}CV_C^2 = \frac{1}{2}0.003 \times (172)^2 = 44.3 \text{ J}$$

d. $V_R = I \times R = 2.5 \times 40 = 100$ V

e. $5\tau_C = 5 \times 0.12 = 0.60$ s

f. At 5 time constants,

$$V_{C \text{ steady state}} = 272 \text{ V}$$

$$I_{C \text{ steady state}} = 0$$

$$W_C = \frac{1}{2}CV_C^2 = \frac{1}{2}0.003 \times (272)^2 = 111 \text{ J}$$

g. $V_R = I_C \times R = 0 \times 40 = 0$ V

5.5 EFFECT OF CAPACITANCE IN AN AC SYSTEM

The capacitive effect is present whether the driving voltage is an alternating-current source or a direct-current source (AC or DC). However, with an AC generator, the driving voltage is continually reversing, causing the voltage across the capacitor to be in a continuous state of delay. Because the voltage across the capacitor is delayed, the current is considered to be ahead of, or leading, the voltage. Figures 5–9a and 5–9b illustrate the current and voltage curves for the two different directions of battery driving voltage, and Figure 5–9c shows the current and voltage curves when an alternating driving voltage is applied.

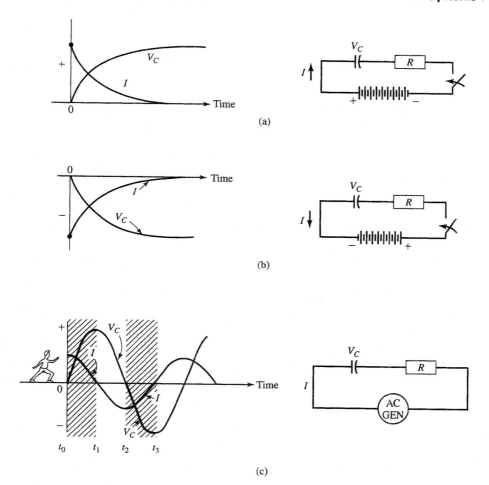

FIGURE 5.9

Effect of an alternating driving voltage on the capacitor current and capacitor voltage: (a, b) curves for two different directions of DC driving voltage; (c) curves for a sinusoidal driving voltage.

From time t_0 to t_1 in Figure 5–9c , the behavior is similar to that shown in Figure 5–9a, and from t_2 to t_3 the behavior is similar to that shown in Figure 5–9b. The difference in curvature between the curves shown in Figure 5–9c and their counterparts in Figures 5–9a and 5–9b is caused by the different driving voltages; the driving voltage of the battery is constant, whereas that of the AC generator alternates in a sinusoidal manner.

The lag or lead of current and voltage waves may be determined by an observer "walking" along the time axis from left to right, observing the positive peaks; for the waves shown in Figure 5–9c, the observer passes the positive peak of the current wave first, and later the positive peak of the voltage wave. Hence, the current wave is said to "lead" the voltage wave.

5.6 DISCHARGING CAPACITORS

If the source voltage in Figure 5–8 is removed, and the capacitor is discharged through the 40-Ω resistor and switch, as shown in Figure 5–10a, the current will be reversed, and the *outrush current* at time zero will be

$$I_0 = -\frac{V_C}{R} = -\frac{272}{40} = -6.8 \text{ A}$$

The voltage and current discharge curves for a capacitor are shown in Figures 5–10b and 5–10c, respectively. After 5 time constants the capacitor will be essentially discharged.

Small capacitors used in radios and audio amplifiers may be discharged by short-circuiting their terminals with a copper wire, called a *jumper*. However, power capacitors should not be discharged in this manner. Power capacitors should be discharged through a resistor. Discharging power capacitors with a short-circuiting jumper could release an amount of energy sufficient to vaporize or explode the shorting conductor, and the mechanical force caused by the magnetic effects of high outrush currents may destroy the capacitor.[4]

---

[4] See Chapter 30 for precautions on discharging and testing power capacitors.

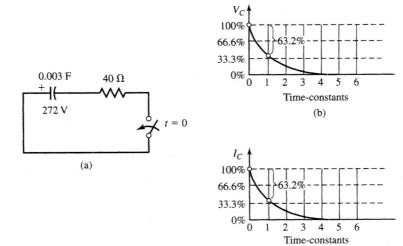

FIGURE 5.10
(a) Capacitor discharge circuit; (b) capacitor voltage curve; (c) capacitor discharge current.

5.7 TESTING SMALL CAPACITORS

> **Warning**
>
> Capacitors should be discharged before and after testing; the residual charge in a capacitor will cause an error in the test results, and may damage the instrument. A charged capacitor also poses a shock hazard.

Small capacitors may be tested easily with a megohmmeter,[5] as shown in Figure 5–11. Unless specified otherwise by the manufacturer, the megohmmeter test voltage for capacitors should approximate the voltage rating of the capacitor.

When a megohmmeter is applied to the two terminals of a good capacitor, as shown in Figure 5–11b, the pointer first deflects sharply to zero and then gradually climbs to the infinity mark (∞) as the capacitor charges. If the pointer deflects to zero and does not climb, the capacitor is shorted. Slight kicks downscale as the pointer climbs indicate leakage current through or across the insulation. The connection for a

---

[5]See Chapter 23 for a discussion on the use of megohmmeters.

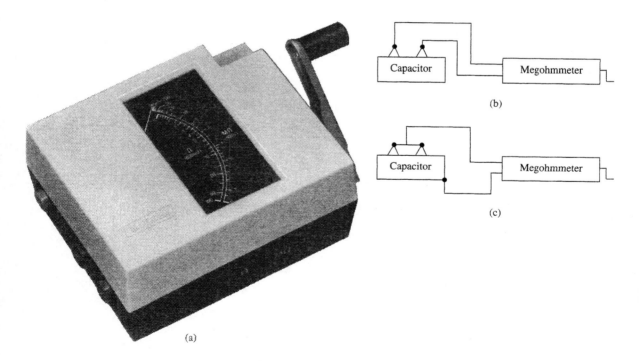

FIGURE 5.11

Testing small capacitors: (a) hand-cranked megohmmeter (courtesy Biddle Instruments); (b) testing for a short circuit; (c) testing for a ground fault.

"ground" test is shown in Figure 5–11c. A good capacitor will indicate ∞; a grounded capacitor will indicate a very low or zero reading.

Electrolytic capacitors, designed for use only in DC circuits, act as capacitors when the applied voltage is in one direction but act as if short circuited when the leads to the capacitor are reversed. Such capacitors will indicate zero ohms when tested in one direction and will climb to a high value of megohms when tested in the opposite direction. Electrolytic capacitors that are designed for use in AC circuits are actually two DC units mounted back to back and enclosed in one housing; such units should indicate a high value of megohms in both directions.

SUMMARY OF EQUATIONS FOR PROBLEM SOLVING

$$C = \frac{q}{V_C} \tag{5-1}$$

$$W_C = \frac{1}{2} C V_C{}^2 \tag{5-3}$$

$$C_{eq.S} = \frac{1}{\dfrac{1}{C_1} + \dfrac{1}{C_2} + \dfrac{1}{C_3} + \cdots + \dfrac{1}{C_n}} \quad \text{or} \quad \frac{1}{C_{eq.S}} = \frac{1}{C_1} + \frac{1}{C_2} + \frac{1}{C_3} + \cdots + \frac{1}{C_n} \tag{5-4}$$

$$C_{eq.P} = C_1 + C_2 + C_3 + \cdots + C_n \tag{5-5}$$

$$I_0 = \frac{E_{bat}}{R} \tag{5-6}$$

$$\tau_C = RC \qquad (5\tau_C \text{ is steady state}) \tag{5-8}$$

REVIEW QUESTIONS

1. What are the defining characteristics of capacitance?
2. List four applications of capacitors.
3. Explain the process by which a capacitor becomes charged.
4. What factors determine the capacitance of a parallel plate capacitor?
5. What are some of the factors that cause a capacitor to lose its charge?
6. What happens to the energy stored in a capacitor when it is disconnected from the supply voltage?
7. Given two capacitors, which connection (series or parallel) will result in a greater total capacitance?
8. What is the recommended safe procedure for discharging a large capacitor?
9. Sketch the voltage and current transients that occur when an uncharged capacitor is connected in series with a resistor and supplied by a DC source.
10. What is meant by the time constant of an electric circuit?
11. Describe the testing procedure for small capacitors.

12. What precautions should be observed before and after testing a capacitor?
13. Sketch the generalized time-constant curves for a capacitor and mark off the time axis in time constants.
14. Sketch the current and voltage waves for a capacitor connected to a sinusoidal source.

PROBLEMS

5–1/1. Determine the electric charge in a 350-μF capacitor when operating at 760 V.

5–2/1. If the electric charge in an 850-μF capacitor is 0.78 coulombs, determine the voltage at the capacitor terminals.

5–3/2. Determine the energy stored in the capacitor in Problem 5–1/1.

5–4/2. Determine the energy stored in the capacitor in Problem 5–2/1.

5–5/3. Determine the equivalent capacitance of the following series-connected capacitors: 2 μF, 18 μF, 60 μF, and 140 μF.

5–6/3. If the capacitors in Problem 5–5/3 are connected in parallel, determine the equivalent capacitance of the parallel combination.

5–7/3. Determine the equivalent capacitance of the following series-connected capacitors: 20 μF, 180 μF, 6 μF, and 110 μF.

5–8/3. If the capacitors in Problem 5–7/3 are connected in parallel, determine the equivalent capacitance of the parallel combination.

5–9/4. A series circuit containing a 20-μF capacitor in series with a 0.5-MΩ resistor and a switch is connected to a 45-V battery. Sketch the circuit and determine (a) the time constant of the circuit; (b) the approximate time it will take, after closing the switch, for the voltage across the capacitor to attain approximately 45 V.

5–10/4. A 300-μF capacitor is connected to a 200-V battery through a 40-Ω resistor. Sketch the voltage-rise and current-decay curves and determine (a) after a lapse of time equal to one time constant, the capacitor voltage and current; (b) the energy stored in one time constant; (c) the energy stored when charging is complete.

5–11/4. A 3-V battery is connected in series with a 4-μF capacitor, an unknown resistor, and a switch. Sketch the circuit. When the switch is closed, the capacitor voltage rises to 63.2 percent of its final value in 10 s. Determine (a) the resistance of the unknown resistor; (b) energy stored in the capacitor when it is completely charged.

5–12/6. A 400-μF capacitor is charged to 1000 V. When discharged through a resistor, the voltage drops to 32.8 percent of its full value in 6 s. Sketch the circuit, and sketch the voltage-discharge curve to an approximate scale. Determine (a) the resistance of the resistor; (b) energy stored in the capacitor before discharge; (c) voltage across the capacitor after 6 s of discharge; (d) energy remaining in capacitor after one time constant of discharge.

6

Current, Voltage, and Impedance in Single-Phase Systems

6.0 INTRODUCTION

The single-phase system is a two-wire alternating-current (AC) system. This system is commonly used in a variety of domestic, commercial, marine, agricultural, and industrial applications including but not limited to lighting, heating, air conditioning, refrigeration, pumps, fans, computers, and copy machines. This chapter introduces the characteristic behavior of inductance and capacitance when connected to an AC source, and provides the mathematical tools necessary to solve problems relating to maintenance and troubleshooting of electrical apparatus.

6.1 ALTERNATING VOLTAGE AND CURRENT

Alternating-current generators, which are designed for lighting and power applications, have sinusoidal current and voltage waves (sine waves) similar to the voltage wave shown in Figure 6–1. One cycle of a sine wave spans 360 degrees, called *electrical degrees*. They are called electrical degrees to differentiate between the space degrees of a circle. The positive and negative half-cycles are called *alternations*, and the maximum value of the wave is called the *amplitude*.

The frequency (f) of a sine wave, such as that shown in Figure 6–1, is the number of complete cycles that the wave makes in 1 second. One cycle per second is defined as 1 hertz (abbreviated Hz). Thus, a 60-Hz wave completes 60 cycles in 1 second.

The period (T) of a sine wave is the time that it takes to complete one cycle. The period is expressed in seconds (s) and is equal to the reciprocal of the frequency. Expressed mathematically,

$$T = \frac{1}{f} \qquad (6\text{–}1)$$

where T = period (s)
f = frequency (Hz).

FIGURE 6.1
Sinusoidal voltage wave.

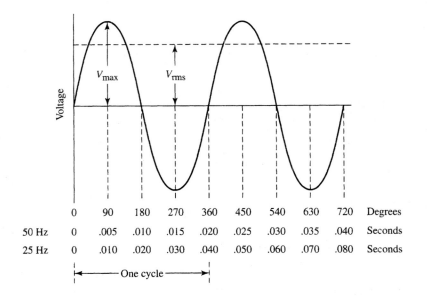

Although there are 360 degrees per cycle, the time it takes to complete one cycle of an alternating wave depends on its frequency. This is shown in Figure 6–1 for 50-Hz and 25-Hz waves. The time required to complete one cycle of a 50-Hz wave is

$$T = \frac{1}{f} = \frac{1}{50} = 0.020 \text{ s}$$

Similarly, the time required to complete one cycle of a 25-Hz wave is

$$T = \frac{1}{f} = \frac{1}{25} = 0.040 \text{ s}$$

Wave forms of alternating voltage and current may be viewed with a handheld graphical multimeter such as that shown in Figure 6–2. This type of instrument provides a clear picture of the waveform, and is very useful in troubleshooting and preventive maintenance operations. Multimeters are also used to measure resistance and frequency, as well as maximum value (called amplitude), average value, and equivalent DC value (rms), of current and voltage waves. They may also be used to troubleshoot diodes and capacitors.

The equation for the sinusoidal voltage wave shown in Figure 6–1 is

$$v = V_{\text{max}} \sin(2\pi \, ft) \qquad (6\text{–}2)$$

where v = voltage at time t (V)
            t = time (s)
        V_{max} = maximum value of the voltage wave (V)
            f = frequency (Hz).

FIGURE 6.2
Digital multimeter (courtesy Fluke
Corporation).

Equivalent DC Value of Alternating Current and Voltage Waves

The equivalent DC value, also called the effective value or root mean square (rms) value, of an alternating-current wave or alternating-voltage wave is commonly used to solve electrical power problems. The rms value is typically found in technical literature and on nameplates of electrical apparatus. Ammeters and voltmeters designed for use in AC power systems indicate the equivalent DC value (rms) of the respective current or voltage wave.[1]

The equivalent DC value (rms) of a sinusoidal wave is equal to its maximum value divided by the square root of 2. That is,

$$V_{rms} = \frac{V_{max}}{\sqrt{2}} = 0.707 \times V_{max}$$

$$I_{rms} = \frac{I_{max}}{\sqrt{2}} = 0.707 \times I_{max}$$

(6–3)

The rms value of an alternating voltage wave is shown in Figure 6–1. The rms values are very useful for calculating current, voltage, and power and, unless otherwise specified, are used throughout the text. However, one must not lose sight of the fact that the actual current is alternating.

---

[1]Ammeters and voltmeters that indicate average values are not recommended for maintenance and troubleshooting applications. Only true-rms measuring instruments should be used.

EXAMPLE 6.1

An rms-responding voltmeter connected to a 60-Hz sinusoidal source indicates 460 V. What is the amplitude of the wave?

Solution
From Eq. (6–3),

$$V_{max} = \frac{V_{rms}}{0.707} = \frac{460}{0.707} = 650.6 \text{ V}$$

6.2 EFFECT OF FREQUENCY ON THE CURRENT IN AN INDUCTOR

As previously discussed, the inductance of a circuit serves to delay any increase or decrease in current but does not prevent or in any way limit the change.[2] When a 0-Hz voltage (DC) is applied to an inductor, the current, although delayed, is limited only by the resistance of the circuit. However, if the circuit is operating at a frequency higher than 0 Hz, the current will be limited in amplitude as well as delayed in time.

The limitation imposed on the amplitude of the current wave, for a given value of inductance, is a direct result of the frequency of reversal of the applied voltage. Current waves at the lower frequencies have the same direction for longer periods of time and therefore attain proportionately higher amplitudes before decreasing than do current waves at the higher frequencies.

The effect of frequency on the current in an inductor is quite similar to the effect of a continually reversing torque applied to a flywheel. In the case of the flywheel, more rapid reversals of the applied torque will result in less motion in each direction. Similarly, in the case of the inductor, higher frequencies of reversal of an applied voltage will result in less current in each direction. This is illustrated in Figure 6–3, where a *constant-voltage*, variable-frequency generator is connected to an "ideal" inductor (an imaginary inductance that has no resistance). Representative voltage waves from 0 Hz to 160 Hz are shown in Figure 6–3b. The corresponding current waves in the inductor are shown in Figure 6–3c. Note that at lower frequencies the current attains higher amplitudes before reversing.

Effect of Increased Inductance

Doubling the inductance doubles the time delay and results in only half the value of current for the same frequency. If there is no inductance in the circuit, then regardless of the frequency, the current will not be delayed and its magnitude will be limited solely by the resistance of the circuit and the amplitude of the driving voltage. Figure 6–4 shows the effect of different values of inductance (L) on the amplitude of the current wave, assuming the frequency and amplitude of the applied voltage wave remain the same.

---

[2]This phenomenon was illustrated with a direct current source in Section 4–3, Chapter 4.

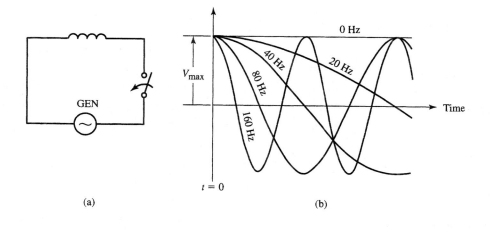

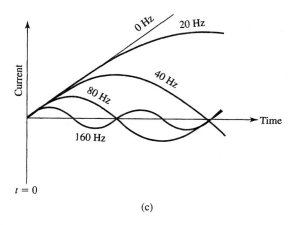

FIGURE 6.3
(a) Alternating-current generator connected to an ideal inductor; (b) representative voltage waves of different frequencies; (c) effect of frequency on the current in an inductor.

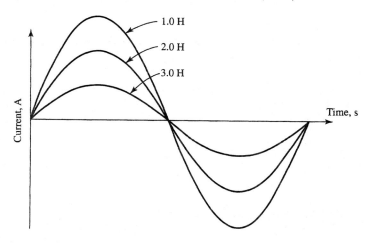

FIGURE 6.4
Effect of increased inductance on the current in an inductor.

6.3 INDUCTIVE REACTANCE

The opposition offered by the inductance of a circuit to a sinusoidal current is called *inductive reactance*. The inductive reactance is proportional to the inductance and to the frequency of reversal of the applied voltage. Expressed mathematically,

$$X_L = 2\pi fL \tag{6-4}$$

where X_L = inductive reactance (Ω)
L = inductance (H)
f = frequency (Hz)
$\pi \approx 3.1416$.

As indicated in Eq. (6–4), the inductive reactance offered by a given inductor decreases with decreasing frequency. If the frequency of the driving voltage is 0 Hz (DC), the inductive reactance will be zero; thus, discounting the initial time delay (five time constants), at 0 Hz, the inductance will offer no opposition to the current. Hence, when used in DC systems, a coil must be designed with adequate internal resistance, by choice of conductor size, or it must be connected in series with an external resistor of such magnitude as to prevent excessive current drain from a DC generator or battery.

The current in an ideal inductance connected to an AC source may be determined by using Ohm's law:

$$I_L = \frac{V_L}{X_L} = \frac{V_L}{2\pi fL} \tag{6-5}$$

where I_L = current in inductor in rms amperes (A)
V_L = voltage applied to inductor in rms volts (V).

EXAMPLE 6.2

A 0.54-H coil of negligible resistance is connected to a 60-Hz sinusoidal driver whose rms voltage is 120 V. The circuit is similar to that shown in Figure 6–3a. Calculate (a) inductive reactance; (b) rms current; (c) current if the rms voltage remains the same but the frequency is changed to 10 Hz; (d) energy stored in the magnetic field of the coil when the 10-Hz current has its maximum value

Solution

a. $X_L = 2\pi fL = 2\pi \times 60 \times 0.54 = 203.6 \; \Omega$

b. $I_L = \dfrac{V}{X_L} = \dfrac{120}{203.6} = 0.59 \; \text{A}$

c. $X_L = 2\pi fL = 2\pi \times 10 \times 0.54 = 33.93 \; \Omega$

$I_L = \dfrac{V}{X_L} = \dfrac{120}{33.93} = 3.54 \; \text{A}$

d. From Eq. (6–3),

$$I_{max} = I_{rms} \times \sqrt{2} = 3.54 \times \sqrt{2} = 5.0 \text{ A}$$

From Eq. (4–11),

$$W_\phi = \frac{1}{2} LI_{max}^2 = \frac{1}{2} \times 0.54 \times 5.0^2 = 6.75 \text{ J}$$

Since inductors have a greater "choking effect" on currents at higher frequencies than on currents at lower frequencies, they are often used in filter circuits to smooth any ripples or irregularities in the current. Inductors used for this purpose are called *choke-coils* or *chokes*.

Large inductors, called *current-limiting reactors*, are used to protect alternating-current generators and other large AC apparatus from damage by limiting short-circuit currents. The reactor places an upper limit on the available short-circuit current that can occur under fault conditions.[3] One type of current-limiting reactor, shown in Figure 6–5, consists of a coil of insulated cable connected in series with the equipment or circuit it is designed to protect. The layers of cable are rigidly clamped between steel-dished heads filled with a mixture of ferromagnetic material and high-strength cement; the ferromagnetic material serves to confine the stray flux. The severe mechanical forces produced by the intense magnetic field can damage the coil if it is not adequately braced. Iron cores are not used, because saturation of the iron at high currents would cause it to lose its effectiveness as a current-limiting inductor.

---

[3]If the available short-circuit current is greater than the interrupting capacity of the circuit breaker, the breaker can explode (see Section 29–6, Chapter 29).

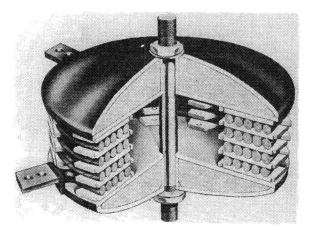

FIGURE 6.5
Current-limiting reactor (courtesy TECO Westinghouse).

6.4 EFFECT OF FREQUENCY ON THE CURRENT IN A CAPACITOR

Figure 6–6 shows the effect of different sinusoidal frequencies on the current in an "ideal" capacitor (an imaginary capacitor that has no resistance). The generator in Figure 6–6a has its voltage V_{max} held constant, but its frequency is adjustable upward from 0 Hz. Representative voltage waves from 0 Hz to 160 Hz are shown in Figure 6–6b.

The current waves that correspond to the different frequencies are shown in Figure 6–6c; higher frequencies cause higher current amplitudes and lower frequencies cause lower current amplitudes. As the frequency approaches 0 Hz the amplitude of the current wave approaches zero. Note the difference in behavior between capacitor circuits and inductor circuits (Figures 6–6c and 6–3c, respectively) when sinusoidal voltages of different frequencies are applied.

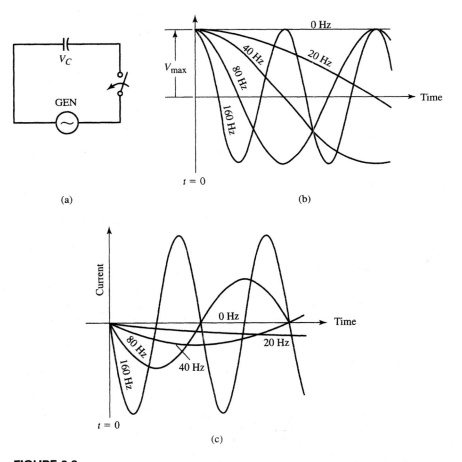

FIGURE 6.6

Effect of frequency on the current in a capacitor: (a) circuit diagram; (b) representative voltage waves of different frequencies; (c) corresponding current waves.

6.5 CAPACITIVE REACTANCE

The opposition offered by capacitance to an alternating current is inversely proportional to the capacitance and to the frequency of reversal of the applied voltage. If the driving voltage is sinusoidal as shown in Figure 6–6, the opposition is called *capacitive reactance* and is expressed mathematically as

$$X_C = \frac{1}{2\pi f C} \tag{6-6}$$

where X_C = capacitive reactance (Ω)
         C = capacitance (F)
         f = frequency (Hz)
         $\pi \approx 3.1416$.

As indicated in Eq. (6–6), the capacitive reactance offered by a given capacitor increases with decreasing frequency. If the frequency of the driving voltage is zero (DC), the capacitive reactance will be infinite, completely blocking the current (discounting the initial charging period).

The current in a capacitor, when connected to an alternating-voltage source, may be determined by using Ohm's law:

$$I_C = \frac{V_C}{X_C} = \frac{V_C}{1/2\pi f C} = 2\pi f C V_C \tag{6-7}$$

where I_C = rms current in capacitor (A)
         V_C = rms voltage applied to capacitor (V)
         C = capacitance (F).

EXAMPLE 6.3

A 250-μF capacitor is connected across a 120-V, 60-Hz system. The circuit is similar to that shown in Figure 6–6a. Determine (a) capacitive reactance; (b) current; (c) energy stored in the capacitor when the voltage across the capacitor has its maximum value. (d) Repeat parts (a), (b), and (c) for a frequency of 6 Hz. (e) Repeat parts (a), (b), and (c) for steady-state conditions with a 120-V DC driver.

Solution

a. $X_C = \dfrac{1}{2\pi f C} = \dfrac{1}{2\pi 60(250 \times 10^{-6})} = 10.61\ \Omega$

b. $I_C = \dfrac{V}{X_C} = \dfrac{120}{10.61} = 11.31\ A$

c. From Eq. (6–3),

$$V_{max} = 120 \times \sqrt{2} = 169.7\ V$$

$$W_{max} = \frac{1}{2}CV_{max}^2 = \frac{1}{2}(250 \times 10^{-6}) \times 169.7^2 = 3.6\ J$$

d. $X_C = \dfrac{1}{2\pi fC} = \dfrac{1}{2\pi 6(250 \times 10^{-6})} = 106.1\ \Omega$

$I_C = \dfrac{V}{X_C} = \dfrac{120}{106.1} = 1.131\ A$

$W_C = \dfrac{1}{2}CV_C^2 = \dfrac{1}{2}(250 \times 10^{-6}) \times 169.7^2 = 3.6\ J$

e. For the DC case $f = 0$, and $V_C = $ DC voltage.

$$X_C = \dfrac{1}{2\pi fC} = \dfrac{1}{2\pi \times 0 \times 250 \times 10^{-6}} = \text{infinity } (\infty)$$

This is, in effect, an open circuit.

$$I_C = \dfrac{V}{X_C} = \dfrac{120}{\infty} = 0\ A$$

$$W = \dfrac{1}{2}CV_C^2 = \dfrac{1}{2}(250 \times 10^{-6}) \times 120^2 = 1.8\ J$$

6.6   PHASE ANGLE OF CURRENT AND VOLTAGE

Figure 6–7a shows the three basic circuit elements connected individually to an AC source. The sine waves in Figure 6–7b illustrate the phase angle relationships between the source voltage (V_T) and and the corresponding current for each type of circuit element. The timescale is for a 50-Hz system. The lag or lead of the current and voltage waves may be determined by the timescale.

An observer "walking" along the time axis from left to right (starting at time zero), *observing the positive peaks* of the voltage and current waves, will pass the positive peak of the capacitor current wave before passing the peak of the applied voltage wave; thus, the capacitor current (I_C) is leading the applied voltage wave by 0.005 seconds or 90°.

For the inductive circuit, the observer passes the positive peak of the voltage wave before passing the peak of the current wave; thus, the wave of inductor current (I_L) is lagging the voltage wave by 0.005 seconds or 90°.

For the resistive circuit, the observer passes the positive peak of the applied voltage wave and the positive peak of the the resistor current wave at the same time. Hence, these waves are *in phase*, rising and falling in unison.

If an AC generator is connected to a series or parallel circuit, containing the three basic circuit elements (inductance, capacitance, and resistance), the current will lag if the overall inductive effect is greater than the overall capacitive effect and will lead if the overall capacitive effect is greater than the overall inductive effect.

If the inductive and capacitive effects are equal, the lagging action of the inductance will be canceled by the leading action of the capacitance, and the current will not

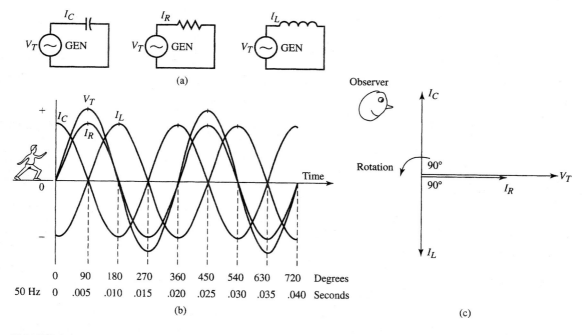

FIGURE 6.7

(a) Basic circuit elements; (b) current waves for resistance, inductance, and capacitance, along with their common driving voltage; (c) phasor diagram corresponding to waves in (b).

lag nor lead the voltage; the current will be in phase with the voltage, rising, falling, and reversing in perfect time with the alternations of the driving voltage. This condition is called *resonance*.

Increasing the resistance of lagging or leading circuits by the introduction of additional resistors makes the respective currents less lagging or less leading, respectively.

Phasors (Rotating Vectors)

Figure 6–7c represents the sine waves as rotating vectors called *phasors*; rotation is counterclockwise (CCW).[4] The phasor diagram provides an easy way to show the magnitude of the waves and their angular position with respect to one another. An observer watching the phasors rotate will note that I_C leads V_T by 90°, I_L lags V_T by 90°, and I_C leads I_L by 180°. Phasor diagrams are used extensively to represent sinusoidal current and sinusoidal voltage in electric power systems. The magnitude of the current or voltage is indicated by the length of the respective phasor.

---

[4]Phasor diagrams in most texts and technical literature do not indicate that the vectors are rotating, but it is implied. The phasors are usually shown "frozen" in place with V_T in the zero-degree position.

6.7 IMPEDANCE

The impedance of an AC circuit is a measure of the overall opposition that the circuit offers to an alternating current. The impedance includes the effects of resistance, as well as any inductance or capacitance that may be present.[5]

Series Impedance

In a series circuit containing resistance, inductance, and capacitance (R-L-C circuit), such as that shown in Figure 6–8a, where the circuit elements are represented as lumped values, the series impedance may be determined from the following equation:

$$Z_S = \sqrt{R^2 + (X_L - X_C)^2} \qquad (6\text{--}8)$$

where
Z_S = impedance of a series circuit (Ω)
R = resistance (Ω)
X_L = inductive reactance (Ω)
X_C = capacitive reactance (Ω).

Parallel Impedance

In a parallel circuit, such as that shown in Figure 6–8b, where the circuit elements are represented as lumped values, the parallel impedance may be determined from the following equation:

$$\frac{1}{Z_p} = \sqrt{\left(\frac{1}{R}\right)^2 + \left(\frac{1}{X_L} - \frac{1}{X_C}\right)^2} \qquad (6\text{--}9)$$

where
Z_P = impedance of a parallel circuit (Ω).

---

[5]Unless otherwise specified in circuit problems, the impedance of the generator and the impedance of the connecting wires are assumed to be negligible.

FIGURE 6.8
(a) Series R-L-C circuit; (b) parallel R-L-C circuit.

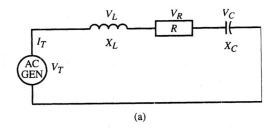

(a)

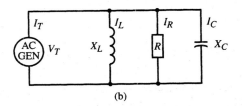

(b)

Note: In those parallel circuits where R, L, or C is not present, the respective $1/R$, $1/X_L$, or $1/X_C$ component must be omitted from the impedance equations. Substituting zero for a missing R, X_L, or X_C in Eq. (6–9), and solving, will result in $Z_P = 0$, which is wrong.

6.8 OHM'S LAW APPLIED TO THE AC CIRCUIT

Applying Ohm's law to an AC circuit is accomplished by substituting the impedance in place of resistance in the Ohm's law equation. Thus, for a sinusoidal AC circuit,

$$I = \frac{V_T}{Z} \qquad\qquad (6\text{–}10)$$

where I = rms current (A)
V_T = rms applied voltage (V)
Z = circuit impedance (Ω)

EXAMPLE 6.4 A 0.2-H coil, whose resistance is 100 ohms, is connected to a 450-V, 60-Hz generator. The connection diagram and equivalent circuit are shown in Figure 6–9. Determine (a) impedance of the coil; (b) circuit current assuming the impedance of the generator and the connecting wires are negligible.

Solution
The coil must be treated as a 0.2-H inductance in series with 100-Ω resistance.

a. $X_L = 2\pi\,fL = 2\pi \times 60 \times 0.2 = 75.4\ \Omega$
Since capacitance is not present, X_C is omitted from the series impedance equation. Thus,

$$Z_S = \sqrt{R^2 + X_L^{\,2}} = \sqrt{100^2 + 75.4^2} = 125\ \Omega$$

b. $I = \dfrac{V_T}{Z_S} = \dfrac{450}{125} = 3.6\ \text{A}$

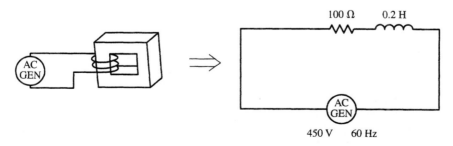

FIGURE 6.9
Connection diagram and equivalent series circuit for Example 6–4.

EXAMPLE 6.5

A 15-Ω resistor, a 1.5-H inductor, and a 3000-μF capacitor are connected in series and supplied by a 120-V, 60-Hz source. Determine (a) the circuit impedance if series connected as shown in Figure 6–10a; (b) the circuit impedance if parallel connected as shown in Figure 6–10b.

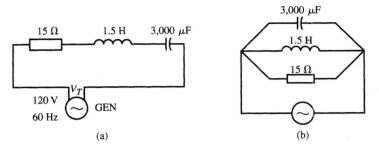

FIGURE 6.10
Circuit for Example 6–5: (a) series connected; (b) parallel connected.

Solution

a. $X_C = \dfrac{1}{2\pi fC} = \dfrac{1}{2\pi 60 \times 3000 \times 10^{-6}} = 0.8842 \ \Omega$

$X_L = 2\pi fL = 2\pi 60 \times 1.5 = 565.487 \Rightarrow 565.5 \ \Omega$

$Z_S = \sqrt{R^2 + (X_L - X_C)^2} = \sqrt{15^2 + (565.487 - 0.8842)^2} = 564.8 \ \Omega$

b. $\dfrac{1}{Z_P} = \sqrt{(1/R)^2 + (1/X_L - 1/X_C)^2} \Rightarrow Z_P = \dfrac{1}{\sqrt{(1/15)^2 + (1/565.487 - 1/0.8842)^2}}$

$Z_P = \dfrac{1}{\sqrt{(0.067)^2 + (0.001768 - 1.13096)^2}} = 0.884 \ \Omega$

6.9 KIRCHHOFF'S VOLTAGE LAW APPLIED TO THE SERIES AC CIRCUIT

A representative series circuit, containing the three different circuit elements, is shown in Figure 6–11. Above each element is its corresponding phasor diagram, and below each element are its corresponding voltage and current waves. *Note that the characteristics of the individual circuit elements remain the same regardless of how they are connected in the circuit—series, parallel, or any other combination.*

At any instant of time, the current in any one element of a series circuit must be identical to the current in all the other elements. Hence, the voltage waves that appear across the three respective elements in Figure 6–11 may be sketched on the same time axis with respect to the common current wave. This is shown in Figure 6–12a. Similarly, the voltage phasors V_L, V_C, and V_R may be sketched on the same phasor diagram with respect to the common current phasor, as shown in Figure 6–12b.

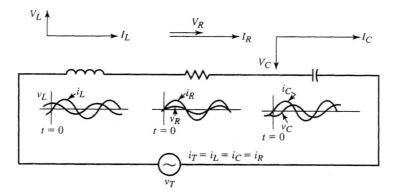

FIGURE 6.11
Series *R-L-C* circuit.

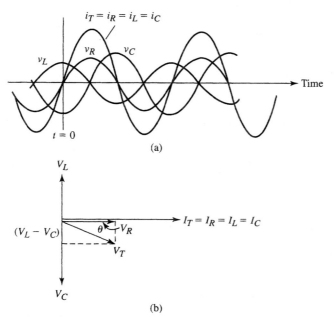

FIGURE 6.12
(a) Voltage drop waves across each element of an *R-L-C* series circuit drawn with respect to the common current wave; (b) corresponding voltage phasors drawn with respect to the common current phasor.

In accordance with Kirchhoff's voltage law for AC circuits, the phasor sum of all voltage drops around any closed loop is equal to the phasor sum of all voltage sources. The voltage drops are V_L, V_C, and V_R, and the only voltage source is V_T. Thus, from the geometry of the phasor diagram in Figure 6–12b,

$$V_T = \sqrt{V_R^2 + (V_L - V_C)^2} \qquad \textbf{(6–11)}$$

The phase angle (θ) between the input voltage (V_T) and and the input current (I_T) is called the *power factor angle*.[6]

---

[6]The relationship between electric power and power factor is presented in Chapter 8.

EXAMPLE 6.6 A coil whose resistance and inductance are 2.4 Ω and 20 mH, respectively, is connected in series with a 450-μF capacitor and a 120-V, 60-Hz source. The equivalent series circuit is shown in Figure 6–13a. Determine (a) inductive reactance; (b) capacitive reactance; (c) impedance; (d) current; (e) voltage drop across each circuit element. (f) Draw the phasor diagram and determine the phase angle between the input current and input voltage.

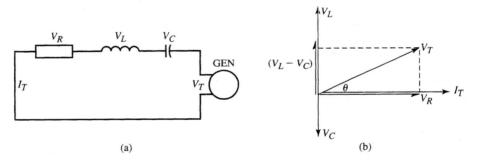

(a)   (b)

FIGURE 6.13
(a) Circuit for Example 6–6; (b) phasor diagram for (a).

Solution

a. $X_L = 2\pi fL = 2\pi \times 60 \times 0.020 = 7.54\ \Omega$

b. $X_C = \dfrac{1}{2\pi fC} = \dfrac{1}{2\pi \times 60 \times 0.00045} = 5.89\ \Omega$

c. $Z_S = \sqrt{R^2 + (X_L - X_C)^2} = \sqrt{2.4^2 + (7.54 - 5.89)^2} = 2.91\ \Omega$

d. $I = \dfrac{V_T}{Z_S} = \dfrac{120}{2.91} = 41.24\ \text{A}$

e. Using Ohm's law,

$$V_R = IR = 41.24 \times 2.4 = 99 \text{ V}$$
$$V_L = IX_L = 41.24 \times 7.54 = 311 \text{ V}$$
$$V_C = IX_C = 41.24 \times 5.89 = 243.1 \text{ V}$$

Note that the voltage across the inductor and the voltage across the capacitor are each greater than the applied voltage. This effect, called *series resonance*, is explained in Chapter 7.

f. From the phasor diagram in Figure 6–13b,

$$\theta = \cos^{-1}\left[\frac{V_R}{V_T}\right] = \cos^{-1}\left[\frac{99}{120}\right] = 34.4°$$

The current lags the applied voltage by 34.4°.

6.10 KIRCHHOFF'S CURRENT LAW APPLIED TO THE PARALLEL AC CIRCUIT

A representative parallel circuit, containing the three different circuit elements, is shown in Figure 6–14. Above each element is its corresponding phasor diagram, and below each element are its corresponding voltage and current waves. Since the voltage in a parallel circuit is the same throughout all parallel components of the circuit, for comparative purposes, the voltage phasor is shown in the same position for all three elements. *Note that the characteristics of the individual circuit elements remain the*

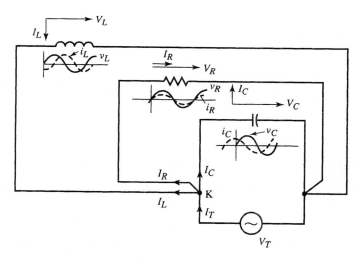

FIGURE 6.14
Parallel *R-L-C* circuit.

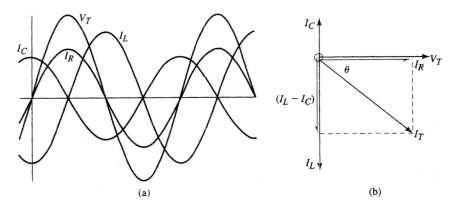

FIGURE 6.15
(a) Current waves in each element of an *R-L-C* parallel circuit drawn with respect to the common voltage wave; (b) corresponding current phasors drawn with respect to the common voltage phasor.

same regardless of how they are connected in the circuit—series, parallel, or any other combination.

At any instant of time, the voltage across any one element of a parallel circuit must be identical to the voltage across all other elements. Hence, the current waves that appear across the three respective elements in Figure 6–14 may be sketched on the same time axis with respect to the common voltage wave. This is shown in Figure 6–15a. Similarly, the current phasors I_L, I_C, and I_R may be sketched on the same phasor diagram with respect to the common voltage phasor. This is shown in Figure 6–15b.

In accordance with Kirchhoff's current law for AC circuits, the phasor sum of all currents entering a junction is equal to the phasor sum of all currents leaving the same junction. The component currents leaving junction K in Figure 6–14 are I_L, I_C, and I_R. The only current entering junction K is I_T. Hence, from the geometry of the phasor diagram in Figure 6–15b,

$$I_T = \sqrt{I_R^2 + (I_L - I_C)^2} \tag{6–12}$$

The phase angle (θ) between the input voltage (V_T) and the input current (I_T) is called the *power factor angle*.

EXAMPLE 6.7   A coil of 0.02 H, a resistor of 10 Ω, and a capacitor of 235 μF are connected in parallel and supplied by a 120-V, 60-Hz generator. The equivalent parallel circuit is shown in Figure 6–16a. Calculate (a) current in each element; (b) line current; (c) circuit impedance. (d) Sketch the phasor diagram and determine the phase angle between the line current and the applied voltage.

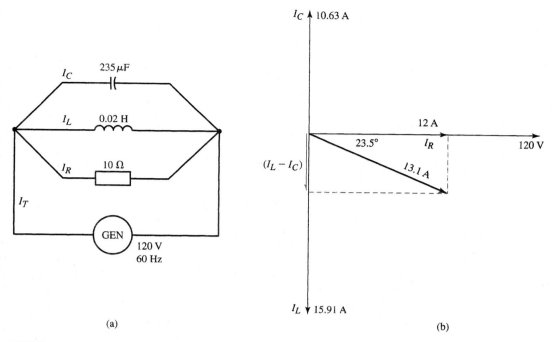

FIGURE 6.16
(a) Circuit for Example 6–7; (b) phasor diagram for (a).

Solution

a. $I_R = \dfrac{V_T}{R} = \dfrac{120}{10} = 12$ A

$X_L = 2\pi fL = 6.28 \times 60 \times 0.02 = 7.539 \Rightarrow 7.54\ \Omega$

$I_L = \dfrac{V_T}{X_L} = \dfrac{120}{7.539} = 15.9154 \Rightarrow 15.91$ A

$X_C = \dfrac{1}{2\pi fC} = \dfrac{1}{2\pi 60(235 \times 10^{-6})} = 11.287 \Rightarrow 11.29\ \Omega$

$I_C = \dfrac{V_T}{X_C} = \dfrac{120}{11.287} = 10.631 \Rightarrow 10.63$ A

b. $I_T = \sqrt{I_R^2 + (I_L - I_C)^2} = \sqrt{12^2 + (15.9154 - 10.631)^2} = 13.112$ A

c. $I_T = \dfrac{V_T}{Z_P} \Rightarrow Z_P = \dfrac{V_T}{I_P} = \dfrac{120}{13.112} = 9.15\ \Omega$

d. From the phasor diagram shown in Figure 6–16b,

$$\theta = \cos^{-1}\left[\frac{I_R}{I_T}\right] = \cos^{-1}\left[\frac{12}{13.112}\right] = 23.76°$$

The total input current lags the applied voltage by 23.76°.

6.11 COMPLEX AC CIRCUITS

Calculating the impedance of series or parallel circuits is straightforward and relatively simple, and is often done when troubleshooting electrical power apparatus. However, a complex impedance made up of other configurations of R, L, and C requires more advanced mathematical methods. In such cases, the impedance may be determined by using the voltmeter-ammeter method and Ohm's law, or specialized impedance measuring instruments. Figure 6–17 illustrates the voltmeter-ammeter method using a multimeter for voltage measurement, and another multimeter with a clamp-on current probe for current measurement.

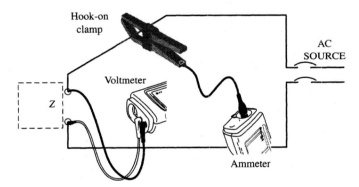

FIGURE 6.17
Voltmeter-ammeter method for determining impedance.

EXAMPLE 6.8 For the circuit in Figure 6–17, assume the ammeter reads 15.8 A, and the voltmeter reads 162.5 V. Determine the impedance of the circuit.

Solution

$$Z = \frac{V}{I} = \frac{162.5}{15.8} = 10.28 \ \Omega$$

SUMMARY OF EQUATIONS FOR PROBLEM SOLVING

For Both Series and Parallel Circuits

$$T = \frac{1}{f} \tag{6-1}$$

$$X_L = 2\pi f L \tag{6-4}$$

$$X_C = \frac{1}{2\pi fC} \tag{6-6}$$

$$V_{rms} = \frac{V_{max}}{\sqrt{2}} = 0.707 \times V_{max} \qquad I_{rms} = \frac{I_{max}}{\sqrt{2}} = 0.707 \times I_{max} \tag{6-3}$$

$$V_R = I_R R \qquad V_L = I_L X_L \qquad V_C = I_C X_C$$

Series Circuit	Parallel Circuit

$$I_T = I_R = I_L = I_C \qquad\qquad V_T = V_R = V_L = V_C$$

$$Z_S = \sqrt{R^2 + (X_L - X_C)^2} \tag{6-8} \qquad \frac{1}{Z_P} = \sqrt{\left(\frac{1}{R}\right)^2 + \left(\frac{1}{X_L} - \frac{1}{X_C}\right)^2} \tag{6-9}$$

$$I_T = \frac{V_T}{Z_S} \qquad\qquad I_T = \frac{V_T}{Z_P}$$

$$V_T = \sqrt{V_R^2 + (V_L - V_C)^2} \tag{6-11} \qquad I_T = \sqrt{I_R^2 + (I_L - I_C)^2} \tag{6-12}$$

$$\theta = \cos^{-1}\left[\frac{V_R}{V_T}\right] \qquad\qquad \theta = \cos^{-1}\left[\frac{I_R}{I_T}\right]$$

REVIEW QUESTIONS

1. What is the relationship between the maximum value of a sine wave and its rms value?
2. What is meant by the period of an alternating wave, and how is it related to the frequency?
3. Explain the effect that the frequency of an alternating voltage wave has on the current in an inductor.
4. Describe the construction details of a current-limiting reactor, and state a specific application.
5. What is inductive reactance, and how does it vary with frequency?
6. Explain the effect that the frequency of an alternating-voltage wave has on the current in a capacitor.

7. What is capacitive reactance, and how does it vary with frequency?
8. Sketch a sinusoidal voltage wave, and then add the individual current waves for a parallel circuit containing a capacitor, an inductor, and a resistor, with their proper phase angles.
9. Sketch a phasor diagram that corresponds to the voltage and current waves in Question 8.

PROBLEMS

6–1/1. The amplitude of a certain 60-Hz source is 650 V. Determine (a) rms voltage; (b) period of the wave. (c) Sketch the wave.

6–2/1. A certain 25-Hz sinusoidal generator develops 240-V rms. Determine (a) maximum value of the voltage wave; (b) period of the wave. (c) Sketch the wave.

6–3/3. A 1.2-H inductor is connected across a 240-V, 25-Hz generator. Determine (a) inductive reactance; (b) rms current; (c) maximum value of the current wave.

6–4/3. A 30-mH inductor is connected across a 108-V, 60-Hz driver. Determine (a) inductive reactance; (b) rms current; (c) amplitude of current wave; (d) peak instantaneous energy stored in the magnetic field.

6–5/3. A 240-V, 60-Hz generator is connected across an inductive reactance of 20 Ω. Sketch the circuit and determine (a) the rms current; (b) the peak value of the sinusoidal current; (c) the inductance.

6–6/5. A 1000-μF capacitor is connected to a 120-V 60-Hz generator. Sketch the circuit and determine (a) the capacitive reactance; (b) the rms current.

6–7/5. A 400-μF capacitor is connected to a 208-V, 25-Hz driver. Sketch the circuit and determine (a) the capacitive reactance; (b) the steady-state current. (c) Calculate the capacitive reactance and (d) the steady current if the 25-Hz driver is replaced by a 208-V battery.

6–8/5. A 240-V, 60-Hz sinusoidal driver is connected to a 500-μF capacitor. Sketch the circuit and determine (a) the capacitive reactance; (b) the steady-state current; (c) the maximum instantaneous voltage across the capacitor.

6–9/8. A coil whose resistance and inductive reactance are 10 Ω and 8 Ω, respectively, is connected in series with a 120-V, 60-Hz source. Determine (a) impedance; (b) current; (c) inductance.

6–10/9. A 10-Ω resistor, a 0.02-H inductance, and a 900-μF capacitor are connected in series and supplied by a 450-V, 50-Hz generator. Sketch the circuit and determine (a) inductive reactance; (b) capacitive reactance; (c) circuit impedance; (d) current; (e) phase angle between input voltage and input current.

6–11/9. A 20-Ω resistor is connected in series with a 1.0-H inductor, a 1500-μF capacitor, and a 208-V, 60-Hz source. Determine: (a) capacitive reactance; (b) inductive reactance; (c) circuit impedance; (d) circuit current; (e) voltage drop across each circuit element; (f) phase angle between input voltage and input current.

6–12/9. A 240-V, 400-Hz generator supplies a series connection of three ideal circuit elements. The elements are a 2.0-Ω resistor, a 3.0-Ω inductive reactance, and a 4.0-Ω capacitive reactance. Sketch the circuit and determine (a) circuit impedance; (b) rms current; (c) voltage drop across the inductance; (d) steady-state current if the driver is replaced by a 240-V battery; (e) steady-state voltage across the capacitor for the conditions in part (d).

6–13/10. A coil of 30 mH, a resistor of 10 Ω, and a capacitance of 350 μF are connected in parallel and supplied by a 120-V, 60-Hz generator. Sketch the circuit and determine (a) the current through each element; (b) the line current; (c) the circuit impedance; (d) the phase angle between input voltage and input current.

6–14/10. A parallel circuit consisting of a 4.0-Ω resistor, a 2.0-Ω capacitive reactance, and a 6.0-Ω inductive reactance is fed from a 30-Hz, 240-V generator. Sketch the circuit and determine (a) circuit impedance; (b) the feeder current.

6–15/10. A 400-V, 50-Hz generator supplies current to a parallel circuit consisting of a 4.0-Ω resistor, a 5.0-Ω inductive reactance, and a 3.0-Ω capacitive reactance. Sketch the circuit and determine (a) input impedance; (b) the feeder current; (c) phase angle between input voltage and input current.

6–16/10. A parallel circuit consisting of a 5.0-Ω capacitive reactance, a 4.0-Ω resistor, and a 2.0-Ω inductive reactance is connected to a 240-V, 60-Hz generator. Sketch the circuit and determine (a) current in each component; (b) feeder current; (c) circuit impedance; (d) values of inductance and capacitance.

6–17/10. An inductor of 20 mH, a capacitor of 450 μF, and a resistor of 5 Ω are connected in parallel and supplied by a 120-V, 60-Hz generator. Determine (a) current through each circuit element; (b) total circuit current; (c) circuit impedance.

6–18/10. A parallel circuit consisting of a 2-Ω resistor, a 15-mH inductor, and a 500-μF capacitor is connected to a 25-Hz, 240-V alternator. Determine (a) circuit current; (b) circuit impedance; (c) current if the frequency is reduced to 0 Hz (direct current).

6–19/10. A 120-V, 60-Hz generator supplies a parallel circuit consisting of a 3-Ω resistor, an inductor whose reactance is 4-Ω, and a capacitor whose reactance is 5-Ω. Determine (a) current drawn by the resistor, (b) current drawn by the capacitor; (c) current drawn by the inductor; (d) total circuit current.

7

Resonance, Harmonics, and their Harmful Effects in Electric Power Systems

7.0 INTRODUCTION

Resonance in electrical power distribution systems can cause current and voltage multiplication that can seriously damage electrical apparatus, blow fuses, and trip circuit breakers.

Another problem that occurs in electrical power systems is the generation of currents and voltages (in the load) whose frequencies are higher than the system frequency. These frequencies, called *harmonics*, are multiples of the system frequency. These can be generated by computer equipment, fluorescent lights, and adjustable speed controllers for induction motors, as well as magnetic saturation effects in transformers and other iron-core apparatus, etc. Harmonic currents, harmonic voltages, and resonance at harmonic frequencies are often the underlying reason for many electrical failures.

This chapter introduces resonance and harmonics, and explains how harmonic resonance can cause excessively high voltage across inductors and capacitors.

7.1 RESONANCE IN ELECTRIC CIRCUITS

Resonance is a condition that occurs in series or parallel AC circuits when the frequency is such that the inductive reactance (X_L) is equal to the capacitive reactance (X_C). A plot of inductive reactance and capacitive reactance versus frequency is shown in Figure 7–1. Note that the capacitive reactance is infinite at a frequency of 0 Hz, decreases with increasing frequency, and approaches zero as the frequency approaches infinity. On the other hand, the inductive reactance is zero at a frequency of 0 Hz, and increases linearly with increasing frequency. The point of intersection of the two curves occurs at a frequency that causes the inductive reactance to equal the capacitive reactance; this is the *resonant frequency*.

FIGURE 7.1
Capacitive reactance and inductive reactance vs. frequency.

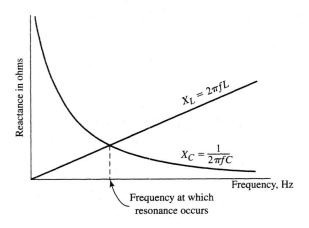

At resonance,

$$X_L = X_C \quad \Rightarrow \quad 2\pi f_r L = \frac{1}{2\pi f_r C}$$

where: f_r = resonant frequency (Hz)
          L = inductance (H)
          C = capacitance (F).

Solving for the resonant frequency,

$$f_r = \frac{1}{2\pi\sqrt{LC}} \qquad\qquad (7\text{–}1)$$

7.2 SERIES RESONANCE

Series resonance is the condition that exists in a series circuit when the frequency is such that the inductive reactance of the series circuit is equal to the capacitive reactance of the series circuit. The behavior of the current in a series *R-L-C* circuit, as the frequency is raised upward from 0 Hz (DC) is shown in Figure 7–2. Note that the rms value of current rises from zero amperes at 0 Hz, reaches its maximum value at the resonant frequency, and then decreases with further increases in frequency, approaching zero amperes as the frequency approaches infinity.

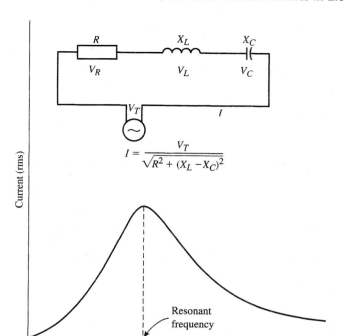

FIGURE 7.2
Current vs. frequency for a series *R-L-C* circuit.

At the resonant frequency, $X_L = X_C$, causing the circuit impedance to be

$$Z_{Sr} = \sqrt{R^2 + (X_L - X_C)^2} = \sqrt{R^2 + (0)^2} = R \qquad (7\text{--}2)$$

where: Z_{Sr} = impedance of series circuit at the resonant frequency (Ω).

When a series circuit is at resonance, the effect is the same as though neither inductance nor capacitance were present. The current under such conditions is dependent solely on the resistance of the circuit.[1] That is,

$$I_{Sr} = \frac{V_T}{Z_{Sr}} = \frac{V_T}{R} \qquad (7\text{--}3)$$

where: I_{Sr} = current in series circuit at the resonant frequency (A)
V_T = input voltage at the resonant frequency (V).

---

[1]Capacitors are sometimes connected in series with transmission lines to offset the inductive reactance of the line and, thus, reduce the voltage drop caused by the inductive reactance.

Rise in Voltage Due to Series Resonance

As shown in Figure 7–2, at series resonance, the current has its maximum rms value, limited only by the circuit resistance. Decreasing the resistance of the circuit increases the current and thus increases the voltage across the inductor (V_L) and the voltage across the capacitor (V_C):

$$V_L = I_{Sr} X_L \tag{7-4}$$

$$V_C = I_{Sr} X_C \tag{7-5}$$

Note: The rise in voltage across the capacitor and the rise in voltage across the inductor may result in destruction of the capacitor, damage to the inductor insulation, or both. Care must always be taken when connecting capacitors and inductors in series. If the particular combination of inductance and capacitance is resonant at the applied frequency, damage to equipment and injury to personnel may occur unless there is sufficient resistance in the circuit or a current-limiting resistor is used.

EXAMPLE 7.1 A series circuit composed of a capacitor rated at 20 μF and 600 V, an inductance rated at 25 mH and 300 V, and a 5-Ω resistor is connected to a 120-V adjustable-frequency generator. Sketch the circuit and determine (a) frequency at which resonance would occur; (b) current at resonance; (c) voltage across the capacitor at resonance; (d) voltage across the inductor at resonance.

Solution
The circuit is shown in Figure 7–3a.

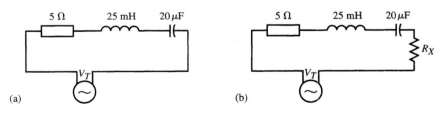

FIGURE 7.3
(a) Circuit for Example 7–1; (b) circuit for Example 7–2.

a. $f_r = \dfrac{1}{2\pi\sqrt{LC}} = \dfrac{1}{2\pi\sqrt{0.025 \times 20 \times 10^{-6}}} = 225.08 \Rightarrow 225$ Hz

b. $I_{Sr} = \dfrac{V_T}{R} = \dfrac{120}{5} = 24$ A

c. $X_C = \dfrac{1}{2\pi \times fC} = \dfrac{1}{2\pi \times 225.08 \times 20 \times 10^{-6}} = 35.355 \Rightarrow 35.4\ \Omega$

$V_C = I_T X_C = 24 \times 35.355 = 848.5$ V

d. $X_L = 2\pi f L = 2\pi \times 225.08 \times 0.025 = 35.355 \Rightarrow 35.4 \ \Omega$

$V_L = I_T X_L = 24 \times 35.355 = 848.5 \ \text{V}$

Note: The excessively high voltage across the capacitor and across the inductor will damage both components.

EXAMPLE 7.2　For the series-resonant circuit in Example 7–1, determine (a) the allowable current in the circuit that will limit the voltage across the inductance to 300 V; (b) the resistance of an external resistor (R_X) to be installed in series with the circuit to limit the current to the value determined in (a).

Solution

a. From Eq. (7–4),

$$I_{Sr} = \frac{V_L}{X_L} = \frac{300}{35.355} = 8.485 \Rightarrow 8.5 \ \text{A}$$

b. The modified circuit with a current-limiting resistor (R_X) in series is shown in Figure 7–3b. Using Ohm's law,

$$I_{Sr} = \frac{V_T}{R + R_X} \Rightarrow 8.485 = \frac{120}{5 + R_X}$$

$$R_X = 9.14 \ \Omega$$

7.3 PARALLEL RESONANCE

Parallel resonance is the condition that exists in a parallel circuit when the frequency is such that the inductive reactance of the parallel circuit is equal to the capacitive reactance of the parallel circuit. The behavior of the current in a parallel *R-L-C* circuit, as the frequency is raised upward from 0 Hz (DC) is illustrated in Figure 7–4.

Note that the rms value of the current decreases from its extremely high value at 0 Hz, reaches a minimum value at the resonant frequency, and then increases with increasing frequency, approaching infinity as the frequency approaches infinity.

Circuit Impedance at Parallel Resonance

The impedance of a parallel *R-L-C* circuit, such as that shown in Figure 7–4, is

$$\frac{1}{Z} = \sqrt{\left(\frac{1}{R}\right)^2 + \left(\frac{1}{X_L} - \frac{1}{X_C}\right)^2}$$

At the resonant frequency, $X_L = X_C$, and the impedance of the parallel circuit becomes

$$\frac{1}{Z_{Pr}} = \sqrt{\left(\frac{1}{R}\right)^2} \Rightarrow Z_{Pr} = R \tag{7–6}$$

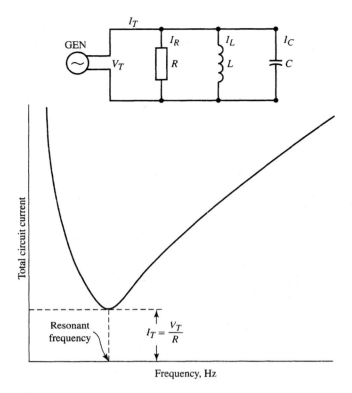

FIGURE 7.4
Current vs. frequency for a parallel *R-L-C* circuit.

The total current drawn by a parallel circuit containing resistance, inductance, and capacitance, such as that shown in Figure 7–4, is given by

$$I_P = \sqrt{I_R{}^2 + (I_L - I_C)^2}$$

If the frequency of the AC generator is adjusted so that $X_L = X_C$, the parallel circuit is at resonance, $I_L = I_C$, and the total circuit current is

$$I_{Pr} = \sqrt{I_R{}^2 + (0)^2} = I_R \qquad (7\text{--}7)$$

where I_{Pr} = total circuit current at parallel resonance (A).

The total current drawn by a parallel *R-L-C* circuit, when at resonance, is that drawn by the resistor alone. If the parallel-connected resistor is removed, as shown in Figure 7–5, the total circuit current drops to zero. However, regardless of the value of the parallel-connected resistance, the individual currents to the inductor and the capacitor are not zero.

Except for the brief initial charging current that occurs when the switch is closed, the current drawn by the capacitor and drawn by the inductor (when at resonance) do not appear at the generator terminals, but oscillate between the inductor and

FIGURE 7.5

Parallel resonance when resistance is negligible.

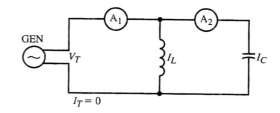

$$I_T = 0$$

the capacitor. Referring to Figure 7–5, ammeter A_1 indicates zero,[2] whereas ammeter A_2 indicates a magnitude of current equal to

$$I_L = I_C = \frac{V_T}{X_L} = \frac{V_T}{X_C}$$

---

[2]Since pure values of inductance and capacitance cannot be obtained in actual practice, ammeter A_1 will indicate some small value, insignificant as it may be.

EXAMPLE 7.3 A 124-Ω resistor, a 25-mH inductance, and a 31.3-μF capacitor are connected in parallel and supplied by a 240-V variable-frequency source that is adjusted to the resonant frequency. Determine (a) resonant frequency; (b) impedance; (c) total circuit current; (d) inductive reactance and capacitive reactance; (e) current in capacitor and in inductor; (f) voltage across the inductor, across the capacitor, and across the resistor.

Solution

The circuit is shown in Figure 7–6.

FIGURE 7.6

Circuit for Example 7–3.

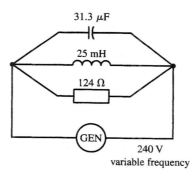

31.3 μF

25 mH

124 Ω

GEN

240 V
variable frequency

a. $f_r = \dfrac{1}{2\pi\sqrt{LC}} = \dfrac{1}{2\pi\sqrt{0.025 \times 31.3 \times 10^{-6}}} = 179.91 \Rightarrow 180$ Hz

b. At resonance $Z_P = R = 124\ \Omega$

c. $I_T = \dfrac{V}{Z_P} = \dfrac{240}{124} = 1.93$ A

d. $X_L = 2\pi f L = 2\pi \times 180 \times 0.025 = 28.262 \Rightarrow 28.3\ \Omega$

$X_C = \dfrac{1}{2\pi f C} = \dfrac{1}{2\pi \times 180 \times 31.3 \times 10^{-6}} = 28.262 \Rightarrow 28.3\ \Omega$

e. $I_C = \dfrac{V_C}{X_C} = \dfrac{240}{28.262} = 8.49$ A $\qquad I_L = \dfrac{V_L}{X_L} = \dfrac{240}{28.262} = 8.49$ A

Note the high current in the L and C branches, and the low total circuit current.

f. The voltage across each circuit element is the source voltage, 240 V.

7.4 HARMONICS IN ELECTRIC POWER APPLICATIONS

The extensive use of nonlinear loads, such as computers, and solid-state control of motors has resulted in severe distortion of the normally sinusoidal current and normally sinusoidal voltage, resulting in overheating of conductors, burnout of transformer windings, and poor performance of some electrical apparatus. The distortion is caused by currents of different frequencies (called *harmonics*) that are generated by nonlinear loads and superimposed on the normally sinusoidal system. An example of a distorted current wave is shown in Figure 7–7.

Analysis of a distorted sinusoidal wave reveals that it is composed of many different frequencies. These frequencies, called *harmonics*, are whole-number multiples of the system frequency. The system frequency is called the fundamental frequency, and is called the 1st harmonic. Thus, if the system frequency is 60 Hz, some of the many possible harmonics are:

2nd harmonic = 2 × 60 = 120 Hz
3rd harmonic = 3 × 60 = 180 Hz
4th harmonic = 4 × 60 = 240 Hz
5th harmonic = 5 × 60 = 300 Hz
6th harmonic = 6 × 60 = 360 Hz
7th harmonic = 7 × 60 = 420 Hz, etc.

A small handheld harmonic analyzer is illustrated in Figure 7–8a. This instrument can provide a direct readout of the magnitude of the harmonics present in a nonsinusoidal wave. The readings are given as a percent of the fundamental.

FIGURE 7.7
Example of a distorted current wave.

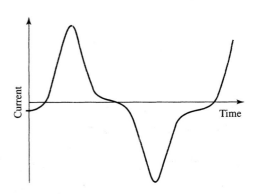

(a)

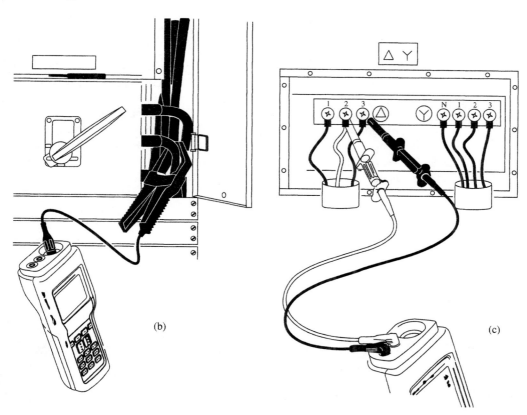

(b)

(c)

FIGURE 7.8

(a) Handheld harmonic analyzer; (b) measuring current harmonics; (c) measuring voltage harmonics (courtesy Fluke Corporation).

131

Harmonic Distortion Factor

The harmonic distortion factor, called *percent total harmonic distortion* (%THD), is a measure of the amount of distortion caused by all harmonics present in the current wave or in the voltage wave, expressed as a percent of the fundamental. The %THD for current is

$$\%\text{THD} = \frac{\sqrt{I_2^2 + I_3^2 + I_4^2 + I_5^2 + \cdots + I_n^2}}{I_1} \times 100 \qquad (7\text{--}8)$$

where $\%\text{THD}$ = percent total harmonic distortion
I_1 = rms current at fundamental frequency (A)
I_2 = rms current at 2nd-harmonic frequency (A)
I_3 = rms current at 3rd-harmonic frequency (A)
I_4 = rms current at 4th-harmonic frequency (A)
I_5 = rms current at 5th-harmonic frequency (A)
I_n = rms current at nth-harmonic frequency (A).

Similarly, the %THD for voltage is:

$$\%\text{THD} = \frac{\sqrt{V_2^2 + V_3^2 + V_4^2 + V_5^2 + \cdots + V_n^2}}{V_1} \times 100 \qquad (7\text{--}9)$$

The harmonic analyzer, shown in Figure 7–8, can provide a direct readout of %THD without going through any of the calculations required in Eqs. (7–8) and (7–9), and is very useful when troubleshooting harmonic problems in distribution systems.[3] Figure 7–8b shows the connections for measuring the harmonic content in the current at the service entrance of a plant; a clamp-on current probe provides a reduced current signal to the meter. Figure 7–8c shows the connections for measuring the harmonic content in the voltage at the terminals of a distribution transformer.

---

[3] Additional features of the harmonic analyzer are presented in Section 8–9, Chapter 8.

EXAMPLE 7.4

A certain electric-arc furnace operating at 660 volts produces the following voltage harmonics.

Harmonic	rms Voltage
Fundamental	660
2nd harmonic	12.4
3rd harmonic	58
4th harmonic	13.2
5th harmonic	66
6th harmonic	9.9
7th harmonic	39.6

Determine the harmonic distortion factor.

Solution

Substituting into Eq. (7–9),

$$\%THD = \frac{\sqrt{12.4^2 + 58^2 + 13.2^2 + 66^2 + 9.9^2 + 39.6^2}}{660} \times 100 = 14.9\%$$

Resonance Overvoltage at Harmonic Frequencies

Overvoltages caused by resonance at harmonic frequencies can cause serious problems in distribution systems and power lines [1]. A discussion of harmonic resonance
that resulted in failure of 13.8-kV switchgear and consequential plant losses is documented in Reference [2].[4]

---

[4]Additional examples of the harmful effects of harmonics and resonance in electric power systems are presented
in Sections 10–7 and 10–9, Chapter 10, and in Section 28–3, Chapter 28.

EXAMPLE 7.5
The series-circuit equivalent of a certain single-phase load connected to two lines of a
460-V, 60-Hz distribution system has a resistance of 6.4 Ω, an inductance of 45 mH,
and a capacitance of 6.25 μF. The equivalent circuit is shown in Figure 7–9. Determine
(a) impedance; (b) current; (c) voltage drop across the inductance and across the
capacitance. (d) Repeat (a), (b), and (c) assuming the circuit is acted on by a 260-V,
5th-harmonic source.

FIGURE 7.9
Circuit for Example 7–5.

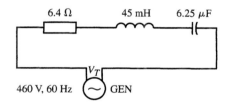

Solution

a. $X_L = 2\pi f_1 L = 2\pi \times 60 \times 0.045 = 16.96 \ \Omega$

$$X_C = \frac{1}{2\pi f_1 C} = \frac{1}{2\pi \times 60 \times 6.25 \times 10^{-6}} = 424.41 \ \Omega$$

$$Z_S = \sqrt{R^2 + (X_L - X_C)^2} = \sqrt{6.4^2 + (16.96 - 424.41)^2} = 407.5 \ \Omega$$

b. $I_T = \dfrac{V_T}{Z_S} = \dfrac{460}{407.5} = 1.13 \ A$

c. $V_L = I_T X_L = 1.13 \times 16.96 = 19.15 \ V$

$V_C = I_T X_C = 1.13 \times 424.4 = 479.09 \ V$

d. $f_5 = 5 \times f_1 = 5 \times 60 = 300 \text{ Hz}$

$X_{L5} = 2\pi f_5 L = 2\pi \times 300 \times 0.045 = 84.82 \ \Omega$

$X_{C5} = \dfrac{1}{2\pi f_5 C} = \dfrac{1}{2\pi \times 300 \times 6.25 \times 10^{-6}} = 84.88 \ \Omega$

$Z_{S5} = \sqrt{R^2 + (X_{L5} - X_{C5})^2} = \sqrt{6.4^2 + (84.82 - 84.88)^2} = 6.4 \ \Omega$

$I_5 = \dfrac{V_{T5}}{Z_{S5}} = \dfrac{260}{6.4} = 40.62 \text{ A}$ $\qquad V_{L5} = I_5 X_{L5} = 40.62 \times 84.82 = 3446 \text{ V}$

$\qquad\qquad\qquad\qquad\qquad\qquad V_{C5} = I_5 X_{C5} = 40.62 \times 84.88 = 3448 \text{ V}$

Note the significant overvoltage and overcurrent caused by the 5th harmonic.

SUMMARY OF EQUATIONS FOR PROBLEM SOLVING

$$f_r = \frac{1}{2\pi\sqrt{LC}} \qquad (7\text{--}1)$$

$$Z_{Sr} = R \qquad (7\text{--}2)$$

$$I_{Sr} = \frac{V_T}{Z_{Sr}} = \frac{V_T}{R} \qquad (7\text{--}3)$$

$$V_L = I_{Sr} X_L \qquad (7\text{--}4)$$

$$V_C = I_{Sr} X_C \qquad (7\text{--}5)$$

$$Z_{Pr} = R \qquad (7\text{--}6)$$

$$I_{Pr} = I_R \qquad (7\text{--}7)$$

$$\%\text{THD} = \frac{\sqrt{I_2^2 + I_3^2 + I_4^2 + I_5^2 + \cdots + I_n^2}}{I_1} \times 100 \qquad (7\text{--}8)$$

$$\%\text{THD} = \frac{\sqrt{V_2^2 + V_3^2 + V_4^2 + V_5^2 + \cdots + V_n^2}}{V_1} \times 100 \qquad (7\text{--}9)$$

SPECIFIC REFERENCES KEYED TO THE TEXT

[1] Currence, E. J., Plizga, J. E., and Nelson, H. N., Harmonic Resonance at a Medium-Sized Industrial Plant, *IEEE Trans. Industry Applications*, Vol. IA-31, No. 4, July/August 1995.

[2] Lemieux, G., Power System Harmonic Resonance—A Documented Case, *IEEE Trans. Industry Applications*, Vol. IA-26, No. 3, May/June 1990.

REVIEW QUESTIONS

1. What is meant by resonance in an electric circuit?
2. Explain why the total circuit current drawn by a series-resonance circuit is equal to that drawn by the resistor alone.
3. Explain how a series-resonance condition can cause a dangerously high voltage across the capacitor and across the inductor.
4. Explain why the total circuit current drawn by a parallel-resonance circuit is equal to that drawn by the resistor alone.
5. What are power system harmonics, how are they generated, and what are some of their adverse effects?
6. Define harmonic distortion factor.

PROBLEMS

7–1/2. A series circuit composed of a 2.21-μF capacitor, a 65-mH inductance coil, and a 10-Ω resistor is connected to a 440-V adjustable-frequency generator. Sketch the circuit and determine (a) resonance frequency; (b) current at resonance; (c) voltage across the capacitor at resonance; (d) voltage across the inductor at resonance.

7–2/2. A series R-L-C circuit consisting of a 6.24-Ω resistor, a 1.5-mH inductor, and a 42-μF capacitor is connected to a 120-V AC source operating at the resonant frequency. Determine (a) resonance frequency; (b) circuit current; (c) voltage across the inductor, voltage across the capacitor, and voltage across the resistor.

7–3/3. A 0.013-H coil, a resistor of 10 Ω, and a capacitor of 175 μF are connected in parallel and supplied by a 120-V, 60-Hz generator. Calculate (a) current through each element; (b) total line current; (c) circuit impedance; (d) current if the frequency is adjusted to cause resonance; (e) resonance frequency.

7–4/3. A resistance of 4.8 Ω, a capacitance of 750 μF, and an inductance of 0.86 H are connected in parallel and supplied by a 480-V, 60-Hz generator. Determine (a) current through each element; (b) total line current; (c) circuit impedance; (d) current if the frequency is adjusted to cause resonance; (e) resonance frequency; (f) line current if the voltage were unidirectional (DC) instead of alternating.

7–5/4. A certain electric apparatus operating at 240 volts produces the following current harmonics:

Harmonic	rms Current
Fundamental	106
2nd harmonic	1.2
3rd harmonic	52.4
4th harmonic	0.8
5th harmonic	22.5
6th harmonic	1.1
7th harmonic	8.3

Determine the harmonic distortion factor.

7–6/4. The harmonic content of a certain load is given here:

Harmonic	rms Voltage
Fundamental	265
2nd harmonic	13.3
3rd harmonic	55.7
4th harmonic	10.6
5th harmonic	31.8
6th harmonic	5.0
7th harmonic	17.2

Determine the harmonic distortion factor.

7–7/4. The equivalent series-circuit parameters of a certain single-phase load connected to two lines of a 60-Hz, 600-V distribution system are $R = 0.652\ \Omega$, 2.48 mH, and 58 μF. Determine (a) circuit impedance; (b) circuit current; (c) voltage across each circuit element; (d) current if the 60-Hz source is replaced by a 410-V, 7th-harmonic source; (e) voltage drop across each circuit element for the conditions in (d).

8

Active, Reactive, and Apparent Power in the Single-Phase System

8.0 INTRODUCTION

When an AC generator supplies energy to a distribution system consisting of heaters, lamps, motors, etc., some of the energy does useful work, but some remains locked in the system, "seesawing" between the generator and the load, and sometimes between loads. An understanding of these characteristics will enable operating engineers and technicians to make necessary adjustments in motor and system loading.

The "health" of an electric motor or distribution system may be determined by analyzing such vital signs as current, voltage, active power (watts), reactive power (vars), and power factor. These values may be measured with handheld power meters, and adjustments made to improve efficiency and power factor, avoid overload, and help manage energy costs.

Principles are introduced using ideal elements, followed by real-life industrial applications.

8.1 ACTIVE AND REACTIVE POWER (kW AND kvar)

Unless otherwise specified, resistance, capacitance and inductance will be assumed to be ideal (pure) elements.

Resistance

When an alternating current passes through a resistance, as shown in Figure 8.1a, the resistance converts the electric energy to heat energy. This type of energy conversion can be found in wire, cabling, lamps, and various heating elements such as space heaters, toasters, and electric ranges. The flow of this energy is unidirectional in that it passes from the generator to the load; it does not return to the generator. The rate of

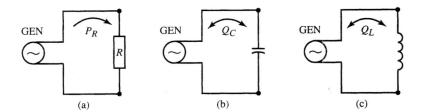

FIGURE 8.1
Circuits showing the direction of energy flow between an AC generator and individual ideal components: (a) resistance; (b) capacitance; (c) inductance.

transfer of this energy as it flows unidirectionally from the generator to the load is called *active power* and is measured with a wattmeter in watts (W). This active power, sometimes called real power or true power, is equal to the voltage across the resistance multiplied by the current in it:

$$P_R = V_R I_R \tag{8-1}$$

where: P_R = active power, expressed in watts (W)
V_R = voltage across the resistor (V)
I_R = current in the resistor (A).

Capacitance

If a capacitor is connected to an AC generator, as shown in Figure 8.1b, the alternating voltage causes the electric charge in the capacitor to alternately build up and decay. The energy accumulated in the capacitor during the buildup of voltage is returned to the generator when the voltage decays. This cyclic exchange of energy between the capacitor and the generator continues undiminished as long as the capacitor is connected to the generator. The rate of energy transfer, as it cycles between generator and capacitor, is called *reactive power*. Reactive power is measured with a varmeter in vars. The term *var* comes from the first letters of the words "volt-amperes reactive." Since reactive power does no useful work it is sometimes called *wattless power*. Capacitor vars are called leading vars because the current to a capacitor leads the voltage across the capacitor. The reactive power drawn by a capacitor is equal to the voltage across the capacitor multiplied by the current in it:

$$Q_C = V_C I_C \tag{8-2}$$

where: Q_C = leading reactive power (var)
V_C = voltage across capacitor (V)
I_C = current in capacitor (A).

Inductance

If an inductor, such as a coil of wire, is connected to an AC generator, as shown in Figure 8.1c, the alternating current causes the magnetic field to buildup and decay. The energy accumulated in the magnetic field of the inductor during the buildup of current is returned to the generator when the current decays. This cyclic exchange of energy between the inductor and the AC generator continues undiminished as long as the inductor is connected to the generator. Inductor vars are called lagging vars because the current to an inductor lags the voltage across the inductor. The rate of transfer of this energy as it cycles between generator and inductor is also called reactive power, and is equal to the voltage across the inductor multiplied by the current in it:

$$Q_L = V_L I_L \qquad\qquad (8\text{–}3)$$

where: $\quad Q_L$ = lagging reactive power (var)
$\quad\quad V_L$ = voltage across inductor (V)
$\quad\quad I_L$ = current in inductor (A).

Mixed Elements

If a circuit includes resistive, inductive, and capacitive loads, as shown in Figure 8.2, where $Q_L > Q_C$, the load with the lagging vars gives up its energy during the same period of time that the load with the leading vars is accepting energy. Hence, some exchange of energy takes place between the inductor and the capacitor, and fewer vars are drawn from the generator.

Lagging vars are defined as having a positive value, and leading vars as having a negative value. Thus, the total (net) vars (Q_T) supplied by the generator represent the difference between the lagging vars and the leading vars:

$$Q_T = Q_L - Q_C \qquad\qquad (8\text{–}4)$$

Resonance

If the frequency is such that the capacitive reactance in Figure 8.2 is equal to the inductive reactance, the capacitive vars will be equal to the inductive vars, and the capacitance and inductance will cycle all energy between one another. No reactive power will be supplied by the generator.

FIGURE 8.2
Circuit that includes all components:
resistive, inductive, and capacitive loads.

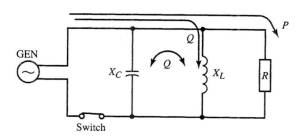

Because of its ability to store energy, a parallel *L–C* circuit is called a *tank circuit*. When the switch is closed, there is an initial transfer of energy from the generator to the tank circuit. Thereafter, the energy cycles back and forth between the inductance and the capacitance, and no additional energy is supplied to the tank circuit. This condition is called *resonance*.[1]

AC Motors

An AC motor converts most of the electrical energy it receives from the generator to mechanical work (active power), and converts a small amount to heat losses in the copper conductors and iron framework (also active power); but the energy used to establish the magnetic field of the motor "seesaws" between the generator and the motor, doing no work at all (reactive power).

8.2 POWER-FLOW DIAGRAM

The power-flow diagram shown in Figure 8.3 shows a prime mover such as a turbine or diesel engine supplying mechanical power to an AC generator. The mechanical power supplied by the prime mover is active power. The generator converts the mechanical power to electrical power. Heating elements and incandescent lamps draw only active power, whereas motors, transformers, and other inductive loads draw both active and reactive power. Note that the active power (kW) is shown as unidirectional, but the reactive power (kvar) is shown as bidirectional:

$$kW = \frac{watts}{1000} \qquad kvar = \frac{vars}{1000}$$

---

[1]More detailed information about resonance in electric power systems is presented in Chapter 7.

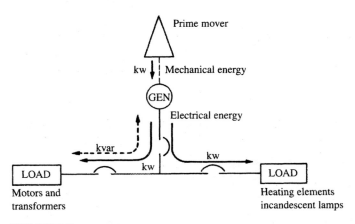

FIGURE 8.3
Power-flow diagram.

8.3 APPARENT POWER AND THE POWER TRIANGLE

The combination of active power and reactive power drawn by a load or combination of loads is called *apparent power* and is expressed in volt-amperes (VA). Apparent power may be calculated by substituting into the following equation:

$$S_T = \sqrt{P_T^2 + Q_T^2} \qquad (8\text{--}5)$$

where: S_T = total apparent power (VA)
P_T = total active power (W)
Q_T = total or net reactive power (var).

The active and reactive components of power in Eq. (8–5) are right-angle components of the power triangle shown in Figure 8.4. The diagonal is the apparent power (S_T), and angle θ is called the power factor angle.

Some useful equations obtained from the geometry of the power triangle are as follows:

$$S_T = \sqrt{P_T^2 + Q_T^2}$$

$$P_T = S_T \times \cos(\theta) \qquad (8\text{--}6)$$

$$Q_T = S_T \times \sin(\theta) \qquad (8\text{--}7)$$

$$\theta = \cos^{-1}\left(\frac{P_T}{S_T}\right) \qquad (8\text{--}8)$$

Note: Active power, reactive power, and apparent power are products of volts times amperes. Thus, it is correct to express all of them in volt-amperes (VA). For example, 20 watts or 20 volt-amperes of active power; 20 vars or 20 volt-amperes of reactive power. Furthermore, when solving power problems, active power, reactive power, and apparent power must all have the same size units. That is, all in VA, all in kVA (kilovolt-amperes), or all in MVA (megavolt-amperes).

FIGURE 8.4
Power triangle.

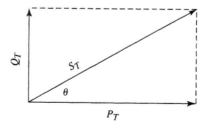

EXAMPLE 8.1   A 60-Hz, 240-V, single-phase source supplies a system that includes a 3-kW resistor heating load, a 0.8-kvar capacitor, and an induction motor that draws 4 kW and 6.2 kvar. Determine (a) total active power; (b) net reactive power; (c) system apparent power;

FIGURE 8.5
(a) Circuit diagram; (b) power-flow diagram;
(c) power triangle for Example 8.1.

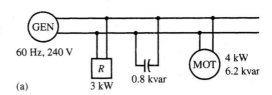

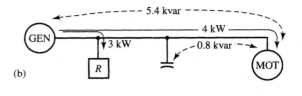

(a)

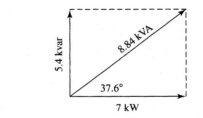

(b)

(c)

(d) power-factor angle. The circuit diagram, power-flow diagram, and power triangle are shown in Figure 8.5.

Solution

a. $P_T = P_{\text{heater}} + P_{\text{motor}} = 3 + 4 = 7\,\text{kW}$

b. $Q_T = Q_L - Q_C = 6.2 - 0.8 = 5.4\,\text{kvar}$

c. $S_T = \sqrt{P_T^2 + Q_T^2} = \sqrt{7^2 + 5.4^2} = 8.84\,\text{kVA}$

d. $\theta = \cos^{-1}\left[\dfrac{P_T}{S_T}\right] = \cos^{-1}\left[\dfrac{7}{8.84}\right] = 37.6°$

8.4 POWER FACTOR

The power factor of a circuit, a motor, or a system is a measure of its effectiveness in utilizing the apparent power it draws from the generator. It is the ratio of the active power input to the apparent power input. Expressed mathematically,

$$F_P = \frac{P_{\text{in}}}{S_{\text{in}}} \tag{8-9}$$

where: P_{in} = active power input (W)
S_{in} = apparent power input (VA)
F_P = power factor, (pu)

The power factor (F_p) in Eq. (8–9) is expressed in decimal form and called per-unit power factor (pu). If the power factor is given in percent, it must be converted to per-unit form by dividing by 100 before substituting into Eq. (8–9). That is,

$$F_P = \frac{\% \text{ power factor}}{100} \tag{8–10}$$

The power factor is also equal to the cosine of the power-factor angle shown in Figure 8–4.[2] That is,

$$F_P = \cos(\theta)$$
$$\theta = \cos^{-1}(F_P) \tag{8–11}$$

The power factor of a load or system can be measured with a power-factor meter, such as that shown in Figure 8.6. The meter can indicate power-factor values ranging from 0.5 to 1.0. Resistors, incandescent lamps, electric ranges, and similar heating equipment draw only active power. For these loads, the apparent power and active power drawn from the generator are one and the same, and the power factor is 1.0. A lagging power factor indicates a predominantly inductive circuit and a leading power factor indicates a predominantly capacitive circuit.[3]

A ceramic capacitor draws very little active power from the generator. (The heat losses in a ceramic capacitor are very small and do not provide a significant wattmeter indication.) Hence, the power factor of a ceramic capacitor is approximately zero:

$$F_{P(cap)} = \frac{P_{cap}}{S_{cap}} \approx \frac{0}{S_{cap}} \approx 0 \tag{8–12}$$

---

[2]Power-factor angle θ is also equal to the phase angle between the input current and the input voltage of a single-phase circuit.

[3]Synchronous motors may be made to operate at lagging, leading, or unity power factor.

FIGURE 8.6
Switchboard-type power-factor meter
(courtesy GE Industrial Systems).

EXAMPLE 8.2

For the 240-V, 60-Hz system shown in Figure 8.7a, determine (a) system active power in kW; (b) system reactive power in kvars; (c) system reactive power in kVA; (d) system power factor. (e) Is the system power factor lagging or leading?

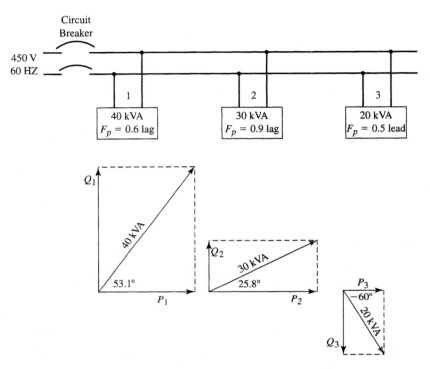

FIGURE 8.7
(a) System diagram and (b) individual power triangles for Example 8.2.

Solution

a. $P_1 = S_1 \times F_{P1} = 40 \times 0.6 = 24$ kW; $\theta_1 = \cos^{-1}(F_P) = \cos^{-1}(0.6) = 53.13°$
   $P_2 = S_2 \times F_{P2} = 30 \times 0.9 = 27$ kW; $\theta_2 = \cos^{-1}(F_P) = \cos^{-1}(0.9) = 25.84°$
   $P_3 = S_3 \times F_{P3} = 20 \times 0.5 = 10$ kW; $\theta_3 = -\cos^{-1}(F_P) = -\cos^{-1}(0.5) = -60°$
   $P_T = P_1 + P_2 + P_3 = 24 + 27 + 10 = 61$ kW

b. The individual power diagrams are shown in Figure 8.7b. Note that the lagging vars have a positive value, and the leading vars have a negative value.
   $Q_1 = S_1 \times \sin\theta_1 = 40 \times \sin(53.13°) = 32.0$ kvar
   $Q_2 = S_2 \times \sin\theta_2 = 30 \times \sin(25.84°) = 13.08$ kvar
   $Q_3 = S_3 \times \sin\theta_3 = 20 \times \sin(-60°) = -17.32$ kvar
   $Q_T = Q_1 + Q_2 + Q_3 = 31.98 + 13.08 - 17.32 = 27.74$ kvar

c. $S_T = \sqrt{P_T^2 + Q_T^2} = \sqrt{61^2 + 27.74^2} = 67$ kVA

d. $F_{P,\,system} = \dfrac{P_T}{S_T} = \dfrac{61}{67} = 0.91$ or 91%

e. Since Q_T has a positive value, the system power factor is lagging $(Q_L > Q_C)$.

8.5 MOTOR EFFICIENCY

When calculating the total load on an electrical distribution system that includes motors, the efficiency of the motor must be taken into consideration. The efficiency of a motor is the ratio of the useful shaft-power output to the total active power input to the motor. Expressed as an equation,

$$\lambda = \frac{P_{out}}{P_{in}} \qquad (8\text{--}13)$$

where: λ = efficiency (pu)
P_{out} = mechanical power output from motor shaft (W)
P_{in} = active power input to motor windings (W).

The efficiency in Eq. (8–13) is expressed in decimal form, called per-unit efficiency (pu). If the efficiency is given in percent, it must be divided by 100 before substituting into Eq. (8–13). That is,

$$\lambda = \frac{\% \text{ efficiency}}{100} \qquad (8\text{--}14)$$

Do not confuse efficiency with power factor!

$$\text{Power factor: } F_P = \frac{P_{in}}{S_{in}} \qquad \text{Efficiency: } \lambda = \frac{P_{out}}{P_{in}}$$

Note: If the shaft power out is expressed in horsepower, it must be converted to watts before substituting into Eq. (8–13).

$$P_{out} = \text{hp} \times 746 \qquad (8\text{--}15)$$

where: hp = shaft horsepower.

EXAMPLE 8.3

A 10-hp motor, operating at rated load, has an efficiency of 95%. Determine (a) kW input; (b) kW input if operating at 25% load with an efficiency of 75%.

Solution

a. $P_{out} = \text{hp} \times 746 = 10 \times 746 = 7460$ W

$$\lambda = \frac{P_{out}}{P_{in}} \Rightarrow P_{in} = \frac{P_{out}}{\lambda} = \frac{7460}{0.95} = 7852.6 \text{ W}$$

b. $P_{out} = hp \times 746 = (10 \times 0.25) \times 746 = 1865$ W

$$\lambda = \frac{P_{out}}{P_{in}} \Rightarrow P_{in} = \frac{P_{out}}{\lambda} = \frac{1865}{0.75} = 2486.7 \text{ W}$$

EXAMPLE 8.4   A 240-V, 60-Hz system supplies a 5-hp motor operating at rated shaft load with an efficiency of 88%, and a power factor of 76%. Determine (a) active power; (b) apparent power; (c) reactive power.

Solution

a. $P_{out} = hp \times 746 = 5 \times 746 = 3730$ W

$$\lambda = \frac{P_{out}}{P_{in}} \Rightarrow P_{in} = \frac{P_{out}}{\lambda} = \frac{3730}{0.88} = 4238.6 \text{ W}$$

b. $F_P = \frac{P_{in}}{S_{in}} \Rightarrow S_{in} = \frac{P_{in}}{F_P} = \frac{4238.6}{0.76} = 5577.1$ VA

c. From Eq. (8–5), $Q = \sqrt{S^2 - P^2} = \sqrt{5577.1^2 - 4238.6^2} = 3625$ var

---

8.6 CALCULATING ACTIVE, REACTIVE, AND APPARENT POWER FROM KNOWN VALUES OF CURRENT, VOLTAGE, AND IMPEDANCE

The following equations are useful for calculating active power, reactive power, and apparent power drawn by the three basic circuit elements (*R*, *L*, and *C*) from the known values of current, voltage, and impedance:

$$P_R = V_R I_R = I_R^2 R = \frac{V_R^2}{R} \tag{8–16}$$

$$Q_L = V_L I_L = I_L^2 X_L = \frac{V_L^2}{X_L} \tag{8–17}$$

$$Q_C = V_C I_C = I_C^2 X_C = \frac{V_C^2}{X_C} \tag{8–18}$$

$$S_T = V_T I_T = I_T^2 Z_T = \frac{V_T^2}{Z_T} \tag{8–19}$$

where: P_R = active power drawn by a resistor (W)
Q_L = reactive power drawn by an inductor (var)
Q_C = reactive power drawn by a capacitor (var)
S_T = apparent power drawn by system (VA)
V_R = voltage across resistor (V)
V_L = voltage across inductor (V)
V_C = voltage across capacitor (V)

V_T = input voltage to system (V)
I_R = current in resistor (A)
I_L = current in inductor (A)
I_C = current in capacitor (A)
I_T = total system current (A).

EXAMPLE 8.5 For the AC system shown in Figure 8.8, determine (a) system active power; (b) system reactive power; (c) system apparent power; (d) system power factor.

FIGURE 8.8
Circuit for Example 8.5.

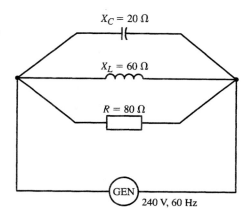

$X_C = 20\ \Omega$

$X_L = 60\ \Omega$

$R = 80\ \Omega$

GEN

240 V, 60 Hz

Solution

a. $P_T = P_R = \dfrac{V_R^2}{R} = \dfrac{240^2}{80} = 720\ \text{W}$

b. $Q_L = \dfrac{V_L^2}{X_L} = \dfrac{240^2}{60} = 960\ \text{var}$

$Q_C = \dfrac{V_C^2}{X_C} = \dfrac{240^2}{20} = 2880\ \text{var}$

$Q_T = Q_L - Q_C = 960 - 2880 = -1920\ \text{var}$

c. $S_T = \sqrt{P_T^2 + Q_T^2} = \sqrt{720^2 + (-1920)^2} = 2050.56 \Rightarrow 2051\ \text{VA}$

d. $F_{P,\,\text{sys}} = \dfrac{P_T}{S_T} = \dfrac{720}{2051} = 0.351$ or 35.1% leading

Other useful formulas for determining the apparent power and the active power, when line current, line voltage, and power factor are given are as follows:

$$S_{\text{in}} = V_{\text{line}} I_{\text{line}} \tag{8–20}$$

$$P_{\text{in}} = V_{\text{line}} I_{\text{line}} F_P \tag{8–21}$$

where: V_{line} = line voltage (V)
I_{line} = line current (A)
S_{in} = apparent power (VA)
P_{in} = active power (W)
F_P = power factor (pu).

EXAMPLE 8.6

A 240-V, 50-Hz, single-phase source supplies a load that draws 48 A at a power factor of 0.65 lagging. Determine (a) apparent power; (b) active power; (c) reactive power; (d) power factor angle. The circuit is shown in Figure 8.9.

FIGURE 8.9
Circuit diagram for Example 8.6.

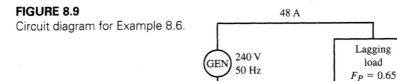

Solution
a. $S_{in} = V_{line}I_{line} = 240 \times 48 = 11{,}520 \text{ VA} = 11.52 \text{ kVA}$
b. $P_{in} = V_{line}I_{line}F_P = 240 \times 48 \times 0.65 = 7488 \text{ W} = 7.488 \text{ kW}$
c. $S = \sqrt{P^2 + Q^2} \Rightarrow Q = \sqrt{S^2 - P^2} = \sqrt{11.52^2 - 7.488^2} = 8.75 \text{ kvar}$
d. $\theta = \cos^{-1}(F_P) = \cos^{-1}(0.65) = 49.5°$

8.7 POWER-FACTOR IMPROVEMENT OF A SINGLE-PHASE SYSTEM

Depending on the relative proportions of heating elements, lamps, induction motors, etc., connected to a single-phase system, the system power factor may be somewhere between 0.35 lagging and 0.90 lagging. Although the lagging kvars drawn by inductive loads do no work, they must be supplied in order to establish a magnetic field.

The operation of electrical power systems at low power factors requires larger generators and associated apparatus. Furthermore, the higher currents drawn at low values of power factor cause greater voltage drops and higher I^2R losses in the connecting lines and apparatus, necessitating conductors of larger cross-sectional area. Utility companies often include power-factor clauses in their rate structures, offering reductions in billing in return for a higher power factor at the load.

If a capacitor is connected to the terminals of a motor, and the capacitor has a kvar rating equal to the kvars drawn by the motor, the seesaw of reactive power will be between the motor and the capacitor. Fewer kilovolt-amperes will be required from the generator, and a smaller size feeder cable may be used. A one-line diagram of power flow between generator, motor, and capacitor for this condition is shown in Figure 8.10.

A synchronous motor may also be used for power-factor improvement; the leading kvars drawn by an overexcited synchronous motor have a capacitive effect on the system.

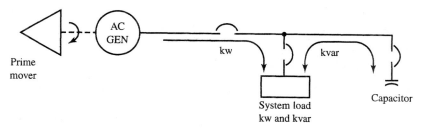

FIGURE 8.10
One-line diagram showing power flow between generator, load, and capacitor.

EXAMPLE 8.7

Figure 8.11a shows a distribution system that includes a 12-kVA load A operating at 0.70 power-factor (pf) lagging, and a 10-kVA load B operating at 0.8 pf lagging, connected in parallel and supplied by a 460-V, 60-Hz source. Determine (a) system active power; (b) system reactive power; (c) system apparent power; (d) system current; (e) system power factor and power-factor angle; (f) kvar rating of a parallel connected capacitor that will raise the system power factor to 90 percent; (g) capacitance of capacitor; (h) total apparent power for the new condition; (i) system current.

FIGURE 8.11
(a) Circuit diagram and (b) power diagram for Example 8.7.

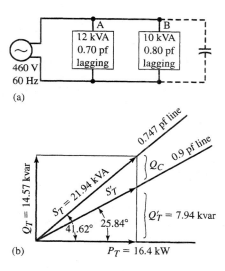

Solution

a. $P_A = S_A \times F_{PA} = 12 \times 0.70 = 8.4 \text{ kW}$
$P_B = S_B \times F_{PB} = 10 \times 0.80 = 8.0 \text{ kW}$
$P_T = P_A + P_B = 8.4 + 8 = 16.4 \text{ kW}$

b. $\theta_A = \cos^{-1}(F_{PA}) = \cos^{-1}(0.70) = 45.57°$ $\theta_B = \cos^{-1}(F_{PB}) = \cos^{-1}(0.80) = 36.87°$
$Q_A = S_A \times \sin \theta_A = 12 \sin(45.57°) = 8.57$ kvar
$Q_B = S_B \times \sin \theta_B = 10 \sin(36.87°) = 6$ kvar
$Q_T = Q_A + Q_B = 8.57 + 6 = 14.57$ kvar, lagging

c. $S_T = \sqrt{P_T^2 + Q_T^2} = \sqrt{16.4^2 + 14.57^2} = 21.937 \Rightarrow 21.94$ kVA

d. $S_T = V_T I_T \Rightarrow I_T = \dfrac{S_T}{V_T} = \dfrac{21.937 \times 1000}{460} = 47.7$ A

e. $F_{P\,sys} = \dfrac{P_T}{S_T} = \dfrac{16.4}{21.937} = 0.7475$ or 74.7% $\theta_{sys} = \cos^{-1}(0.7475) = 41.62°$

f. The system power-factor angle corresponding to a power factor of 0.90 is

$$\theta'_{sys} = \cos^{-1}(0.90) = 25.84°$$

A power diagram showing the original conditions at 74.7% power factor, and the new conditions for 90% power factor is shown in Figure 8.11b. The new conditions are indicated by primes ('). *Note:* Installing a capacitor does not change the total active power.[4] Thus, from Figure 8.11b, after installing the required capacitance, the total reactive power supplied by the generator is

$$\tan(25.84°) = \frac{Q'_T}{16.4} \Rightarrow Q'_T = 7.94 \text{ kvar}$$

The amount of kvars supplied to the distribution system by the capacitor is

$$Q_C = Q_T - Q'_T = 14.57 - 7.94 = 6.63 \text{ kvar}$$

g. When solving power problems that involve current or voltage, components P, Q, and S must be in their respective basic units (watts, vars, and VA). Thus, converting kvars to vars, and substituting into Eq. (8–18),

$$Q_C = \frac{V_C^2}{X_C} \Rightarrow X_C = \frac{V_C^2}{Q_C} = \frac{460^2}{6630} = 31.93 \text{ } \Omega$$

$$X_C = \frac{1}{2\pi f C} \Rightarrow C = \frac{1}{2\pi f X_C} = \frac{1}{2 \times \pi \times 60 \times 31.93} = 83 \text{ } \mu\text{F}$$

h. $S'_T = \sqrt{P_T^2 + Q'_T^2} = \sqrt{16.4^2 + 7.94^2} = 18.22$ kVA

i. Converting kVA to VA, and substituting into Eq. (8–19),

$$S'_T = V_T I'_T \Rightarrow I'_T = \frac{S'_T}{V_T} \Rightarrow \frac{18.22 \times 1000}{460} = 39.6 \text{ A}$$

Note: The 83-μF capacitor reduced the system current from 47.7 A to 39.6 A.

---

[4] The active power drawn by a capacitor is negligible.

EXAMPLE 8.8 What would be the power factor of the system in Example 8.7 if 20 kvar of capacitance were used?

Solution

The power diagram for this condition is shown in Figure 8.12. Note that the overall system kvars will be negative. Hence, the power factor will be leading.

$$Q'_T = Q_L - Q'_C = 14.57 - 20 = -5.43 \text{ kvar}$$
$$S'_T = \sqrt{P_T^2 + Q'^2_T} = \sqrt{16.4^2 + (-5.43)^2} = 17.28 \text{ kVA}$$
$$F_P' = \frac{P_T}{S'_T} = \frac{16.4}{17.28} = 0.95 \text{ leading}$$

FIGURE 8.12
Power diagram for Example 8.8.

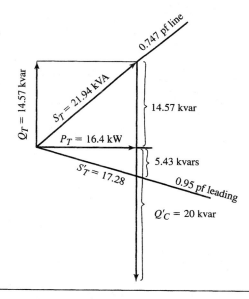

8.8   THE CAPACITOR: A TWO-EDGED SWORD

Capacitors can provide significant benefits in electrical distribution systems. Power-factor improvement by the addition of capacitors is based on an engineering study of the economics of the particular installation. The rate of return on capacitor investment depends on the type of power factor clause imposed by the utility and the extent to which the power factor deviates from unity. Power-factor improvements up to 90 percent or 95 percent are generally economically justified.

However, if improperly applied, capacitors can cause trouble. Too much, or improperly connected, capacitance could result in overvoltage, blown fuses, damaged motor windings, and broken motor shafts [1].

The selection of capacitor size, and its location, must be based on the particular distribution system, including transformers, motors, motor control, and capacitor switching.

8.9 POWER MEASUREMENT IN A SINGLE-PHASE SYSTEM

Power measurement in the two-wire system is best done with handheld multifunction power meters, such as that shown in Figure 8.13a. The meter measures active power, reactive power, apparent power, and power factor without having to break the circuit

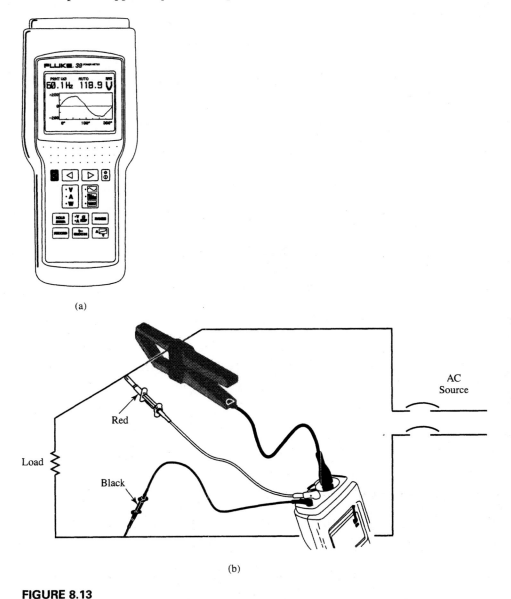

(a)

(b)

FIGURE 8.13
(a) Handheld multifunction power meter and (b) connections for power measurement (courtesy Fluke Corporation).

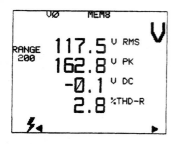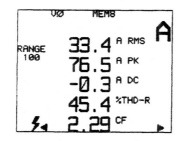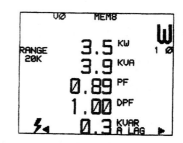

FIGURE 8.14
Text screens of current, voltage, and power (courtesy Fluke Corporation).

to connect the current coil. Current data are obtained via a clamp-on current probe, and voltage data are obtained directly from the source via the red and black test clips as illustrated in Figure 8.13b. The data may be viewed as text on separate current, voltage, and power screens as shown in Figure 8.14. Note that the **V** and **A** screens display rms values, peak values (PK), a DC component if present, and percent total harmonic distortion (%THD). The **W** screen displays active power (KW), reactive power (kvar), apparent power (KVA), power factor (PF), and displacement power factor (DPF). Displacement power factor is the power factor measured at the fundamental frequency. Lag or lead of current is indicated by (ALAG) or (ALEAD), respectively, and the reactive component of power at the fundamental frequency is also indicated.

Waveforms of voltage, current, and power corresponding to the text screens in Figure 8.14 are shown in Figure 8.15. The text above the waveform indicates the frequency. The different screen modes shown in Figures 8.14 and 8.15 are selected by pressing different keys.

Some multifunction power meters, such as that shown in Figure 8.13, can store and recall text data and multiple waveforms. If an optional interface is used, the meter can communicate with a printer or a computer.

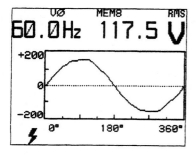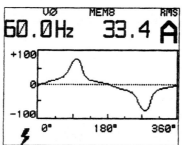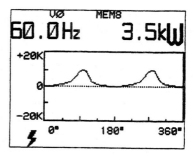

FIGURE 8.15
Waveform screens of voltage, current, and power (courtesy Fluke Corporation).

SUMMARY OF EQUATIONS FOR PROBLEM SOLVING

$$P_R = V_R I_R \tag{8-1}$$

$$Q_C = V_C I_C \tag{8-2}$$

$$Q_L = V_L I_L \tag{8-3}$$

$$Q_T = Q_L - Q_C \tag{8-4}$$

$$S_T = \sqrt{P_T^2 + Q_T^2} \tag{8-5}$$

$$P_T = S_T \times \cos(\theta) \tag{8-6}$$

$$Q_T = S_T \times \sin(\theta) \tag{8-7}$$

$$\theta = \cos^{-1}\left(\frac{P_T}{S_T}\right) \tag{8-8}$$

$$F_P = \frac{P_{\text{in}}}{S_{\text{in}}} \tag{8-9}$$

$$\lambda = \frac{P_{\text{out}}}{P_{\text{in}}} \tag{8-13}$$

$$P_{\text{out}} = \text{hp} \times 746 \tag{8-15}$$

$$P_R = V_R I_R = I_R^2 R = \frac{V_R^2}{R} \tag{8-16}$$

$$Q_L = V_L I_L = I_L^2 X_L = \frac{V_L^2}{X_L} \tag{8-17}$$

$$Q_C = V_C I_C = I_C^2 X_C = \frac{V_C^2}{X_C} \tag{8-18}$$

$$S_T = V_T I_T = I_T^2 Z_T = \frac{V_T^2}{Z_T} \tag{8-19}$$

$$S_{\text{in}} = V_{\text{line}} I_{\text{line}} \tag{8-20}$$

$$P_{\text{in}} = V_{\text{line}} I_{\text{line}} F_P \tag{8-21}$$

SPECIFIC REFERENCE KEYED TO THE TEXT

[1] Zucker, M., and Erhart, J., Capacitors Near Loads? The Engineering Viewpoint, *IEEE Trans. Industry Applications*, Vol. IA–21, March/April 1985.

REVIEW QUESTIONS

1. Differentiate between active power, reactive power, and apparent power.
2. List three types of electrical power apparatus that draw lagging vars, and list two types that draw leading vars.
3. Sketch an energy flow diagram showing an induction motor, a capacitor, a resistance heater, a generator, and a prime mover. Assume the reactive power drawn by the capacitor is less than the reactive power drawn by the motor.
4. What is the difference between power factor and efficiency?
5. Sketch the power diagram of an induction motor, showing active, reactive, and apparent power components. What is the mathematical relationship between them?
6. How can a capacitor be used to improve the power factor of a distribution system? Illustrate with a diagram.
7. What is the disadvantage of operating a system at low power factor?
8. What ill effects can be caused by misapplication of capacitors?
9. If a system is operating at a low lagging power factor, will raising the power factor to unity increase or decrease the current? Explain.

PROBLEMS

8–1/3. A 120-V, 60-Hz source supplies a 4-kW lighting load, and an induction motor that draws 10.2 kW and 7.4 kvar. Determine the apparent power supplied by the source.

8–2/3. The apparent power drawn by a certain motor is 30.4 kVA, and its active power is 24.6 kW. Sketch the power triangle and determine its reactive power.

8–3/4. A 460-V, 60-Hz, 546-kVA load has a power factor of 82%. Sketch the power triangle and determine (a) active power; (b) reactive power; (c) power-factor angle.

8–4/4. The active and reactive components of power drawn by a certain load are 2360 W and 1854 var. Sketch the power triangle and determine (a) apparent power; (b) power factor; (c) power-factor angle.

8–5/4. An electric circuit draws 100 kW at 0.8 power factor lagging from a 240-V 60-Hz source. Sketch the circuit and the power triangle, and determine (a) apparent power input; (b) kvar input.

8–6/4. A 440-V 60-Hz source supplies 20 kVA to a load whose power factor is 70%. Sketch the circuit and the power triangle, and determine (a) kW; (b) kvars; (c) power-factor angle.

8–7/4. A 450-V, 60-Hz generator supplies power to a 150-kVA 0.80-pf lagging load and a 100-kVA 0.75-pf lagging load. Sketch the circuit and the power diagram and determine (a) the total active power; (b) the total reactive power; (c) the total apparent power; (d) the overall system power factor.

8–8/4. A 120-V, 60-Hz generator supplies power to two parallel loads. One load draws 10 kVA at unity power factor, and the other draws 40 kVA at 0.60 power factor lagging. Sketch the circuit and the power diagram and determine (a) the total active power; (b) the total reactive power; (c) the total apparent power; (d) the system power factor.

8–9/4. A capacitor, an electric resistance heater, and an impedance are connected in parallel to a 120-V, 60-Hz system. The capacitor draws 50 var, the heater draws 100 W, and the impedance draws 26 VA at a power factor of 0.74 lagging. Sketch the circuit and determine (a) the system active power; (b) the system reactive power; (c) the system apparent power; (d) the system power factor.

8–10/5. A 240-V, 50-hp motor operating at 1/2 rated load has an efficiency of 71.4%. Determine the kW input to the motor.

8–11/5. If the efficiency and power factor of a 460-V, 60-Hz, 100-hp motor operating at rated load are 92% and 86%, respectively, determine (a) active power input; (b) apparent power input; (c) reactive power input.

8–12/5. A 20-hp 450-V motor operating at rated load has an efficiency of 85% and a power factor of 0.76 lagging. The motor is in parallel with a 30-kVA load B whose power factor angle is 35° lagging. The two loads are supplied from a 450-V, 60-Hz source. Sketch the circuit and power diagram and determine (a) the active power and reactive power drawn by the motor; (b) the active power and reactive power drawn by the 30-kVA load; (c) the total active power supplied by the system; (d) the total reactive power supplied by the system; (e) the total apparent power supplied by the system; (f) system power factor.

8–13/5. A 3-hp motor draws 28 A from a 60-Hz, 120-V generator. The motor has an efficiency of 86%. Determine (a) active power supplied to motor; (b) power factor; (c) motor reactive power.

8–14/6. A parallel circuit consisting of a 3.0-Ω resistor, a 10-Ω capacitive reactance, and an ideal inductor whose inductive reactance is 6.0 Ω is connected to a 120-V, 60-Hz source. Sketch the circuit and determine (a) current through each branch; (b) circuit impedance; (c) total active power input; (d) total reactive power input; (e) apparent power input; (f) feeder current.

8–15/6. Two parallel-connected single-phase loads A and B are supplied by a 440-V, 60-Hz generator. Load A draws an apparent power of 10 kVA at 0.80 pf lagging and load B draws an apparent power of 7 kVA at 0.65 pf lagging. Sketch the circuit and determine (a) system active power; (b) system reactive power; (c) system apparent power; (d) system power factor; (e) feeder current; (f) kWh of energy supplied by the generator in 30 min.

8–16/7. A 10-kVA load operating at 0.8 power factor lagging is connected in parallel with a 20-kVA load operating at 0.7 power factor lagging. Calculate the kVA rating of a parallel-connected capacitor required to correct the system power factor to 0.95 lagging.

8–17/7. A 50-kVA, 460-V, 50-Hz generator supplies energy to a parallel circuit containing a 20-kVA 0.8-pf lagging load and a 10-kVA unity power-factor load.

Sketch the power diagram and determine (a) kvar rating of a parallel-connected capacitor that will change the system power factor to 0.95 leading; (b) capacitance of the capacitor; (c) voltage rating of the capacitor.

8–18/7. A 220-V, 1.5-hp, single-phase, 60-Hz induction motor draws 7.6 A when operating at rated conditions. The efficiency at rated conditions is 85%. Calculate (a) kVA input; (b) kW input; (c) power factor; (d) capacitance of a parallel connected capacitor that will cause the system to operate at unity power factor; (e) total input current after the capacitor is installed; (f) capacitor current.

9

Current, Voltage, and Power in the Three-Phase System

9.0 INTRODUCTION

The three-phase system, invented by Nikola Tesla in 1883, is the most common method for the generation, transmission, distribution, and utilization of large quantities of electrical power. In its simplest form, it consists of a three-phase generator, three distribution lines, and a three-phase load such as a three-phase motor. Three-phase motors weigh less and occupy less space than other machines of equal horsepower and speed ratings.

This chapter introduces the current, voltage, and power relationships in the three-phase system. Three-phase power measurement, power-factor improvement, and the significance of phase sequence are included.

9.1 BASIC THREE-PHASE SYSTEM

The basic three-phase system consists of three sinusoidal voltage sources, called *phases,* that are equal in amplitude and equal in frequency, but reach their maximum voltages at different instants of time.[1] Figure 9.1a illustrates the sine waves of a representative three-phase source, and Figure 9.1b illustrates the corresponding phasor diagram. The three voltages are 120 degrees apart.

An observer walking along the time axis, or observing the phasors as they rotate counterclockwise (CCW), will note that the phase sequence is phase A followed by phase B, followed by phase C.

---

[1] The three voltage sources that make up a three-phase system can be obtained electronically or by means of rotating generators.

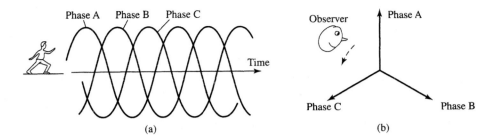

FIGURE 9.1
(a) Voltage waves of the respective phases; (b) phasor diagram.

Wye-Connected Voltage Source

A wye-connected source is shown in Figures 9.2a and 9.2b; one terminal of each of the three AC voltage sources is connected to a common junction, called the *neutral*. The three voltage sources (or phases) must be connected in a manner that will cause the instantaneous positive direction of voltage rise (e^+) of each source to be in the same direction; that is, all toward the common junction, or all away from the common junction, as shown in Figure 9.2a. A reversal of any one of the three voltage sources invalidates the required 120-degree phase relationship.

A fourth cable, called the *common* or *neutral line,* is often brought out from the common junction, as shown in Figure 9.2b. The voltage between any line and the neutral is called the *phase voltage* (V_{phase}); and the voltage between any two lines is called the line-to-line voltage, or simply the line voltage (V_{line}).

The voltage between any two lines in Figure 9.2b can be determined by traversing the circuit from one line to the other, adding the voltages along the way (phasor sum). Thus, the line voltage measured from line A to line B in Figure 9.2b can be determined by adding the voltage from a to a′ of phase A to the voltage from b′ to b of phase B. Using subscript notation,

$$\mathbf{E}_{\text{a to b}} = \mathbf{E}_{\text{a to a}'} + \mathbf{E}_{\text{b}' \text{to b}}$$

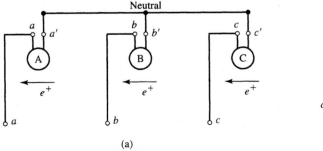

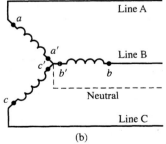

FIGURE 9.2
(a, b) Circuit diagrams for a wye-connected source.

Or, in simplified form, with the word "to" implied, the voltage from a to b is

$$\mathbf{E}_{ab} = \mathbf{E}_{aa'} + \mathbf{E}_{b'b} \qquad (9\text{--}1)$$

Similarly,

$$\mathbf{E}_{bc} = \mathbf{E}_{bb'} + \mathbf{E}_{c'c} \qquad (9\text{--}2)$$

$$\mathbf{E}_{ca} = \mathbf{E}_{cc'} + \mathbf{E}_{a'a} \qquad (9\text{--}3)$$

A phasor diagram of the the component phasors is illustrated in Figure 9.3a. Note that the voltage measured from a primed terminal to an unprimed terminal is the negative of the measurement from an unprimed terminal to a primed terminal.

$$\mathbf{E}_{a'a} = -\mathbf{E}_{aa'} \qquad \mathbf{E}_{b'b} = -\mathbf{E}_{bb'} \qquad \mathbf{E}_{c'c} = -\mathbf{E}_{cc'}$$

Figure 9.3b shows the graphical determination of the line voltages represented by Eqs. (9–1), (9–2) and (9–3). The three line voltages are equal in magnitude but displaced from each other by 120°, and the set of line-voltage phasors is displaced from the set of phase-voltage phasors by 30°. From the geometry of the phasor diagram, the line-to-line voltage of a wye-connected source is equal to the line-to-neutral voltage times $\sqrt{3}$. Expressed mathematically,

$$V_{\text{line Y}} = \sqrt{3}\, V_{\text{phase}} = 1.732\, V_{\text{phase}} \qquad (9\text{--}4)$$

The neutral line serves three functions:

1. It provides a means for obtaining a lower value single-phase voltage in a four-wire three-phase distribution system.

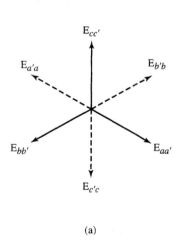

(a)

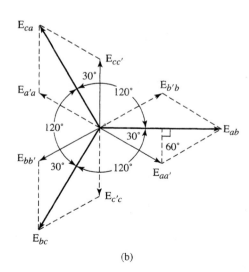

(b)

FIGURE 9.3
Wye-connected source: (a) component phasors; (b) graphical determination of line voltages.

2. It provides a means for safety grounding of the electrical system.[2]

3. It provides a path for third-harmonic currents and their multiples.[3]

Delta-Connected Voltage Source

Figures 9.4a and 9.4b illustrate the three AC voltage sources connected in series to form a closed loop. The three phases are connected in a manner that will cause the instantaneous positive direction of voltage rise (e^+) of the three individual phases to be from a' to a, b' to b, and c' to c, respectively, when each in turn is ascending in the positive direction. Thus, the instantaneous positive direction of voltage rise of the three phases will be all clockwise as shown in Figure 9.4a or all counterclockwise. Any other closed-loop arrangement of the three voltage sources will not constitute a valid delta connection and may result in damage to the voltage source.

Although the three phases of the delta-connected source form a closed loop, the phasor sum of the three source voltages around the delta loop is equal to zero. This is shown in Figure 9.5, where the phasor sum of $V_{a'a} + V_{b'b}$ is equal and opposite to $V_{c'c}$.

Because the phasor summation of the source voltages around the delta loop is zero, current cannot circulate around the delta loop. However, each of the three phases is still capable of supplying current to external loads. As noted in Figure 9.4b for a delta-connected source, the line voltage is equal to the phase voltage. Expressed mathematically,

$$V_{\text{line}\Delta} = V_{\text{phase}\Delta} \tag{9-5}$$

---

[2] See solidly grounded distribution systems in Section 28–6, Chapter 28.

[3] See harmonic heating of delta–wye transformer banks in Section 10–9, Chapter 10.

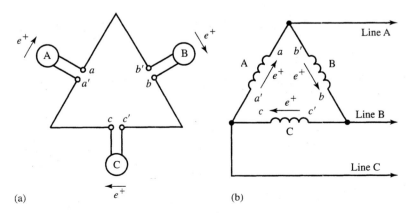

(a) (b)

FIGURE 9.4

(a) Closed loop formed by a delta-connected source; (b) output line connections.

FIGURE 9.5

Phasor summation of voltages in the loop formed by a delta source.

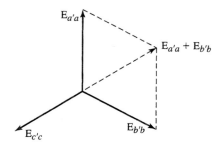

9.2 CURRENT AND VOLTAGE IN THREE-PHASE, THREE-WIRE SYSTEMS WITH BALANCED LOADS

A three-phase load is considered balanced if the load connected to each phase has the same impedance. A three-phase motor is a balanced three-phase load.

Balanced Wye-Connected Load

A three-phase source supplying a balanced wye-connected load is illustrated in Figure 9.6. Each line feeds into one impedance. Hence, the line current for a wye-connected load is equal to the phase current. Expressed mathematically,

$$I_{\text{line Y}} = I_{\text{phase Y}} \qquad (9\text{–}6)$$

A neutral line is not needed when supplying a balanced wye-connected load. The voltage between any line and the neutral junction of a balanced wye-connected load is equal to the line voltage divided by the $\sqrt{3}$:

$$V_{\text{phase Y}} = \frac{V_{\text{line Y}}}{\sqrt{3}} = 0.577 \; V_{\text{line Y}} \qquad (9\text{–}7)$$

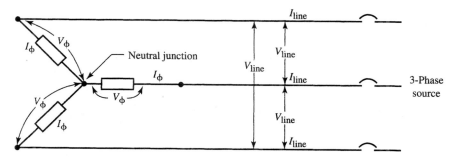

FIGURE 9.6

Three-phase source supplying a balanced wye-connected load.

where: V_ϕ = phase voltage (V)
             V_{line} = line voltage (V)
             I_ϕ = phase current (A)
             I_{line} = line current (A)
             Z_ϕ = load impedance per phase (Ω).

Impedance is always given on a per-phase basis.

Balanced Delta-Connected Load

A three-phase source supplying a delta-connected load is shown in Figure 9.7. Note that each line feeds into two impedances. Hence, the line current to a delta-connected load is greater than the current in the individual phases. In a balanced delta-connected load, the line current is equal to $\sqrt{3}$ times the current in one phase:

$$I_{line\Delta} = \sqrt{3}\, I_{\phi\Delta} = 1.732\, I_{\phi\Delta}$$

The current and voltage relationships for balanced three-phase loads, such as those shown in Figures 9.6 and 9.7, are given here:

Connection	Voltage	Current	Ohm's Law
Wye	$V_{lineY} = \sqrt{3}\, V_{\phi Y}$	$I_{lineY} = I_{\phi Y}$	$I_\phi = \dfrac{V_\phi}{Z_\phi}$
Delta	$V_{\phi\Delta} = V_{line\,\Delta}$	$I_{line\Delta} = \sqrt{3}\, I_{\phi\Delta}$	$I_\phi = \dfrac{V_\phi}{Z_\phi}$

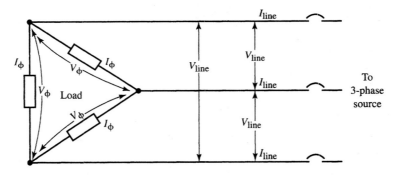

FIGURE 9.7
Three-phase source supplying a balanced delta-connected load.

EXAMPLE 9.1   A balanced delta-connected load is supplied from a 460-V, 60-Hz, three-phase source. Each phase of the balanced delta-connected load consists of a coil whose resistance and inductive reactance are 6 Ω and 9 Ω, respectively. Determine (a) impedance of each phase; (b) phase voltage; (c) phase current; (d) line current.

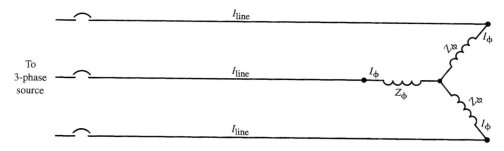

FIGURE 9.8
Circuit for Example 9.1.

Solution
The circuit is shown in Figure 9.8.

a. $Z_\phi = \sqrt{R^2 + X_L^2} = \sqrt{6^2 + 9^2} = 10.82 \ \Omega$

b. $V_\phi = 460$ V

c. $I_\phi = \dfrac{V_\phi}{Z_\phi} = \dfrac{460}{10.82} = 42.53$ A

d. $I_{line} = \sqrt{3} \times I_\phi = \sqrt{3} \times 42.53 = 73.7$ A

**EXAMPLE
9.2**   Repeat Example 9.1, assuming the three branches of the delta are connected in a wye as shown on Figure 9.9.

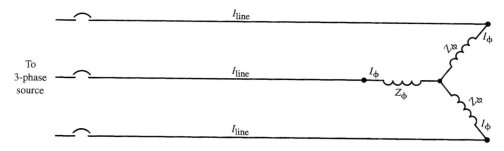

FIGURE 9.9
Circuit for Example 9.2.

a. $Z = 10.82 \ \Omega$

b. $V_\phi = \dfrac{V_{line}}{\sqrt{3}} = \dfrac{460}{\sqrt{3}} = 265.58$ V

c. $I_\phi = \dfrac{V_\phi}{Z_\phi} = \dfrac{265.58}{10.82} = 24.55$ A

d. For a wye connection, $I_{\text{line}} = I_\phi = 24.55$ A

9.3 FOUR-WIRE, THREE-PHASE SYSTEM

A four-wire three-phase system uses a wye-connected source with a neutral line brought out as shown in Figure 9.10.

Representative wye, delta, and single-phase loads are shown connected to the source. The voltage across impedance Z_5 is from line to neutral, which is the phase voltage (V_ϕ); the voltage across each Z_1 impedance in the wye-connected load is the phase voltage (V_ϕ); and the voltage across each Z_2 impedance in the delta-connected load is the line voltage (V_{line}).

9.4 THREE-PHASE POWER AND POWER FACTOR

For a balanced three-phase (3ϕ) load, the respective three-phase active power, reactive power, and apparent power are equal to three times the corresponding single-phase (1ϕ) values. Thus, building on the power relationships developed in Chapter 8 for the single-phase system,

$$P_{3\phi} = 3P_{1\phi} \tag{9–8}$$

$$Q_{3\phi} = 3Q_{1\phi} \tag{9–9}$$

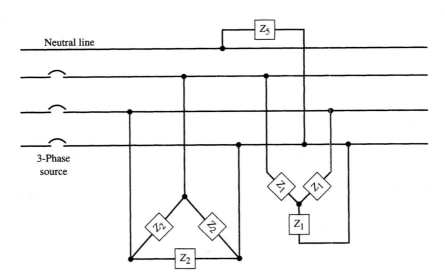

FIGURE 9.10
Four-wire three-phase system.

$$S_{3\phi} = 3S_{1\phi} \tag{9-10}$$

$$S_{3\phi} = \sqrt{P_{3\phi}^2 + Q_{3\phi}^2} \tag{9-11}$$

where: $S_{3\phi}$ = three-phase apparent power (VA)
$Q_{3\phi}$ = three-phase reactive power (var)
$P_{3\phi}$ = three-phase active power (W).

The active and reactive components of power in Eq. (9–11) are right-angle components of the power triangle shown in Figure 9.11. The diagonal is the apparent power, and angle θ is the power-factor angle. From the geometry of the power triangle,

$$P_{3\phi} = S_{3\phi} \times \cos(\theta) \tag{9-12}$$

$$Q_{3\phi} = S_{3\phi} \times \sin(\theta) \tag{9-13}$$

$$\theta = \cos^{-1}\left(\frac{P_{3\phi}}{S_{3\phi}}\right) \tag{9-14}$$

$$F_P = \frac{P_{in}}{S_{in}} = \frac{P_{3\phi}}{S_{3\phi}} = \frac{P_{1\phi}}{S_{1\phi}} = \cos\theta \tag{9-15}$$

$$\theta = \cos^{-1}(F_P) \tag{9-16}$$

where: F_P = power factor, per unit (pu).

In Eqs. (9–11) through (9–15), the units used must all have the same dimensions: W, VA, and var; or kW, kVA, and kvar; or MW, MVA, and Mvar; respectively.

The power factor in Eqs. (9–15) and (9–16) is expressed in decimal form, called per-unit power factor (pu). If the power factor is given in percent, it must be divided by 100 before substituting into those equations. That is,

$$F_P = \frac{\% \text{ power factor}}{100} \tag{9-17}$$

Note: The power factor (F_P) of a balanced three-phase load is equal to the power factor of one phase.

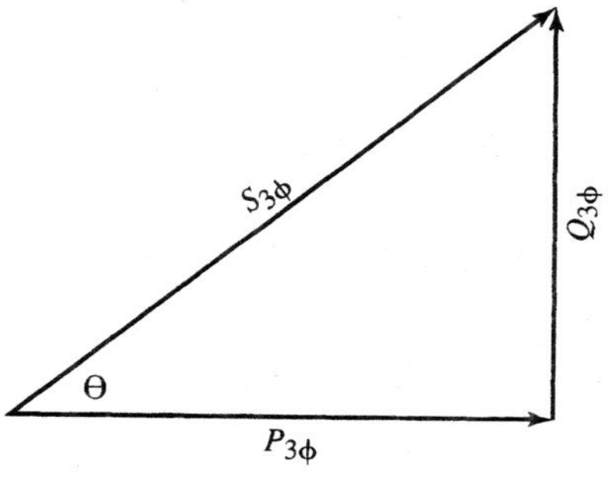

FIGURE 9.11
Power triangle.

Motor Efficiency

The efficiency of a motor is equal to the ratio of the shaft power output to the electrical power input. Expressed mathematically,

$$\lambda = \frac{P_{\text{shaft}}}{P_{\text{in}}} \qquad (9\text{--}18)$$

where: λ = efficiency (pu)
P_{shaft} = shaft output power (W)
P_{in} = input power (W).

If the shaft power out is expressed in horsepower (hp), it must be converted to watts before substituting into Eq. (9–18), where $P_{\text{shaft}} = \text{hp} \times 746$. If the efficiency is given in percent, it must be divided by 100 to convert to per-unit form before substituting into Eq. (9–18). That is,

$$\lambda = \frac{\% \text{ efficiency}}{100} \qquad (9\text{--}19)$$

EXAMPLE 9.3   The circuit in Figure 9.12a shows a 460-V, three-phase, 60-Hz source supplying power to an induction motor. The motor, rated at 60 hp, 460 V, 60 Hz, and 890 revolutions per minute (rpm), is operating at rated load with an efficiency of 92.7% and a power factor of 83.5% lagging. Determine (a) active power in; (b) apparent power in; (c) power factor angle; (d) reactive power in.

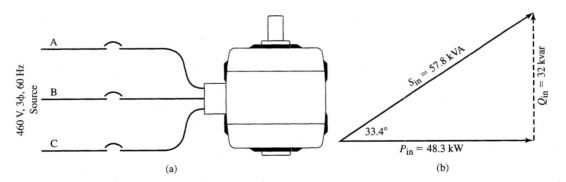

FIGURE 9.12
(a) Circuit for Example 9.3; (b) associated power triangle.

Solution
a. $P_{\text{shaft}} = \text{hp} \times 746 = 60 \times 746 = 44{,}760 \text{ W}$

From Eq. (9–18), $P_{\text{in}} = \dfrac{P_{\text{shaft}}}{\lambda} = \dfrac{44{,}760}{0.927} = 48{,}285 \text{ W} = 48.3 \text{ kW}$

b. From Eq. (9–15), $S_{in} = \dfrac{P_{in}}{F_P} = \dfrac{48{,}285}{0.835} = 57{,}826$ VA $= 57.8$ kVA

c. $\theta = \cos^{-1}(F_P) = \cos^{-1}(0.835) = 33.384° \Rightarrow 33.4°$

d. From the power triangle in Figure 9.12b, $\tan(\theta) = \dfrac{Q_{3\phi}}{P_{3\phi}}$

$Q_{in} = P_{in} \times \tan(\theta) = 48{,}285 \times \tan(33.384°) = 48{,}285 \times 0.659$
$= 31{,}819$ var $\Rightarrow 32$ kvar

9.5 CALCULATING ACTIVE POWER AND APPARENT POWER FROM KNOWN VALUES OF CURRENT, VOLTAGE, AND POWER FACTOR IN A THREE-PHASE SYSTEM

The power relationships in a balanced three-phase load, expressed in terms of line current, line voltage, and power factor, are as follows:

$$P_{3\phi} = \sqrt{3} \times V_{line} \times I_{line} \times F_P \qquad (9\text{–}20)$$

$$S_{3\phi} = \sqrt{3} \times V_{line} \times I_{line} \qquad (9\text{–}21)$$

Note: All equations involving current and/or voltage must use only W, VA, and var as appropriate; kW, kVA, and kvar must not be used in these equations.

EXAMPLE 9.4

A certain three-phase load draws 12 A when operating at 8 kW from a 460-V, 60-Hz system. Determine (a) apparent power; (b) reactive power; (c) power factor.

Solution
a. $S_{3\phi} = \sqrt{3} \times V_{line}I_{line} = \sqrt{3} \times 460 \times 12 = 9560.9$ VA $\Rightarrow 9.56$ kVA
b. From Eq. (9–11):

$$Q_{3\phi} = \sqrt{S_{3\phi}^2 - P_{3\phi}^2} = \sqrt{9560.9^2 - 8000^2} = 5235.5 \text{ var} \Rightarrow 5.24 \text{ kvar}$$

c. $F_P = \dfrac{P_{3\phi}}{S_{3\phi}} = \dfrac{8000}{9560.9} = 0.8367 \Rightarrow 83.7\%$

EXAMPLE 9.5

Figure 9.13a shows a three-phase, 2300-V, 60-Hz source supplying power to a system containing a 300-kW unity power factor load and a 200-hp induction motor operating at rated shaft horsepower. The motor is 92% efficient, and has a power factor of 89.4%. Determine (a) total active power; (b) total reactive power; (c) total apparent power; (d) system power factor; (e) input current from generator.

Solution
All values of kW, kvar, and kVA are for a three-phase system.

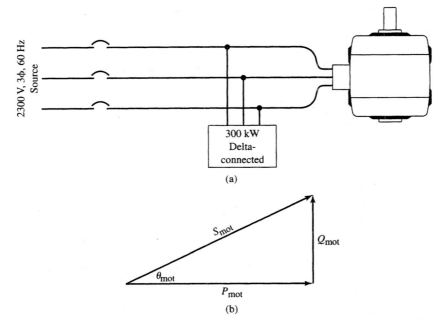

FIGURE 9.13

(a) Circuit for Example 9.5; (b) associated power triangle.

a. For the motor,

$$P_{shaft} = hp \times 746 = 200 \times 746 = 149,200 \text{ W} \Rightarrow 149.2 \text{ kW}$$

$$P_{mot,\,in} = \frac{P_{shaft}}{\lambda} = \frac{149.2}{0.92} = 162.174 \text{ kW}$$

Total active power input:

$$P_T = 300 + 162.174 = 462.174 \text{ kW}$$

b. $\theta_{mot} = \cos^{-1}(F_{P,\,mot}) = \cos^{-1}(0.894) = 26.62°$

From the motor power triangle shown in Figure 9.13b,

$$Q_{mot} = P_{mot,in} \times \tan(\theta_{mot}) = 162.174 \times \tan(26.62°) = 81.28 \text{ kvar}$$

Total reactive power input:

$$Q_T = Q_{mot} = 81.28 \text{ kvar}$$

c. $S_T = \sqrt{P_T^2 + Q_T^2} = \sqrt{462,174^2 + 81,280^2} = 469,267 \text{ W} \Rightarrow 469.3 \text{ kVA}$

d. $F_{P.sys} = \dfrac{P_T}{S_T} = \dfrac{462.174}{469.267} = 0.9848 \Rightarrow 98.5\%$

e. $S_T = \sqrt{3} \times V_{line}I_{line} \Rightarrow I_{line} = \dfrac{S_T}{\sqrt{3} \times V_{line}} = \dfrac{469,267}{\sqrt{3} \times 2300} = 117.8 \text{ A}$

9.6 POWER-FACTOR IMPROVEMENT IN THE THREE-PHASE SYSTEM

The calculation for the kvar of capacitance necessary for power-factor correction in a three-phase system is very similar to that for a single-phase system and is discussed in detail in Chapter 8. However, in a three-phase system, wye-connected or delta-connected capacitors must be used.

EXAMPLE 9.6

A balanced three-phase load, shown in Figure 9.14a, draws 153 A from a 460-V, 60-Hz source, and has a power factor of 60%.

Part 1: Determine (a) apparent power of load; (b) active power of load; (c) reactive power of load; (d) power-factor angle of load.

Part 2: We want to install a three-phase capacitor bank to correct the system power factor to 95%. Determine (e) new power-factor angle of system; (f) new system vars; (g) kvar rating of a three-phase capacitor bank required to make this correction; (h) capacitance of each branch of the bank if the capacitors are delta connected; (i) capacitance of each branch of the bank if the capacitors are wye connected.

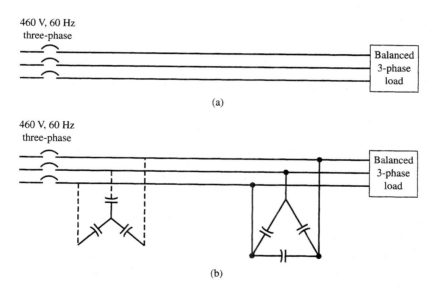

FIGURE 9.14

Circuits for Example 9.6; (a) original conditions; (b) with capacitor bank installed.

Solution

Figure 9.14b shows how the respective delta and wye capacitor banks would be connected.

a. $S_{\text{load}} = \sqrt{3} \times V_{\text{line}} I_{\text{line}} = \sqrt{3} \times 460 \times 153 = 121{,}902 \text{ VA} \Rightarrow 121.9 \text{ kVA}$

b. $P_{\text{load}} = \sqrt{3} \times V_{\text{line}} I_{\text{line}} \times F_P = \sqrt{3} \times 460 \times 153 \times 0.60 = 73{,}141.04 \text{ W} \Rightarrow 73.1 \text{ kW}$

c. From Eq. (9–11),

$$Q_{load} = \sqrt{S_{load}^2 - P_{load}^2} = \sqrt{121,902^2 - 73,141^2} = 97,521.4 \text{ var} \Rightarrow 97.5 \text{ kvar}$$

d. The load power-factor angle is

$$\theta_{load} = \cos^{-1}(F_{P\,load}) = \cos^{-1}(0.60) = 53.13°$$

e. The desired power-factor angle is

$$\theta'' = \cos^{-1}(F_P) = \cos^{-1}(0.95) = 18.195°$$

The power diagram, showing the initial conditions at 60% power factor and the desired conditions at 95% power factor, is given in Figure 9.15.

f. The system vars (Q'') when corrected to 95% power factor will be

$$Q'' = P_{load} \times \tan(\theta'') = 73,141 \times \tan(18.195°) = 73,141 \times 0.329$$
$$= 24,040.5 \text{ var} \Rightarrow 24.04 \text{ kvar}$$

g. Referring to Figure 9.15, the three-phase capacitive var required to reduce the system var to 24.04 kvar is

$$Q_C = Q_{load} - Q'' = 97,521.4 - 24,040.5 = 73,481.1 \text{ var} \Rightarrow 73.5 \text{ kvar}$$

$$Q_C/\text{phase} = \frac{73,481.1}{3} = 24,493.7 \text{ var}$$

h. For the delta (Δ) bank,

$$Q_C/\text{phase} = \frac{V_{phase\Delta}^2}{X_{C\Delta}} \Rightarrow X_{C\Delta} = \frac{V_{phase\Delta}^2}{Q_C/\text{phase}} = \frac{460^2}{24,493.7} = 8.64 \ \Omega$$

The capacitance per phase for the delta bank is

$$C_\Delta = \frac{1}{2\pi f X_{C\Delta}} = \frac{1}{2\pi \times 60 \times 8.64} = 0.00030705 \text{ F} \Rightarrow 307 \ \mu\text{F}$$

FIGURE 9.15
Power triangles for Example 9.6.

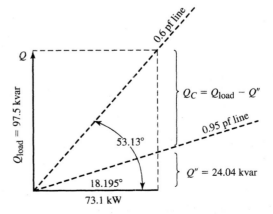

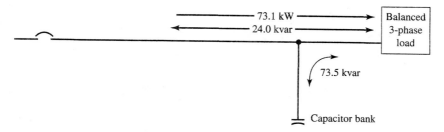

FIGURE 9.16
Power-flow diagram for Example 9.6.

i. For the wye (Y) bank,

$$X_{CY} = \frac{V_{phase}^2}{Q_C/phase} = \frac{(460/\sqrt{3})^2}{24,493.7} = 2.879 \ \Omega$$

The capacitance per phase for the wye bank is

$$C_Y = \frac{1}{2\pi f X_{CY}} = \frac{1}{2\pi \times 60 \times 2.879} = 0.000921 \ F \Rightarrow 921 \ \mu F$$

A one-line diagram illustrating the flow of power is shown in Figure 9.16. Note that the installation of the capacitor bank does not change the active and reactive power drawn by the load. However, some of the reactive power formerly supplied by the source is now supplied by the capacitor bank.

9.7 POWER MEASUREMENT IN A THREE-PHASE, THREE-WIRE SYSTEM

Maintenance testing of operating three-phase motors and other three-phase power apparatus is best done with clamp-on power meters, such as that previously shown in Figure 8.13, Chapter 8. The big advantage of power clamp-on meters is that only one instrument is needed to measure volts, amperes, watts, volt-amperes, vars, power factor, and harmonics. Once connected, simple switching at the handheld instrument gives the desired readings. Another big advantage of clamp-on power meters is that the circuit wiring does not have to be disturbed.

Power Measurement in Three-Phase Balanced Loads

Figure 9.17 shows the clamp-on connections for measuring power drawn by a three-wire, three-phase *balanced load*. Only one reading is required. The instrument automatically calculates and displays a three-phase readout. A three-phase induction motor is an example of a balanced load.

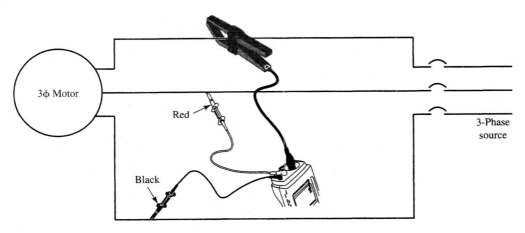

FIGURE 9.17
Clamp-on power meter connections for measuring balanced three-phase power.

Power Measurement in Three-Phase Unbalanced Loads

Figure 9.18 shows the proper connections required for power measurements in a three-phase, three-wire unbalanced load. The arrow on the clamp-on current probe (not shown) must point toward the load. Two readings are required, and the total three-phase power is equal to the algebraic sum of the two readings. This is known as the *two-wattmeter method* and may be used for both balanced and unbalanced loads. *Note:* Three wattmeters would be required to measure the power in a three-phase, four-wire system, which includes single-phase loads.

To use the two-wattmeter method, connect the clamp-on wattmeter as indicated by W_1 in Figure 9.18, and record the reading. Then disconnect the wattmeter, reconnect it as indicated by W_2 in Figure 9.18, and record the second reading. The total three-phase power is the algebraic sum of the two wattmeter readings:

$$P_{3\phi} = W_1 + W_2$$

- If the power factor is 1.0 (unity), both readings will be positive and equal.
- If the power factor is less than unity, but greater than 0.5, both readings will be positive, but one reading will be higher than the other.
- If the power factor is 0.5, one reading will be zero, and the other reading will indicate total three-phase power.
- If the power factor is less than 0.5, one reading will have a negative sign associated with it.

EXAMPLE 9.7

Assume that the two readings of the clamp-on wattmeter in Figure 9.18 indicate 560 kW and −180 kW. Determine the three-phase power.

Solution

$$P_{3\phi} = W_1 + W_2 = 560 + (-180) = 380\ kW$$

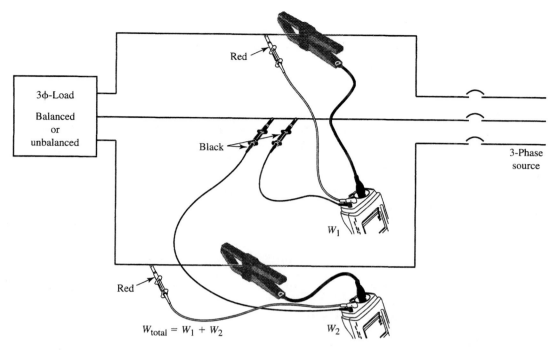

FIGURE 9.18
Clamp-on power meter connections for unbalanced or balanced three-phase power.

9.8 PHASE SEQUENCE

The phase sequence, or phase rotation of a three-phase source, is the order or sequence in which the voltage waves reach their maximum positive values. This is illustrated in Figure 9.19, where an observer traveling the time axis notes that the sequence is

$$V_a, V_b, V_c, V_a, V_b, V_c, V_a, V_b, V_c, \cdots$$

For convenience, this may be simplified by denoting only the subscript in the voltage sequence. Thus, using the subscripts, the sequence is written as

abcabcabc   or simply   abc,   or cab,   or bca

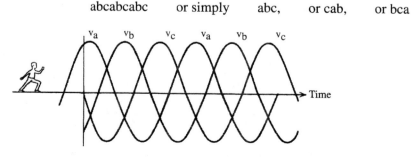

FIGURE 9.19
Determining the phase sequence of the three voltage waves.

The phase sequence of the source can have a profound effect on the performance of the load. For example, reversing the phase sequence of the driving voltage to a three-phase motor will reverse its direction of rotation; reversing the phase sequence of the driving voltage applied to an unbalanced three-phase load could cause major changes in the magnitudes and phase angles of the line currents; and reversing the phase sequence of a three-phase generator that is to be paralleled with another three-phase generator can cause extensive damage to both machines.

When analyzing circuit diagrams involving three-phase loads, it is convenient to define the phase sequence at the load, reading from top to bottom or left to right as applicable. For example, in Figure 9.20a, reading repetitively from top to bottom, the phase sequence at the load is abcabcabc, or simply abc. Interchanging any two of the three cables causes the sequence at the load to be reversed. This is shown in Figures 9.20b, 9.20c, and 9.20d and is indicated in the following tabulation:

Figure	Cables Interchanged	Phase Sequence
9.20b	a and b	bacbacbac or cba
9.20c	b and c	acbacbacb or cba
9.20d	c and a	cbacbacba or cba

Note that the three-phase system has only two possible phase sequences: abc or cba.

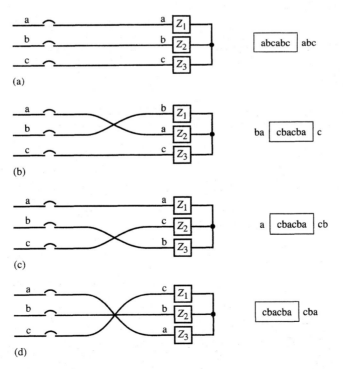

(a)

(b)

(c)

(d)

FIGURE 9.20
Changing phase sequence of a three-phase system by interchanging any two cables.

SUMMARY OF EQUATIONS FOR PROBLEM SOLVING

Load	Voltage	Current	Ohm's Law
Wye	$V_{\text{line}Y} = \sqrt{3}\, V_{\phi Y}$	$I_{\text{line}Y} = I_{\phi Y}$	$I_\phi = \dfrac{V_\phi}{Z_\phi}$
Delta	$V_{\phi\Delta} = V_{\text{line}\Delta}$	$I_{\text{line}\Delta} = \sqrt{3}I_{\phi\Delta}$	$I_\phi = \dfrac{V_\phi}{Z_\phi}$

$$P_{3\phi} = 3P_{1\phi} \tag{9-8}$$

$$Q_{3\phi} = 3Q_{1\phi} \tag{9-9}$$

$$S_{3\phi} = 3S_{1\phi} \tag{9-10}$$

$$S_{3\phi} = \sqrt{P_{3\phi}{}^2 + Q_{3\phi}{}^2} \tag{9-11}$$

$$P_{3\phi} = S_{3\phi} \times \cos(\theta) \tag{9-12}$$

$$Q_{3\phi} = S_{3\phi} \times \sin(\theta) \tag{9-13}$$

$$\theta = \cos^{-1}\!\left(\frac{P_{3\phi}}{S_{3\phi}}\right) \tag{9-14}$$

$$F_P = \frac{P_{\text{in}}}{S_{\text{in}}} = \frac{P_{3\phi}}{S_{3\phi}} = \frac{P_{1\phi}}{S_{1\phi}} = \cos\theta \tag{9-15}$$

$$\theta = \cos^{-1}(F_P) \tag{9-16}$$

$$\lambda = \frac{P_{\text{shaft}}}{P_{P_{\text{in}}}} \tag{9-18}$$

$$\lambda = \frac{\%\ \text{efficiency}}{100} \tag{9-19}$$

$$P_{3\phi} = \sqrt{3} \times V_{\text{line}} \times I_{\text{line}} \times F_p \tag{9-20}$$

$$S_{3\phi} = \sqrt{3} \times V_{\text{line}} \times I_{\text{line}} \tag{9-21}$$

REVIEW QUESTIONS

1. What is a three-phase system, and what are its advantages over a single-phase system?
2. Compare the current and voltage relationships in a wye connection with those in a delta connection.
3. What is meant by a balanced three-phase load?
4. How is power factor corrected in a three-phase system?
5. What means is available for measuring power in a three-phase system without breaking into the circuit? How is this done?

6. Explain what is meant by phase sequence.
7. Make a sketch showing how the phase sequence may be reversed at the terminals of a three-phase induction motor.
8. What effect does reversing the phase sequence have on a three-phase induction motor?

PROBLEMS

9–1/2. A wye-connected resistor bank of 10 Ω per leg is supplied by a three-phase, 460-V, delta-connected, 60-Hz generator. Sketch the circuit and determine (a) current in each resistor; (b) line current; (c) voltage across each resistor.

9–2/2. A delta-connected resistor bank of 25 Ω per leg is supplied by a 208-V, wye-connected, 60-Hz generator. Sketch the circuit and determine (a) current in each resistor; (b) line current; (c) voltage across each resistor.

9–3/2. A three-phase generator has a voltage rating of 260 volts per phase. What is the line voltage when it is (a) wye-connected and (b) delta-connected? Sketch the circuits.

9–4/2. A 380-V, 50-Hz, three-phase source supplies a wye-connected load whose resistance and inductive reactance per phase are 6 Ω and 15 Ω, respectively. Determine (a) impedance per phase; (b) phase current; (c) line current.

9–5/4. The instrumentation of a generator system indicates an output of 350 kW at 0.7 power factor from a 450-volt 60-hertz generator. Determine the apparent power supplied by the generator.

9–6/4. A three-phase induction motor draws 36 kW and 20 kvar from a 460-V, 50-Hz, 3-phase source. Sketch the circuit and determine (a) apparent power; (b) power factor.

9–7/4. A balanced three-phase load draws 30 A at 0.86 power factor lagging from a 240-V, three-phase source. Sketch the circuit and determine (a) apparent power; (b) active power; (c) reactive power.

9–8/4. A 3-phase, 30-hp, 1765-rpm, 460-V induction motor operating at three-quarter load has an efficiency of 91% and a power factor of 87.4%. Sketch the circuit and determine (a) active power; (b) apparent power; (c) reactive power; (d) line current.

9–9/4. A three-phase, 460-V, 60-Hz, 10,000-kVA generator supplies the following three-phase loads: 6 kVA at 0.8 pf lagging, 8 kVA at 0.8 pf lagging, and 20 kVA at unity pf. Sketch a one-line diagram and determine (a) system active power; (b) system reactive power; (c) system apparent power; (d) system power factor.

9–10/4. A 2300-V, 60-Hz, delta-connected generator supplies a system consisting of the following parallel connected three-phase loads: 10 kVA at 0.8 pf lagging, 6 kVA at 0.7 pf lagging, and a 4 kVA wye-connected resistor load. Sketch a one-line diagram and determine (a) system active power; (b) system reactive power; (c) system apparent power; (d) system power factor.

9–11/5. A three-phase induction motor draws 20 amperes from a 2400-volt, three-phase, 60-hertz generator. Sketch the circuit and determine the kVA supplied.

9–12/5. A three-phase, 60-hertz motor draws 25 amperes at 0.75 pf from a 460-volt system. Sketch the circuit and determine (a) the apparent power drawn by the motor: (b) the active power; (c) the reactive power.

9–13/5. The instrumentation for a given three-phase load indicates power of 24,500 watts, voltage of 550 volts, and current of 28.6 amps. Determine (a) the apparent power delivered to the load; (b) the reactive power; (c) the power factor.

9–14/6. A 100-kVA, balanced three-phase load is operating at 0.65 pf lagging from a 450-V, 25-Hz three-phase supply. Sketch a one-line diagram and determine (a) active power drawn by the load; (b) reactive power drawn by the load; (c) line current. Sketch the power diagram and (d) determine the kvar rating of a delta-connected capacitor bank required to obtain a system power factor of 0.90 lagging; (e) capacitance of each capacitor in the delta bank; (f) new line current.

9–15/6. A 4000-V, 60-Hz, three-phase source supplies a 700-hp, 1775-rpm, 60-Hz, three-phase induction motor. The motor is operating at rated horsepower at an efficiency of 93.6% and a power factor of 88.5%. Sketch the circuit and determine (a) kvar rating of a capacitor bank required to obtain unity power factor; (b) capacitance of each branch of the bank if delta connected; (c) line current at unity power factor.

9–16/6. A 100-kVA balanced three-phase load is operating at 0.65 power factor lagging from a 460-V, 60-Hz source. Sketch a one-line diagram and determine (a) active power; (b) reactive power; (c) line current; (d) kvar rating of a capacitor bank required to obtain a system power factor of 90% lagging; (e) capacitance of the bank if wye connected.

10

Transformers in Power Applications: Principles, Operation, and Maintenance

10.0 INTRODUCTION

Transformers are used extensively in power applications to raise or lower voltage in a distribution system. They are also used to isolate one circuit from another and to prevent excessive current from damaging a motor when starting, etc. Although a transformer has no moving parts to wear out, it must be monitored for excessive heating due to overload, internal shorts, harmonic currents, and signs of insulation deterioration.

10.1 PRINCIPLE OF TRANSFORMER ACTION

When two coils are arranged with respect to one another, so that all or part of the flux caused by current in one coil passes through the window of the other coil, as illustrated in Figure 10.1, the combination is called a *transformer*. The coil that is connected to the AC source is called the *primary* and the other coil is called the *secondary*. The AC source causes an alternating magnetic flux to be generated in the primary, some of which passes through the window of the secondary. That part of the magnetic flux that passes through both windings is called the *mutual flux* (Φ_M), and that part of the magnetic flux that does not pass through the secondary window is called the *leakage flux* (Φ_{leak}).

In practical transformers, such as those illustrated in Figures 10.2a and 10.2b, the iron core contains the flux and guides it from one coil to the other. The result is less leakage and, hence, a greater amount of flux passing through the secondary window. The transformer in Figure 10.2a is called a core-type transformer, and the transformer in Figure 10.2b is called a shell-type transformer. The core-type transformer is simpler in construction, and the wider spacing between primary and secondary windings gives it an advantage in high-voltage applications. However, the shell type has the advantage of less leakage flux.

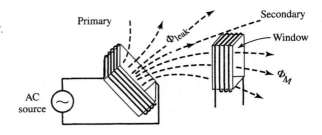

FIGURE 10.1
Elementary transformer.

Eddy-Currents and Hysteresis

The alternating flux that generates a voltage in the secondary winding also generates heat in the iron core. The heat is generated by circulating currents (called *eddy currents*) induced in the iron core by the alternating flux, and by "molecular friction" as the magnetic domains in the iron are twisted back and forth. The power loss due to molecular friction, called *hysteresis loss* (P_{hys}), is proportional to the frequency and the 1.6 power of the maximum flux density. The power loss due to eddy currents, called *eddy-current loss* (P_{eddy}), is proportional to the square of the maximum flux density, and the square of the frequency. Expressed mathematically,

$$P_{hys} \propto f B_{max}^{1.6} \qquad \qquad \textbf{(10–1)}$$

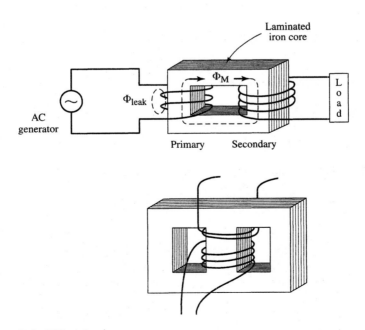

FIGURE 10.2
(a) Core-type transformer; (b) shell-type transformer.

$$P_{eddy} \propto f^2 B^2_{max} \tag{10-2}$$

where: P_{hys} = power loss due to hysteresis (W)
P_{eddy} = power loss due to eddy currents (W)
f = frequency (Hz)
B_{max} = amplitude (peak value) of flux density wave in teslas (T).

To reduce eddy currents, each of the laminations illustrated in Figure 10.2 is insulated with a coating of varnish or oxide on one or both sides. Any damage to the laminated core that causes the laminations to make electrical contact with each other will increase the eddy currents and result in overheating of the core. The laminations are stamped out of low-loss silicon steel to minimize hysteresis heating.

10.2 TRANSFORMER TURN RATIO

Transformers may be used to either step up or step down voltage. Hence, it is convenient to refer to the two windings as the high-voltage-side or high-side (HS) winding and the low-voltage-side or low-side (LS) winding.

The turn ratio of a transformer is the ratio of the number of turns in the high-voltage winding to the number of turns in the low-voltage winding, assuming no shorted turns. The turn ratio is essentially equal to the voltage ratio obtained from high-side voltage measurements and low-side voltage measurements, *with no load connected to the secondary*. Expressed mathematically,

$$a = \frac{N_{HS}}{N_{LS}} = \frac{E_{HS}}{E_{LS}} \tag{10-3}$$

where: a = turn ratio
E_{HS} = no-load voltage at high-voltage terminals (V)
E_{LS} = no-load voltage at low-voltage terminals (V)
N_{HS} = number of turns of wire in high-voltage winding
N_{LS} = number of turns of wire in low-voltage winding.

EXAMPLE 10.1   The turn ratio of a certain transformer is 4 to 1 (also written as 4:1). If the low-side voltage (measured at no load) is 240 V, determine the voltage at the high side.

Solution

$$\frac{E_{HS}}{E_{LS}} = \frac{N_{HS}}{N_{LS}} \Rightarrow E_{HS} = E_{LS} \times \frac{N_{HS}}{N_{LS}} = 240 \times 4 = 960 \text{ V}$$

Nameplate Voltage Rating

The voltage ratings on the nameplate of a transformer are the voltages obtained when no load is connected to the secondary. The nameplate voltage ratings are based on the turn ratio. With rated nameplate voltage impressed on one winding, all other nameplate voltages shall be correct within ±0.5% of the nameplate markings [1].

Voltage Ratio under Load

The voltage ratio when under load is the ratio of high-side voltage to low-side voltage under specified conditions of load; it is not the turn ratio [2].

When a load is connected to the secondary, as illustrated in Figure 10.2a, voltage drops (IR drops) will occur within the primary and secondary windings due to the resistance of the windings, and additional voltage drops will occur due to increased leakage flux. Since the leakage flux and IR drops are both proportional to the load current, increases in load current will result in lowered secondary voltage.[1]

However, when operating transformers at rated and below-rated load, the internal voltage drops (due to leakage flux and winding resistance) are small, and the turn ratio may be used as a rough approximation of the voltage ratio under load. Expressed mathematically,

$$a = \frac{N_{HS}}{N_{LS}} \approx \frac{V_{HS}}{V_{LS}} \qquad (10\text{--}4)$$

where: V_{HS} = voltage under load at high-side terminals (V)
V_{LS} = voltage under load at low-side terminals (V).

Tap Changers

The construction details of a single-phase distribution transformer equipped with a manually operated tap changer are illustrated in Figure 10.3. The tap changer is used to obtain slightly different turn ratios, so that adjustments can be made to compensate for small differences in system voltages.

A wide variety of tap changers are in use today and each one's operating handle is positioned in a different way. Hence, it is imperative that the maintenance engineer read the operating instructions before changing taps on this type of transformer.

10.3 TRANSFORMER NO-LOAD CONDITIONS

With no load connected to the secondary, the current in the primary is just enough to establish the magnetic flux needed for transformer action and to supply the hysteresis and eddy-current losses in the iron core. This no-load current, called the *excitation current* or exciting current, varies between 1 and 2 percent of rated current in large

---

[1] A capacitive load that causes series resonance with the transformer inductance will cause an increase in voltage.

FIGURE 10.3
Transformer with tap changer for changing
the turn ratio.

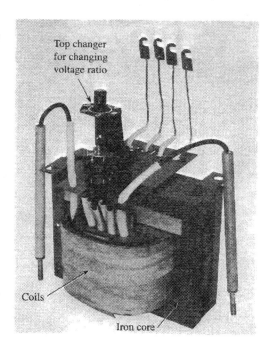

power transformers and can be as high as 6 percent of rated current in small distribution transformers.

The component of excitation current that establishes the magnetic flux in the transformer core is called the *magnetizing current* (I_M). The magnetizing current and, hence, the magnetic flux are proportional to the applied voltage and are inversely proportional to the frequency. Expressed mathematically, and assuming the iron core is not saturated with flux,

$$I_M \propto \frac{V}{f} \qquad (10\text{--}5)$$

$$\phi \propto \frac{V}{f} \qquad (10\text{--}6)$$

where: I_M = magnetizing current (A)
            ϕ = magnetic flux (Wb)
            V = applied voltage (V)
            f = frequency (Hz).

Damaging Effects of Overexcitation

Operating a transformer significantly above its rated primary voltage or below its rated frequency will result in a higher magnetizing current and, hence, a higher flux in the transformer core. This is called *overexcitation*. Excessive overexcitation causes

FIGURE 10.4
No-load current with magnetic saturation.

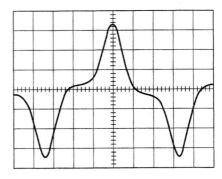

saturation of the iron core. When this occurs, the iron cannot hold any more flux. The excess flux will spill into the surrounding space, including the metal frame, copper conductors, and transformer case. Eddy currents caused by this "stray flux" cause hot spots that can blister paint, char insulation, and sometimes melt metal. Furthermore, each time the iron saturates, the inductance of the primary winding drops to a very low value, causing a high pulse of primary current. This behavior is illustrated in Figure 10.4. The high current pulses result in additional heating of the transformer windings.[2]

Transformer In-Rush Current

When the switch to the primary winding of an unloaded transformer is closed, there may be an in-rush of current many times greater than rated full-load current of the transformer. The magnitude of the in-rush current depends on the instant of time that the circuit breaker is closed, and on the amount of the residual magnetism left in the iron core from some previous excitation.

If the switch is closed at the instant the voltage wave is at its peak value, there will be no in-rush. The current will be the rated no-load current; this is illustrated in Figure 10.5a. However, if the switch is closed at the instant the voltage wave passes through zero, the resulting in-rush current will be approximately double the rated current as illustrated in Figure 10.5b.

If the doubled current causes the iron core to saturate, the inductance of the primary winding will drop to a very low value and the in-rush current will be significantly larger.

10.4 TRANSFORMER LOAD CONDITIONS

An elementary circuit diagram of a two-winding transformer supplying a load is illustrated in Figure 10.6. Voltages V_{HS} and V_{LS} are the respective high-side and low-side voltages, as measured at the terminals of the transformer, when it is carrying a load.

---

[2] See Section 4.10, Chapter 4, for the effect of magnetic saturation on inductance.

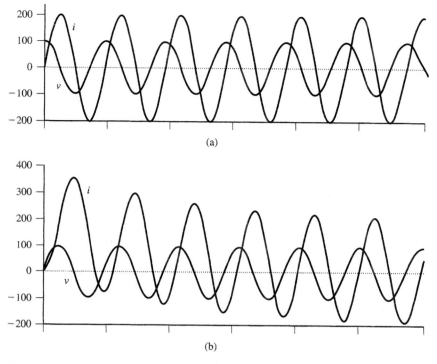

(a)

(b)

FIGURE 10.5
Transformer in-rush current: (a) switch closed when voltage wave is passing through its maximum value; (b) switch closed at the instant voltage wave passes through zero.

The efficiency of distribution and power transformers varies from 96 to 99 percent. Thus, for practical purposes, when calculating currents and voltages for power input and power output, it may be assumed that the power input to the transformer is equal to the power output. That is,

$$P = V_{HS}I_{HS}F_P = V_{LS}\,I_{LS}F_P \qquad (10\text{--}7)$$

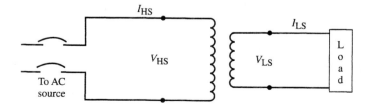

FIGURE 10.6
Elementary circuit diagram of a two-winding transformer.

Similarly,

$$S = V_{HS}I_{HS} = V_{LS}I_{LS} \qquad \text{(10–8a)}$$

Expressing Eq. (10–8a) as a ratio, and substituting Eq. (10–4),

$$\frac{I_{LS}}{I_{HS}} = \frac{V_{HS}}{V_{LS}} = \frac{N_{HS}}{N_{LS}} \qquad \text{(10–8b)}$$

where: V_{HS} = voltage under load at high-side terminals (V)
V_{LS} = voltage under load at low-side terminals (V)
I_{HS} = high-side current (A)
I_{LS} = low-side current (A)
P = active power (W)
S = apparent power (VA)
F_P = power factor.

EXAMPLE 10.2

A 20-kVA, 60-Hz, 480–240-V transformer is supplying a 20-kVA, 0.82 power factor load at 240 V. Neglecting losses, determine (a) low-side current; (b) high-side current. The circuit is similar to that in Figure 10.6.

Solution

a. $S = V_{LS}I_{LS} \Rightarrow I_{LS} = \dfrac{S}{V_{LS}} = \dfrac{20{,}000}{240} = 83.3 \text{ A}$

b. $V_{HS}I_{HS} = V_{LS}I_{LS} \Rightarrow I_{HS} = \dfrac{V_{LS}I_{LS}}{V_{HS}} = \dfrac{240 \times 83.3}{480} = 41.7 \text{ A}$

or $S = V_{HS}I_{HS} \Rightarrow I_{HS} = \dfrac{S}{V_{HS}} = \dfrac{20{,}000}{480} = 41.7 \text{ A}$

10.5 AUTOTRANSFORMERS

A transformer that accomplishes voltage transformation with a single winding is called an *autotransformer*. This is illustrated in Figure 10.7, where part of the total winding is common to both the primary and secondary circuits; the common section is drawn with dark lines.

Autotransformers require less copper and are more efficient than two-winding transformers. Their major disadvantage is the lack of isolation between the primary and secondary circuits. Autotransformers should only be used in applications where lack of isolation between the high-voltage side and the low-voltage side will not present a safety hazard. However, note that if used as a step-down transformer, and the winding that is common to both primary and secondary is accidentally opened, the full primary voltage will appear across the secondary terminals.

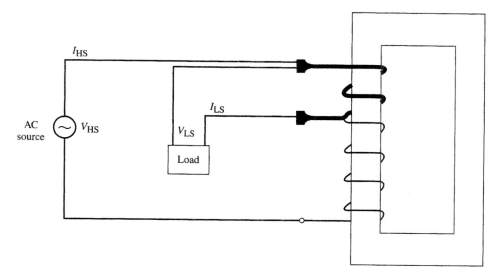

FIGURE 10.7
Basic autotransformer circuit.

In power applications, autotransformers are used for reduced-voltage starting of AC motors, as booster transformers on distribution systems to compensate for voltage drops along the line, and for voltage step up or step down for high-voltage transmission lines.

Percent Voltage Tap

The output voltage of an autotransformer is usually expressed as a percent of the input voltage, called the *percent tap*. Standard autotransformers, used for starting AC motors of 50 hp and below, are equipped with 65 percent and 80 percent voltage taps. For motors above 50 hp, the standard taps are 50 percent, 65 percent, and 80 percent.

The power relationships for the two-winding transformer, as expressed in Eqs. (10–7) and (10–8), are directly applicable to the autotransformer.

EXAMPLE 10.3

A certain autotransformer operating in the step-down mode supplies 140 A from a 20% tap to a unity power factor load. The voltage input to the autotransformer is 480 V at 60 Hz. Neglecting losses, determine (a) low-side voltage; (b) apparent power input; (c) high-side current. The circuit diagram is illustrated in Figure 10.8.

Solution

a. $V_{LS} = V_{HS} \times \dfrac{\% \text{ tap}}{100} = 480 \times 0.20 = 96$ V

b. Because the power input must equal the power output (neglecting losses),

$$S = V_{LS}I_{LS} = 140 \times 96 = 13{,}440 \text{ VA} = 13.44 \text{ kVA}$$

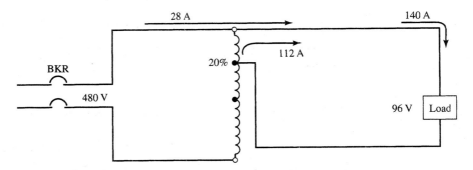

FIGURE 10.8
Circuit for Example 10.3.

c. $S = V_{HS}I_{HS} \Rightarrow I_{HS} = \dfrac{S}{V_{HS}} = \dfrac{13,440}{480} = 28 \text{ A}$

Note: Although 140 A is drawn by the load, only 28 A is conducted from the supply line. The remaining 112 A is supplied by transformer action at 96 V, as illustrated in Figure 10.8.

10.6 TRANSFORMER TERMINAL MARKINGS AND SINGLE-PHASE CONNECTIONS

In order to parallel transformers successfully or connect them properly in three-phase arrangements, the polarity of the respective primary and secondary terminals must be known. High-voltage terminals are designated by the letter H, and low-voltage terminals by the letter X, with corresponding subscripts denoting the same instantaneous polarity. The standard terminal markings for a transformer are illustrated in Figure 10.9 [3].

Terminals H_1 and X_1 have the same instantaneous polarity; likewise H_2 and X_2. Thus, at any given instant during most of each half-cycle, when the primary current is entering H_1, the secondary current is leaving X_1 as though the two leads form a continuous line. The lowest and highest subscript numbers mark the respective ends of the full winding, and the intermediate numbers mark the fractions of windings or taps (if present), as illustrated in Figures 10.9c and 10.9f.

Additive and Subtractive Polarity

The position of the terminals with respect to the high-voltage and low-voltage sides of the transformer affects the voltage stress on the external leads, especially in high-voltage transformers. In Figures 10.9a, 10.9b, and 10.9c, the terminals with the same instantaneous polarity are opposite each other, and if accidental contact of two adjacent terminals occurs, one from each winding, the voltage across the other ends will be the difference between the high and low voltages. Hence, this arrangement of terminals is called *subtractive polarity*.

Subtractive polarity

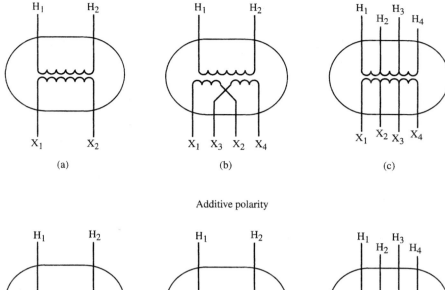

(a) (b) (c)

Additive polarity

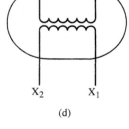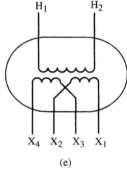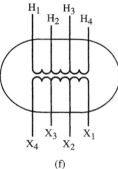

(d) (e) (f)

FIGURE 10.9
Standard terminal markings for transformers.

However, if the terminals are arranged as illustrated in Figures 10.9d, 10.9e, and 10.9f, and if accidental contact of two adjacent terminals occurs, one from each winding, the voltage across the other ends will be equal to the sum of the high and low voltages. This arrangement of terminals is called *additive polarity* and has the disadvantage of producing high-voltage stresses between windings. For this reason, and with few exceptions, transformer terminals are arranged for subtractive polarity.

Connecting Transformers in Parallel

Regardless of additive or subtractive polarity, when transformers are to be connected in parallel, all terminals with the same letter and subscript markings must tie together. This is illustrated in Figures 10.10a, 10.10b, and 10.10c. Furthermore, the turn ratios of both must be equal.

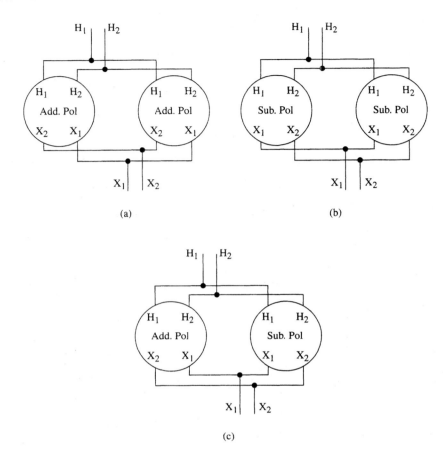

FIGURE 10.10
Paralleling connections for single-phase transformers.

10.7 HARMONIC HEATING OF TRANSFORMERS

If a transformer supplies circuits containing nonlinear loads, such as computers, light dimmers, electronic speed control, etc., the transformer current will be nonsinusoidal. A current wave for a representative nonlinear load is shown with a heavy line in Figure 10.11. Analysis of this particular nonsinusoidal wave shows it to be made up of component waves, which are all sinusoidal but of different frequencies and different phase angles. The component waves are called *harmonics* and are multiples of the source frequency (called the *fundamental frequency*). The harmonic components in Figure 10.11 are the fundamental (sometimes called the first harmonic), the 3rd, the 5th, and the 7th.

The harmonic components of load current cause heating of the transformer core, as well as heating of the conductors, above and beyond that caused by rated current and rated frequency of the transformer.

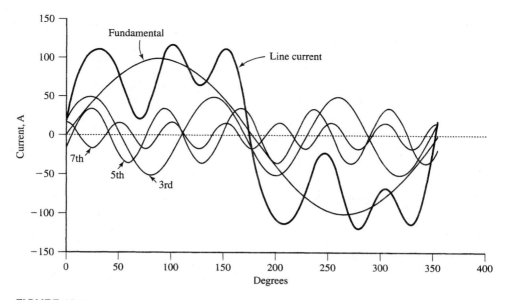

FIGURE 10.11
Line-current wave and its components when nonlinear loads are present.

The K-Factor Rating

The K-factor rating of a transformer is a measure of the transformer's ability to withstand the total heating effect caused by all the harmonic components in the load current plus the fundamental current. The transformer K-factor needed for a particular load may be measured with a handheld clamp-on harmonic meter, such as the one illustrated in Figure 7.8, Chapter 7. When clamped around the conductor supplying the load, the harmonic meter will read the required K-factor.[3] Transformers with the required K-factor rating or higher will not overheat when carrying that particular nonsinusoidal load. Some of the K-factor ratings, standardized by Underwriters Laboratories Inc. (UL), are K1, K-4, K-13, K-20, K-30, and K-50.

10.8 THREE-PHASE CONNECTIONS OF SINGLE-PHASE TRANSFORMERS

Single-phase transformers can be used to transform three-phase power by connecting them in wye–wye, delta–wye, wye–delta, delta–delta, or open-delta arrangements, respectively, as illustrated in Figure 10.12. The wye–wye connection has some applications in high-voltage transmission, but is not very popular because of poor voltage regulation. The wye–delta connection is good for stepping down high transmission line voltages. The delta–wye is most commonly used in industrial distribution systems.

---

[3] The basis for the K-factor rating is given in Reference [4].

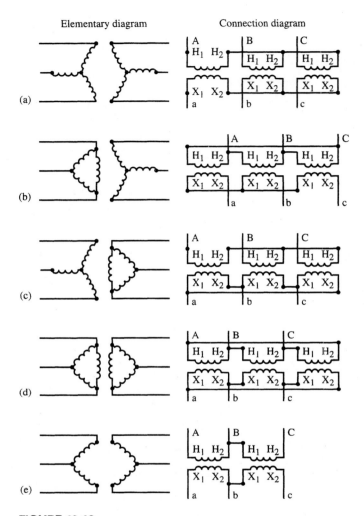

Elementary diagram   Connection diagram

FIGURE 10.12
Three-phase connections of single-phase transformers: (a) wye–wye; (b) delta–wye; (c) wye–delta; (d) delta–delta; (e) open-delta (V connection).

The delta–delta bank has the advantage of being able to operate continuously with reduced load if one of the three transformers is disconnected from the circuit. This open-delta connection, also called a V connection, is illustrated in Figure 10.12e. The open-delta connection provides a convenient means for inspection, maintenance, testing, and replacing of delta-connected transformers, one at a time, with only a momentary power interruption. Note that, for safety reasons, the circuit breaker in the primary circuit must be opened before disconnecting a transformer.

Disconnecting one transformer of a delta–delta bank converts the circuit illustrated in Figure 10.12d to the circuit illustrated in Figure 10.12e. Although two transformers supply the same balanced three-phase voltage as do the three transformers in the delta–delta bank, its kVA rating is reduced to 58 percent of the delta–delta rating.

$$S_{\text{open}-\text{delta}} = 0.58 \times S_{\Delta-\Delta} \qquad (10\text{–}9)$$

Two transformers in an open-delta connection may also be used initially to provide three-phase service, with the expectation of adding a third (for Δ–Δ operation) if the projected increase in load occurs.

EXAMPLE 10.4

A delta–wye transformer bank composed of three single-phase transformers reduces a 4160-V supply to 460 V for use at a load center. Determine (a) the line current on the high-voltage side if the line current to the load is 240 amperes; (b) current in each delta branch; (c) current in each wye branch.

Solution
The circuit is illustrated in Figure 10.13.

a. $\sqrt{3}V_{\text{line}} I_{\text{line}} = \sqrt{3}V_{\text{line}} I_{\text{line}} \Rightarrow \sqrt{3} \times 4160 I_{\text{line}} = \sqrt{3} \times 460 \times 240$
  high side low side high side
$I_{\text{line high-side}} = 26.5 \text{ A}$

b. $I_{\Delta,\,\text{phase}} = \dfrac{I_{\Delta,\,\text{line}}}{\sqrt{3}} = \dfrac{26.5}{\sqrt{3}} = 15.3 \text{ A}$

c. $I_{Y,\,\text{phase}} = I_{Y,\,\text{line}} = 240 \text{ A}$

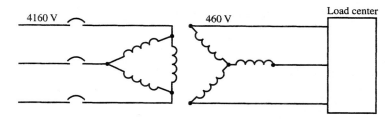

FIGURE 10.13
Circuit for Example 10.4.

EXAMPLE 10.5

Three 100-kVA, 4160–480-V, single-phase transformers are connected delta–delta. Determine (a) apparent power rating of the bank; (b) high-side current rating of the bank; (c) allowable apparent power rating of the bank if due to an emergency one transformer is removed for repair; (d) allowable high-side current for the emergency conditions.

Solution

a. $S_\Delta = 3 \times S_{1\text{-phase}} = 3 \times 100 = 300 \text{ kVA}$

b. $S = \sqrt{3} V_{\text{line}} I_{\text{line}} \Rightarrow I_{\text{line}} = \dfrac{S}{\sqrt{3} V_{\text{line}}} = \dfrac{300,000}{\sqrt{3} \times 4160} = 41.64 \text{ A}$

c. $S_{\text{open-delta}} = 300 \times 0.58 = 174 \text{ kVA}$

d. $S = \sqrt{3} V_{\text{line}} I_{\text{line}} \Rightarrow I_{\text{line}} = \dfrac{S}{\sqrt{3} V_{\text{line}}} = \dfrac{174,000}{\sqrt{3} \times 4160} = 24.15 \text{ A}$

10.9 DELTA–WYE DISTRIBUTION SYSTEM

Almost all commercial and industrial distribution systems are fed by transformer banks with a delta primary and a wye secondary. Large power loads are fed from the three-phase line, and single-phase loads are connected between secondary lines and neutral. This is illustrated in Figure 10.14, with clamp-on meters measuring the current in the line and neutral conductors.

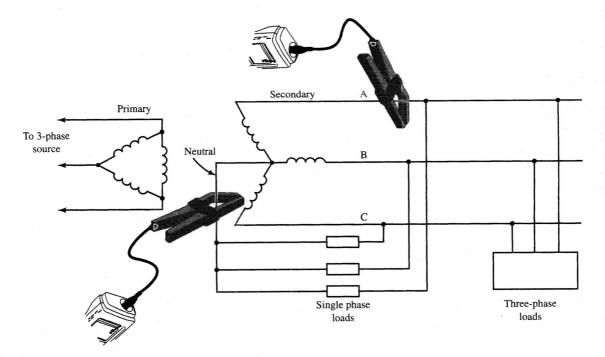

FIGURE 10.14
Delta–wye transformer supplying single-phase and three-phase loads.

If all of the single-phase loads that are connected to each leg of the three phases are identical and linear, such as resistance heaters, incandescent lamps without dimmers, and single-phase motors, *the system is balanced and there will be no current in the neutral line.*

However, if the single-phase circuits supply nonlinear loads, such as computers, light dimmers, solid-state speed controls, and other electronic equipment, the load current in lines A, B, C, and the neutral will contain harmonics. The harmonics that are particularly troublesome in the delta–wye system are the 3rd harmonic and its odd multiples, called *triplen harmonics*. These include the 3rd, 9th, 15th, 21st, etc. Thus, for a 60-Hz source, the triplen harmonics will include 180 Hz, 540 Hz, 900 Hz, etc.

Effect of Load-Produced Triplen Harmonics on Neutral Conductors

Many older distribution systems designed for use with essentially balanced linear loads have neutral conductors that are smaller in diameter than the phase conductors. This was justified because in the older systems the neutral carried a relatively small unbalanced current. However, the proliferation of computers and other solid-state loads causes significant triplen harmonic currents in the neutral, which in many cases may be greater than the respective phase currents. The respective triplen harmonics, from all three phases, are arithmetically additive in the neutral. That is, the third harmonic currents in lines A, B, and C are additive in the neutral line (they do not cancel each other). Similarly, the 9th harmonic currents in lines A, B, and C are additive in the neutral line, etc. As a result, older distribution systems are beginning to suffer from overheated neutrals, neutral burnout, and voltage drops in the neutral that cause computers to malfunction [5, 6]. In many cases the only solution is to completely rip out the old wiring and rewire with larger neutral conductors.

Distribution systems designed specifically for use with nonlinear loads must have neutral conductors that are larger in diameter than the phase conductors, or must use a separate neutral conductor for each phase and have it connected directly to the wye junction of the transformer.

Figure 10.15 shows a five-conductor armored cable, called Super-Neutral Cable®, that has a larger diameter neutral conductor for use in predominantly nonlinear systems.[4] The cable includes three line conductors, one ground conductor, and a much larger neutral conductor.

Effect of Triplen Harmonics on Delta–Wye Transformers

Load-produced triplen harmonic currents in the three wye-connected secondaries of a delta–wye transformer generate triplen harmonic currents in the three primary windings. These triplen harmonic currents circulate around the closed delta loop, generating heat. Thus, when operating with rated load current in the delta primary, the added heating effect caused by triplen harmonic currents circulating in the delta will cause

---

[4] Super-Neutral Cable® is a registered trademark of AFC Co.

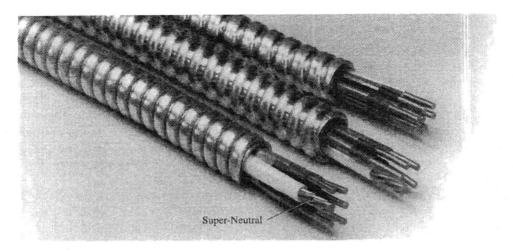

FIGURE 10.15
Super-Neutral Cable® (courtesy AFC Incorporated).

excessive heating of and possible damage to the primary windings. Note that triplen harmonic currents *do not appear in the three lines supplying the delta* and, therefore, are not easily measured by maintenance technicians.

Damaging Effects of Harmonics on Distribution Systems

Although an infinite number of harmonics make up the current wave in a nonlinear load, only a few harmonics cause big problems in the distribution system. The problems are due to resonance between the inductance of the transformer and the capacitors used for power-factor correction. The most severe resonance problems usually involve the 5th and 7th harmonics

An excellent example of the damaging effects caused by resonance at harmonic frequencies in distribution systems is documented in Reference [7]; parallel resonance at the 5th harmonic occurred between the system inductive reactance and a 20-MVA, 13.8-kV capacitor bank installed for power-factor improvement. The resultant over-voltage caused destruction of a utility circuit breaker, resulting in a fire. Reference [8] explores the application of filters (also called wave traps) that can be used to reduce adverse harmonic effects on a distribution system. The reference includes both hypothetical and actual cases.

10.10 THREE-PHASE CONNECTIONS OF SINGLE-PHASE TRANSFORMERS WHEN TERMINALS ARE UNMARKED

If the terminals of the three transformers are unmarked, it is necessary to measure the voltage at the open ends of the secondary after each connection is made. A voltmeter that has a range equal to double the expected voltage of the low-voltage winding should be used.

Procedure

First connect the three high-side windings in wye or delta, as desired. Then connect one secondary terminal of each of any two low-side windings together, as illustrated in Figure 10.16a. Excite the high-voltage windings with rated or below-rated voltage. Measure, with a voltmeter, the voltage of one secondary and then the voltage across the open ends of the two transformer secondaries just connected. The voltage will be equal to that of one transformer, or 1.73 times the voltage of one transformer.

If a wye-connected secondary is desired, the voltage across the open ends should be equal to 1.73 times the voltage across one transformer. If it is not equal to this value, reverse one of the secondaries just connected. Connect one terminal of the third transformer secondary to the junction point of the other two, as illustrated in Figure 10.16b. Measure the line-to-line voltage across the three remaining open ends. All three line-to-line voltages should equal the value of one transformer secondary times 1.73. If the voltage is not correct, reverse the secondary connections of the third transformer. The transformer secondary is now correctly wye connected.

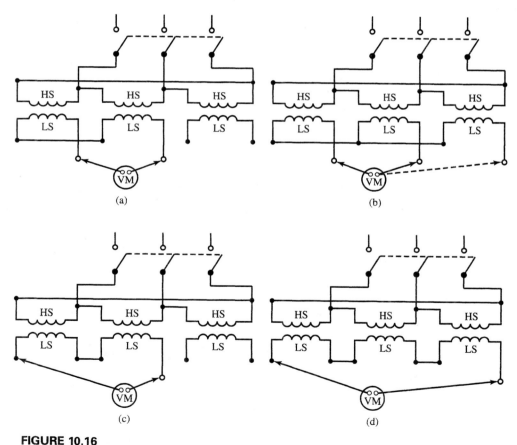

FIGURE 10.16

(a, b) Testing for a wye-connected secondary; (c, d) testing for a delta-connected secondary.

If a delta-connected secondary is desired, connect any two secondaries in series, as illustrated in Figure 10.16c, so that the voltage across the open ends will be equal to the voltage across one transformer. If it is not equal to this value, reverse one of the secondaries just connected. Then connect the third secondary in series with the other two, as illustrated in Figure 10.16d. The voltage across the remaining open ends will be equal to zero,[5] or twice the voltage of one transformer secondary. If the voltage is zero, the remaining ends may be closed; if it is equal to double the voltage of one secondary, reverse the third transformer secondary. With zero volts across the open ends, the delta may be closed.

10.11 INSTRUMENT TRANSFORMERS

Instrument transformers are used to transform high currents and high voltages to low values for instrumentation and control. An instrument potential transformer (PT), used to step down voltage, is illustrated in Figure 10.17a. The PT has two windings and is similar in construction to the standard two-winding transformer shown in Figure 10.2b.

An instrument current transformer (CT), used to step down current for ammeters, is illustrated in Figure 10.17b. The basic construction of a current transformer is illustrated in Figure 10.17c. It consists of a section of insulated copper bus or rod that acts as the primary, around which is a ferromagnetic doughnut. The secondary is wrapped around the doughnut and connected to an ammeter, relay, or other instrumentation circuit. Current transformers used in power applications have 5-A secondaries and are connected to 5-A ammeters; the ammeters are calibrated so that a full-scale reading represents rated CT primary current.

Circuits representing the two types of instrument transformers and their methods of connection to a circuit are illustrated in Figure 10.18. The instrument or relay connected to the secondary is called the *burden*.

In those applications involving currents in excess of 2000 A, the return conductor, L1 in Figure 10.18a, should be kept some distance away from the current transformer; this will prevent the relatively large magnetic flux about L1 from entering the iron core, causing an error in the instrument reading.

The relative instantaneous polarities of the transformer terminals are indicated with ± markings, or H1 and X1 markings, and are generally supplemented with a bright paint marker.

For instrument transformers, H1 always denotes the primary polarity lead or terminal and X1 the secondary polarity lead or terminal. At the instant that current enters the polarity terminal of the primary coil, current leaves the polarity terminal of the secondary coil, as though the two leads formed a continuous line (see Figure 10.18b).

The secondary winding of a current transformer must always be connected to a burden, or it must be short-circuited at the terminals with a shorting switch as illustrated in Figure 10.18a. Before servicing instruments and relays connected to the secondary of

---

[5] If the primary is wye connected and the neutral is not connected to the generator, a third harmonic voltage will prevent a zero indication. However, it will be less than that of a single transformer secondary.

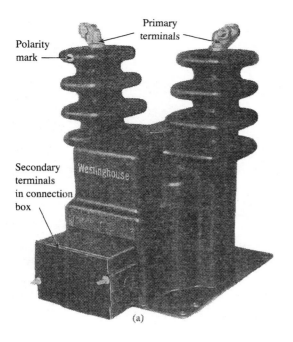

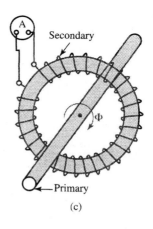

(c)

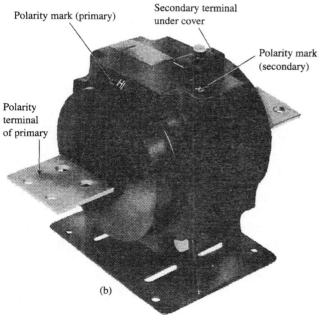

(b)

FIGURE 10.17

(a) Potential transformer (courtesy TECO Westinghouse); (b) current transformer (courtesy GE Industrial Systems); (c) basic construction of a current transformer.

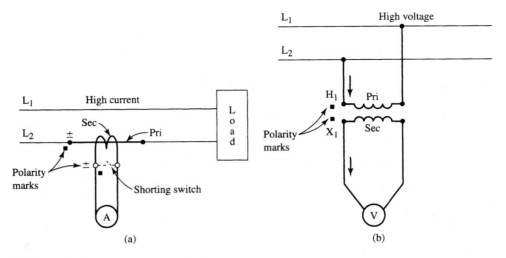

FIGURE 10.18
Proper polarity connections for (a) current transformers; (b) potential transformers.

a current transformer, the secondary must be shorted. *Although short-circuiting a current transformer does not harm it, short-circuiting a potential transformer will burn it out.*

EXAMPLE 10.6   For the circuit illustrated in Figure 10.18a, determine the current in the ammeter if the CT ratio is 600:5 and the load current is 450 A.

Solution

$$\frac{I_P}{I_S} = \text{CT ratio} \implies \frac{450}{I_S} = \frac{600}{5}$$

$$I_S = 3.75 \text{ A}$$

Opening the secondary circuit of a current transformer while current is in the primary may cause dangerously high voltages at the secondary terminal; it may also permanently magnetize the transformer iron, introducing errors in the transformer ratio.

To demagnetize a current transformer, disconnect both primary and secondary, and connect the secondary in series with a 120-V, 600-W heating element, a slide-wire autotransformer, and a 120-V, 60-Hz source as illustrated in Figure 10.19. A clamp-on ammeter can be used to measure the current. With the autotransformer slide set for maximum output voltage, close the switch, and then gradually adjust the slide-wire to obtain minimum current. Open the switch, disconnect, and reinstall the current transformer.

Polarity Check

A polarity check should always be made on new, rebuilt, or repaired instrument transformers to make sure that the polarity markings are indeed correct. Extensive damage

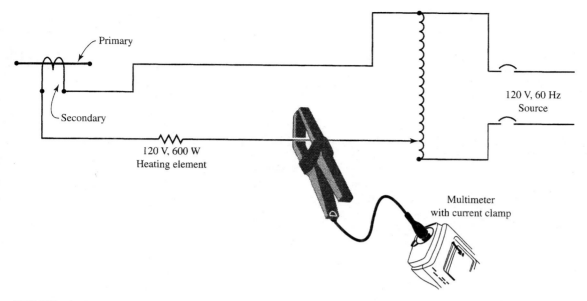

FIGURE 10.19
Demagnetizing circuit for a current transformer.

can be caused to AC generators and distribution systems that utilize protective relaying if the instrument transformers have incorrect polarity.

To make a polarity check of a potential transformer, first de-energize the system, and disconnect the primary and secondary terminals. Then connect a DC analog-type voltmeter to the high-side winding with the positive terminal of the voltmeter connected to the polarity terminal of the high side, as illustrated in Figure 10.20a. Connect a battery and push-button to the low side, with the positive terminal of the battery con-

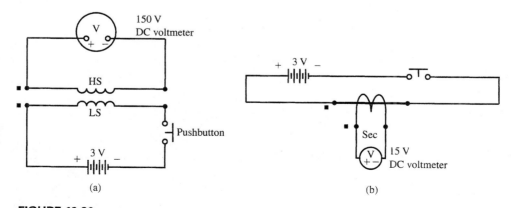

FIGURE 10.20
Polarity check of (a) potential transformer; (b) current transformer.

nected to the polarity terminal of the low side. The polarity check is made by depressing and releasing the push-button quickly. If the polarity is correct, the deflection on pushing the button will be upscale and on releasing the button will be downscale. To prevent the iron from being permanently magnetized, no more than two D or C cells (for a total of 3 V) should be used. A polarity check of a current transformer can be made in a similar manner; the test circuit is illustrated in Figure 10.20b.

10.12 MAINTENANCE OF TRANSFORMERS

Although transformers require less attention than most other electrical power apparatus, some routine maintenance is required. Figure 10.21 shows damage due to neglected maintenance; deterioration of the insulation resulted in a severe short circuit, rendering the transformer a total loss.

To help reduce accidental outages, periodic inspections should be made. The inspector should check for connections that are rusty or discolored (indicating excessive heat), accumulation of dirt on high-voltage bushings, rust, and the accumulation of refuse from birds, squirrels, etc., on transformer lids. Dirty bushings or the accumulation of conducting refuse on transformer lids can cause a flashover at the high-

FIGURE 10.21
Damage to a transformer caused by lack of maintenance (courtesy Mutual Boiler and Machinery Insurance Co.).

voltage terminals. These inspections should include checks of temperature, liquid level, and leaks in liquid-filled transformers.

Maintenance of Dry-Type Transformers

Dry-type transformers are air cooled and designed for installation in dry locations. Hence, care should be exercised to prevent the entrance of water by splashing from open windows or from leaky or broken steam lines. Dry-type transformers should be located at least 12 inches away from walls, so that the free circulation of air around and through the transformer is not impaired. They should be sheltered from dust and chemical fumes. Dust settling on the windings, core, and enclosing case reduces heat dissipation and results in overheating.

Periodic inspections of the windings and core of distribution and power transformers should be made at least once a year to detect incipient failures. To do this, the transformer should be disconnected from the line, the covers removed, and inspection made for accumulations of dirt, corrosion, loose or discolored connections, discoloration caused by excessive heat, and carbonized paths (called tracking) caused by electron creepage over the insulation surfaces.

Maintenance of Liquid-Filled Transformers

A liquid-filled transformer requires more attention than does the dry type. The liquid used for the insulating medium is either mineral oil or silicone oil.[6] These liquids should be checked at least once a year for the presence of moisture and sludge. The accumulation of sludge on transformer coils and in cooling ducts reduces the heat-transfer capability, causing higher operating temperatures. Moisture may be detected by testing samples of liquid at high voltage to determine its dielectric strength. Recommended procedures for cleaning and drying liquid-filled transformers and dry-type transformers are discussed in Chapter 26.

10.13   TESTING FOR SHORTED TURNS

Turn-to-turn shorts in transformer windings can be caused by vibration, moisture, accumulated dirt, rapidly fluctuating loads, sustained overloads, and voltage surges due to switching or lightning. Shorted turns in the primary or secondary increase the input current to the primary, which raises the temperature and shortens the life of the transformer.

A few shorted turns in the primary (high side) of a step-down transformer decrease the turn ratio, resulting in a higher secondary voltage, which could damage or shorten the useful life of the connected equipment. Shorted turns in the secondary will result in a higher turn ratio and, hence, a lower secondary voltage.

---

[6] Insulating liquids, called askerels or PCBs, were used in earlier construction. However, because of their hazardous nature they are no longer manufactured.

The turn ratio of a transformer is essentially equal to the ratio of high-side voltage to low-side voltage measured at no load, and should be equal to the ratio of high-side to low-side voltages stamped on transformer nameplates. However, the IEEE and NEMA standards on power and distribution transformers allow a manufacturing tolerance (difference) of 0.5 percent between the nameplate voltage ratio and a measured voltage ratio [1, 2, 9]. Defining the percent difference mathematically,

$$\% \text{ difference} = \frac{|\text{TVR} - \text{NVR}|}{\text{NVR}} \times 100 \qquad (10\text{--}10)$$

$$\text{TVR} = \frac{V_{\text{HS}}}{V_{\text{LS}}} \qquad (10\text{--}11)$$

where: TVR = test voltage ratio
NVR = nameplate voltage ratio
V_{HS} = test voltage applied to high side (V)
V_{LS} = voltage measured at low side (V).

A voltage-ratio test made on all newly installed or repaired power and distribution transformers provides baseline data for evaluating data obtained from future maintenance tests.

Two-Voltmeter Voltage-Ratio Test for Shorted Turns

The circuit diagram for this test is illustrated in Figure 10.22. Digital-type voltmeters should be used and positioned so that both instruments can be read at the same time. There must be no load connected to the transformer, and the applied test voltage should be at rated frequency and equal to or less than rated voltage. A convenient and relatively safe testing voltage would be somewhere between 8 and 24 volts.

FIGURE 10.22
Voltmeter test for determining the turn ratio of a transformer.

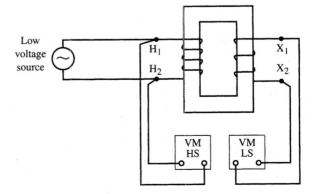

EXAMPLE 10.7   The nameplate voltage ratings for a certain 10-kVA, 60-Hz transformer are 2300 V and 230 V. Data obtained from a voltage-ratio test of the transformer, using a low-

voltage 60-Hz source, are 24.1 V and 2.21 V. Determine (a) nameplate voltage ratio (NVR); (b) test voltage ratio (TVR); (c) percent difference with respect to nameplate voltage ratio.

Solution

a. $\text{NVR} = \dfrac{2300}{230} = 10$

b. $\text{TVR} = \dfrac{24.1}{2.21} = 10.905$

The higher ratio indicates shorted turns in the secondary.

c. $\text{Percent difference} = \dfrac{\text{TVR} - \text{NVR}}{\text{NVR}} \times 100 = \dfrac{10.95 - 10}{10} \times 100 = 9.05\%$

This is a significant error; the voltage ratio should be within 0.5% of the nameplate ratio. The transformer should be repaired or replaced.

Turn-Ratio Test Set

A more accurate determination of the turn ratio may be made with specially designed self-contained turn-ratio test sets such as that illustrated in Figure 10.23. The test set consists of a reference transformer having an adjustable ratio of 0.1 to 129.9, a voltmeter, an ammeter, a null detector, and a self-contained hand-cranked AC generator.

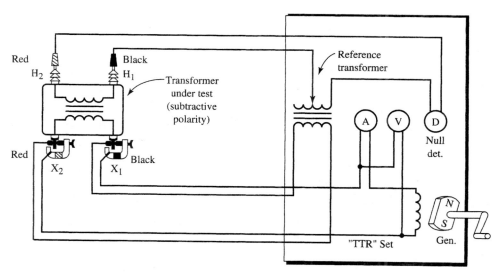

FIGURE 10.23
Turn-ratio test set (courtesy Biddle).

The hand-cranked 8-V AC generator provides for safety and portability. When connected properly, and used in accordance with the manufacturer's instructions, the test set will provide a direct readout of the turn ratio. Note that the turn-ratio test set must not be connected to an energized transformer; it will be damaged and the operator injured.

Excitation-Current Test

The magnitude of the excitation current will indicate whether or not the winding has shorted turns or perhaps shorted laminations. To check the excitation current, disconnect the secondary from the load and, using a clamp-on ammeter, measure the primary current at rated voltage and rated frequency. If the no-load current (exciting current) is higher than the manufacturer's specifications, the transformer is shorted. Note that the ammeter indication will read high whether the short is in the primary or the secondary windings.

Testing between Windings and between Windings and Ground

An insulation test of the transformer windings to ground, and between windings, will help to determine the relative conditions of the insulation.[7]

10.14 INTERPRETING TRANSFORMER NAMEPLATE DATA

Typical transformer nameplate data include voltage rating, kVA rating, frequency, number of phases, temperature rise, cooling class, percent impedance, and name of manufacturer. The nameplates of large power transformers also include basic impulse level (BIL), diagrams for three-phase operation, and tap-changing information.

Voltage Ratings

The high-side and low-side voltage ratings are *no-load* values. Full-load values depend on the power factor of the connected load and are therefore not given. Voltage ratings include a winding designator such as a long dash (—), slant (/), cross (X), or wye (Y), to indicate how the voltages are related to each other [2]. These NEMA standard markings indicate the following:

Long-dash (—): Indicates voltages are from *different* windings.

Slant (/): Indicates voltages are from the *same* winding (taps).

Cross (X): Indicates that voltages may be obtained by reconnecting a two-part winding in series or multiple (parallel). This type of winding is not suitable for three-wire operation.

Wye (Y): Indicates voltages in a wye-connected winding.

---

[7] For details on testing and evaluating electrical insulation, see Chapters 24 and 25.

The following examples indicate how the winding designators are used in single-phase and three-phase applications:

Single Phase

240/120: 240-V winding with a center tap.

240 × 120: Two-part winding that may be connected in series for 240 V or connected in parallel for 120 V.

240—120: A 240-V winding and a separate 120-V winding.

Three Phase

4160—480Y/277: A 4160-V delta-connected winding, and a separate 480-V wye-connected winding with an available neutral connection. *Note:* The voltage of the delta winding is always given first.

Miscellaneous Definitions

Frequency: Rated frequency of the transformer.

kVA: Rated apparent power of the transformer.

Percent impedance: The percent impedance of the transformer measured at the indicated temperature. Temperature affects resistance and, thus, affects impedance.

Percent impedance voltage: The voltage drop within the transformer when operating at rated current, as a percent of rated voltage.

Temperature rise: The maximum allowable temperature rise of the transformer based on an ambient temperature of 30°C.

Class: The insulating medium and the method of cooling.

BIL: The basic impulse level (BIL) of a transformer, or any other apparatus, is a measure of the *transient voltage stress* that the insulation can withstand without damage. The BIL rating of a transformer indicates that it was tested using an impulse voltage that rises to its peak value in 1.2 μs, and then decays to 50 percent peak voltage after a total of 50 μs has elapsed. The impulse test simulates a lightning surge induced in a transmission line, with the surge voltage modified by a lightning arrester [2].

SUMMARY OF EQUATIONS FOR PROBLEM SOLVING

$$a = \frac{N_{HS}}{N_{LS}} = \frac{E_{HS}}{E_{LS}} \qquad (10\text{–}3)$$

$$a = \frac{N_{HS}}{N_{LS}} \approx \frac{V_{HS}}{V_{LS}} \qquad (10\text{–}4)$$

$$P = V_{HS} I_{HS} F_P = V_{LS} I_{LS} F_P \qquad (10\text{--}7)$$

$$S = V_{HS} I_{HS} = V_{LS} I_{LS} \qquad (10\text{--}8a)$$

$$\frac{I_{LS}}{I_{HS}} = \frac{V_{HS}}{V_{LS}} = \frac{N_{HS}}{N_{LS}} \qquad (10\text{--}8b)$$

$$S_{\text{open-delta}} = 0.58 \times S_{\Delta\text{-}\Delta} \qquad (10\text{--}9)$$

$$\% \text{ difference} = \frac{|\text{TVR} - \text{NVR}|}{\text{NVR}} \times 100 \qquad (10\text{--}10)$$

$$\text{TVR} = \frac{V_{HS}}{V_{LS}} \qquad (10\text{--}11)$$

Wye	**Delta**	**Open-Delta**
$V_{\text{line}} = \sqrt{3} \times V_{\text{phase}}$	$V_{\text{line}} = V_{\text{phase}}$	$V_{\text{line}} = V_{\text{phase}}$
$I_{\text{line}} = I_{\text{phase}}$	$I_{\text{line}} = \sqrt{3} \times I_{\text{phase}}$	$I_{\text{line}} = I_{\text{phase}}$

GENERAL REFERENCES

Hubert, Charles I., *Electric Machines: Theory, Operation, Applications, Adjustment, and Control*, 2nd ed. Prentice Hall, Upper Saddle River, NJ, 2001.

Heathcote, Martin J., *JSP Transformer Book: A Practical Technology of the Power Transformer*, 12th ed. Oxford, Boston, 1998.

SPECIFIC REFERENCES KEYED TO THE TEXT

[1] American National Standards Institute, *General Requirements for Dry-Type Distribution, and Power Transformers, Including Those with Solid Cast and/or Resin Encapsulated Windings*, ANSI/IEEE C57.12.01–1998.

[2] National Electrical Manufacturers Association, *Dry-Type Transformers for General Applications*, Standards Publication No. ST 20–1992.

[3] American National Standards Institute, *Terminal Markings and Connections for Distribution and Power Transformers*, C57.12.70–1993.

[4] American National Standards Institute, *Recommended Practice for Establishing Transformer Capability When Supplying Non-Sinusoidal Load*, ANSI/IEEE C57.110–1998.

[5] Gruz, Thomas M., A Survey of Neutral Currents in Three-Phase Computer Power Systems, *IEEE Trans. Industry Applications*, Vol. 1A-26, No. 4, July/August 1990.

[6] Moravek, J., and Lethert, E., Field Study of Harmonic Loading in Modern Electrical Systems, *Electrical Design and Installation*, March 1991.

[7] Lemieux, G., Power System Harmonic Resonance—A Documented Case, *IEEE Trans. Industry Applications*, Vol. IA-26, No. 3, May/June 1990.

[8] Stratford, Ray P., Harmonic Pollution on Power Systems—A Change in Philosophy, *IEEE Trans. Industry Applications*, Vol. IA-16, No. 5, September/October 1980.

[9] American National Standards Institute, *General Requirements for Liquid Immersed Distribution, Power, and Regulating Transformers*, ANSI/IEEE C57. 12.00–1993.

REVIEW QUESTIONS

1. What is the function of a transformer? How is it constructed? Why is the core laminated?
2. Using suitable sketches, explain the principle of transformer operation. What is leakage flux?
3. Why are some transformers provided with tap changers? What precautions should be observed when operating a manually operated tap changer?
4. Explain why increasing the load on the secondary of a transformer causes the primary current to increase.
5. What is an autotransformer? State several applications of autotransformers.
6. Differentiate between subtractive and additive polarity as it applies to transformers.
7. What maintenance advantage does a delta–delta connection of single-phase transformers have over other three-phase arrangements? Explain.
8. Sketch the circuits for wye–wye, wye–delta, delta–wye, delta–delta, and open-delta connections of single-phase transformers.
9. What are harmonics, and how do they affect transformers?
10. What are triplen harmonics?
11. What is meant by the K-factor rating of a transformer?
12. State the procedure for determining the correct connections for three single-phase transformers with unmarked terminals that are to be connected in a wye–delta format.
13. State the procedure for determining the correct connections for three single-phase transformers with unmarked terminals that are to be connected in a delta–delta format.
14. Will a delta–delta bank of single-phase transformers continue to supply three-phase power when an open occurs in one transformer secondary? Explain.
15. Sketch a circuit showing the correct connections for current and potential transformers.
16. Describe a test that can be used to check the polarity of a current transformer and a potential transformer.
17. Why should the secondary terminals of a current transformer always be closed, either by a burden or a short circuit? What will happen if the secondary terminals of a potential transformer are short-circuited?
18. Describe a test that can be used to check the turn ratio of a transformer.
19. What should an electrical inspector look for when making an external check of transformers?

PROBLEMS (Neglect Losses and Voltage Drops within the Transformer.)

10–1/2. A 440-V supply is connected to the 880-turn primary winding of a transformer. (a) What is the voltage across the 220-turn secondary? (b) What are the volts per turn (V/t) of each winding?

10–2/2. A bell-ringing transformer has 400 turns of wire in the primary winding. If 120 volts applied to the primary results in 12 volts at the secondary, how many turns are in the secondary?

10–3/4. A 4160—460-V transformer operating at rated secondary voltage supplies 200 A at 460 V to a resistor load. Sketch the circuit and determine the primary current.

10–4/4. A 450—110-V transformer is operating at rated voltage. A clamp-on ammeter indicates a primary current of 25 A. Sketch the circuit and determine the secondary current.

10–5/4. A transformer with a 600-turn primary and a 30-turn secondary has a 600-V, 60-Hz input and a 10-Ω resistor load connected to the secondary. Sketch the circuit and determine (a) voltage at load; (b) current in load; (c) primary current.

10–6/4. A transformer with a 4:1 turn ratio has a high side input of 400 V at 50 Hz, and supplies a 60-Ω load. Sketch the circuit and determine (a) voltage at load; (b) load current; (c) input current.

10–7/4. A transformer with a 400-turn primary has a 1380-V, 60-Hz input. Its 4140-V output is connected to a 200-Ω load. Sketch the circuit and determine (a) load current; (b) input current; (c) number of turns in secondary.

10–8/5. An autotransformer with 80 and 65 percent taps is connected to a 230-volt, 60-hertz supply voltage. Sketch the circuit and determine the two tap voltages.

10–9/5. A 1200-V generator supplies the high-side voltage for an autotransformer. A 25-Ω load is connected to an 80% tap. Sketch the circuit and determine (a) secondary current; (b) primary current.

10–10/5. The primary of an autotransformer is supplied from a 600-V, 60-Hz source. Determine the percent tap required to supply 50 A to a 10-Ω load.

10–11/5. An autotransformer with a 20% tap supplies 80 A to a resistor load from a 450-V supply line. Sketch the circuit and determine (a) secondary voltage (tap voltage); (b) primary current; (c) power supplied to the transformer.

10–12/8. A wye–delta transformer bank composed of three single-phase transformers reduces a 2300-V three-phase supply to 450-V three-phase supply for use at a load center. Sketch the circuit and determine (a) turn ratio of the transformers; (b) line current on the high-voltage side if the line current to a balanced three-phase load is 300 A.

10–13/8. Three 4160—277-V single-phase transformers are connected delta–wye for step-down operation and supply a balanced 300-kVA load. Sketch the circuit and determine (a) line voltage at the load; (b) high-side line current; (c) high-side phase current; (d) low-side line current; (e) low-side phase current.

10–14/8. A delta–delta bank consisting of three single-phase transformers reduces a 460-volt, three-phase supply to 115 volts, three-phase supply for lighting circuits. The bank rating is 150 kVA. Sketch the circuit and determine (a) turn ratio of the transformers; (b) line current on the high-voltage side if the line current to the load is 650 A; (c) total apparent power input.

10–15/8. Referring to the delta–delta bank in Problem 10–14, assume one transformer (primary and secondary) is disconnected for maintenance, and the same three-phase load remains. Sketch the circuit and determine (a) rating of the open-delta bank; (b) the amount of load in kVA that must be removed in order to prevent overheating the transformer.

10–16/13. The nameplate voltage ratings for a certain 50-kVA, 60-Hz transformer are 2400 V and 480 V. Data obtained from a two-voltage turn-ratio test of the transformer, using a 16-V, 60-Hz source, are 16.2 V and 3.0 V. Determine (a) turn ratio from nameplate data; (b) turn ratio from test; (c) percent error with respect to the nameplate turn ratio.

10–17/13. The nameplate voltage ratings for a certain 100-kVA, 60-Hz transformer are 4160 V and 600 V. Data obtained from a two-voltage turn-ratio test of the transformer, using a 24-V, 60-Hz source, are 24.5 V and 3.54 V. Determine (a) turn ratio from nameplate data; (b) turn ratio from test; (c) percent error with respect to the nameplate turn ratio.

10–18/13. The nameplate voltage ratings for a certain 50-kVA, 60-Hz transformer are 13,800 V and 2400 V. Data obtained from a two-voltage turn-ratio test of the transformer, using a 120-V, 60-Hz source, are 120.8 V and 19.5 V. Determine (a) turn ratio from nameplate data; (b) turn ratio from test; (c) percent error with respect to the nameplate turn ratio.

11

Three-Phase
Induction Motors

11.0 INTRODUCTION

The induction motor, invented by Nikola Tesla in 1888, has become the "workhorse" of industry. Induction motors range in size from a fraction of a horsepower to more than 100,000 horsepower. They are more rugged, require less maintenance, and are less expensive than other motors of equal power and speed ratings. This chapter illustrates the general construction details of induction motors, explains how induction-motor torque is developed, and how speed and direction of rotation may be changed. Guidance in the application of motors for specific loads is provided.

11.1 INDUCTION-MOTOR CONSTRUCTION

A cutaway view of a three-phase squirrel-cage induction motor is shown in Figure 11.1a. The stationary member, called the *stator*, has overlapping coils arranged in phase groups that are connected in wye or delta.

The rotating member, called the *rotor*, is illustrated in Figure 11.1b. The rotor is manufactured with a slotted laminated steel core in which molten aluminum is cast to form a one-piece "cage" consisting of the rotor conductors, end rings, and fan blades. There is no insulation between the iron core and the aluminum conductors because none is required; the current induced in the rotor is contained within the circuit formed by the conductors and the end rings.

The purpose of the skewed slots of the rotor is to eliminate some of the slot noise produced when a slot of the rotor passes over a slot of the stator; a skewed rotor slot passes gradually rather than abruptly across the straight slot of the stator. In addition, skewed slots prevent "dead spots" by eliminating positions of minimum reluctance. If the slots of the stator were parallel to the slots of the rotor and the rotor stopped in a position in which the teeth of the rotor were lined up with the teeth of the stator, the

FIGURE 11.1
(a) Cutaway view of a three-phase induction motor (courtesy Siemens Energy and Automation Co.).

strong magnetic attraction caused by the low-reluctance path could prevent the rotor from starting.

There is no electrical connection to the rotating member. The energy to do work is transferred electromagnetically across the air gap between the stator and the rotor. Hence, the gap is made quite small so as to offer minimum reluctance to the flux.

So that all stator coils can be equally loaded, the air gap between rotor and stator must be uniform all around the circumference of the rotor. An unequal air gap could cause burned-out coils in the region of the smaller air gap; the region with a smaller air gap has lower reluctance; thus, more loading takes place in those coils.

11.2 INDUCTION-MOTOR ACTION[1]

The windings of the three phases of the stator are spaced and connected in a manner that causes the development of a rotating magnetic field when the stator is connected to a three-phase supply voltage. As illustrated in Figure 11.1a and Figure 11.2a, the

[1] For an in-depth discussion of induction motor theory, see Reference [1].

Aluminum
conductors

Aluminum
blades for
cooling

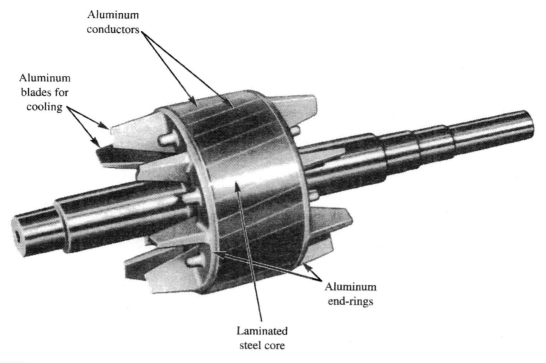

Aluminum
end-rings

Laminated
steel core

FIGURE 11.1 (cont'd)
(b) Squirrel-cage rotor (courtesy Allis Chalmers Mfg. Co.).

coils of the three phases are staggered in an overlapping arrangement so that the magnetic field contribution of each, for the development of a given pole, is nonconcentric.
Furthermore, because the current in each phase of the three-phase supply attains its
respective maximum value at a different instant in time, the centerline of magnetic
flux shifts from A to B to C, assuming this to be the phase sequence of the applied
voltage. The shifting magnetic field has the same effect on the squirrel-cage rotor as
that produced by a magnet sweeping around the rotor, as illustrated in Figure 11.2b.

The rotating magnetic field set up by the three-phase current in the stator passes
through the many windows formed by the aluminum conductors (bars) in the squirrel-
cage rotor. This behavior is simulated in Figure 11.2c, where the moving magnets represent the "rotating poles" set up in the stator and a representative window of the stationary squirrel-cage rotor is in the process of being swept by the clockwise rotation of
these "rotating poles" (stator flux).

At the instant shown, there are three magnetic lines directed downward through
the window, and one line directed upward, for a net flux of two lines in the downward
direction. As the flux rotation proceeds, the net downward flux through the window
is reduced to zero and then increases in the upward direction. For the instant shown,
the net flux is in the downward direction but is steadily decreasing. The decreasing
flux through the window induces a current in the associated rotor loop, that is, in a

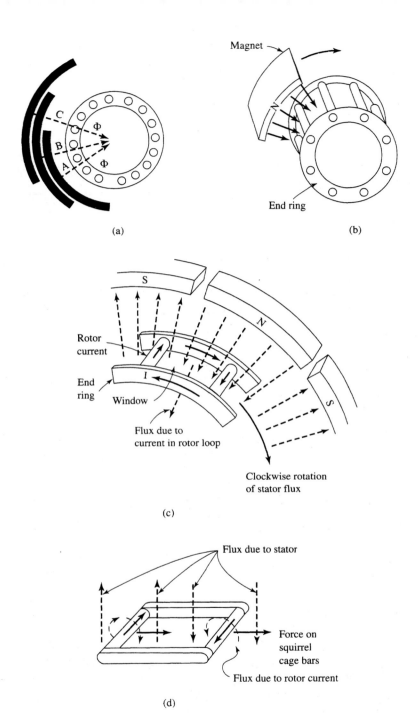

FIGURE 11.2

(a) Coils of three phases that make up one pole; (b) magnetic field sweeping around the rotor; (c) rotating stator flux sweeping rotor bars; (d) mechanical force produced by interaction of the magnetic field of the stator with the magnetic field produced by the induced rotor current.

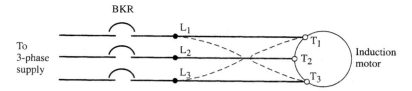

FIGURE 11.3
Reversing the direction of rotation of a three-phase induction motor.

direction to delay the change in the flux (Lenz's law). The result, shown in Figure 11.2c, is a clockwise (CW) induced current that produces an additional downward contribution of flux.

The interaction of the magnetic field caused by the induced current in the rotor bars, with the magnetic field of the stator, produces a mechanical force on the bars that is in the same direction as the rotating flux. The direction of the mechanical force may be verified by observing where flux bunching occurs at the rotor bars.[2] This is illustrated in Figure 11.2d for a representative rotor loop. Thus, the rotor will rotate in the same direction as the rotating field of the stator.

Reversing a Three-Phase Induction Motor

To reverse the direction of rotation of a three-phase induction motor, it is necessary to reverse the direction of rotation of the stator field. This is accomplished by reversing the phase sequence of the applied voltage. To do this interchange any two of the three motor leads where they connect to the three-phase source. This is shown by the broken lines in Figure 11.3; motor lead T_1 is switched from line L_1 to line L_3, and motor lead T_3 is switched from line L_3 to line L_1.

11.3 PREDETERMINING THE DIRECTION OF ROTATION OF A THREE-PHASE INDUCTION MOTOR

For many motor applications, rotating the shaft in the wrong direction can cause serious problems. For example, circulating pumps operating in reverse may cause failure of a heating system, forced-air fans for boilers may blow smoke and fire into an engine room, fire pumps will fail to work, and some driven equipment may be seriously damaged if rotated for even a few degrees in the wrong direction.

Fortunately, a phase-sequence indicator along with a motor-rotation indicator can be used to predetermine the direction of rotation of a three-phase motor without having to apply voltage to the machine. It is also a time-saver, since temporary hookups to determine the direction of rotation are not required. These instruments are shown in Figures 11.4a and 11.4b, respectively.

---

[2] Flux bunching is discussed in Section 3–5, Chapter 3.

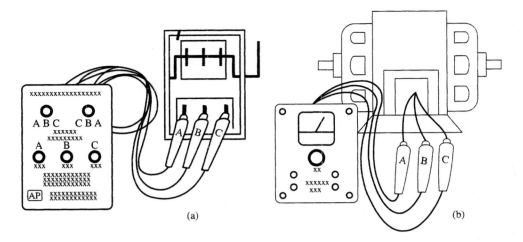

FIGURE 11.4
(a) Connecting the phase-sequence indicator; (b) connecting the motor-rotation indicator (courtesy Associated Research Inc.).

Using the Phase-Sequence Indicator

1. Open the breaker or disconnect switch on the motor control panel.
2. Connect the phase-sequence indicator to the control panel, and then close the breaker or disconnect switch, as shown in Figure 11.4a.
3. If the phase-sequence indicator is properly connected, the ABC or CBA light will indicate the phase sequence, and all three phase lights will glow to show that all three phases are present. If there is an open phase, or open connection to the phase-sequence indicator, only one of the three lights will glow and it will indicate which phase is open.
4. If no phase is open, and the ABC lamp lights, mark the output terminals of the control panel with the letters corresponding to the A, B, C markings on the instrument leads. If the BCA lamp lights, interchange any two of the instrument leads so that the instrument indicates sequence ABC. Mark the output terminals of the control panel with the letters corresponding to the A, B, C markings on the instrument leads.

Using the Motor-Rotation Indicator

1. Connect the motor-rotation indicator to the motor leads, as shown in Figure 11.4b. Hold the ON button down, and manually rotate the motor shaft about one-quarter turn in the desired direction of rotation. Note whether the first deflection is to ABC or to CBA. If ABC shows first, mark the motor leads to correspond with the instrument leads of the motor-rotation indicator. If CBA shows first, interchange leads A and C of the motor-rotation indicator. Make one final test, and if ABC shows first, mark the motor leads to correspond with the terminal markings of the instrument leads.

2. Connect the motor to the starter (A to A, B to B, C to C), and it will run in the desired direction.

11.4 FACTORS AFFECTING INDUCTION-MOTOR SPEED

The speed of rotation of an induction motor depends on the shaft load and on the speed of the rotating flux. The speed of the rotating stator flux (called *synchronous speed*) is determined by the frequency of the supply voltage and the number of poles in the stator winding.

How Poles Affect Speed

The number of poles in a stator winding can be determined by the span of a stator coil. If a coil spans a distance equal to one-fourth of the circumference, it is a four-pole winding; if it spans one-eighth of the circumference, it is an eight-pole winding; etc. This is illustrated in Figure 11.5 for four-pole and eight-pole windings. Note that the centerline of flux shifts 60° (A to C) for the four-pole winding and 30° (A to C) for the eight-pole winding. Because of the greater angle swept, in the same period of time (assuming that the frequency is the same), the four-pole winding has a higher synchronous speed than does the eight-pole winding. Since poles are always in pairs (every north has a corresponding south), the number of poles in a stator must be an even number (2, 4, 6, 8, etc.).

How Frequency Affects Speed

The effect of frequency on the speed of the rotating flux is explained by reference to Figures 11.5a and 11.5b. Increasing the frequency shortens the time for the flux to shift from the centerline of coil A to the centerline of coil C, resulting in a proportionately faster sweep around the rotor. The effect is a higher rotor speed.

FIGURE 11.5
Span of stator coils in a (a) four-pole winding; (b) eight-pole winding.

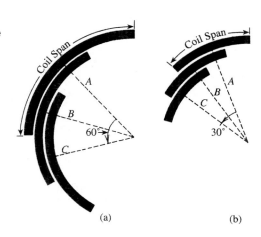

(a)   (b)

The mathematical relationship between speed, frequency, and the number of poles in an induction motor is

$$n_s = \frac{120\,f}{P} \tag{11-1}$$

where: n_s = synchronous speed (rpm)
f = frequency (Hz)
P = number of poles in stator winding.

The synchronous speeds for some commonly used pole arrangements are listed in Table 11.1 for 50-Hz and 60-Hz systems.

Note that the speed of the rotor, when operating at rated or below-rated load, is somewhat less than the synchronous speed. Thus, if the rotor speed and frequency are known, the synchronous speed and the number of stator poles can be determined from Table 11.1. For example, if the nameplate speed of a certain induction motor is 1750 rpm, and the frequency is 60 Hz, the synchronous speed is 1800 rpm, and the stator has four poles.

Slip

The difference between the speed of the rotating flux and the speed of the rotor is called the *slip speed*, and the ratio of slip speed to synchronous speed is called the *slip*. Expressed in equation form:

$$n = n_s - n_r \tag{11-2}$$

$$s = \frac{n_s - n_r}{n_s} \tag{11-3}$$

where: s = slip (decimal form)
n = slip speed (rpm)
n_r = rotor speed (rpm)
n_s = synchronous speed (rpm).

TABLE 11.1
Synchronous Speeds of Some Commonly Used Motors

Poles	50 Hz Synchronous Speed (rpm)	60 Hz Synchronous Speed (rpm)
2	3000	3600
4	1500	1800
6	1000	1200
8	750	900
10	600	720
12	500	600

If slip is given in percent, it must be divided by 100 before substituting into any equations. That is,

$$s = \frac{\% \text{ slip}}{100} \tag{11-4}$$

EXAMPLE 11.1

A certain two-pole, 10-hp, 450-V, 50-Hz, three-phase induction motor, operating at rated frequency and rated voltage has a speed of 2930 rpm. Determine (a) synchronous speed of the rotating flux; (b) slip speed; (c) slip.

Solution

a. $n_s = \dfrac{120f}{P} = \dfrac{120 \times 50}{2} = 3000$ rpm

b. $n = n_s - n_r = 3000 - 2930 = 70$ rpm

c. $s = \dfrac{n_s - n_r}{n_s} = \dfrac{3000 - 2930}{3000} = 0.023$ or 2.3%

11.5 INDUCTION-MOTOR BEHAVIOR DURING ACCELERATION AND LOADING

The current generated in the rotor is due to the relative motion between the rotating stator flux and the rotor. As the rotor accelerates, the relative motion between the rotating flux and the rotating rotor decreases, resulting in less rotor current and thus less torque.

If there is no load on the motor shaft, the rotor will accelerate to a speed just short of synchronous speed. The rotor cannot reach synchronous speed while operating as an induction motor. Induction-motor action can only occur when there is relative motion between the rotor and the rotating magnetic field of the stator. The no-load current, called the excitation or exciting current, ranges from a low of about 10 percent rated current, to a high of over 75 percent rated current. Motors with small air gaps draw less exciting current.

Figures 11.6a and 11.6b demonstrate the respective torque-speed characteristic and corresponding current-speed characteristic for a representative induction motor. Normal operation of induction motors occurs between no-load and rated load as shown in Figure 11.6. The flow of energy is electromagnetically from stator to rotor, and electromechanically from rotor to shaft. An increase in shaft load causes a decrease in rotor speed, which causes an increase in rotor current, and therefore a proportional increase in stator current; the increased current causes an increase in developed torque. The process continues until breakdown occurs.[3]

The four significant points in Figure 11.6a are breakdown torque, locked-rotor torque, rated torque, and pull-up torque, as discussed next.

---

[3] For a more detailed analysis of induction-motor behavior, see Reference [1].

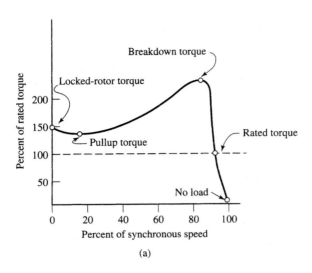

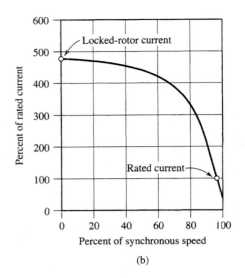

(a)   (b)

FIGURE 11.6
(a) Representative torque-speed characteristic; (b) current-speed characteristic of an
induction motor.

Breakdown Torque

The breakdown torque is the maximum torque that the machine can develop as a result
of increased shaft loads without causing an appreciable drop in speed.

Locked-Rotor Torque

The locked-rotor torque (starting torque) is the torque produced by the motor at the
instant that voltage is applied to the stator. The combined inertia of the rotor and
driven equipment prevents an instantaneous start when voltage is applied; the conditions are the same as if the rotor were locked in place. Hence, the expression "locked-
rotor" torque.

Rated Torque

The rated torque is the shaft torque when operating at rated horsepower at rated voltage and rated frequency.

Pull-Up Torque

The pull-up torque is the minimum torque developed by the motor during the period
of acceleration from rest to the speed at which breakdown torque occurs. If the pull-
up torque is less than the load torque on the shaft, the motor will not accelerate past
the pull-up point. The presence of significant dips in the torque-speed characteristic

of an induction motor may indicate a defective design, a damaged rotor, or improper repair of a damaged stator. Undesirable bumps and dips in the motor torque-speed characteristic during acceleration may cause the rotor to lock-in at some very low crawling speed.

11.6 INDUCTION-MOTOR IN-RUSH CURRENT

Induction motors of almost any horsepower may be started by connecting them across full voltage, and most are started that way. However, in many cases the high in-rush current associated with full-voltage starting can cause large voltage dips in the distribution system. This may cause lights to dim or flicker, unprotected control systems to drop out due to low voltage, unprotected computers to go off line or lose data, etc. Furthermore, the impact torque that occurs when starting at full voltage can, if high enough, damage gears and other components of the driven equipment. As shown in Figure 11.6b, the in-rush current is very high at locked rotor and remains very high during most of the acceleration period.

If the motor and driven equipment have low inertia, rapid acceleration quickly reduces the high starting current before overheating of the winding can occur. However, motor-load combinations with high inertia accelerate slowly, permitting the current to remain high for a longer period of time. The methods commonly used for reducing in-rush current are reduced-voltage starting using autotransformers, current limiting through wye–delta connections of stator windings, series impedance, and solid-state control.[4]

11.7 NEMA-DESIGN SQUIRREL-CAGE MOTORS AND THEIR APPLICATIONS

Squirrel-cage motors are standardized for specific applications by the National Electrical Manufacturers Association (NEMA) and are available in five basic designs: A, B, C, D, and E.

Representative torque-speed characteristics of NEMA-design induction motors are shown in Figure 11.7. The curves illustrate the relative differences in motor performance.[5] The design F motor (an obsolete standard) is included because many of these motors are still in use. The different characteristics are determined by the resistance and reactance of the rotor winding.

The *design B* motor has the broadest field of application and serves as the basis for comparison with other designs. It is used to drive centrifugal pumps, fans, blowers, and machine tools. It has a relatively high efficiency, even at light loads, and a relatively high power factor at full load.

---

[4] See Chapter 22.

[5] Values of locked-rotor torque, breakdown-torque, and pull-up torque for specific motors are given in References [1] and [2].

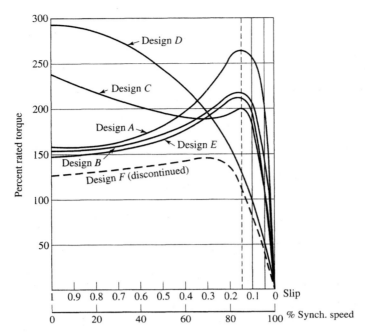

FIGURE 11.7
Comparison of torque-speed characteristics for representative NEMA-design induction motors.

The *design A* motor has essentially the same characteristics as the design B one, except for a somewhat higher breakdown torque. However, since its starting current is higher, its field of application is limited.

The *design C* motor has a higher locked-rotor torque, but a lower breakdown torque than design B. The higher starting torque makes it suitable for driving plunger pumps, vibrating screens, conveyers, and compressors without unloading devices. The starting current and slip at rated torque are essentially the same as for design B.

The *design D* motor has a very high locked-rotor torque and a high slip. Its principal field of application is in high-inertia loads such as flywheel-equipped punch presses, elevators, and hoists.

The *design E* motor is a high-efficiency motor that is used to drive centrifugal pumps, fans, blowers, and machine tools. However, except for isolated cases, the locked-rotor torque, breakdown torque, and pull-up torque of a design E motor are somewhat lower than that of design B motors for the same power and synchronous speed ratings. Furthermore, the locked-rotor current (starting current) of design E motors is significantly higher than that for design B motors for the same power and synchronous speed ratings.

The obsolete design F motor had much lower locked-rotor torque and much lower breakdown torque than design B motors. It was used when starting current limitations were severe, and both starting and maximum torque requirements were low; for example, fans, centrifugal pumps and compressors with unloading devices.

TABLE 11.2

Minimum Critical Torques in Percent of Rated Torque for 60-hp
Four-Pole Motors from NEMA Tables

NEMA Design	Locked Rotor	Breakdown	Pull-Up
B	140	200	100
E	120	180	90

The Upgrading Problem [3]

Before replacing a design B motor with a design E motor of the same horsepower and synchronous speed ratings, be sure to check the NEMA tables to see if the design E motor has sufficient torque to start and accelerate the load. Table 11.2 provides a comparison of the significant torque points for design B and design E 4–pole motors, both rated at 60 hp.

In Table 11.2, note that for the same horsepower and speed ratings (60 hp, 1800 rpm), the design E motor has lower minimum critical torques than the design B motor. This may cause problems. For the given load, the motor must be able to develop sufficient locked-rotor torque to start, sufficient pull-up torque to accelerate, and sufficient breakdown torque to handle any peak loads. It would also be wise to check with the manufacturer of the motor for their recommendations.

11.8 INTERPRETATION OF INDUCTION-MOTOR NAMEPLATE DATA

Nameplate data offer very pertinent information on the limits, operating range, and general characteristics of electrical apparatus. Interpretation of data and adherence to their specifications are vital to successful operating, servicing, and life of such equipment. Correspondence with the manufacturer should always be accompanied by the complete nameplate data of the apparatus. Figure 11.8 illustrates a typical nameplate for an induction motor.

The nameplate lists the rated operating conditions of the motor as guaranteed by the manufacturer. If the motor, represented by the nameplate in Figure 11.8, was supplied with a three-phase power source of exactly 460 volts, 60 Hz, and operated in an 40°C environment with a shaft load of exactly 150 hp, it would run exactly at the rated rpm, current, and efficiency specified by the manufacturer. In reality, motors seldom run exactly as specified by the manufacturer. Unless the motor is operated in a controlled environment (i.e., laboratory), it will be very unlikely that the rated horsepower, ambient temperature, and utilization voltages (voltage at the apparatus) will correspond exactly to the motor nameplate.

Although the system frequency most often matches the rated frequency of the motor, instances do occur, especially in isolated generator systems (offshore drilling rigs or ships), where the frequency is subject to change.

The nameplate acts as a guide to motor applications. Satisfactory performance is assured if the applied voltage is the approximately rated voltage, the frequency is the

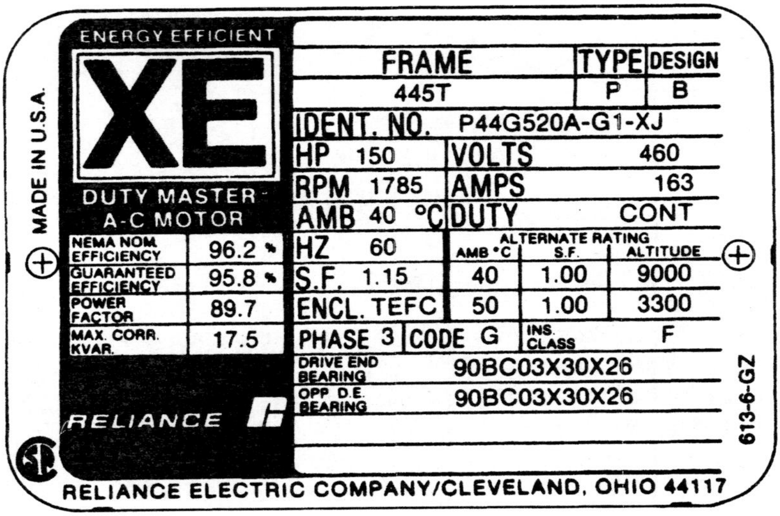

FIGURE 11.8
Induction-motor nameplate (courtesy Reliance Electric Co.).

approximately rated frequency, the shaft load does not exceed the service factor rating, and the temperature of the ambient is within the limits indicated on the nameplate. The horsepower rating for each speed of a multispeed motor is based on the type of industrial application: constant horsepower, constant torque, or variable torque. The relationship between shaft torque, shaft horsepower, and shaft speed is given here:

$$HP = \frac{Tn_r}{5252} \tag{11–5}$$

where: HP = shaft power output (hp)
  T = shaft torque (lb-ft)
  n_r = shaft speed (rpm).

Nominal Efficiency

The nominal efficiency indicated on the nameplate is the average efficiency of a large number of motors of the same design. The manufacturer guarantees that if the motor is operating at rated nameplate conditions, a certain minimum efficiency is to be expected.

Design Letter

The design letter indicates the NEMA-design characteristics of the machine and thereby directs the reader to tables of minimum values of locked-rotor torque, breakdown torque, and pull-up torque that may be expected from the machine. These tables are available in References [1] and [2].

Service Factor

The service factor (S.F.) of a motor is a multiplier that, when multiplied by the rated power, indicates the permissible loading, provided that the voltage and frequency are maintained at the value specified on the nameplate. However, it should be noted that if induction motors are operated at a service factor greater than 1.0, the efficiency, power factor, and speed will be different from those at rated load.

Insulation Class

The letter designating insulation class specifies the maximum allowable temperature rise above the temperature of the cooling medium for motor windings, based on a maximum ambient temperature of 40°C.[6] All winding temperatures are to be determined by winding resistance measurement. The motor whose nameplate is shown in Figure 11.8 has Class F insulation, a service factor of 1.15, and is rated for continuous duty.

---

[6]A list of insulation classes and maximum allowable winding-temperature rises for different classes of insulation systems is given in Chapter 23.

EXAMPLE 11.2

For the motor whose nameplate is given in Figure 11.8, determine (a) rated torque; (b) synchronous speed.

Solution

a. $\text{HP} = \dfrac{Tn_r}{5252} \Rightarrow T = \dfrac{5252 \times \text{HP}}{n_r} = \dfrac{5252 \times 150}{1785} = 441.3 \text{ lb–ft}$

b. The rated speed and rated frequency stamped on the motor nameplate are 1785 rpm and 60 Hz, respectively. From Table 11.1, Section 11–4, the synchronous speed for this motor is 1800 rpm, and the stator has four poles.

Code Letter

The code letter provides a means for determining the expected locked-rotor in-rush current to the stator when starting the motor with rated voltage and rated frequency applied directly to the stator terminals. The code letter directs the reader to a table of locked-rotor kVA per horsepower, from which the in-rush current can be calculated (see Table 11.3 and *read the footnote*).

EXAMPLE 11.3

Determine the expected range of locked-rotor current for the motor whose nameplate is shown in Figure 11.8.

Solution

The code letter on the nameplate is G. From Table 11.3, the range of locked-rotor kVA/hp is 5.6–6.299.

Minimum expected starting kVA = 5.6 kVA/hp × 150 hp = 840 kVA

TABLE 11.3

NEMA Code Letters for Locked Rotor kVA per Horsepower

Code Letter	kVA/hp[a]	Code Letter	kVA/hp[a]
A	0.0–3.15	K	8.0– 9.0
B	3.15–3.55	L	9.0–10.0
C	3.55–4.0	M	10.0–11.2
D	4.0–4.5	N	11.2–12.5
E	4.5–5.0	P	12.5–14.0
F	5.0–5.6	R	14.0–16.0
G	5.6–6.3	S	16.0–18.0
H	6.3–7.1	T	18.0–20.0
J	7.1–8.0	U	20.0–22.4
		V	22.4 and up

Reprinted with permission from NFPA 70, National Electrical Code, Copyright © 1999 National Fire Protection Association, Quincy, MA 02269. This reprinted material is not the complete and official position of the NFPA on the referenced subject, which is represented only by the standard in its entirety. National Electrical Code® and NEC® are trademarks of The National Fire Protection Association, Inc., Quincy, MA.
[a]Locked kVA per horsepower range includes the lower figure up to, but not including, the higher figure. For example, 3.14 is designated by letter A, and 3.15 by letter B.

$$S = \sqrt{3}V_{line}I_{line} \Rightarrow I_{line} = \frac{S}{\sqrt{3}V_{line}} = \frac{840,000}{\sqrt{3} \times 460} = 1054 \text{ A}$$

Maximum expected starting kVA = 6.299 kVA/hp × 150 hp = 945 kVA

$$S = \sqrt{3}V_{line}I_{line} \Rightarrow I_{line} = \frac{S}{\sqrt{3}V_{line}} = \frac{945,000}{\sqrt{3} \times 460} = 1186 \text{ A}$$

Operators of electrical machinery should make periodic checks on the load that each machine is carrying. Exceeding nameplate values shortens the useful life of a machine. Clamp-on multimeters, discussed in Chapter 9, are very convenient instruments for making periodic checks of motor current, applied voltage, power, power factor, and frequency.

Motor temperature readings are also very helpful in detecting impending trouble. If the temperature rise of the machine increases for the same load condition, it may mean that the air passages and ventilating screens are clogged with dirt, the flow of water to the coolers is cut off, the strip heaters have been left on, etc. If embedded temperature detectors are not installed, thermometers may be secured to the stationary parts of the windings on open machines, or they may be fastened to the hottest part of the motor frame, if the machine is totally enclosed.

11.9 MULTISPEED INDUCTION MOTORS

The speed of a squirrel-cage induction motor operating from a fixed frequency system can be changed only by changing the number of poles in the stator. Such machines are called *multispeed motors* [1,2,4].

There are three types of special-purpose multispeed squirrel-cage induction motors, each type designed for a specific type of application: constant-torque motors, variable-torque motors, and constant-horsepower motors. The horsepower listing of multispeed motors always applies to the highest speed. Figure 11.9 shows the NEMA standard terminal markings and table of connections for single-winding multispeed induction motors [2]. Terminals L_1, L_2, and L_3 are the line terminals. *Note that terminals that are not used must be individually insulated to prevent short-circuits or grounds.*

If a two-speed motor with a speed ratio of 2:1 is desired from a single stator winding, each of the three phases (A, B, and C) is provided with a center tap, as shown by the dark dot in Figure 11.9. The lower speed is obtained by reconnecting the stator winding so that all poles have the same polarity (all north or all south). As a consequence of this connection, opposite polarity poles (called consequent poles) form between the poles established by the stator winding, thus doubling the number of poles.

Constant-Torque Motor

A constant-torque multispeed motor, whose circuit and connections are shown in Figure 11.9a, is designed to be capable of delivering approximately the same torque with every speed connection. Hence, the rated horsepower for the different speed connections will vary directly with the synchronous speed. For example, the horsepower rating for an 1800-rpm connection would be double the horsepower rating at the 900-rpm connection. Multispeed motors of this type are used for constant-torque loads such as conveyers, compressors, reciprocating pumps, printing presses, and similar loads. *However, note that, although the motor can develop the same torque at different speed connections, it will not deliver the same torque unless the load demands it.*

Variable-Torque Motor

A variable-torque multispeed motor, whose circuit and connections are shown in Figure 11.9b, is designed to be capable of developing torque in direct proportion to the synchronous speed connection. Such motors could develop double the torque at the double-speed connection, half the torque at the half-speed connection, etc. Multispeed motors of this type are used for fans, centrifugal pumps, or other loads with similar characteristics. The power requirement of a variable-torque motor is directly proportional to the square of the speed. Hence, higher speed connections can draw significantly more power from the system. *However, note that, although the motor can develop torque in direct proportion to the different speed connections, it will not do so unless the load demands it.*

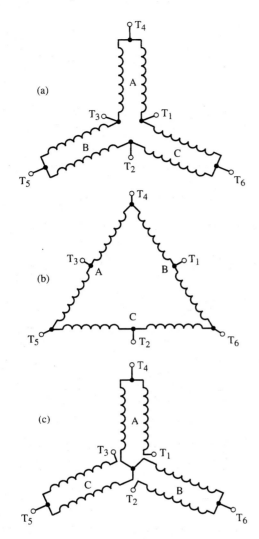

Two-speed single-winding (constant torque)

Speed	L₁	L₂	L₃	Tie together
Low	T₁	T₂	T₃	-------------
High	T₆	T₄	T₅	T₁ T₂ T₃

Two-speed single-winding (variable torque)

Speed	L₁	L₂	L₃	Tie together
Low	T₁	T₂	T₃	-------------
High	T₆	T₄	T₅	T₁ T₂ T₃

Two-speed single-winding (constant horsepower)

Speed	L₁	L₂	L₃	Tie together
Low	T₁	T₂	T₃	T₆ T₄ T₅
High	T₆	T₄	T₅	-------------

FIGURE 11.9
Winding connections and standard terminal markings for single-winding multispeed motors.

Constant-Horsepower Motors

A constant-horsepower, multispeed motor, whose circuit and connections are shown in Figure 11.9c, is designed to be capable of delivering approximately the same rated horsepower with every speed connection. Thus, a constant-horsepower motor can develop proportionately higher torque at the lower speed connections, and proportionally lower torque at the higher speed connections. For example, the torque that the motor can develop at a 900-rpm connection would be twice that which can be developed at an 1800-rpm connection. Multispeed motors of this type are used for lathes and other machine tools that often require a constant rate of doing work. *However, note that, although the motor can develop the same horsepower at all speed connections, it will not deliver the same horsepower unless the load demands it.*

Speed Ratios Higher than 2 to 1

For speed ratios higher than 2:1, the stator must have two or more independent windings, each with a different number of poles, and only one winding energized at a time. The terminals for the lowest speed winding are marked with single-digit numbers (T_1, T_2, T_3, T_4, etc.); the terminal marks for the second-speed winding have 10 added to them (T_{11}, T_{12}, T_{13}, T_{14}, etc.); the terminals for the third-speed winding have 20 added to them (T_{21}, T_{22}, T_{23}, T_{24}, etc.). Examples of three-speed and four-speed squirrel-cage induction motors are shown in Figure 11.10. Terminals T_3, T_7 and T_{13}, T_{17} of one of the windings in Figure 11.10c is open when the other winding is energized to prevent circulating current by transformer action. The same is true for terminals T_5, T_7, and T_{15}, T_{17} in Figure 11.10d. *Note that terminals that are not used must be individually insulated to prevent short circuits or grounds.*

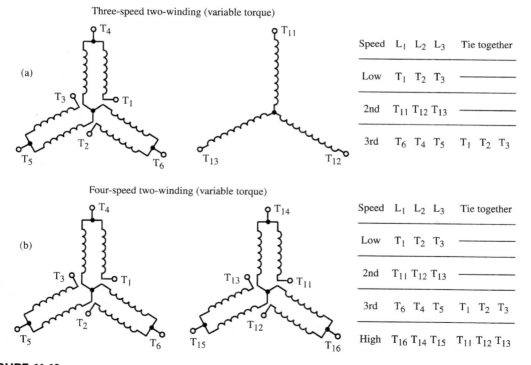

Three-speed two-winding (variable torque)

(a)

Speed	L_1	L_2	L_3	Tie together
Low	T_1	T_2	T_3	————
2nd	T_{11}	T_{12}	T_{13}	————
3rd	T_6	T_4	T_5	T_1 T_2 T_3

Four-speed two-winding (variable torque)

(b)

Speed	L_1	L_2	L_3	Tie together
Low	T_1	T_2	T_3	————
2nd	T_{11}	T_{12}	T_{13}	————
3rd	T_6	T_4	T_5	T_1 T_2 T_3
High	T_{16}	T_{14}	T_{15}	T_{11} T_{12} T_{13}

FIGURE 11.10
Connection diagrams for representative two-winding multispeed induction motors.

Four-speed two-winding (constant torque)

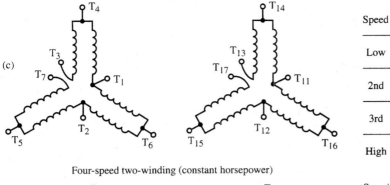

Speed	L_1	L_2	L_3	Tie together
Low	T_1	T_2	T_3	T_3 T_7
2nd	T_{11}	T_{12}	T_{13}	T_{13} T_{17}
3rd	T_6	T_4	T_5	T_1 T_2 T_3 T_7
High	T_{16}	T_{14}	T_{15}	T_{11} T_{12} T_{13} T_{17}

Four-speed two-winding (constant horsepower)

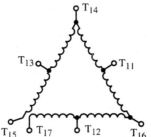

Speed	L_1	L_2	L_3	Tie together
Low	T_1	T_2	T_3	T_4 T_5 T_6 T_7
2nd	T_{11}	T_{12}	T_{13}	T_{14} T_{15} T_{16} T_{17}
3rd	T_6	T_4	T_5	T_5 T_7
High	T_{16}	T_{14}	T_{15}	T_{15} T_{17}

FIGURE 11.10 (cont'd)

11.10 WOUND-ROTOR INDUCTION MOTOR

A wound-rotor induction motor, shown in Figure 11.11a, is an adjustable-speed machine that uses a wound rotor in place of a squirrel cage.

The stator construction is the same as that for the squirrel-cage motor. The rotor, shown in Figure 11.11b, uses insulated coils that are set in slots. The overlapping coils are connected in series or parallel arrangements to form phase groups, and the phase groups are connected in a wye or delta. The rotor circuit is completed through a set of slip rings, carbon brushes, and an external wye-connected rheostat, as shown in Figure 11.11a.

The principle of wound-rotor motor operation is the same as for a squirrel-cage motor. The stator develops a rotating magnetic field that sweeps past the rotor, and the interaction of the magnetic field of the stator with that produced by the current in the rotor results in the development of motor torque.

An equivalent circuit for a wound-rotor motor is shown in Figure 11.12. The three-phase rheostat is composed of three rheostats connected in a wye, and a common lever is used to adjust all three arms simultaneously. The rheostat serves to change the torque-speed characteristic of the machine. When starting, all rheostat resistance should be inserted in the rotor circuit. This provides a relatively high starting torque and low stator current.

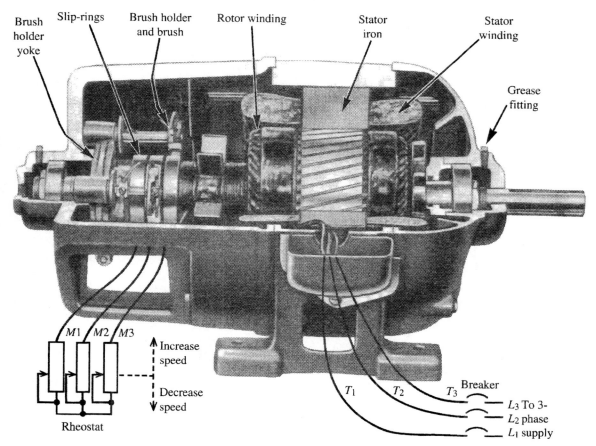

FIGURE 11.11

(a) Wound-rotor induction motor; (b) wound rotor (courtesy Louis Allis Co.).

Reducing the rheostat resistance, assuming the same shaft load, causes an increase in motor speed. Speed reduction through rheostat control can be obtained only if the machine is loaded. The speed of the machine can also be changed by varying the frequency or by changing the number of stator poles. If the motor is to be used without a rheostat, the three rotor terminals M_1, M_2, and M_3 in Figures 11.11 and 11.12 should be connected together; this can be done at the motor connection box.

To reverse the motor, interchange any two of the stator connections; interchanging the rotor connections does not reverse rotation, nor does it have any effect on the motor performance.

Typical speed-torque curves for different rheostat settings are shown in Figure 11.13. Curve R_1 is with all of the rheostat resistance shorted out, and curve R_9 is with all of the rheostat resistance in. The speed for each rheostat setting for a given load can be found by following the corresponding torque line across the graph. Furthermore, as seen from the curves, an increase in the resistance in the rotor circuit does not cause an

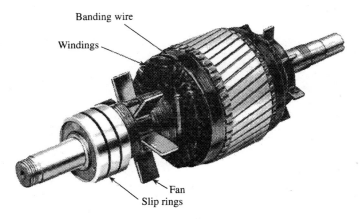

Banding wire

Windings

Fan

Slip rings

FIGURE 11.11 (cont'd)

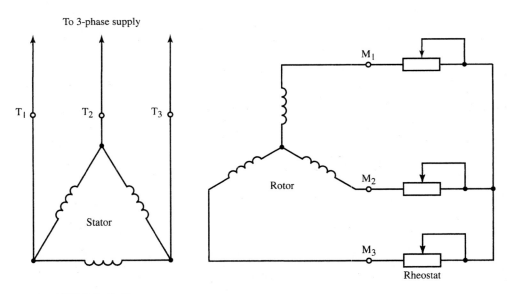

To 3-phase supply

T_1 T_2 T_3

Stator

Rotor

M_1

M_2

M_3

Rheostat

FIGURE 11.12
Rotor and stator circuits for a wound-rotor induction motor.

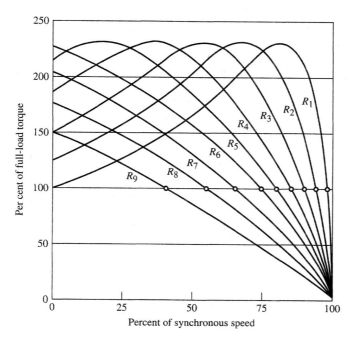

FIGURE 11.13
Torque-speed curves for wound-rotor induction motors with different rheostat settings.

increase in the maximum torque (breakdown torque) that the machine can develop; it merely shifts the position of maximum torque further to the left.

Wound-Rotor Motor Applications

The speed-torque characteristics of wound-rotor motors make them adaptable to loads requiring constant-torque variable-speed drives, high starting torques, and relatively low starting currents. Blowers, compressors, stokers, and hoists are some of its applications. However, prolonged operation at speeds below 50 percent of synchronous speed should be avoided, unless forced ventilation is provided; if cooling is not provided, the machine will overheat when operated at slow speed.

11.11 INDUCTION MOTOR AS AN INDUCTION GENERATOR

Induction generators have the same basic construction as squirrel-cage induction motors. In fact, all induction motors can be operated very effectively as induction generators by driving them at a speed greater than synchronous speed. Induction generators are suitable for operation by wind turbines, hydraulic turbines, steam turbines, and gas engines powered by natural gas or biogas. They can range in size from a few kilowatts to 10 MW or higher, and are used extensively in cogeneration operations.

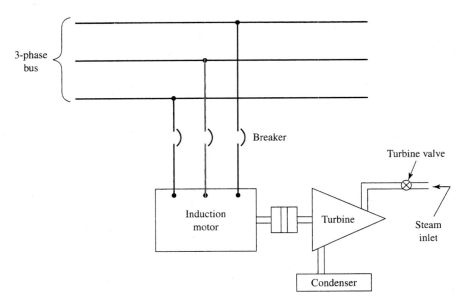

FIGURE 11.14
Induction motor coupled to a turbine for induction generator operation.

Cogeneration is the sequential production of two forms of energy, usually steam for process operations and electricity for plant use and for sale to utilities.

Figure 11.14 shows an induction motor connected to a three-phase system, with its shaft mechanically coupled to a steam turbine. Assume the turbine valve is closed so that no steam enters the turbine, and the induction motor is started from the electrical systems and accelerates the turbine to somewhat less than the synchronous speed of the rotating stator flux.

Gradually opening the turbine valve causes a gradual buildup of turbine torque, adding to that developed by the induction motor, resulting in an increase in rotor speed. When the speed of the turbine-motor set reaches the synchronous speed of the stator, the slip becomes zero, and no motor torque is developed. At zero slip, the induction machine is neither a motor nor a generator; it is "floating" on the bus. Increasing the speed of the turbine causes the induction machine to act as a generator, feeding electric power into the system.[7]

---

[7] For additional information on the operation of induction generators, see Reference [1].

SUMMARY OF EQUATIONS FOR PROBLEM SOLVING

$$n_s = \frac{120f}{P} \qquad (11\text{--}1)$$

$$n = n_s - n_r \qquad (11\text{--}2)$$

$$s = \frac{n_s - n_r}{n_s} \qquad\qquad (11\text{--}3)$$

$$s = \frac{\% \text{ slip}}{100} \qquad\qquad (11\text{--}4)$$

$$\text{HP} = \frac{T n_r}{5252} \qquad\qquad (11\text{--}5)$$

SPECIFIC REFERENCES KEYED TO THE TEXT

[1] Hubert, Charles I., *Electric Machines: Theory, Operation, Applications, Adjustment, and Control*, 2nd ed. Prentice Hall, Upper Saddle River, NJ, 2001.
[2] National Electrical Manufacturers Association, *Motors and Generators*, NEMA Standards Publication No. MG 1–1998.
[3] DeDad, John, Design-E Motor: You May Have Problems, *Electrical Construction and Maintenance*, September 1999, pp. 36, 38.
[4] Heredos, F. P., Selection and Application of Multi-Speed Motors, *IEEE Trans. Industry Applications*, Vol. IA–23, No. 2, March/April 1987.

REVIEW QUESTIONS

1. Explain how a three-phase stator produces a rotating magnetic field.
2. Explain how current is generated in the rotor of a squirrel-cage rotor.
3. Explain why the squirrel-cage rotor revolves in the same direction as the rotating magnetic field of the stator.
4. Explain why interchanging any two of the three line leads to a three-phase squirrel-cage motor causes it to run in the opposite direction.
5. Explain why a two-pole machine runs at a higher speed than a four-pole machine when both are connected to the same three-phase service.
6. What two methods are used to change the speed of a three-phase squirrel-cage induction motor?
7. What is the purpose of a consequent-pole connection?
8. What factors determine whether or not a squirrel-cage induction motor can be started at full line voltage?
9. Define breakdown torque, locked-rotor torque, and pull-up torque as they pertain to squirrel-cage induction motors. Which NEMA-design motor develops the greatest locked-rotor torque?
10. Sketch a generalized torque-speed characteristic of a squirrel-cage motor. Mark and label the four significant points.
11. State the difference in construction details between the squirrel-cage rotor and the wound rotor as used in induction motors.
12. What are the advantages of the wound-rotor motor over the squirrel-cage motor? State an application for a wound-rotor motor.
13. State three methods that can be used for adjusting the speed of a wound-rotor motor.

14. State the correct procedure for (a) starting a wound-rotor induction motor; (b) reversing a wound-rotor motor.

15. On a single sheet of graph paper, sketch and label the torque-speed characteristics of design A, B, C, D, and E motors. State an application for each.

16. What are the three basic types of multispeed motors? State an application for each.

17. What information is provided by the code letter on an induction-motor nameplate?

PROBLEMS

11–1/4. Determine the synchronous speed of a 10-hp, 450-volt, four-pole induction motor when operating from a 60-hertz line.

11–2/4. A 25-hp induction motor has a synchronous speed of 3600 rpm, and a slip of 2.8% when operating at rated load from a 2300-volt, 60-hertz supply. Determine (a) number of poles in its stator winding; (b) slip speed; (c) rotor speed.

11–3/4. A 30-hp, four-pole, 60-Hz, 460-V, three-phase induction motor operating at rated conditions has a speed of 1725 rpm. Determine (a) speed of rotating flux; (b) slip speed; (c) slip.

11–4/8. A 50-hp, 208-V, 60-Hz, three-phase induction motor operating at rated conditions has a speed of 1780 rpm. Determine the rated torque.

11–5/8. A 150-hp, 460-V, 60-Hz, three-phase induction motor runs at 3560 rpm when operating at rated load. Determine the rated torque.

11–6/8. A three-phase induction motor operating at 580 rpm delivers 145 lb-ft of shaft torque to a pump. Determine the shaft horsepower.

11–7/8. A 10-hp fuel-oil service pump is driven by a three-phase, 1719 rpm, 440-V, 60-Hz induction motor, with code letter G. Determine the expected range of locked-rotor current.

11–8/8. A 100-hp, 440-V, 60-Hz, three-phase induction motor is used to drive a circulating pump. The motor runs at 710 rpm, and has code letter G. Determine the expected range of locked-rotor current.

11–9/8. The main condensate pump on a ship is driven by a 440-V, 450-hp, three-phase induction motor. The motor runs at 1755 rpm and has code letter E. Determine the expected range of locked-rotor current.

11–10/9. A 40-hp, three-phase, 460-V, 60-Hz, two-speed motor has synchronous speeds of 3600 rpm and 900 rpm. (a) Determine the number of poles that correspond to each speed. (b) Could there be a consequent-pole connection?

11–11/9. A 15-hp, three-phase, 460-V, 60-Hz, two-speed motor has rated speeds of 1775 rpm and 880 rpm. (a) Determine the number of poles that correspond to each speed. (b) Could there be a consequent-pole connection?

11–12/9. A 7.5-hp, three-phase, 440-V, 60-Hz, two-speed motor has rated speeds of 1170 rpm and 585 rpm. (a) Determine the number of poles that correspond to each speed. (b) Could there be a consequent-pole connection?

12

Synchronous Motors

12.0 INTRODUCTION

Synchronous motors are useful in applications in which constant speed is essential or in which the power factor of the system must be kept at a high level. Furthermore, the efficiencies of large low-speed (500 rpm and below) synchronous motors are higher than that of induction motors.

Large machines, which are in continuous service for months at a time, will operate more economically when driven by synchronous motors. These types of motors are typically used to drive compressors and centrifugal pumps. The motor's high efficiency coupled with power factor improvement makes it particularly adaptable for that type of service.

When synchronous motors are to drive reciprocating compressors, flywheels are used in conjunction with a damper winding to smooth the large current fluctuations caused by the torque pulsations of the compressor. One very interesting application is the 44-MW (59,000-hp), 10-kV, 60-Hz, 50-pole, 144-rpm synchronous motors used to drive the *Queen Elizabeth-II* passenger ship. A solid-state volts/hertz drive-circuit provides speed control through frequency adjustment.

This chapter introduces the principles of synchronous motor behavior and presents the correct way to start, stop, and reverse synchronous motors.

12.1 SYNCHRONOUS MOTOR CONSTRUCTION

The stator of a large three-phase synchronous motor used for ship propulsion is shown in Figure 12.1a. It is similar in construction to the stator of a three-phase induction motor. When energized from a three-phase supply, it develops a rotating field in the same manner as described for induction motors.

241

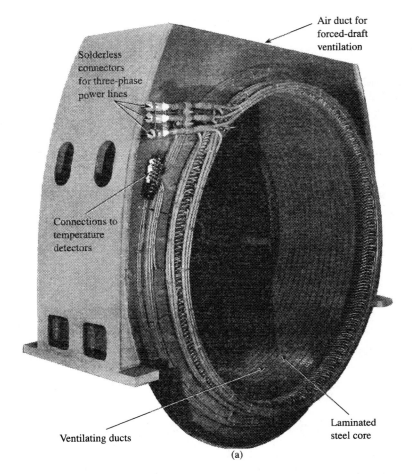

Air duct for
forced-draft
ventilation

Solderless
connectors
for three-phase
power lines

Connections to
temperature
detectors

Laminated
steel core

Ventilating ducts

(a)

FIGURE 12.1
(a) Synchronous motor stator; (b) corresponding rotor (courtesy TECO Westinghouse).

The corresponding rotor, shown in Figure 12.1b, consists of a set of electromagnets projecting radially outward from a steel spider. The magnets (called field poles) are bolted or keyed to the spider, and the spider is keyed to the shaft. The magnet coils (also called field coils or field windings) are connected in series or series-parallel and in a manner that provides alternate north and south poles. The number of rotor poles is equal to the number of stator poles.

The rotor circuit terminates at slip rings (also called collector rings). Graphite brushes press against the slip rings to provide the connection between the field windings and a DC source. A squirrel-cage winding, consisting of copper bars embedded in the pole faces, is used to accelerate the rotor to near synchronous speed. The squirrel-cage winding is also called the amortisseur winding, damper winding, or starting winding.

FIGURE 12.1 (cont'd)

A salient-pole rotor with a shaft-mounted DC generator, called an exciter, is shown in Figure 12.2. Direct current from the exciter armature is supplied to the field windings by means of carbon brushes (not shown) riding on the commutator that connects to carbon brushes riding on the slip rings. The field-pole structure for the exciter armature in Figure 12.2 is not shown.

12.2 STARTING, STOPPING, AND REVERSING A SYNCHRONOUS MOTOR

An elementary circuit diagram that illustrates the minimum rotor and stator connections required for starting and running a synchronous motor is shown in Figure 12.3. To start the motor, first make sure that the DC circuit breaker that supplies the magnets is open, then close the three-phase breaker. This energizes the three-phase stator, causing a rotating magnetic field that sweeps the squirrel-cage bars of the rotor.

The resultant induction-motor torque accelerates the rotor to a speed somewhat less than the synchronous speed of the rotating flux. At this speed the slip is very small and the rotating flux of the stator moves very slowly relative to the revolving rotor. Direct current supplied by the exciter is then applied to the field windings (magnets), forming alternate north and south poles that "lock" in rotational synchronism with the

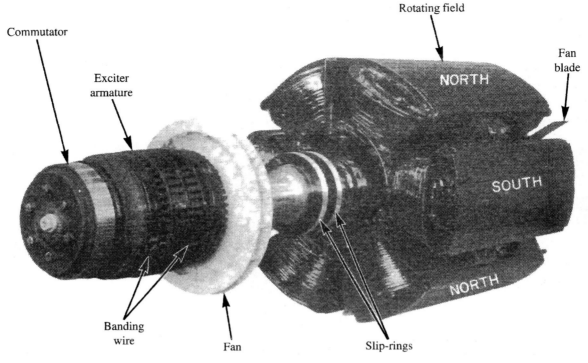

FIGURE 12.2
Salient-pole rotor with a shaft-mounted exciter (courtesy GE Industrial Systems).

corresponding opposite poles of the rotating flux; the slip is zero, all induction-motor action ceases, and the magnets are "dragged around" at the synchronous speed of the stator flux. Expressed mathematically,

$$n_r = n_s = \frac{120 f_s}{P} \tag{12–1}$$

where: n_r = rotor speed (rpm)
n_s = synchronous speed (rpm)
f_s = frequency of applied voltage (Hz)
P = number of poles (fixed for a synchronous motor).

Field-Discharge Resistor

When starting as an induction motor, the magnetic field of the stator sweeps the magnet coils (as well as the squirrel-cage winding) and induces a very high voltage in the magnet coils. This induced voltage, which may be as high as 35,000 volts, could cause flashover across the slip rings and cause failure of the magnet insulation. The varistor,

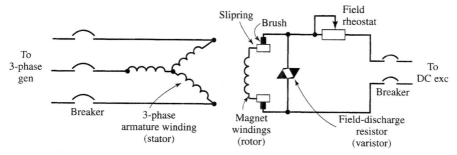

FIGURE 12.3
Minimum rotor and stator connections required for starting and running a synchronous motor.

shown in Figure 12.3, effectively short-circuits the slip rings when the voltage is high, reducing the voltage to a safe value. The varistor also serves as a field-discharge resistor when the field circuit is opened.

Precautions When Starting a Synchronous Motor

If the rotor magnets are energized before the machine reaches its maximum speed as an induction motor, the rotor may not synchronize and severe vibration will occur; every time a pole of the rotating flux passes a rotor pole, the alternate attraction and repulsion will occur. Such out-of-step operation, called *pole slipping,* causes cyclic current surges and torque pulses at slip frequency in the stator windings. Similarly, if the resistance of the field rheostat is set too high for the particular load on the motor shaft, the magnet poles will not be strong enough to hold the rotor in synchronism, and pole slipping will occur.

The squirrel-cage winding is designed for starting duty. If operated as an induction motor for more than a short period of time, severe overheating may damage the squirrel-cage winding. Furthermore, the high stator current associated with operation as an induction motor, if sustained, will cause overheating of the stator windings.

Stopping a Synchronous Motor

To stop a synchronous motor when on manual control, as shown in Figure 12.3, first de-energize the magnets by opening the DC breaker, and then open the three-phase breaker. The field-discharge resistor, shown in Figure 12.3, dissipates the energy stored in the magnetic field when the DC breaker is opened. The energy stored in the magnetic field of the magnets is converted to heat energy in the resistor and in the resistance of the field coils, and dissipated to the atmosphere. A broken (open) discharge resistor results in arcing and burning at the breaker tips when the breaker is opened, and the high voltage induced in the magnet windings may cause serious damage to its insulation.

Reversing a Synchronous Motor

The direction of rotation of a three-phase synchronous motor is reversed by first stopping it, interchanging any two of the three AC line leads at the stator, and then restarting. Interchanging two of the AC line leads reverses the phase rotation, causing the stator flux to rotate in the opposite direction.[1] Reversing the direction of current in the field windings will not change the direction of rotation.

Changing the Speed of a Synchronous Motor

The speed of a synchronous motor is the synchronous speed of the rotating field, and it is changed by changing the frequency of the three-phase source. Changing the speed by changing the number of poles is not done; it would require more slip rings, and a more complicated and more costly control system.

12.3 SYNCHRONOUS MOTOR TORQUE

The torque load that a synchronous motor can handle is dependent on the hold-in strength of the poles. Hence, to increase the hold-in torque, it is necessary to increase the direct current supplied to the rotor magnets.

Although the poles of the rotor are locked in rotational synchronism with the poles of the rotating flux, a certain amount of "elasticity" permits variations of the rotor position in relation to the flux of the stator for different shaft loads, while still going at the same speed. The synchronous motor behaves very much like the nonslip rubber-disk coupling shown in Figure 12.4; increased load on the driven shaft causes it to lag the driving shaft by some small angle, while both are still rotating at the same speed.

The sketches in Figure 12.5 are simulated stroboscopic views showing the position of the rotor magnets relative to the poles of the rotating magnetic field of the stator for different load conditions. With the rotor running at synchronous speed, and no load on the shaft, the rotor poles will be directly in line with the corresponding stator

---

[1] A motor-rotation indicator, discussed in Section 11–3, Chapter 11, for induction motors, may also be used to predict the direction of rotation of synchronous motors.

FIGURE 12.4
Nonslip rubber-disk coupling.

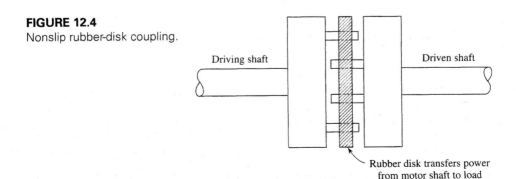

Driving shaft

Driven shaft

Rubber disk transfers power
from motor shaft to load

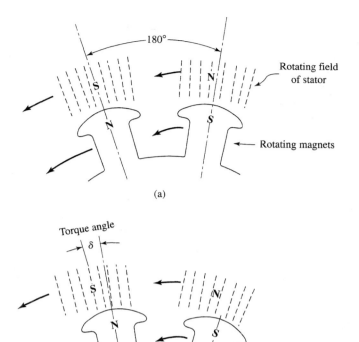

FIGURE 12.5
Synchronous motor running (a) with no load on the shaft; (b) with load on the shaft.

poles of opposite polarity; this is shown in Figure 12.5a. Increasing the shaft load causes the rotor poles to lag slightly behind the stator poles, while still rotating at synchronous speed. This mechanical angle of lag, called the torque angle, load angle, or power angle, is indicated by the Greek letter delta (δ) in Figure 12.5b. The torque angle is zero at no-load, and increases with increases in shaft load. The torque angle is between 20 and 30 electrical degrees when operating at rated load.[2]

12.4 SYNCHRONOUS MOTOR COUNTER-ELECTROMOTIVE FORCE

As the rotor turns, the magnets sweep the stator conductors, generating a voltage in the stator coils. This voltage, called a counter-emf or cemf, is in opposition to the applied voltage. The magnitude of the cemf is constant for all load conditions, as long as the

---

[2] In electrical machines, the measurement of circular arcs is in electrical degrees, where 180 electrical degrees is defined as the angle of circular arc measured between the centers of adjacent poles of opposite polarity, as shown in Figure 12.5a.

current in the magnets and the speed of the rotor are not changed. By adjusting the magnet current (field current), the cemf may be made equal to, less than, or greater than the applied stator voltage. The part played by the cemf in synchronous motor behavior may be analyzed by phasor diagrams and stroboscopic views for different load conditions.

Figure 12.6a shows the no-load stroboscopic view of a rotor pole as it sweeps past a stator conductor; the strobe light shows the instantaneous position of a rotor pole with respect to the stator conductor. At no load, the cemf (E) is almost equal and opposite the applied voltage (V), as shown in Figure 12.6b, and the stator current will be very low.

As the machine is loaded, the rotor changes its relative position with respect to the rotating flux of the stator, lagging by angle δ, as shown in Figure 12.6c. The corresponding phasor diagram is shown in Figure 12.6d. Note that the cemf phasor is now lagging its no-load position by δ degrees. Although the magnitude of the cemf did not change, the lagging angle makes the cemf less effective in its opposition to the applied voltage; the opposition is now $E \cos \delta$. The reduced opposition permits more current to enter the stator. The increased current strengthens the stator flux, enabling the rotor to remain in synchronism with the rotating flux of the stator, thus enabling it to carry the additional load.

As additional load is placed on the machine, the rotor magnets continue to increase their lag relative to the rotating flux. This increases the angle of lag of the cemf phasor and thus increases the magnitude of the stator current. Except for the transient conditions whereby the rotor assumes a new position in relation to the rotating flux, the speed of the machine does not change with additional loading. Finally, as load continues to increase, a point is reached at which a further increase in δ fails to cause a corresponding increase in motor torque, and the rotor pulls out of synchronism. This point of maximum torque occurs at a power angle of approximately 60° for the salient-pole machines shown in Figures 12.1, 12.2, and 12.7.

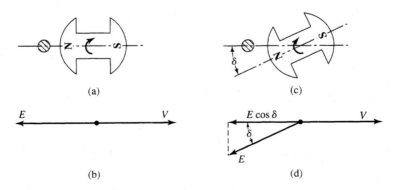

(a)   (c)

(b)   (d)

FIGURE 12.6
Position of cemf phasor, relative to the applied voltage phasor, for a synchronous motor.

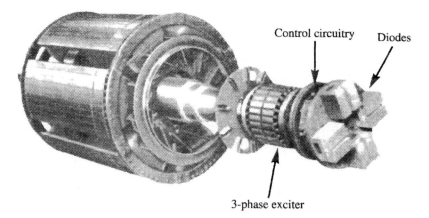

Control circuitry   Diodes

3-phase exciter

FIGURE 12.7
Salient-pole rotor equipped with a brushless excitation system (courtesy Dresser Rand/Electric Machinery Co.).

Pull-Out Torque

The value of torque load that causes the rotor to pull out of synchronism is called the *pull-out torque*. If the rotor pulls out of synchronism, pole slipping will occur and the motor will vibrate severely. The DC breaker must be opened, the field rheostat adjusted to a higher magnet current, and the rotor resynchronized. Increasing the magnet current increases the hold-in strength of the magnets, permitting higher torque loads to be carried without pulling out of synchronism.

Damping Action

When operating at synchronous speed, the squirrel-cage winding acts as a damper to smooth the effects of pulsating loads, hunting of the power supply, and the sudden applications of load. The damping action is explained by Lenz's law. A pulsating torque causes oscillation of the magnets about their normal position with respect to the rotating flux. This causes relative motion between the squirrel-cage bars and the rotating flux of the stator. The resultant induced current in the squirrel-cage winding sets up a torque in opposition to the hunting action.

12.5 BRUSHLESS EXCITATION SYSTEM

Conventional synchronous motors, such as that shown in Figure 12.2, have slip rings, a commutator, and carbon brushes to conduct current to the rotating components. These moving and sliding parts are subject to wear and, thus, require periodic maintenance.[3]

---

[3] See Chapter 19 for a discussion of maintenance of commutators, brushes, and slip rings.

A salient-pole rotor equipped with a brushless excitation system, shown in Figure 12.7, does not have slip rings, commutator, or brushes. Brushless excitation is provided by a small three-phase generator winding (called an exciter armature), a three-phase rectifier, control circuitry, and magnets. All of these components are mounted on the same shaft. The circuit for the brushless excitation system in Figure 12.7 is shown in Figure 12.8.

Direct current for the exciter field is supplied from the three-phase source through a rectifier, which converts the alternating current to direct current. A frequency-sensitive solid-state control circuit using silicon-controlled rectifiers (SCRs) monitors the frequency of the emf induced in the rotor magnet winding by the rotating flux of the stator. The frequency of the emf in the rotor magnet winding is the same as that

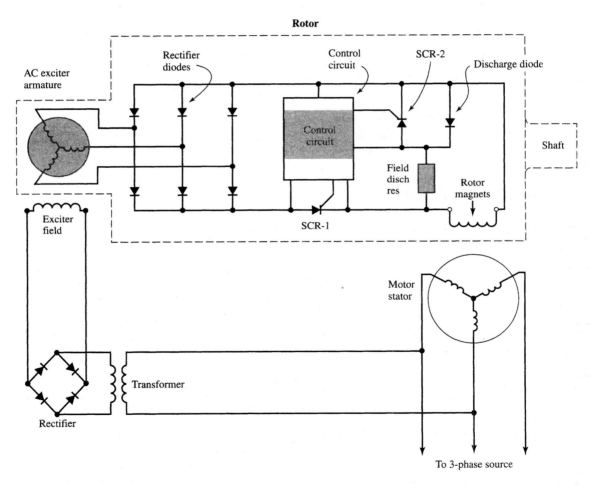

FIGURE 12.8
Circuit diagram for the rotor in Figure 12.7 (courtesy Dresser Rand. Electric/Machinery Co.).

induced in the squirrel-cage winding, and it is a function of the frequency of the applied stator voltage and the slip. That is,

$$f_r = sf_s \qquad \qquad \text{(12–2)}$$

where: f_r = frequency of the voltage induced in the rotor (Hz)
                f_s = frequency of the applied stator voltage (Hz)
                s = slip.

At locked rotor, the frequency of the current induced in the magnet windings is equal to the applied stator frequency. The frequency-sensitive circuit, sensing the line frequency, closes SCR-2 and opens SCR-1; opening SCR-1 blocks the exciter current to the field windings, and closing SCR-2 connects the discharge resistor across the magnet windings. At near-synchronous speed, the frequency of the voltage generated in the magnets is very low (approaching zero), and the control circuit opens SCR-2 and closes SCR-1; opening SCR-2 disconnects the discharge resistor, and closing SCR-1 admits current to the magnet windings. The solid-state control circuit is programmed to close SCR-1 at an instant that will ensure that the rotor poles will be facing stator poles of opposite polarity, thus preventing pole slipping.

When the stator is disconnected from the three-phase source, the motor slows down, SCR-1 opens, and the energy stored in the magnet windings discharges into the discharge resistor via the discharge diode.

12.6   POWER FACTOR OF A SYNCHRONOUS MOTOR

In addition to adjusting the hold-in strength of the magnets, changing the amount of direct current in the magnet windings causes a change in the power factor of the stator. Low values of magnet current result in less hold-in strength and a lagging power factor; high values of magnet current result in greater hold-in strength and a leading power factor. This characteristic of a synchronous motor is a useful fringe benefit when power-factor adjustment of a system is required. When the power factor is leading, the machine is said to be overexcited; when lagging, it is said to be underexcited; and when at unity power factor the machine is said to be normally excited.

Values of magnet current (field current) that is higher or lower than that required to obtain unity power factor results in higher stator current, which in turn increases the temperature of the stator windings.

In industrial applications where both synchronous motors and induction motors are used, the synchronous motor is usually operated at a leading power factor to compensate for the lagging power factor of the induction motors.

When a synchronous motor is operated without load, merely for the purpose of improving the power factor of a system, the machine is called a synchronous condenser. Machines designed specifically for this application are built without external shafts, and the large ones are hydrogen cooled.

GENERAL REFERENCE

[1] Hubert, Charles I., *Electric Machines: Theory, Operation, Applications, Adjustment, and Control,* Prentice Hall, Upper Saddle River, NJ, 2001.

REVIEW QUESTIONS

1. What is the difference in construction details between a three-phase squirrel-cage motor and a three-phase synchronous motor?
2. State the correct procedure for (a) starting a three-phase synchronous motor, (b) stopping the motor, and (c) reversing the motor.
3. Explain the purpose of a field-discharge resistor that is connected across the field circuit of a synchronous motor.
4. How is the speed of a synchronous motor adjusted?
5. What is cemf, and how does it affect synchronous motor operation?
6. What can cause a synchronous motor to fail to pull into synchronism?
7. Assuming that a synchronous motor is operating at rated load and 1.0 pf, what effect does increasing the field current to the magnets have on the torque angle and on the power factor?
8. What damaging effects can be produced when a synchronous motor pulls out of synchronism? Assume that the stator and rotor circuits are still energized.
9. What remedial action should be taken when a synchronous motor pulls out of synchronism?
10. Explain why the stator current increases when the shaft load is increased.
11. Explain why the increase in stator current, with increased loading, enables the rotor to remain in synchronism.

13

Operational Problems of Three-Phase Motors

13.0 INTRODUCTION

This chapter explains the problems that can afflict three-phase motors (squirrel-cage, wound-rotor, and synchronous) due to improper operating procedures and poor power quality (off-standard voltage, off-standard frequency, and unbalanced voltage). Low voltage at the motor terminals may be caused by voltage drops in distribution circuits, decreased voltage supplied by a utility due to system faults or overload, etc. Low frequency is caused by overloaded generators, and it slows down the prime mover. Suggestions for remedial action that could extend the useful life of the machine are discussed. The importance of baseline data and condition checks, as an integral part of a preventive maintenance program, are emphasized.

13.1 EFFECT OF OFF-STANDARD VOLTAGE AND OFF-STANDARD FREQUENCY ON LOCKED-ROTOR TORQUE AND BREAKDOWN TORQUE

The locked-rotor and breakdown torque developed by an induction motor, and the locked-rotor torque of a synchronous motor are proportional to the square of the voltage and inversely proportional to the square of the frequency.[1] The following equation provides a convenient method for determining the locked-rotor torque (and breakdown torque) of an induction motor when operating at off-rated conditions:

$$\frac{T_{LR2}}{T_{LR\ rated}} = \frac{\left[\dfrac{V_2}{f_2}\right]^2}{\left[\dfrac{V_{rated}}{f_{rated}}\right]^2} \quad \Rightarrow \quad T_{LR2} = T_{LR\ rated}\left[\frac{V_2 f_{rated}}{V_{rated}\ f_2}\right]^2 \qquad (13\text{--}1)$$

---

[1] This relationship between voltage and torque is true only for a sinusoidal voltage wave. If the motor is fed from a solid-state control, the voltage would be nonsinusoidal, and the torque would be less.

where: $T_{LR\,rated}$ = locked-rotor torque at rated conditions (lb-ft)
T_{LR2} = locked-rotor torque at off-rated conditions (lb-ft)
V_{rated} = rated voltage (V)
V_2 = off-rated voltage (V)
f_{rated} = rated frequency (Hz)
f_2 = off-rated frequency (Hz).

EXAMPLE 13.1

A three-phase, 60-Hz, 2300-V, 200-hp, design B, 1185-rpm induction motor is to be started from a system that is experiencing a 10% drop in voltage. The locked-rotor torque at rated voltage is 120% rated torque. Determine the locked-rotor torque (a) at rated voltage; (b) at the reduced operating voltage; (c) determine the percent decrease in locked-rotor torque.

Solution

a. The relationship between shaft horsepower, shaft torque and shaft speed is

$$\text{HP} = \frac{T n_r}{5252} \qquad (13\text{--}2)$$

where: T = shaft torque (lb-ft)
HP = shaft power (hp)
n_r = rotor speed (rpm).

Note: All percent values must be divided by 100 before entering into equations. From Eq. (13–2),

$$T_{rated} = \frac{5252 \times \text{HP}}{n_r} = \frac{5252 \times 200}{1185} = 886.4 \text{ lb-ft}$$

$$T_{LR\,rated} = \frac{120}{100} \times T_{rated} = 1.2 \times 886.4 = 1063.7 \text{ lb-ft}$$

b. The frequency has not changed, and $V_2 = 2300 \times 0.90 = 2070$ V. Thus,

$$T_{LR2} = T_{LR\,rated} \times \left[\frac{V_2 f_{rated}}{V_{rated} f_2} \right]^2 = 1063.7 \times \left[\frac{2070 \times 60}{2300 \times 60} \right]^2 = 861.6 \text{ lb-ft}$$

c. Percent decrease $= \dfrac{T_{LR2} - T_{LR\,rated}}{T_{LR\,rated}} \times 100 = \dfrac{861.6 - 1063.7}{1063.7} \times 100 = -19\%$

As illustrated in Example 13.1, a 10 percent decrease in voltage caused a 19 percent decrease in starting torque (locked-rotor torque). A supply voltage that is too low may prevent a heavily loaded induction motor from starting, or if running, may cause it to stop if a heavy load is applied. A higher than rated supply voltage will result in higher locked-rotor torque and higher breakdown torque.

In the case of a synchronous motor, if starting with too low a voltage, the rotor may not reach sufficient speed on its squirrel-cage winding to enable it to pull into synchronism when the magnets are energized. The result will be severe torque pulses and high current pulses as pole slippage occurs.

13.2 EFFECT OF OFF-STANDARD VOLTAGE AND OFF-STANDARD FREQUENCY ON LOCKED-ROTOR CURRENT

At locked-rotor, the motor acts as an impedance that is essentially inductive reactance. Thus, the current at locked rotor is directly proportional to the applied voltage and inversely proportional to the frequency. Expressed mathematically,

$$\frac{I_{LR2}}{I_{LR\ rated}} = \frac{V_2/f_2}{V_{rated}/f_{rated}} \quad \Rightarrow \quad I_{LR2} = I_{LR\ rated} \times \left[\frac{V_2 f_{rated}}{V_{rated} f_2}\right] \qquad (13\text{--}3)$$

where: $I_{LR\ rated}$ = locked-rotor current at rated conditions (A)
I_{LR2} = locked-rotor current at off-rated conditions (A)
V_{rated} = rated voltage (V)
V_2 = off-rated voltage (V)
f_{rated} = rated frequency (Hz)
f_2 = off-rated frequency (Hz).

EXAMPLE 13.2   A certain three-phase, 240-V, 60-Hz, 150-hp, 1750-rpm, design B motor draws a locked-rotor current of 2722 A and has a locked-rotor torque of 110% rated torque. Assume very heavy loads on the system caused the frequency and voltage of the system to drop to 59 Hz and 230 V, respectively. Determine for the new conditions (a) the locked-rotor current; (b) the locked-rotor torque.

Solution

a. $I_{LR2} = I_{LR\ rated}\left[\dfrac{V_2 f_{rated}}{V_{rated} f_2}\right] = 2722 \times \left[\dfrac{230 \times 60}{240 \times 59}\right] = 2653$ A

b. From Eq. (13–2),

$$T_{rated} = \frac{5252 \times HP}{n_r} = \frac{5252 \times 150}{1750} = 450.2 \text{ lb-ft}$$

$$T_{LR\ rated} = \frac{110}{100} \times T_{rated} = 1.1 \times 450.2 = 495.2 \text{ lb-ft}$$

$$T_{LR2} = T_{LR\ rated} \times \left[\frac{V_2 f_{rated}}{V_{rated} f_2}\right]^2 = 495.2 \times \left[\frac{230 \times 60}{240 \times 59}\right]^2 = 470.3 \text{ lb-ft}$$

13.3 EFFECT OF OFF-STANDARD VOLTAGE AND OFF-STANDARD FREQUENCY ON THE PERFORMANCE CHARACTERISTICS OF A RUNNING MACHINE

The general effect of voltage and frequency variation on the performance characteristics of a running machine is illustrated in Figure 13.1. The starting torque and starting current (locked-rotor torque and locked-rotor current, respectively) are also illustrated.

Operating below Rated Voltage

The increased current that results from operating a loaded motor at reduced voltage causes increased heat losses (I^2R effect) in stator and rotor conductors. This condition results in a temperature rise in the conductor insulation, thus shortening insulation life.[2] The reduction in core losses and, hence, lower iron temperature (due to a lower

---

[2] See 10-degree half-life rule for electrical insulation in Section 23–6, Chapter 23.

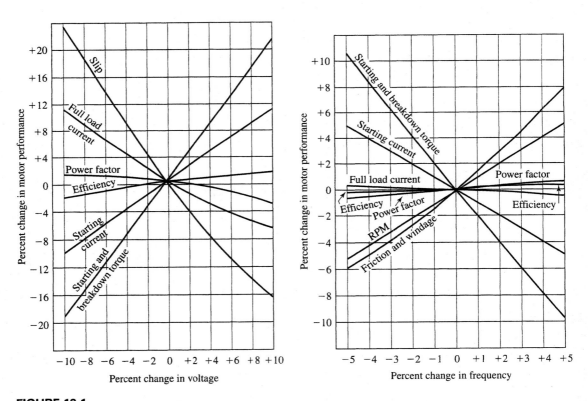

FIGURE 13.1

General effect of voltage and frequency variations on induction-motor characteristics (courtesy GE Industrial Systems).

operating voltage) do not offset the increased heating of the conductors. However, if the motor is lightly loaded, and the increase in current due to a decrease in voltage is below rated current, no overheating will occur.

For those synchronous motors that obtain the DC field current from rectifiers supplied by the same three-phase source, a drop in AC voltage will cause a proportional drop in DC voltage. The resultant reduction in magnet current reduces the hold-in strength of the magnets, which may cause the motor to pull out of synchronism.

Operating above Rated Voltage

If the line voltage is increased above the rated voltage of the motor, the stator current will decrease with increasing voltage until the applied voltage is high enough to cause significant magnetic saturation of the iron. If this occurs, there will be a steep rise in the exciting current, which added to the load component of current, results in over-heating of the windings. Significant magnetic saturation may occur at voltages above 110 percent rated voltage.

Assuming a constant shaft-power load, and neglecting internal losses, the electrical power input to the motor is equal to shaft power output. That is,

$$\sqrt{3}V_{line} \times I_{line} \times F_P = HP_{shaft} \times 746 \qquad (13\text{–}4)$$

With a constant shaft-power load, the electrical power input at off-standard voltage will be approximately equal to the electrical power input at rated voltage. Thus, by neglecting changes in windage and friction due to small changes in speed, we get

$$\sqrt{3}V_2 I_2 F_P \approx \sqrt{3}V_{rated}I_{rated}F_P \qquad (13\text{–}5)$$

Assuming an insignificant change in power factor,

$$I_2 \approx \frac{V_{rated}\,I_{rated}}{V_2} \qquad (13\text{–}6)$$

where: I_2 = current at off-rated voltage (A)
      I_{rated} = rated current (A)
      V_{rated} = rated voltage (V)
      V_2 = off-rated voltage (V).

Effect of Off-Standard Frequency on the Line Current in a Running Machine

A reduction in frequency increases the excitation current (no-load component of motor current), thus increasing the total motor current. If the frequency is low enough to cause significant magnetic saturation of the iron, a steep rise in the exciting current will occur, resulting in overheating of the windings.

Percent Heating Caused by Overcurrent

The percent heating of motor windings due to overcurrent, caused by off-standard voltage, off-standard frequency, or overload, can be determined from

$$\% \text{ heating} \approx \left[\frac{I_2}{I_{\text{rated}}} \right]^2 \times 100 \qquad (13\text{--}7)$$

where: I_2 = current at off-standard voltage.

EXAMPLE 13.3

A three-phase, 460-V, 60-Hz, 50-hp, 1775-rpm squirrel-cage motor driving a constant rated shaft horsepower load has a line current of 59.8 A. Because of a serious system problem, the utility is forced to lower the voltage by 10%. Determine (a) the new voltage; (b) the approximate motor current; (c) the stator heating at the lower voltage as a percent of stator heating that occurs at rated current.

Solution

a. $V_2 = 460 \times (1 - 0.10) = 414$ V

b. From Eq. (13–6),

$$I_2 \approx \frac{V_{\text{rated}} I_{\text{rated}}}{V_2} = \frac{460 \times 59.8}{414} = 66.4 \text{ A}$$

Or using the curve in Figure 13.1, a 10% decrease in voltage results in an increase in motor current of approximately 11%. Thus,

$$I_{\text{motor}} = 1.11 \times 59.8 = 66.4 \text{ A}$$

c. $\% \text{ heating} \approx \left[\dfrac{I_2}{I_{\text{rated}}} \right]^2 \times 100 = \left[\dfrac{66.4}{59.8} \right]^2 \times 100 = 123.45 \Rightarrow 123\%$

In this example, a 10% decrease in line voltage caused a 23% increase in conductor heating.

13.4 NEMA CONSTRAINTS ON VOLTAGE AND FREQUENCY

NEMA-rated induction motors are expected to carry *rated load* as long as variations in applied voltage and applied frequency do not exceed the following constraints [1]:

1. A voltage variation of up to ±10 percent rated voltage while operating at rated frequency;
2. A frequency variation of up to ±5 percent rated frequency, while operating at rated voltage;
3. A combined variation in voltage and frequency, with the sum of the absolute values of the respective variations not exceeding 10 percent providing the frequency does not exceed ±5 percent of rated frequency.

Note: In accordance with constraint 3, a 9 percent rise in system voltage, accompanied by a 4 percent drop in system frequency would cause a combined variation of 9% + 4% = 13%. Adding algebraically 9% + (−4%) = 5% is incorrect! The calculation must be an addition of the absolute values. Off-standard frequency and off-standard voltage each have an adverse effect on motor performance, and the adverse effects of a decrease in frequency do not offset the adverse effects of an increase in voltage, and vice versa. This is illustrated in Figure 13.1, where for starting current and starting torque, the effect of an increase in frequency is not offset by a decrease in voltage.

13.5 EFFECT OF UNBALANCED LINE VOLTAGES ON INDUCTION-MOTOR PERFORMANCE [2]

If the three line voltages supplied to a three-phase induction motor are not equal in magnitude, they not only cause unequal phase currents in the rotor and stator windings, but the percent current unbalance may be six to ten times larger than the percent voltage unbalance. The resultant increase in I^2R losses will overheat the insulation, shortening its life. Unbalanced voltages also cause a decrease in locked-rotor torque and breakdown torque. Thus, in those applications where there is only a small margin between the locked-rotor torque and the load torque, severe voltage unbalance may prevent the motor from starting. The full-load speed of running motors is reduced slightly by voltage unbalance.

A system with perfectly balanced line voltages can become unbalanced if single-phase loads are added to the system without regard to proper load balancing. Phases that carry more load have larger voltage drops in their connecting lines. Reconnecting some of the single-phase loads to other phases may be all that is necessary to balance the voltages. Errors in transformer tap connections may also produce a voltage unbalance.

Percent Voltage Unbalance

Percent voltage unbalance is defined by NEMA as 100 times the maximum line voltage deviation from the average value of the three line-to-line voltages, divided by the average voltage. Expressed as an equation,

$$\% \text{ VUB} = \frac{V_{\text{max dev.}}}{V_{\text{avg}}} \times 100 \qquad \textbf{(13–8)}$$

$$V_{\text{avg}} = \frac{V_{\text{ab}} + V_{\text{bc}} + V_{\text{ca}}}{3} \qquad \textbf{(13–9)}$$

where: % VUB = percent voltage unbalance
V_{avg} = average line-to-line voltage (V)
$V_{\text{max dev.}}$ = maximum voltage deviation between any line voltage and V_{avg}
$V_{\text{ab}}, V_{\text{bc}}, V_{\text{ca}}$ = line-to-line voltages (V)

Voltage measurements, taken for the purpose of determining voltage unbalance, should be made as close as possible to the motor terminals, and the readings should be taken with a digital voltmeter for greater accuracy.

Increase in Motor Temperature Caused by Unbalanced Voltage

When operating at rated load, the percentage increase in motor temperature ($\%\Delta T$), due to unbalanced line voltages, is approximately equal to twice the square of the percentage voltage unbalance [3]. Expressed mathematically,

$$\%\Delta T = 2 \times (\%\text{VUB})^2 \qquad (13\text{--}10)$$

A graph of Eq. (13–10) is shown in Figure 13.2.

The expected maximum allowable temperature rise of a motor caused by voltage unbalance can be determined from the following equation:

$$T_{\text{rise,unb}} = T_{\text{rise,rated}} \times \left[1 + \frac{\%\Delta T}{100}\right] \qquad (13\text{--}11)$$

where: $T_{\text{rise,unb}}$ = temperature rise due to voltage unbalance
$T_{\text{rise,rated}}$ = rated maximum allowable temperature rise[3]
$\%\Delta T$ = percent increase in motor temperature.

---

[3] The rated maximum allowable temperature rise for induction motors is given in Table 23.2, Chapter 23.

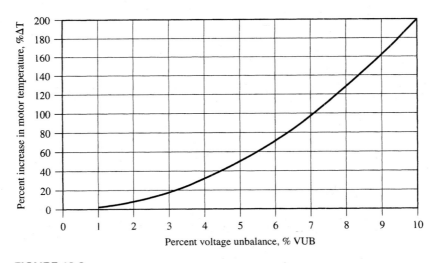

FIGURE 13.2
Graph showing the percent increase in motor temperature vs. percent voltage unbalance.

EXAMPLE 13.4

A 150-hp, design B, 460-V, 60-Hz, four-pole, totally enclosed, nonventilated induction motor with class F insulation and a service factor of 1.15 is to be operated at rated power from an unbalanced three-phase system. If the three line-to-line voltages are 460 V, 425 V, and 440 V, determine (a) the percent voltage unbalance; (b) the percent increase in motor temperature when operating at rated load; (c) the rated maximum temperature rise if operating at rated load in 40°C ambient and balanced voltages; (d) the expected maximum temperature rise if operating with the specified unbalanced voltages.

Solution

a. $V_{avg} = \dfrac{460 + 425 + 440}{3} = 441.67 \text{ V}$

The voltage deviations from the average are:

$$|460 - 441.67| = 18.33 \text{ V}$$
$$|425 - 441.67| = 16.67 \text{ V}$$
$$|440 - 441.67| = 1.67 \text{ V}$$

$$\% \text{ VUB} \approx \frac{V_{max\ dev.}}{V_{avg}} \times 100 = \frac{18.33}{441.67} \times 100 = 4.15\%$$

b. $\%\Delta T = 2 \times (\% \text{ VUB})^2 = 2 \times (4.15)^2 = 34.5\%$

Note: $\% \Delta T$ may also be approximated from Figure 13.2.

c. From Table 23.2, Chapter 23, the rated temperature rise for a motor with Class F insulation and a 1.15 service factor is 115°C.

d. $T_{rise,unb} = T_{rise,rated} \times \left[1 + \dfrac{\%\Delta T}{100} \right] = 115 \times \left[1 + \dfrac{34.5}{100} \right] = 154.6°C$

Derating Motor to Prevent Overheating

In those applications where voltage unbalance cannot be corrected, the motor should be derated (operated at a lower horsepower). The derating curve, shown in Figure 13.3, should be used to determine the required derating. A 1 percent unbalance will not

FIGURE 13.3

Motor derating curve for unbalanced line voltages (courtesy National Electrical Manufacturers Association; from NEMA Standards Publication MG1-1998, copyright 1999 by NEMA).

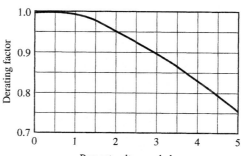

cause significant problems. However, operating a motor with a voltage unbalance greater than 5 percent is not recommended.

EXAMPLE 13.5

For the motor in Example 13.4, determine (a) the derating factor required to prevent overheating; (b) the allowable horsepower.

Solution

a. The percent unbalanced voltage (%VUB) is 4.15%. The corresponding derating factor obtained from Figure 13.3 is approximately 0.82.
b. The motor load should be limited to 150 × 0.82 = 123 hp.

13.6 UNBALANCED MOTOR LINE CURRENTS

Periodic monitoring of motor line currents should be performed to determine if the motor load exceeds the nameplate value and if there is any current unbalance. If the line currents are unequal, there is either a fault in the motor or the line voltages are unbalanced.

The cause of current unbalance may be determined by "rolling" the line leads at the motor terminals and measuring the three line currents at the motor, as shown in Figure 13.4. When rolling the leads, all three motor leads must be shifted. If all three leads are *not* shifted, the phase sequence will change and the motor direction will be reversed.

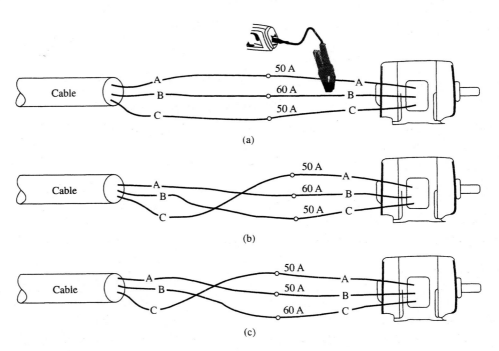

FIGURE 13.4
Rolling line leads to determine cause of unbalanced currents.

Before rolling, mark the motor leads and line leads A, B, and C, as shown in Figure 13.4a, and using a clamp-on meter, measure and record the current in each motor lead. Then shut off the power, roll the leads as shown in Figure 13.4b, restart the motor, and again measure the current in each motor lead. If the motor lead that had the high current before rolling still has the high current, as shown in Figure 13.4b, the motor is defective and should be serviced. However, if a different motor lead has the high current after rolling, as shown in Figure 13.4c, the problem is an unbalanced line voltage.

13.7 EFFECT OF THE NUMBER OF STARTS ON MOTOR LIFE

The high in-rush current associated with every start, or attempted start, causes severe thermal and mechanical stresses on rotor and stator components. The effects of these stresses are cumulative and adversely affect the service life of the machine. Large motors (above 50 hp) and motors with high inertia loads are particularly susceptible to damage caused by frequent starting. For this reason, manufacturers of large motors often provide information on the permissible number of starts and the required waiting period between starts. The number of starts and cooling periods between starts, are based on the inertia (Wk^2) of the motor and load, the type of load, and the temperature of the machine [1].

Operating and maintenance personnel should avoid unnecessary starts; repeated attempts at starting a motor after tripping due to fault or overload can cause extensive damage to the machine. The reason for tripping or failure to start must be determined and corrected before attempting a restart.

13.8 PERILS OF RECLOSING OUT OF PHASE [4,5]

When the stator of an induction motor is disconnected from the line (circuit opened), the closed circuit formed by the rotor bars and end rings prevents the quick collapse of rotor current and associated rotor flux. The residual current and residual flux decay at a rate determined by the inductive time constant of the rotor. In an elapsed time equal to one time constant, the rotor current and flux will have decayed by only 63.2 percent. Rotor time constants of 0.3 seconds are not uncommon.

The flux in the revolving rotor induces a three-phase voltage in the stator windings. This voltage appears at the stator terminals. However, because of the decreasing speed, the frequency of this residual voltage will be less than the system frequency. Thus, when the stator is disconnected from the line, the residual voltage will be cycling in and out of phase with the system voltage.

If, after a power interruption, the motor is reconnected to the power line with the residual voltage out of phase with the line voltage, the in-rush current will be higher than if the motor were started from a stopped position. The worst possible condition would be reclosing immediately after a power interruption, with the residual voltage almost equal to the line voltage and nearly 180° out of phase. In this situation, the in-rush current would be almost double the locked-rotor value and severe damage to the motor could occur. If the machine is large, a power blackout may occur.

Reclosing with high residual voltage and out of phase is likely to occur when switching rapidly from a failed power supply to a standby or emergency supply.

In the case of a synchronous motor, the magnets make the situation worse. Rapid reclosing to another power source may cause damage to the stator coils, twist the shaft, or even rip the motor from its base plate [5].

Rapid and safe reclosing to alternate power supplies may be accomplished with an in-phase monitor, which measures the phase angle between the source voltage and the motor residual voltage and automatically initiates reclosing when the phase angle approaches zero.

Permanently connecting a voltmeter across the motor terminals and manually reclosing when the residual voltage drops below 20 percent or less of rated voltage is another method that will enable a relatively quick restart while avoiding an abnormally high in-rush current. If reclosed at 20 percent voltage and 180° out of phase, the net voltage applied to the stator windings will be only 120 percent rated voltage.

Note that in those induction-motor applications where capacitors for power-factor improvement are connected directly across the motor terminals, the residual voltage will take a much longer time to decay; capacitors cause the motor to act as a self-excited induction generator.[4]

13.9 OPERATING 60-Hz MOTORS ON A 50-Hz SYSTEM

Operating an induction motor significantly below rated frequency, such as operating a 60-Hz motor at 50 Hz, causes a significant decrease in magnetizing reactance and, because of magnetic saturation effects, an out-of-proportion increase in magnetizing current. The net result is severe overheating of the motor windings. To prevent overheating, a reduction in applied frequency must be accompanied by a reduction in applied voltage. Simply stated, the ratio of volts per hertz must be kept constant. General-purpose, three-phase, 60-Hz, NEMA-design induction motors with two, four, six, or eight poles are capable of operating satisfactorily from 50-Hz systems, provided the horsepower rating is reduced to 5/6 of its 60-Hz rating, and the voltage supplied at 50 Hz is 5/6 of the 60-Hz voltage rating. Expressed mathematically,

$$HP_{50} = \frac{5}{6} \times HP_{60} \qquad V_{50} = \frac{5}{6} \times V_{60} \qquad (13–12)$$

When operating in this manner, overheating will not occur, and the locked-rotor torque and breakdown torque at 50 Hz will be essentially the same as for 60-Hz operation [1].

---

[4] For information on induction generators, see Section 11–11, Chapter 11, and Reference [6].

EXAMPLE 13.6 A three-phase, 200-hp, six-pole, 460-V, 60-Hz induction motor is to be operated from a 50-Hz source. Determine (a) the allowable horsepower rating; (b) the correct operating voltage for the reduced frequency.

Solution

$$HP_{50} = \frac{5}{6} \times HP_{60} = \frac{5}{6} \times 200 = 167 \text{ hp}$$

$$V_{50} = \frac{5}{6} \times V_{60} = \frac{5}{6} \times 460 = 383 \text{ V}$$

Electric power systems in the United States, Canada, and most of the Americas operate at 60 Hz, whereas Europe, Asia, and Africa operate at 50 Hz. Solid-state frequency converters, or motor-generator sets, can be used to provide compatibility between U.S. and other international power standards for operation of electrical and electronic equipment. Frequency converters are used for testing American-made products for export as well as domestic applications of imported equipment. It is also useful for supplying shore power to a ship whose frequency does not match the shore frequency.

Rotary frequency converters, similar to that shown in Figure 13.5, provide true line isolation, and thus protect the load from switching transients, voltage fluctuations, and power line noise. The output waveform is clean and free of sharp peaks.

FIGURE 13.5
Rotary frequency changer (courtesy Georator Corporation).

FIGURE 13.6
Single-phasing fault occurring in a three-phase induction motor stator.

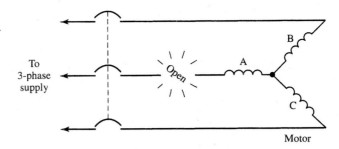

13.10 DAMAGING EFFECTS OF AN OPEN PHASE

A fault condition, called single phasing, occurs when one line of a three-phase supply opens while the motor is running, as shown in Figure 13.6. Although less torque is developed and severe and possibly damaging vibration may occur, the motor will continue to run if it is not heavily loaded. However, once stopped, a three-phase motor will not start with one line open, even if lightly loaded. It will make growling noises, overheat, and smoke.

If single phasing occurs while a three-phase induction motor is operating, and the shaft load is not changed, the power input when operating in the single phase will be equal to the power input when operating in three phases. That is,

$$V_{\text{line}} I_{1\phi} \times F_P = \sqrt{3} V_{\text{line}} I_{\text{line}} \times F_P$$

Solving for $I_{1\phi}$

$$I_{1\phi} = \sqrt{3} \times I_{\text{line}} = 1.73 I_{\text{line}} \qquad \text{(13–13)}$$

As indicated in Eq. (13–13), a 73 percent increase in line current will occur, in each of the remaining two lines, if an open phase occurs while the motor is running. If the motor is at or near rated load, rapid heating of the windings will occur, and unless thermal-overload devices or phase-failure devices disconnect the machine from the power line, the excessively high temperature will burn out the windings [7,8].

If a synchronous motor is running at rated load when single phasing occurs, it will probably lose synchronism and have to be stopped. However, if the synchronous motor is lightly loaded when single phasing occurs, the motor may not pull out of synchronism. The resultant unbalanced voltage will cause excessively high current in the squirrel-cage winding and may burn it out.

13.11 BASELINE DATA FOR THREE-PHASE MOTORS

Baseline data provide a fixed reference for comparison with future readings. If periodic maintenance tests show significant deviations from baseline values, it may indicate that cleaning, drying, or repair is required, or that operational problems such as undervoltage, overvoltage, unbalanced voltage, wrong frequency, overload, inadequate cooling, improper operation, or misapplication exist.

Preinstallation

Before installing a new or reconditioned motor, the following baseline measurements should be made:

1. Using a megohmmeter, measure and record the insulation resistance between the motor windings and ground (framework of machine).[5]
2. Using a low resistance ohmmeter, measure and record the resistance between motor terminals.
3. Record the date, time of measurement, ambient temperature, and humidity.

On Initial Start-Up

While the motor is operating at its rated speed and normal load, the following baseline measurements should be made:

1. Using a clamp-on ammeter, measure and record the steady-state current in each of the three motor lines. Make sure that the jaws close tightly and that the jaws are not too close to energized transformers. Also separate the three leads sufficiently to prevent stray flux from causing an error.
2. Using a digital voltmeter (for greater accuracy), measure and record the line-to-line voltage between the three motor terminals: a to b, b to c, and c to a. Measurements should be made at the motor. If the voltage is measured at the starter, allowance must be made for the voltage drop in the lines between the motor and the starter.
3. Assuming the three line currents in step 1 are equal, shut down the motor, and using a clamp-on ammeter with peak-hold capabilities, measure and record the motor starting current (in-rush current) in one of the lines.
4. Measure and record the vibration level at each bearing.[6]
5. If convenient, measure and record the no-load current.

13.12 CONDITION CHECK OF THREE-PHASE MOTORS: LOOK, LISTEN, SMELL, TEST, AND RECORD

The following condition check is the most important part of a preventive maintenance program:

1. Inspect motor for accumulation of dirt, check ventilation, check air filter on motor or filter in the air supply. Clean if dirty. Record condition and action taken.

---

[5] See Chapter 24 for insulation resistance measurement and its interpretation.

[6] See Chapter 21 for vibration measurement.

2. Be alert for unusual noises and odors.
3. Measure and record bearing temperature. Lubricate only as specified by manufacturer.
4. Measure and record bearing vibration.
5. Measure and record current in all three motor lines.
6. Using a digital voltmeter, measure and record the three line-to-line voltages.
7. On shutdown, measure and record the insulation resistance to ground and the machine temperature. Check for loose connections and tighten if necessary.
8. Compare condition check data with baseline data, note any discrepancies, and report results to the plant engineer.

SUMMARY OF EQUATIONS FOR PROBLEM SOLVING

$$T_{LR2} = T_{rated} \left[\frac{V_2 f_{rated}}{V_{rated} f_2} \right]^2 \tag{13-1}$$

$$HP = \frac{T n_r}{5252} \tag{13-2}$$

Neglecting losses
$$\begin{cases} I_{LR2} = I_{LR\ rated} \times \left[\frac{V_2 f_{rated}}{V_{rated} f_2} \right] & \text{(13-3)} \\ \sqrt{3} V_{line} \times I_{line} \times F_p = HP_{shaft} \times 746 & \text{(13-4)} \\ \sqrt{3} V_2 I_2 F_P \approx \sqrt{3} V_{rated} I_{rated} F_P & \text{(13-5)} \end{cases}$$

$$I_2 \approx \frac{V_{rated} I_{rated}}{V_2} \tag{13-6}$$

$$\% \text{ heating} \approx \left[\frac{I_2}{I_{rated}} \right]^2 \times 100 \tag{13-7}$$

$$\% \text{ VUB} = \frac{V_{max\ dev.}}{V_{avg}} \times 100 \tag{13-8}$$

$$V_{avg} = \frac{V_{ab} + V_{bc} + V_{ca}}{3} \tag{13-9}$$

$$\% \Delta T = 2 \times (\% VUB)^2 \tag{13-10}$$

$$T_{rise,\ unb} = T_{rise,\ rated} \times \left[1 + \frac{\% \Delta T}{100} \right] \tag{13-11}$$

$$HP_{50} = \frac{5}{6} \times HP_{60} \qquad V_{50} = \frac{5}{6} \times V_{60} \tag{13-12}$$

$$I_{1\phi} = \sqrt{3} \times I_{line} = 1.732 I_{line} \tag{13-13}$$

SPECIFIC REFERENCES KEYED TO THE TEXT

[1] NEMA Standards Publication No. MG 1-1998, *Motors and Generators.*

[2] Woll, R. F., Effect of Unbalanced Voltages on the Operation of Polyphase Induction Motors, *IEEE Trans. Industry and General Applications,* Vol. IGA-11, No. 1, January/February 1975.

[3] Brighton, R. J., Jr., and Ranade, P. R., Why Overload Relays Do Not Always Protect Motors, *IEEE Trans. Industry and General Applications,* Vol. IGA-8, No. 6, November/December 1982.

[4] Gill, J. D., Transfer of Motor Loads Between Out of Phase Sources, *IEEE Trans. Industry and General Applications,* Vol. IGA-15, No. 4, July/August 1979.

[5] Hauck, T. R., *Motor Reclosure and Bus Transfer, IEEE Trans. Industry and General Applications,* Vol IGA-6, No. 3, May/June 1970.

[6] Hubert, Charles I., *Electric Machines: Theory, Operation, Applications, Adjustment, and Control,* 2nd ed., Prentice Hall, Upper Saddle River, NJ, 2001.

[7] Griffith, M. S., A Penetrating Gaze at One Open Phase: Analyzing the Polyphase Induction Motor Dilemma, *IEEE Trans. Industry Applications,* Vol. IA-13, No. 6, November/December 1977.

[8] *IEEE Guide for AC Motor Protection,* IEEE Std. 588-1976.

REVIEW QUESTIONS

1. What are some of the causes of low voltage and low frequency in a distribution system? Explain.
2. How do off-rated frequency and off-rated voltage affect locked-rotor torque and breakdown torque? Explain.
3. How do off-rated frequency and off-rated voltage affect the locked-rotor current. Explain.
4. What effect does operating below nameplate voltage have on the operating temperature of a running machine? Assume the motor is driving a constant power load.
5. What are the NEMA constraints on voltage and frequency for three-phase induction motors?
6. What can cause a distribution system, with perfectly balanced voltages, to become unbalanced? Explain.
7. What very serious effect can unbalanced three-phase voltages have on the performance of an induction motor?
8. If, for some reason, voltage unbalance cannot be corrected, can induction motors still be used? Explain.
9. What test can be used to determine if unbalanced current in the lines to a three-phase induction motor is caused by a defective motor or by unbalanced voltages? Describe the test.
10. Why do manufacturers of large motors place a limitation on the number of times a motor can be started and on the required waiting time between starts?

11. What precautions should be taken in order to safely reclose a motor to a standby power supply after the primary supply fails?

12. Under what conditions can a 60-Hz motor be safely operated from a 50-Hz supply?

13. What causes single phasing, and what are its harmful effects?

14. What is meant by baseline data, and what is their application in a preventive maintenance program?

PROBLEMS

13–1/1. A three-phase, 60-Hz, 2300-V, 350-hp, design E, 1185-rpm induction motor is to be started from a system that is experiencing a 15% drop in voltage. The locked-rotor torque at rated voltage is 75% rated torque. Determine the locked-rotor torque (a) at rated voltage; (b) at the reduced operating voltage. (c) Determine the percent decrease in locked-rotor torque.

13–2/1. A three-phase, 60-Hz, 230-V, 75-hp, design A, 875-rpm induction motor is to be started from a system that is experiencing a 10% drop in voltage. The locked-rotor torque at rated voltage is 125% rated torque. Determine the locked-rotor torque (a) at rated voltage; (b) at the reduced operating voltage. (c) Determine the percent decrease in locked-rotor torque.

13–3/1. A three-phase, 460-V, 1765-rpm, 60-Hz, design C motor is used to drive a compressor. The motor locked-rotor torque is 363 lb-ft, and the compressor requires 300 lb-ft to start. If a utility dimout causes the utilization voltage[7] to drop to 440 V, will the compressor start? Show all work.

13–4/1. A 40-hp, three-phase, 60-Hz, 460-V, 3530-rpm, design B motor is used to drive a centrifugal pump. The locked-rotor torque of the motor is 126 lb-ft. The pump requires a minimum starting torque of 65 lb-ft. However, for adequate acceleration the locked-rotor torque must be at least 75 lb-ft. Determine the minimum voltage that would provide satisfactory acceleration.

13–5/2. A certain three-phase, 220-V, 60-Hz, 200-hp, 1775-rpm, design B motor draws a locked-rotor current of 2900 A. Assume that very heavy loads on the system caused the frequency and voltage of the system to drop to 55 Hz and 200 V, respectively. Determine the locked-rotor current for the new conditions.

13–6/2. A certain three-phase, 440-V, 60-Hz, 150-hp, 1750-rpm, design B motor draws a locked-rotor current of 1085 A. Assume that very heavy loads on the system caused the frequency and voltage of the system to drop to 59 Hz and 410 V, respectively. Determine the locked-rotor current for the new conditions.

13–7/2. A certain 460-V, 60-Hz, three-phase induction motor has a locked-rotor current of 362 A. Determine the locked-rotor current if the utilization voltage is 440 V.

---

[7] The utilization voltage is the actual voltage applied to the motor or other apparatus.

13–8/2. A certain 2300-V, 60-Hz, three-phase induction motor has a locked-rotor current of 58 A. Determine the locked-rotor current if the utilization voltage is 2400 V.

13–9/2. A certain 575-V, 60-Hz, three-phase induction motor has a locked-rotor current of 720 A. Determine the locked-rotor current if the utilization voltage and frequency are 560 V and 58 Hz, respectively.

13–10/3. A three-phase, 15-hp, 440-V, 1750-rpm, design B motor driving a constant horsepower load has a line current of 18.5 A. If the system voltage drops to 396 V, determine (a) approximate motor current; (b) percent increase in heating.

13–11/3. A three-phase, 30-hp, 440-V, 1750-rpm, design B motor driving a constant horsepower load has a line current of 39 A. If the system voltage drops to 410 V, determine (a) approximate motor current; (b) percent increase in heating.

13–12/5. A 25-hp, design B, 230-V, 60-Hz, four-pole, totally enclosed nonventilated induction motor, with Class B insulation and a service factor of 1.15, is to be operated at rated power from an unbalanced three-phase system. If the three line-to-line voltages are 232 V, 218 V, and 215 V, determine (a) the percent voltage unbalance; (b) the percent increase in motor temperature when operating at rated load; (c) the rated maximum temperature rise if operating at rated load in a 40°C ambient, and balanced voltages; (d) the expected maximum temperature rise if operating with the specified unbalanced voltages.

13–13/5. A 100-hp, design B, 460-V, 60-Hz, six-pole induction motor, with Class F insulation and a service factor of 1.0, is to be operated at rated power from an unbalanced three-phase system. If the three line-to-line voltages are 434 V, 456 V, and 460 V, determine (a) the percent voltage unbalance; (b) the percent increase in motor temperature when operating at rated load; (c) the rated maximum temperature rise if operating at rated load in a 40°C ambient, and balanced voltages; (d) the expected maximum temperature rise if operating with the specified unbalanced voltages.

13–14/5. A 10-hp, design B, 240-V, 60-Hz, two-pole induction motor, with Class F insulation and a service factor of 1.15, is to be operated at rated power from an unbalanced three-phase system. If the three line-to-line voltages are 220 V, 219 V, and 241 V, determine (a) the percent voltage unbalance; (b) the percent increase in motor temperature when operating at rated load; (c) the rated maximum temperature rise if operating at rated load in a 40°C ambient, and balanced voltages; (d) the expected maximum temperature rise if operating with the specified unbalanced voltages.

13–15/5. The three line-to-line voltages supplied to a 25-hp, 460-V, 60-Hz, design B motor are 460 V, 440 V, and 430 V. The motor is totally enclosed, nonventilated, with a 1.0 service factor and Class F insulation. Determine (a) average voltage; (b) percent voltage unbalance; (c) derating factor; (d) permissible load.

13–16/5. The three line-to-line voltages measured at the terminals of a three-phase 20-hp, design C motor are 450 V, 430 V, and 472 V. Determine (a) average voltage; (b) percent voltage unbalance; (c) derating factor; (d) derated horsepower.

13–17/5. A 200-hp, 2300-V, 1190-rpm, three-phase, 60-Hz, design B motor is overheating when operating at rated load. Voltage measurements of the three line-to-line voltages are 2300 V, 2320 V, and 2200 V. Determine (a) average voltage; (b) percent voltage unbalance; (c) derating factor; (d) maximum permissible load required to prevent overheating.

13–18/9. A three-phase, 100-hp, six-pole, 460-V, 60-Hz motor is to be operated from a 50-Hz source. Determine (a) the allowable horsepower rating; (b) the required operating voltage.

13–19/9. A three-phase, 50-hp, two-pole, 240-V, 60-Hz motor is to be operated from a 50-Hz source. Determine (a) the allowable horsepower rating; (b) the required operating voltage.

13–20/9. A three-phase, 75-hp, four-pole, 220-V, 50-Hz motor is to be operated from a 60-Hz source. Determine (a) the allowable horsepower rating; (b) the required operating voltage.

14

Synchronous Generators: Principles and Operational Problems

14.0 INTRODUCTION

Synchronous generators, also called alternators or AC generators, are the principal source of electrical power throughout the world. They range in size from a fraction of a kVA to 1500 MVA. The bulk of electric power is generated with high-speed steam turbines driving cylindrical rotor machines, and low-speed hydraulic turbines driving salient-pole machines. Diesel-driven generators and gas-turbine-driven generators are used for large and small isolated loads, for utility standby power to handle peak demand, and for applications in remote pumping stations, ship propulsion, drilling rigs, and the like.

This chapter begins with an introduction to alternating current generators, their construction, and principles of operation. This is followed by the general procedure for safe and efficient operation, paralleling, and shutdown of synchronous generators. Methods for power-factor control and for avoiding potential operational problems are also presented.

14.1 GENERATION OF AN ALTERNATING VOLTAGE

In its simplest form, an AC generator, also called alternator, consists of a set of magnet poles, a coil of wire, called an armature coil, and a prime mover, as shown in Figure 14.1.

As the magnet poles rotate, the flux through the window of the armature coil is made to vary with time. A graph of this variation for one revolution of the prime mover is shown in Figure 14.2a. The direction and relative magnitude of flux through the window of the coil for different positions of the magnet are indicated by arrows; the greater the flux, the longer the arrow. The apparent sinusoidal variation of the flux through the window, as the magnet rotates, is obtained by proper shaping of the field poles.

273

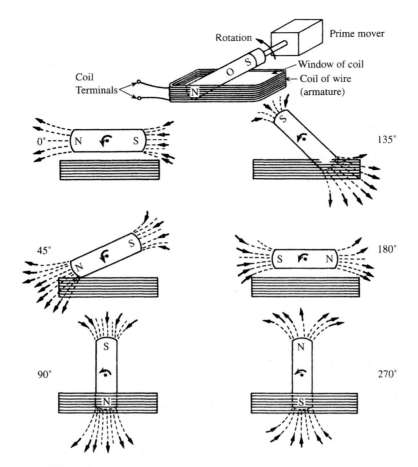

FIGURE 14.1
Variation of flux through the window of a coil as the magnet poles rotate.

The change in flux through the window, as the magnet rotates (increasing, decreasing, and reversing), generates an alternating voltage within the coil. The magnitude of this magnetically induced voltage is proportional to the number of turns of wire in the coil and the *rate of change* of flux through the window. The mathematical relationship that expresses this behavior is called Faraday's law:

$$v = -N\left(\frac{d\phi}{dt}\right) \quad\quad (14\text{--}1)$$

where: v = generated voltage (V)
N = number of turns of wire in coil
$\left(\dfrac{d\phi}{dt}\right)$ = *rate of change* of flux through window of coil.

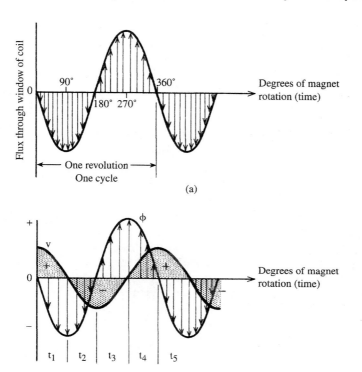

Flux through window of coil

90°   360°

180° 270°

Degrees of magnet
rotation (time)

← One revolution →
One cycle

(a)

φ

v

Degrees of magnet
rotation (time)

t_1 t_2 t_3 t_4 t_5

(b)

Time interval	Flux through window	Direction of generated voltage as determined from Lenz's law
t_1 to t_2	Increasing negatively	Positive
t_2 to t_3	Decreasing negatively	Negative
t_3 to t_4	Increasing positively	Negative
t_4 to t_5	Decreasing positively	Positive

FIGURE 14.2

(a) Graph showing variation of flux through the window of the coil for one revolution of the prime mover; (b) voltage and flux curves.

The minus sign (−) indicates that the voltage generated will be in a direction to oppose any change in flux through the window of the coil (Lenz's law).

The magnitude of the generated voltage does not depend on the amount of flux passing through the window of the coil but on the *rate of change* of flux through the window. Referring to Figure 14.2a, as the magnet rotates, to the 90° and 270° positions, the flux through the window attains its maximum amount and is about to decrease; at these brief instants of time, the flux is neither increasing nor decreasing, there is no change in flux through the window, and no voltage is generated.

At 0°, 180°, and 360°, the slope of the flux curve at these points is at its steepest value, indicating that the flux through the window is changing at its greatest rate;

hence, at the instants of time corresponding to these instantaneous positions of the revolving magnet (even though the flux through the window is zero), the voltage generated in the coil has its maximum value.

The direction of the generated voltage depends on the direction of the flux through the window and whether it is increasing or decreasing. This is shown in Figure 14.2b; the accompanying table illustrates the use of Lenz's law for determining the direction (polarity) of the generated voltage.

Increasing the speed of rotation of the magnet increases the rate of change of flux through the window, and a higher voltage is induced. The same increase in voltage may be accomplished without changing the speed by using a stronger magnet. A stronger magnet rotating at the same speed causes a greater rate of change of flux through the window and hence a higher voltage. Still another means of obtaining a higher voltage without changing the strength of the magnet or changing the speed is to increase the number of turns of wire in the armature coil. Practical generators use DC electromagnets whose strength is adjusted by varying the current to the magnet coils. Direct current is used to provide poles of fixed polarity. The small generator used to supply the current to the magnets is called an exciter generator or exciter. The exciter builds up its own voltage from residual magnetism.

The voltage generated in an AC machine can be conveniently expressed in terms of speed, flux, and a machine constant:

$$E = n\Phi K_G \qquad \qquad \textbf{(14–2)}$$

where: $\quad n$ = rpm
$\qquad \quad \Phi$ = flux per pole, in webers (Wb)[1]
$\qquad \quad K_G$ = generator constant.

The generator constant includes such design factors as the number of conductors, arrangement of armature winding, and the number of poles. If the generated voltage for a given speed and flux is known, the voltage for some other speed and flux can be determined by substituting into the following equation:

$$\frac{E_2}{E_1} = \frac{n_2\Phi_2}{n_1\Phi_1} \qquad \qquad \textbf{(14–3)}$$

Subscript 1 denotes the original combination of voltage, speed, and flux; subscript 2 denotes the new combination.

---

[1] *Note:* 10^6 maxwells (magnetic lines of force) equal one weber (Wb).

EXAMPLE 14.1 A single-phase generator running at 3600 rpm generates 450 volts. If the speed is reduced to 3400 rpm and the field current is not changed, determine the new voltage.

Solution
Substituting into Eq. (14–3),

$$\frac{E_2}{450} = \frac{3400}{3600} \frac{\Phi_2}{\Phi_1}$$

However, in this example Φ_2 is equal to Φ_1, therefore

$$E_2 = 450 \times \frac{3400}{3600} = 425 \text{ V}$$

The voltage output of an AC generator rises, falls, and reverses with time in a sinusoidal manner, as shown in Figure 14.2b. Examination of this voltage wave shows it to consist of repetitive cycles, each composed of a positive and a negative alternation. The number of cycles that occur in 1 second, called the frequency, depends on the rotational speed of the magnets and the total number of magnet poles in the field structure. Expressed mathematically,

$$f = \frac{Pn}{120} \tag{14-4}$$

where: f = cycles per second or hertz
P = total number of magnet poles
n = speed of rotating member (rpm).

EXAMPLE 14.2 A single-phase, two-pole synchronous generator running at 3600 rpm generates a voltage of 460 V. If the speed is reduced to 3400 rpm and the field current is not changed, determine (a) voltage; (b) frequency.

Solution
a. Since the magnetic field strength is not changed, $\Phi_1 = \Phi_2$.

$$\frac{E_2}{E_1} = \frac{n_2\Phi_2}{n_1\Phi_1} \Rightarrow \frac{E_2}{460} = \frac{3400\ \Phi_2}{3600\ \Phi_2} \Rightarrow E_2 = 434.4 \text{ V}$$

b. $f = \dfrac{Pn}{120} = \dfrac{2 \times 3400}{120} = 56.7 \text{ Hz}$

Because the generation of a voltage depends on a changing flux through the window of a coil, it does not matter whether the armature coils are stationary and the magnet poles rotate, or the magnets are stationary and the armature coils rotate. Elementary circuit diagrams and one-line diagrams for rotating-armature and rotating-field generators are shown in Figure 14.3. In each case, it is the armature that supplies current to the load. Two, three, or more output terminals may be provided, depending on the type of armature winding and output connections desired. In Figure 14.3a, a turbine is used to drive both the exciter generator and the generator armature; in Figure 14.3b, the turbine is used to drive the exciter and the rotating magnets of the generator. The one-line diagrams, shown in Figure 14.3c and 14.3d,

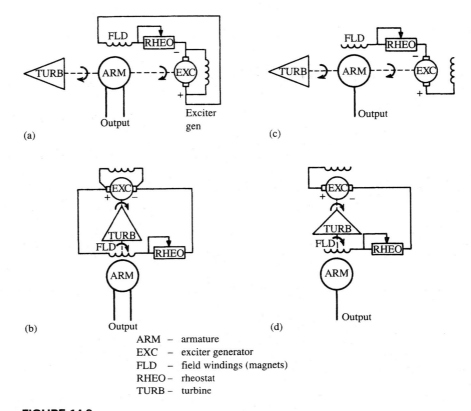

ARM – armature
EXC – exciter generator
FLD – field windings (magnets)
RHEO – rheostat
TURB – turbine

FIGURE 14.3
(a,b) Elementary circuit diagrams; (c,d) corresponding one-line diagrams for rotating-armature and rotating-field machines.

are shorthand drawings that show the interconnection of circuit components. The return wires that complete the circuit are deliberately not shown. One-line diagrams are particularly useful in complex systems as an aid to understanding the mechanics of plant operation.

14.2 THREE-PHASE GENERATORS

Except for very small generators such as those used for home emergency lighting, most AC generators are three-phase machines. A three-phase generator has three separate but identical armature windings that are acted on by one system of rotating magnets. Each winding, called a *phase*, consists of a set of armature coils.

Figure 14.4 shows the construction details for a six-pole, three-phase rotating-field generator. The armature coils are tied securely to a steel bracing ring to prevent the large mechanical forces caused by the armature current from pulling the coils out of shape.

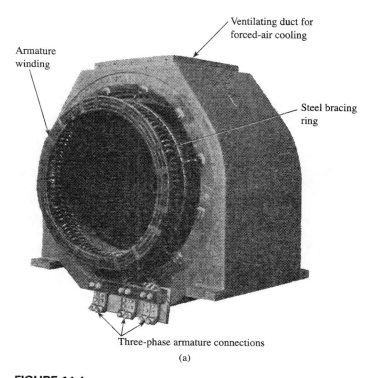

Ventilating duct for forced-air cooling

Armature winding

Steel bracing ring

Three-phase armature connections

(a)

FIGURE 14.4

Three-phase AC generator with a six-pole field. (a) Stationary armature; (b) rotating field and exciter armature (courtesy GE Industrial Systems).

The magnet coils are connected in series and terminate at two slip rings. Carbon brushes, not shown, are pressed against the slip rings and provide the connecting link between the magnet coils and the small DC generator called an exciter. The exciter is mounted on the same shaft as the rotating magnet member. The ventilating duct at the top of the stator and the fans on the rotor provide forced-draft cooling. The field frame for the exciter is not shown.

Figure 14.5 shows the armature and field members of a two-pole, three-phase generator. The field structure is called a cylindrical rotor because it is in the form of a cylinder. The field structure of the machine in Figure 14.4 is called a salient-pole rotor.

Although the three phases are acted on by the same rotating magnets and therefore have the same maximum values of voltage, the spacing and connections of the coils cause the three line-to-line voltages to reach their respective positive maximums 120° apart.

Figures 14.6a and 14.6b illustrate the current and voltage relationships for wye-connected and delta-connected generators, each supplying the same line current and the same line voltage to identical loads. The neutral, or common connection, shown

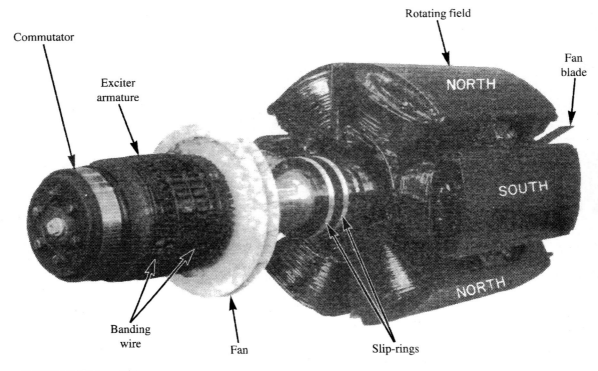

Commutator

Exciter
armature

Rotating field

Fan
blade

NORTH

SOUTH

NORTH

Banding
wire

Fan

Slip-rings

FIGURE 14.4 (cont'd)

with a broken line in the wye arrangement, is not always provided. The voltage and current relationships illustrated in the figures are summarized in Table 14.1.

EXAMPLE 14.3   A three-phase, 2300-V, 60-Hz synchronous generator is supplying a line current of 4600 A. Determine the phase current if the generator is (a) wye connected; (b) delta connected.

a. For the wye connection: $I_{\text{phase}} = I_{\text{line}} = 4600 \text{ A}$

b. For the Delta connection: $I_{\text{phase}} = \dfrac{4600}{\sqrt{3}} = 2659 \text{ A}$

14.3   BRUSHLESS EXCITATION SYSTEM

Figure 14.7a shows the construction details for the rotating member of a self-excited brushless generator system; the stationary members are not shown. The rotor differs from its counterpart in Figure 14.4 in that it has no slip rings or brushes. The generator field, exciter armature, and rectifier assembly are all part of the rotating member. The exciter armature is a small rotating three-phase armature whose output current is rectified (changed from AC to DC) by means of semiconductor diodes and fed to the field

(a)

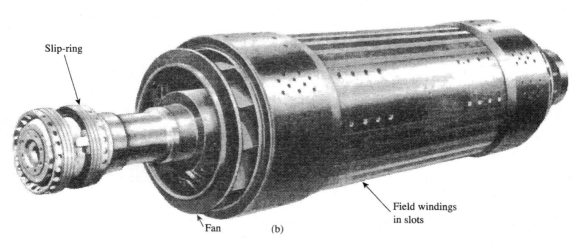

Slip-ring

Fan   (b)

Field windings
in slots

FIGURE 14.5
Three-phase AC generator with a two-pole cylindrical field rotor. (a) Stationary armature
called the stator; (b) rotating field called the rotor (courtesy TECO Westinghouse).

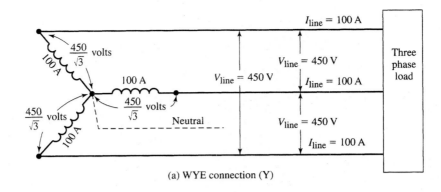

(a) WYE connection (Y)

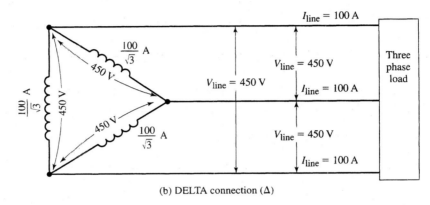

(b) DELTA connection (Δ)

FIGURE 14.6
Current and voltage relationships for wye- and delta-connected generators.

of the main generator. A heat sink absorbs the heat generated by the flow of electrons through the diode.

Operating Principle of the Brushless Generator

The very simplified circuit shown in Figure 14.7b is used to explain the principle of operation. When the prime mover is started, the residual magnetism in the iron poles of the exciter field generates a small three-phase voltage in the rotating exciter arma-

TABLE 14.1
Voltage and Current Relationships in a Three-Phase Generator

Connection	Current	Voltage
Wye (Y)	$I_{line} = I_{phase}$	$V_{line} = \sqrt{3}V_{phase}$
Delta (Δ)	$I_{line} = \sqrt{3}I_{phase}$	$V_{line} = V_{phase}$

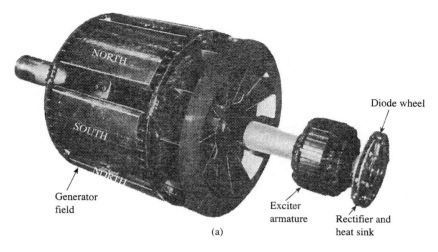

(a)

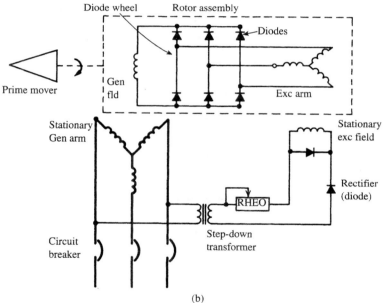

(b)

FIGURE 14.7

Brushless generator system: (a) rotor assembly; (b) elementary diagram showing a simplified manual control circuit (courtesy GE Industrial Systems).

ture. The exciter armature supplies a small direct current (rectified by the diode wheel) to the rotating field magnets of the generator, which in turn induces a small three-phase voltage in the generator armature. This low-output AC voltage is converted to DC, via a step-down transformer and diode rectifier, and fed to the exciter field winding.

The flux generated by current in the exciter field coils adds to the residual magnetism in the pole iron of the exciter. Thus a chain reaction occurs; the increase in exciter flux causes a greater generator output voltage, which causes an even greater exciter field current and corresponding increase in exciter flux, etc., until a voltage regulator (not shown) takes control and holds the exciter current at a predetermined value. The discharge diode in parallel with the exciter field provides an essentially smooth DC current in the exciter field even though the current in the rectifier occurs in pulsed form. The complete brushless excitation system includes a means for restoring residual magnetism for initial start-up or after a prolonged idle period. This process is called *flashing the field*.

14.4 VOLTAGE AND FREQUENCY ADJUSTMENT

To prevent the dimming of lights and the malfunctioning of motors and other electrical equipment, the output voltage and frequency of an AC generator must be held relatively constant as electrical loads are switched on and off. Unfortunately, when a load is connected to a generator, two undesirable reactions take place. First, the prime mover slows down, causing a lowering of both the frequency and the generated voltage. Second, the passage of current through the impedance of the armature winding causes a further drop in output voltage. The reduction in prime-mover speed is caused by the development of a countertorque. When a load is connected to the generator terminals, the current in the armature conductors sets up a magnetic field that interacts with the field of the rotating magnets. The mechanical force produced by this action is in opposition to the driving torque of the prime mover (Lenz's law).

An elementary circuit diagram and its one-line diagram counterpart for a generator and distribution system are shown in Figure 14.8. The AC generator and exciter are driven by a single prime mover. The generator feeds a distribution bus, which in turn supplies various connected loads. A current transformer (CT) is used to provide a reduced but proportional current to the ammeter. Changes in prime-mover speed caused by the application of a load are corrected by an automatic speed governor that uses electronic or mechanical sensors to detect a change in speed and then automatically increases or decreases the energy input to the prime mover. Similarly, an automatic voltage regulator uses electrical sensors to detect a change in voltage and then automatically adjusts the resistance in the field circuit to raise or lower the voltage.

14.5 INSTRUMENTATION AND CONTROL OF SYNCHRONOUS GENERATORS

Figure 14.9 illustrates a functional diagram for a two-generator system, showing the minimum instruments, switches, and adjustable controls that an operating engineer must be acquainted with when operating AC generators singly or in parallel. The con-

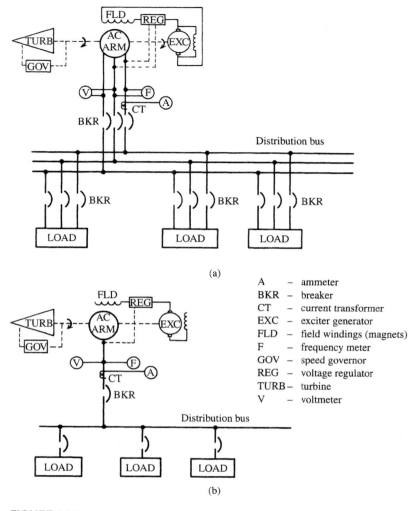

(a)

A	–	ammeter
BKR	–	breaker
CT	–	current transformer
EXC	–	exciter generator
FLD	–	field windings (magnets)
F	–	frequency meter
GOV	–	speed governor
REG	–	voltage regulator
TURB	–	turbine
V	–	voltmeter

(b)

FIGURE 14.8

(a) Elementary diagram of a generator and distribution system; (b) corresponding one-line diagram.

necting lines indicate what each control device operates and where each meter gets its signal. Panels 1 and 3 are identical generator panels and contain the instruments and controls for the particular machine.

Panel 2 in Figure 14.9 contains the instruments and controls necessary for paralleling. As should be expected, the wattmeter and power-factor meters each require both current and voltage signals, and the synchroscope requires voltage signals from each generator.

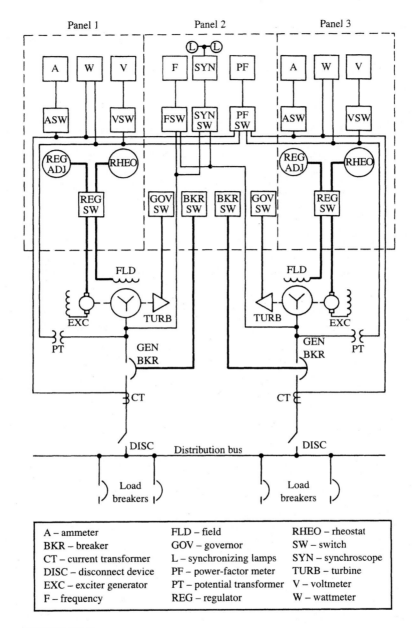

A – ammeter	FLD – field	RHEO – rheostat
BKR – breaker	GOV – governor	SW – switch
CT – current transformer	L – synchronizing lamps	SYN – synchroscope
DISC – disconnect device	PF – power-factor meter	TURB – turbine
EXC – exciter generator	PT – potential transformer	V – voltmeter
F – frequency	REG – regulator	W – wattmeter

FIGURE 14.9
Functional diagram for a two-generator control system.

14.6 SINGLE-GENERATOR START-UP (NO OTHER GENERATOR ON THE BUS)

For those situations where manufacturer's instructions are not available, the following procedure may be used as a guide for single-generator operation:

1. Make sure that the generator breaker is open.
   Warning: The circuit breaker must always be opened before opening or closing the disconnect switch or links. Wear rubber gloves and safety goggles when working behind the switchboard.

2. Close the disconnect switch if one is used (also called isolation switch or disconnect links). These are generally located in the rear of the switchboard. Disconnect links, illustrated in Figure 14.10, must be closed with an insulated wrench. The bolt should be turned until moveable link A makes solid contact with stationary link B. The purpose of the "disconnect" is to allow complete isolation of the breaker from the bus for the purposes of test, maintenance, or repair.

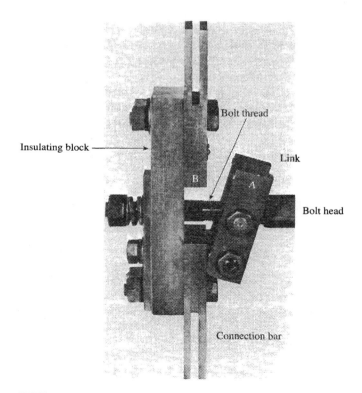

FIGURE 14.10
Switchboard disconnect links (courtesy GE Industrial Systems).

3. Switch the voltage regulator to "manual."[2]
4. Start the turbine or engine and bring it up to speed.
5. Using the governor-control switch, adjust the frequency to the rated name-plate frequency.
6. Adjust the manual voltage control to obtain rated voltage.
7. Switch the voltage regulator to automatic and adjust to rated voltage.
8. Close the circuit breaker manually or by turning the breaker-closing switch to "close."

14.7 PARALLELING SYNCHRONOUS GENERATORS

Multiple generators operating in parallel provide economy of operation and flexibility for scheduling routine maintenance. As increases in oncoming load approach the rated load of the machine or machines already on the bus, additional machines may be paralleled to share the load. Similarly, as the bus load decreases, one or more machines can be taken off the line to allow the remaining units to operate at higher efficiencies.

Phase Sequence

The phase sequence of new or refurbished generators should be checked before paralleling to see if they have the same phase sequence as the bus. Phase sequence may be determined with commercially available phase sequence indicators[3] or with a small, properly fused, three-phase induction motor of the same voltage rating as the bus voltage. When using a motor for this purpose, mark the three leads, connect them to the bus terminals of the circuit breaker, and note the direction of rotation. Then transfer the three motor leads to the corresponding generator terminals of the breaker and again note the direction of rotation. If the motor rotation is the same as before, the incoming machine has the same phase sequence as the bus. Any attempt to parallel a generator with another already on the bus, but with opposite phase sequence, may black out the plant, and may cause serious damage to the machines and associated equipment.

The dangers associated with incorrect phase sequence are explained with the aid of Figure 14.11. Figure 14.11a shows generator 2 connected to the system bus, and generator 1 with its breaker open. Both machines are operating at rated voltage, have the same phase sequence, and their respective voltages are in phase. The three-phase voltage waves for generator 1 and generator 2 are shown in Figure 14.11b and 14.11c, respectively. As indicated by the two sets of curves, at any given instant of time (for example, time t_1) voltages b, a, and c of generator 1 will be equal to and in the same direction as corresponding voltages b', a', and c' of generator 2. Thus, if breaker 1 is closed, there will be no interchange of energy between machines; both machines will feed energy to the system bus.

---

[2] Raising the voltage by manual control, before switching to automatic mode, is a precautionary measure to avoid excessive current in the regulator and field during start-up.

[3] See Section 11.3, Chapter 11, for information on commercially available phase-sequence indicators.

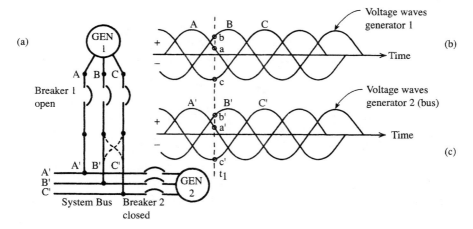

FIGURE 14.11
Dangers associated with incorrect phase sequence.

However, if any two of the three cables from the bus to breaker 1 are interchanged, for example, B' and C' shown with broken lines in Figure 14.11c, then the voltages corresponding to the top and the bottom of each pole of breaker 1 will be, respectively, a to a', b to c', c to b'. For this condition, at time t_1, a positive voltage b of generator 1 will connect to the same breaker pole as a negative-voltage c' of generator 2. Similarly, a negative-voltage c will connect to the same breaker pole as a positive-voltage b'. If the breaker is closed for this condition, a violent interchange of energy between machines will result, and the circuit breaker overload device will trip the breaker.

Aside from the impossibility of operating AC generators in parallel with opposite phase sequence, if the three-phase power lines supplied to a factory, or to a ship for shore power, inadvertently had two of the three incoming power cables reversed, every three-phase motor in the plant would run in the reverse direction. This can have disastrous consequences.

Basic Considerations for Successful Paralleling

Assuming correct phase sequence, the voltage of the incoming machine should be adjusted to obtain a value approximately equal to the bus voltage. Although a few volts higher is preferred, a few volts lower will not adversely affect the operation.

The frequency of the incoming machine should be adjusted to be a fraction of a hertz higher than the bus frequency; approximately 0.1 Hz is preferred. This will ensure it will take on the load at the instant it is paralleled. If the frequency of the incoming machine is slightly lower than the bus frequency, the incoming machine will be driven as a motor by the other generators on the bus, at the instant the circuit breaker is closed. Synchronous generators in parallel are locked in rotational synchronism, whether motoring or generating.

The no-load speed (and hence the no-load frequency) of a synchronous generator is changed by remote control from the generator panel. A control switch on the generator panel actuates a servomotor that raises or lowers the no-load speed setting of the governor.

Although it is possible to adjust the frequency of the incoming machine to the exact frequency as the bus and to be in phase with the bus, the incoming machine will "float" on the line when paralleled. It will not be driven as a motor, nor will it take load. Hence, there is no advantage to this procedure.

Since a generator is paralleled for the specific purpose of taking some of the bus load, it is advantageous to adjust it to a fraction of a hertz higher than the bus frequency prior to paralleling. With its frequency higher, the voltage wave of the incoming machine will slide slowly past the voltage wave of the bus. This is illustrated in Figure 14.12; the voltage wave representing the incoming machine is shown slightly higher than the bus voltage, and its frequency slightly higher than the bus frequency.

The incoming machine should be paralleled with the bus at an instant when its voltage wave coincides with that of the bus (0° in Figure 14.12). Note how the error angle changes with time when the voltage of the incoming machine has a different frequency than that of the bus.

Synchroscope

A synchroscope, shown in Figure 14.13, indicates the instantaneous angle of phase displacement (error angle) between two voltage waves. If the voltage waves have different frequencies, as shown in Figure 14.12, the error angle will always be changing, causing the pointer to revolve. When properly connected, a synchroscope indicates the

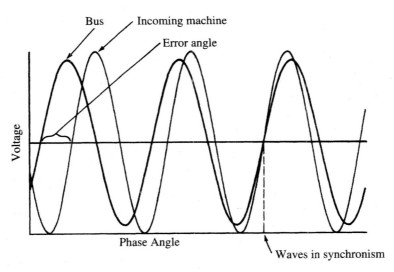

FIGURE 14.12
Voltage waves of bus and incoming machine.

FIGURE 14.13
Synchroscope (courtesy GE Industrial Systems).

electrical speed (frequency) of the incoming machine with respect to the bus. If the synchroscope revolves in the direction marked SLOW, the incoming machine has a lower frequency than that of the bus. If the synchroscope revolves in the direction marked FAST, the incoming machine has a higher frequency than that of the bus. If the frequency of the incoming machine is equal to the bus frequency, the pointer will not revolve, and the position of the pointer will indicate the error angle. If this happens, the prime mover of the incoming machine should have its speed increased by adjusting its governor control. The incoming machine should be paralleled with the bus by closing its breaker at the instant the synchroscope pointer enters the zero-degree position while revolving slowly in the FAST direction. Once paralleled, the synchroscope pointer will no longer revolve, but will stay in the 0° position.

Due to the slight delay caused by human reaction time and breaker closing time, it is good practice to start the breaker closing operation one or two degrees before the synchroscope pointer reaches the 0° position. Assuming a 60-Hz system, a breaker with a 15-cycle closing time takes 15/60 or 1/4 second for the closing mechanism to complete its operation. A 30-cycle closing device takes 30/60 or 1/2 second for closing to occur.

If the breaker is closed at an angle other than 0° (12 o'clock on the synchroscope), a crosscurrent will occur between the incoming generator and the system bus. The severity of this crosscurrent depends on how far out of synchronism the machines are when the breaker is closed; the greater the error angle, the higher the crosscurrent. The worst condition corresponds to an error angle of 180° and has the same effect as a short circuit at double voltage. If the bus is already heavily loaded, the entire system may be blacked out. Furthermore, the high transient current associated with an extreme out-of-phase attempt at synchronization causes excessive oscillating torques on both the stator and rotor windings, which can result in immediate damage or, at best, a decrease in its service life.

Under no circumstances, emergency or otherwise, should the breaker be closed when the error angle exceeds 10°. Closing the generator breaker when the error angle is greater than 10° will cause a high transient synchronizing current between the incoming generator and the bus, which may cause automatic tripping of the generator breaker.

Synchronizing Lamps

Synchronizing lamps provide a means for checking the synchroscope and can be used as a secondary method for determining the correct instant for paralleling if the synchroscope is defective. The lamp connections are shown in Figure 14.14.

Assuming the incoming machine and the bus have the same phase sequence, but slightly different frequencies, both lamps will go bright and dark in unison, and both will be dark when the synchroscope is at approximately 0 degrees.

If the cables from the generator to the breaker are wrongly connected (as previously shown with broken lines in Figure 14.11a), the phase sequence at breaker 1 will be wrong, and the synchronizing lamps will not go bright and dark together.

The significant correspondence between the synchroscope and the synchronizing lamps is the 0°-lamps-dark relationship. When the lamps are dark, the synchroscope pointer must be at 0°, or the synchroscope is defective; the lamps are always correct, assuming that they are not burned out. When using lamps for synchronizing, the duration of the dark period should be timed, and the breaker closed midway through the dark period. Timing is important, because the dark period may extend over 15° or more of error angle. Lamps used for this purpose are generally of the nonfrosted type. One disadvantage of using synchronizing lamps is that they do not indicate whether the incoming machine is fast or slow.

Both lamps must have the same voltage and wattage ratings. The voltage rating must be as specified by the manufacturer. Because in all but a few applications the

FIGURE 14.14
Synchronizing lamp circuit.

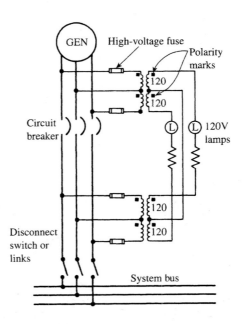

generator voltage is greater than 250 volts, potential transformers are connected between the generator and the synchroscope (or lamps) to reduce this voltage to the rated synchroscope (or lamp) value of 120 or 240 volts. Accidental reversing of the connections in the primary or secondary of the transformer will cause the synchroscope and the synchronizing lamps to give false indications; the synchroscope will indicate 0° and the lamps will be dark when the generator is 180° out of phase with the bus. Hence, all polarity connections of repaired or replaced potential transformers must be carefully checked to avoid this very serious but easily made error.

14.8 TRANSFER OF ACTIVE POWER BETWEEN SYNCHRONOUS GENERATORS IN PARALLEL

The transfer of active power between synchronous generators in parallel must be accomplished by adjustment of the governor controls of the respective prime movers. The machine that is to take load must have more energy admitted to its prime mover, and the machine that is to lose load must experience a reduction in the energy to its prime mover. This is accomplished by adjustment of the no-load speed setting of the respective prime movers' governors.

Assuming the bus load is to be divided equally between the incoming machine and the machine already on the bus, turn the governor no-load speed setting of the incoming machine to "raise," and that of the machine on the bus to "lower" until the kW load is divided equally between machines. If the machines are not identical, the kW load should be divided between machines in proportion to their kW ratings.

If both governor controls are close enough, they may be operated simultaneously. However, if they are too far apart, the transfer should be made in small increments. Turn the governor control of the incoming machine to "raise," and hold it there until a small amount of load is transferred; then turn the governor control of the other machine to "lower" until another equal increment is transferred. Repeat until the desired load transfer is accomplished. While making this transfer, keep an eye on the frequency meter to avoid excessive changes in frequency.

Prime-Mover Governor Characteristics

A typical prime-mover governor characteristic, shown in Figure 14.15, is a plot of prime-mover speed (or generator frequency) vs. active power (kW). Although usually drawn as a straight line, the actual characteristic has a slight curve. The drooping characteristic shown in the Figure 14.15 provides inherent stability of operation when paralleled with other machines. Machines with zero droop, called *isochronous machines,* are inherently unstable when operated in parallel with other isochronous machines; they are subject to unexpected load swings, unless electrically controlled with solid-state regulators. The no-load speed setting (and hence the no-load frequency setting) of a synchronous generator can be changed by remote control from the generator panel by using a remote-control governor switch, shown as GOV SW in Figure 14.9. The switch actuates a servomotor that repositions the no-load speed setting of the

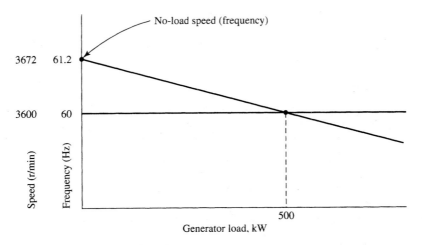

FIGURE 14.15
Typical prime-mover governor characteristic.

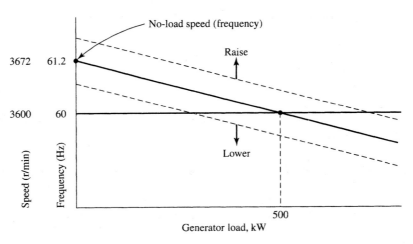

FIGURE 14.16
Curves for different no-load speed settings.

governor, raising or lowering the characteristic without changing its slope. Curves for different no-load speed settings are shown with broken lines in Figure 14.16.

Governor-Speed-Droop

The governor-speed-droop (GSD) expressed in terms of frequency is

$$GSD = \frac{f_{nl} - f_{rated}}{f_{rated}} \times 100 \qquad (14-5)$$

where: GSD = governor-speed-droop (%)
             f_{nl} = frequency at no load (Hz)
             f_{rated} = frequency at rated load (Hz).

EXAMPLE 14.4   A 2300-V, 1200-kW, diesel-driven synchronous generator is operating at rated load and a frequency of 60 Hz. The governor-speed-droop is 3%. Determine the frequency if the circuit breaker is tripped (no load).

Solution
From Eq. (14–5),

$$f_{nl} = \frac{GSD \times f_{rated}}{100} + f_{rated} = \frac{3 \times 60}{100} + 60 = 61.8 \text{ Hz}$$

Figure 14.17 graphically illustrates the transfer of load between two turbine generators with identical droop characteristics. Machine B carries a load of 150 kW at 60 Hz. Machine A has just been paralleled and is carrying no load. The initial conditions are indicated by subscript 1 and the solid lines. To transfer some load from machine B to machine A without permanently changing the frequency of the system requires adjustment of both governors.

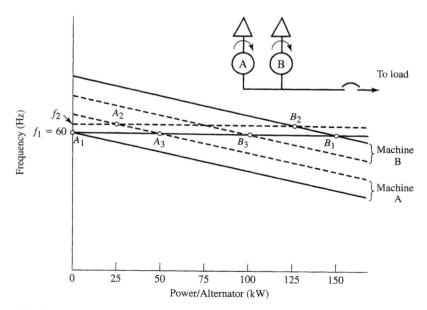

FIGURE 14.17
Graphical illustration of the transfer of power between two turbine generators.

Assume that one-third of the load on machine B is to be transferred to machine A; that is, machine B is to carry 100 kW and machine A is to carry 50 kW. The governor control switch of machine A is turned to RAISE and held there until one-half the amount of load that is to be transferred is shifted, namely, 25 kW. This adjustment raises the no-load speed setting, which raises the entire characteristic of machine A without changing its droop. Because 25 kW was removed from machine B, it also increases in speed. Its characteristic, however, has not been shifted. The system is now operating at some higher frequency f_2, and the characteristic of machine B intersects the new and higher frequency line at 125 kW, as denoted by subscript 2.

To complete the transfer of load, the governor control switch of machine B is moved to LOWER and held there until the remaining 25 kW is transferred. The characteristic of machine B is thereby lowered, and the frequency of the system comes back to its original 60-Hz value. This is indicated by subscript 3, where machine A now takes 50 kW and machine B takes 100 kW. Throughout the transition period, the slope of the curves did not change. The entire characteristic of each machine was moved either up or down.

During the transition period, when more energy was admitted to the prime mover of machine A, the energy balance of the system was disturbed; more energy was fed into the turbine generators than was going out. The total input was greater than the output, resulting in an increase in system speed and frequency. Prime-mover A increased its speed because of the direct increase in energy input brought about by raising the no-load speed setting of its governor. Prime-mover B increased its speed because of a reduction in its load. Except for the brief interval when the power angles are changing, both machines maintain identical speeds. Synchronous machines with the same number of field poles, operating in parallel, cannot run at different rotational speeds; if one increases in speed, so must the other. Like two engines driving a bull gear through identical pinions and flexible couplings, both machines are locked in rotational synchronism; but the engine with the greater fuel input will do more work than the other.

In actual practice, the division of active power (kW) between machines is accomplished with very little transient change in frequency, by making small governor adjustments (alternately on each machine) until the desired transfer is accomplished.

If the governor control switches are located on a common synchronizing panel (panel 2 in Figure 14.9), both controls can be operated simultaneously, holding one on RAISE and the other on LOWER until the desired transfer is accomplished. However, this can be a little tricky.

Constant-Frequency System

Isochronous machines are sometimes paralleled with droop machines to provide a constant-frequency system, as shown in Figure 14.18. In this system, generator A has a drooping governor characteristic and generator B is isochronous. The loads on machines A and B are 40 kW and 120 kW, respectively, and are indicated in Figure 14.18 by subscript 1. To transfer 60 kW from generator B to generator A, the governor

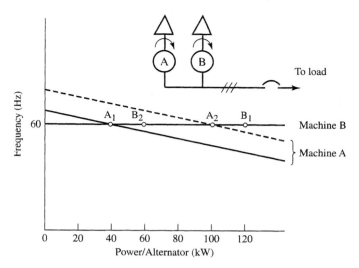

FIGURE 14.18
Isochronous machine paralleled with a droop machine.

control switch of generator A should be turned to RAISE and held there until the entire transfer is completed. The governor control switch of the isochronous machine (generator B) should not be used, since adjustment of its governor would affect the frequency of the system. The transfer of load must be made by adjusting only the governor control of the machine with droop. The final conditions are indicated by subscript 2.

For a given no-load speed setting of its governor, a generator with a drooping characteristic can carry only one value of load at a given frequency. An isochronous machine, however, can carry any value of load from no load to rated load at the given frequency. Thus, if an isochronous machine is in parallel with a droop machine, any increase or decrease in bus load will be absorbed by the isochronous machine.

Inherent Instability of Two or More Isochronous Machines in Parallel

When two or more generators with isochronous governors are in parallel, their operation is very unstable unless controlled by special-purpose electronic governors. The slightest change in frequency may cause a considerable interchange of load between machines. This is illustrated in Figure 14.19. Since no two governors are identical in the true sense of the word, it may be assumed that generator A has zero droop and generator B has a very minute droop. The initial conditions are illustrated by subscript 1, where both generators are shown as having equal shares of the bus load. If a slight decrease in system speed occurs, due perhaps to very slight irregularities in the isochronous governing mechanism, a considerable portion of the bus load will be thrown on generator B, and generator A will then be lightly loaded or driven as a motor; the new condition is shown by subscript 2. Likewise, a slight increase in speed

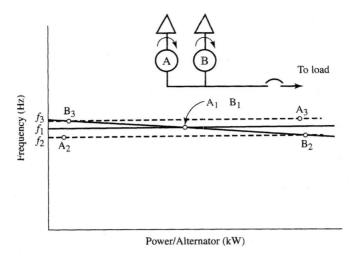

FIGURE 14.19
Inherent instability of isochronous machines in parallel.

would shift the load in the other direction, as indicated by subscript 3. The parallel operation of isochronous machines requires fast-acting electronic governors. Mechanical governors do not respond fast enough to avoid wild load swings.

14.9 TRANSFER OF REACTIVE POWER BETWEEN SYNCHRONOUS GENERATORS IN PARALLEL

Although active power may be divided equally between synchronous machines in parallel by using governor control, the division of reactive power requires adjustment of the internal voltage of each machine. To do this, the DC field excitation of the machine that took some load should be increased, and that of the other machine decreased, until the power-factor meters or var meters of each generator read the same lagging values.

If on manual control, division of reactive power between machines is accomplished by adjusting the respective field rheostats, turning that of the incoming machine in the "raise voltage" direction, and those already on the bus in the "lower voltage" direction until the power factors are equalized.

Machines that are equipped with automatic voltage regulators do not require manual adjustment of their field rheostats to balance the reactive power. A compensating device in the regulator automatically adjusts the kvar distribution and hence the power factor of each machine. If they are not equal, the voltage-adjusting control of the voltage regulators may be used to provide equalization.

Note: Changing the field excitation of an AC generator, when in parallel with others, does not appreciably alter the division of active power (kW) between machines. Active power transfer must be done by governor adjustment.

In factories, hotels, or large ships, where two or or more synchronous generators are operated in parallel to supply the connected load, it is good operating practice to balance both the kW and the kvar loads in proportion to the kW ratings of the machines. By doing so, transient overcurrents and short circuits in the distribution system are shared by the generators, reducing the probability of one generator tripping. The tripping of one generator due to overcurrent will throw the entire load on the remaining unit or units, which may cause them to trip and black out the plant.

If the generator panels are not equipped with a power-factor meter or varmeter, the balancing of kvars may be made through observation of the AC ammeters. The field excitation of the machine that took some load should be increased, and that of the other machines decreased, until the AC ammeters of all machines indicate the same. If carried too far, it will result in an unbalanced situation in the other direction.

14.10 DIVISION OF ONCOMING LOAD BETWEEN SYNCHRONOUS GENERATORS IN PARALLEL

The division of *oncoming* bus load between synchronous generators in parallel is determined automatically by the droops of the respective governors.

Machines with Identical Governor Droops

Machines with identical governor droops will divide all increases in bus load and all decreases in bus load equally between them, regardless of the power ratings of the individual machines and the number of machines in parallel. If, after balancing the bus kW between two turbogenerators with supposedly identical governor droops, the switching of load on or off the bus causes an unbalanced condition, it is an indication that the governor droops of the two machines are not the same. One or both require droop adjustment, which should be done by experienced personnel.

Machines with Dissimilar Governor Droops

Synchronous generator sets with dissimilar governor characteristics do not divide increases or decreases in bus load equally. The machine with the least droop assumes a greater portion of the change in bus load. This is shown in Figure 14.20, where three paralleled generators, each with a different governor droop, are taking equal shares of the bus load at frequency f_1. The broken lines show the effect that different governor droops have on the distribution of oncoming bus load; the new lower frequency f_2 caused by the oncoming bus load causes the machine with the least droop to take a greater share of the oncoming load. Thus, if generators with different power ratings are to be operated in parallel, the respective governor droops should be adjusted so that load distribution between generators will be in proportion to their respective power ratings.

In some applications the parallel operation of machines with dissimilar governor droops is desirable. For example, if one machine is adjusted to have zero droop, and

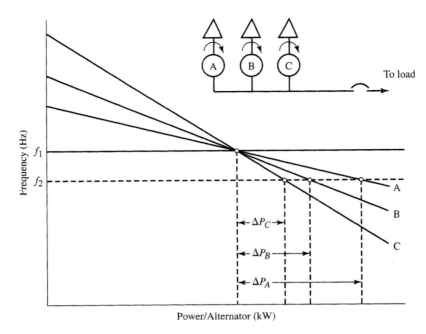

FIGURE 14.20
Machines with dissimilar governor droops.

the others in parallel with it to have a 2 or 3 percent droop, the droop machines may be loaded for optimum operating efficiency. The machine with droop can carry only one value of load at a specific frequency for a given no-load speed setting of its governor. Hence, all fluctuations in the bus load will be absorbed by the machine with zero droop; it will take all the additional oncoming load and will also lose any load that may be disconnected from the bus. Any transfer of active power between such machines must be accomplished by adjusting only the governor control of the machine with droop. Adjusting the governor control of the zero-droop machine will change the frequency of the system.

14.11 MOTORIZATION OF SYNCHRONOUS GENERATORS

If the energy input to the prime mover of a generator is decreased to a point where the energy admitted is insufficient to keep the machine running in synchronism with the bus, the other machines on the bus will feed into the machine and run it as a synchronous motor at the same speed and in the same direction as before. This is known as *motorization* or *motoring*. Although motorization is not harmful to the generator, it does result in an energy loss in the form of an additional kW load on the bus, and in some cases may cause serious damage to the prime mover.

A reverse-power relay, also called a power-directional relay or wattmeter relay, shown in Figure 14.21, is provided to trip the machine off the bus if motorization

FIGURE 14.21
Reverse-power relay (courtesy GE
Industrial Systems).

occurs. A motorized generator can be detected by the reversed reading of its wattmeter and, if detected early enough, can be corrected by admitting more energy to its prime mover.

14.12 REMOVING A SYNCHRONOUS GENERATOR FROM THE BUS

To remove a synchronous generator from the bus, transfer all the active power (kW) to the remaining machine by adjusting the governor controls of both. Trip the circuit breaker of the outgoing machine, switch its voltage regulator to manual, and then shut down the prime mover. Transferring load by tripping the breaker of the outgoing machine, instead of through governor adjustment, should be avoided; it exerts a severe impact force on the mechanical parts of the machine that take the load. If a generator is to be out of service for an extended period of time, the disconnect switch or links (if present) should be opened after generator shutdown and kept open until the generator is ready to go back on line. This prevents damage to the machine in the event that the breaker is inadvertently closed.

14.13 LOSS OF SYNCHRONISM

If a generator pulls out of synchronism with the other machines on the bus, exceedingly heavy currents will circulate between the armature of the affected machine and the bus. Although this fault current has a magnitude closely approaching that of the short-circuit value, it occurs in periodic surges. The effect is identical to that of a synchronous motor that pulls out of synchronism. As each rotor pole slips past a pole of the rotating flux set up in the stator by the bus voltage, a large pulse of stator current and a torque reversal are produced.

The high voltage induced in the field windings by the extremely high current pulsations may cause a flashover at the collector rings, and the violent vibration caused by the torque reversals may cause severe mechanical damage. To prevent serious damage, an out-of-synchronism machine should be removed immediately from the bus by tripping its circuit breaker.

Loss of synchronism may be caused by a low value of field current, a sluggish governor, overload, or attempts to parallel with a high phase angle difference (error angle) between the bus voltage and the voltage of the incoming machine.

14.14 LOSS OF FIELD EXCITATION

The loss of field excitation to a generator operating in parallel with others causes it to lose load and overspeed. Unless tripped off the bus in a short time, the machine may be damaged by excessively high temperature; high armature current is caused by the high voltage differential between the armature and the bus. Furthermore, high currents induced in the field iron and field windings by the armature current will cause rapid heating of the apparatus.

Voltage failure of an AC generator may be caused by an open in the field circuit, an open in the field rheostat, or failure of the exciter generator. A voltage check of the exciter generator will determine if it is at fault. To test the rheostat, short-circuit its terminals and observe the generator voltage. A buildup of voltage indicates an open rheostat. If the exciter is operating at rated voltage and the AC generator fails to build up when the rheostat is short-circuited, the trouble is in the field circuit or the wires connecting the exciter to the field.

14.15 OPERATING OUTSIDE OF THE NAMEPLATE RATING

Synchronous generators are rated on the basis of supplying three-phase power to balanced loads. An excessive imbalance, caused by large single-phase loads, produces eddy currents along the surface of the rotor, in the slot wedges, and in the slot teeth, resulting in overheating of the rotor.

Most generators can be operated with a voltage variation of 5 percent above or below rated voltage and still safely carry rated kVA at rated power factor. If the generator is to be operated at reduced frequency, a proportional decrease from rated voltage and rated kVA is required. Although a reduction in operating frequency of up to 5 percent of rated is not harmful, the manufacturer should be consulted.

Under certain conditions, for example, the heavy air-conditioning load in exceptionally hot weather, it sometimes becomes necessary to operate a generator in excess of its rated kVA. However, because the increased load causes increased heating of the armature and field windings, additional cooling must be provided to prevent damage to the machine. In the case of self-contained air or gas recirculating systems, an increase in the flow of cooling water or other coolant will help keep the temperature down. Under no circumstances should the temperature of the air or gas be allowed to reach or go below the dew point; temperatures below the dew point cause condensation of any entrapped moisture on the windings of the machine.

14.16 SYNCHRONIZING SUMMARY

The procedure for paralleling synchronous generators depends on the particular installation, which ranges from fully automatic, to semiautomatic, to manual. Most modern plants have automatic synchronizers that correctly synchronize a generator with the bus. However, nonautomatic plants and plants with defects in the automatic synchronizing equipment require manual synchronization. If manufacturer's instructions for paralleling are not available, the following procedure may be used as a guide:

1. Make sure that the breaker for the incoming machine is open.
2. If applicable, close the disconnect switch, or if links, close with an insulated wrench. Wear rubber gloves and safety goggles when working behind the switchboard.
3. Switch the voltage regulator to manual, and set it to the minimum volts position.
4. Start the turbine or engine, and bring it up to speed.
5. Using the governor-control switch, adjust the frequency of the incoming machine to approximately 1/10 Hz higher than the bus frequency.
6. Adjust the manual voltage control of the incoming machine to obtain a voltage approximately equal to or slightly higher than the bus voltage.
7. Switch the voltage regulator from manual to automatic and, if necessary, readjust the voltage of the incoming machine to equal the bus voltage.
8. Switch the synchroscope to the incoming machine, and using the governor-control switch, adjust the frequency of the incoming machine until the synchroscope pointer revolves slowly in the "fast" direction.
9. Close the circuit breaker at the instant the synchroscope pointer passes into the zero (12 o'clock) position. This parallels the incoming machine with the bus. *Note:* It is good practice to start closing the breaker one or two degrees before the 0° position, thus ensuring that the actual closing occurs at about 0°.
10. Turn the synchroscope switch to "off."
11. Assuming the bus load is to be divided equally between the incoming machine and the machine already on the bus, turn the governor no-load speed setting of the incoming machine to "raise" and that of the machine on the bus to "lower" until the kW load is divided equally between machines. If the machines are not identical, the kW load should be divided between the machines in proportion to their kW ratings.
12. Adjust the automatic voltage control of the incoming machine in the "raise voltage" direction, and that of the machine that gave up kW load in the "lower voltage" direction, until the power-factor meter indication (or var meter indication) of each generator reads the same lagging values. If the machines are not identical, the kvar load should be divided between the machines in proportion to their kW ratings.

SUMMARY OF EQUATIONS FOR PROBLEM SOLVING

$$\frac{E_2}{E_1} = \frac{n_2 \Phi_2}{n_1 \Phi_1} \tag{14-3}$$

$$f = \frac{Pn}{120} \tag{14-4}$$

Connection	Current	Voltage
Wye (Y)	$I_{line} = I_{phase}$	$V_{line} = \sqrt{3} V_{phase}$
Delta (Δ)	$I_{line} = \sqrt{3} I_{phase}$	$V_{line} = V_{phase}$

$$\text{GSD} = \frac{f_{nl} - f_{rated}}{f_{rated}} \times 100 \tag{14-5}$$

GENERAL REFERENCE

Hubert, Charles I., *Electric Machines: Theory, Operation, Applications, Adjustment, and Control*, 2nd ed. Prentice-Hall, Upper Saddle River, NJ, 2001.

REVIEW QUESTIONS

1. What adjustments can be made to an operating AC generator to change the generated voltage?
2. What adjustment can be made to an operating AC generator to change its frequency?
3. Explain the operation of a brushless excitation system used with some AC generators.
4. When a relatively heavy load is connected to an AC generator, the frequency and voltage output of the machine are reduced. Explain why this happens. What equipment is provided to minimize this effect, and how does it work?
5. What is the purpose of generator disconnect links when a generator has a circuit breaker? Where are the links located, and how are they opened and closed? What cautions must be observed?
6. Outline the steps required for paralleling an AC generator with another one already on the bus. Assume that the machine on the bus is operating at 460 V, 60 Hz, and has a 300-kW load at 0.8 pf lagging. Include the correct procedure for dividing the active and reactive components of the bus load equally.
7. Describe the procedure for dividing the kW load between two generators in parallel when the governor controls are too far apart to be operated simultaneously.
8. Why is it good practice to start the breaker-closing operation 1° or 2° before the 0° position?

9. Why is it good practice to balance both the kW and kvar load in proportion to the kW ratings of the machines?

10. Assume that two identical generators are operating in parallel and sharing a small bus load. Outline the correct procedure for removing one of the generators from the bus.

11. What is motorization? Is it harmful? How can it be detected and corrected?

12. What damage can occur if an out-of-synchronization machine is not immediately removed from the bus?

13. Is there any advantage to adjusting the frequency of an incoming machine to equal the bus frequency before paralleling? Explain.

14. Interpret each of the following synchroscope indications: (a) rotating slowly clockwise; (b) rotating slowly counterclockwise; (c) stationary at 90°.

15. What is the maximum error angle permissible when paralleling generators? What bad effect does a large error angle have on the machines? Explain.

16. If a synchroscope indicates that conditions are right for paralleling but the synchronizing lamps indicate otherwise, which is correct? Explain.

17. State the correct procedure for paralleling when using synchronizing lamps. Are there any restrictions on the type, voltage, and wattage ratings of lamps when used as synchronizing indicators?

18. What effect do different governor droop characteristics have on the division of oncoming load between generators? Explain with the aid of a sketch.

19. Two generators A and B are in parallel, taking equal shares of the bus load. The governor of prime-mover A has zero speed droop, and the governor of B has a 2 percent speed droop. If an additional 100-kW load is connected to the bus, what percentage of the additional load will be taken by each machine?

20. Generator A is connected to the bus and is supplying the total bus load of 400 kW at a lagging power factor of 0.8. Generator B is paralleled with generator A, and the bus load is divided equally between machines using the governor controls. Is the power factor of each machine 0.8? Explain.

21. Assume that a damaged AC generator has new armature coils installed, the terminals are connected to the switchboard, and the machine is apparently ready to be paralleled with the bus. State in detail the procedure that you would follow to determine whether or not the phase sequence is correct.

22. What effect does an opposite phase-sequence power supply have on the operation of three-phase motors?

23. What will happen if a generator with the wrong phase sequence is paralleled with the bus?

24. State the possible reasons why an AC generator might fail to build up voltage.

25. What will happen if a generator in parallel with others loses its field excitation?

PROBLEMS

14–1/1. A 60-Hz, 450-V three-phase generator operates at 1800-rpm. Assume that a defect in the excitation system that supplies the field current causes the output voltage to drop to 400 V. (a) At what speed must the generator operate to obtain 450 V? (b) What effect does this new speed have on the frequency?

14–2/1. Determine the rated operating speed for a 450-V, three-phase, six-pole, 60-Hz, 1200-kVA generator.

14–3/1. Determine the operating speed for a three-phase, 12-pole, 450-V, 60-Hz, 600-kW generator.

14–4/1. Determine the frequency of a four-pole, 600-kW, three-phase generator operating at 1750 rpm.

14–5/1. A 450-V, two-pole, 60-Hz, three-phase generator is operating at a reduced speed of 3000 rpm. Determine the frequency at this lower speed.

14–6/1. Determine the number of poles in a generator that generates 4160 V and 60 Hz at 600 rpm.

14–7/1. If an AC generator develops 200 V at 60 Hz, what will be its voltage and frequency when both the speed and the strength of the magnetic field are doubled?

14–8/2. Calculate the phase current of a delta-connected generator that supplies 320 A of line current at 4160 V to a three-phase induction motor. Sketch the circuit.

14–9/2. What is the phase current of a wye-connected generator that supplies 50 A to a three-phase resistor load? Sketch the circuit.

14–10/2. A three-phase generator has a voltage rating of 260 V per phase. What is the line voltage when it is (a) wye-connected; (b) delta-connected? Sketch the circuits.

14–11/2. A three-phase generator has a line voltage rating of 460 V. What is the phase voltage when it is (a) wye-connected; (b) delta-connected? Sketch the circuits.

14–12/8. A 600-V, 1000-kW, diesel-driven synchronous generator is operating at rated load and a frequency of 60 Hz. The governor speed droop is 2%. Determine the frequency if the circuit breaker is tripped.

14–13/8. A 240-V, 500-kW, diesel-driven synchronous generator is operating at rated load and a frequency of 60 Hz. The governor speed droop is 2.5%. Determine the frequency if the circuit breaker is tripped.

15

Troubleshooting and Emergency Repair of Three-Phase Motors

15.0 INTRODUCTION

The objective of a good preventive maintenance program is to reduce the number of avoidable breakdowns. However, even with the best maintenance program, problems will arise and electrical equipment will fail. This chapter provides a methodical commonsense approach to identifying and troubleshooting some of the more common problems associated with AC machinery. Suggestions for emergency repairs are also provided to keep equipment safely operating until a replacement is available or an overhaul is scheduled.

This section will not cover difficulties of a mechanical nature such as brush tension, slip-ring wear, bearing wear, or vibration because these topics are discussed in other chapters. In addition, electrical apparatus does not by itself reverse the connections to its field, armature, or control system. Unless equipment tampering is suspected, it is assumed that connections within the motor controller and motor are correct.

The nameplate data of the machine should always be checked against the actual operating values of voltage, current, frequency, temperature, speed, etc. Poor performance can often be traced to operating above or below the rated nameplate values.

If a motor does not start (assuming the cables between the motor and control were not disconnected or severed) and no humming/growling noises or smoke is coming from the motor, then the fault lies in the control system.

The branch-circuit fuses or circuit breaker, thermal overload relay, and small fuses on the motor control panel should be the first things checked. If rated voltage is available at the starter and if the fuses and breakers check out OK, push the start button to energize the starter and then measure the output voltage at starter terminals T_1, T_2, and T_3. If no output voltage is present, the fault is within the starter.[1]

---

[1] See Chapter 22 for troubleshooting and maintenance of motor controls.

Troubleshooting a three-phase motor should begin with a troubleshooting chart such as that shown later in Table 15.2. The chart lists the most likely causes of specific motor symptoms, and refers the reader to the pertinent book sections that could aid in diagnosis and repair.

15.1 LOCATING SHORTED STATOR COILS

Shorted coils in the stator of an AC machine are generally indicated by higher than normal current. A shorted coil that is not identified by burning, overheating, or discoloration of its insulation may be located with a growler.

A growler, shown in Figure 15.1, consists of a coil of wire wound around an iron core and connected to a source of alternating current. When placed on the stator core, the growler coil acts as the primary of a transformer, and the stator coils act as the secondary. A *feeler,* fashioned from a thin strip of steel, or hacksaw blade with the teeth ground off, is used as a short detector.

The alternating magnetic flux set up by the growler passes through the window of the armature coil, generating an alternating voltage in the coil. If the coil is shorted, an alternating current induced in the coil will cause an alternating magnetic field to encircle the shorted conductors.

With the growler energized, the feeler is moved from slot to slot. When the feeler is moved over a slot containing the shorted coil, the alternating magnetic field will alternately attract and release the feeler, causing it to vibrate in synchronism with the

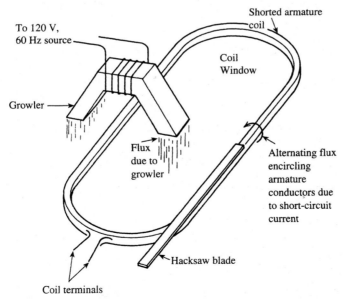

FIGURE 15.1
Principle of growler action.

FIGURE 15.2
Testing stator coils for short circuits using a portable growler (courtesy United States Merchant Marine Academy).

alternating current. A strong vibration of the feeler accompanied by a growling noise indicates that the coil is shorted.

A growler with a built-in feeler in its center is shown in Figure 15.2. The built-in feeler is particularly handy for very small stators, where there is insufficient room for a separate feeler. The growler is moved from slot to slot, and vibrates with a very loud growling noise when the growler is directly over a slot that contains the shorted coil. If there is sufficient room, a separate feeler will provide a higher degree of sensitivity.

If a growler is not available, the application of reduced voltage to the stator, approximately 25 percent of rated nameplate voltage, will cause the shorted coil to become noticeably warmer than the others. A test stand with a three-phase variable voltage supply is recommended. The rotor must be removed for this test.

15.2 LOCATING GROUNDED STATOR COILS

If a stator is grounded, as indicated by a megger insulation test, the grounded coil may be located by applying reduced voltage between the motor terminals and ground, as illustrated in Figure 15.3. The rheostat should be adjusted to permit sufficient current to flow through the defective coil to cause overheating and smoking at the point of

FIGURE 15.3
Circuit connections for smoking out a ground.

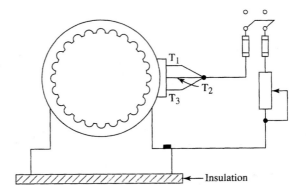

←— Insulation

grounding. Before making this test, the stator should be insulated from ground with dry wood blocks or other heavy insulating material, and an approved fire extinguisher should be present at the test site.

15.3 EMERGENCY REPAIR OF A SHORTED OR GROUNDED STATOR COIL

An emergency repair of a shorted or grounded stator may be made by cutting the defective coil in half. Depending on the size, the stator coil can be cut with a diagonal type of pliers, bolt cutter, hacksaw, or hammer and chisel. Figure 15.4 illustrates how this cut can be made. This procedure is illustrated in Figure 15.5.

The *ends of the shorted coils* should be disconnected, properly insulated, and tied down to prevent movement. The coil leads should be disconnected, and a jumper should be used to complete the circuit. The success of such emergency repairs depends to a good extent on the number of coils in the stator. Cutting one coil out of the circuit of a machine that has only a few coils has a serious effect on its operation; whereas a large machine, such as a 10,000-hp synchronous propulsion motor of an electric-drive ship, will operate satisfactorily at reduced load, even though several coils may be cut out of the circuit. However, whenever possible, the manufacturer's recommendations for emergency repairs should be followed.

15.4 LOCATING AND EMERGENCY REPAIR OF OPEN STATOR COILS

Single-Circuit Stators

If the open occurs in a single-circuit stator, as illustrated in Figure 15.6, the machine will not start. It will hum and have symptoms identical to those produced by a blown fuse in one of the three line leads. An open that occurs while the machine is carrying

FIGURE 15.4
Cutting out a defective coil as part of an emergency repair of a shorted stator
(courtesy United States Merchant Marine Academy).

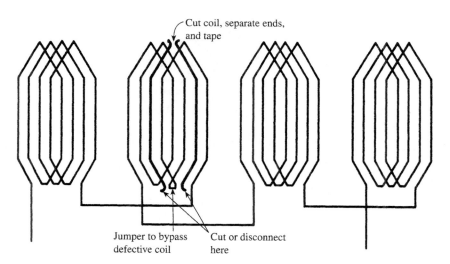

FIGURE 15.5
Emergency repair of a shorted or grounded stator.

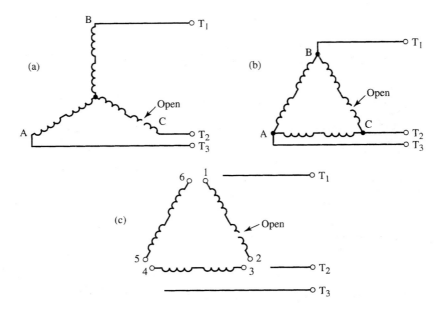

FIGURE 15.6
Single-circuit, three-phase stator connections showing representative open coils: (a) wye connection; (b) delta connection; (c) delta with all connections opened.

load causes arcing and generally burns the insulation in the vicinity of the fault. However, if the open is not apparent by visual inspection, a series of tests must be made to determine the exact location of the fault.

A preliminary test should be made to determine the type of winding, wye or delta. To do this, megohmmeter readings should be taken between (T_1 and T_2), (T_2 and T_3), and (T_3 and T_1). A delta connection will indicate zero ohms for all combinations, whereas a wye connection will indicate zero ohms for one combination only, T_1 and T_3 of Figure 15.6a.

To locate the phase that contains the open in a delta-connected stator, the connections at A, B, and C must be opened, as shown in Figure 15.6c, and the nine leads (including the three line leads) must be carefully marked so that they may be reconnected in the same manner after the emergency repair is made. The six stator leads should then be tested with a megohmmeter for continuity; the two stator leads that indicate infinity, when tested with each of the other five stator leads, connect to the phase that contains the open (leads 1 and 2 in Figure 15.6c).

The actual location of the open may be determined by testing the faulty phase with a megohmmeter and needle-point probes, as illustrated in Figure 15.7. The needle points easily penetrate the insulation. One probe should be connected to one end of the phase, and the other probe should be shifted from junction to junction until the

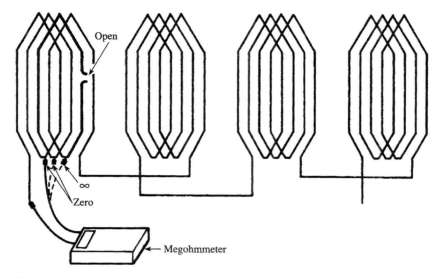

FIGURE 15.7
Tracking down an open coil in one phase of a stator that has four coils per pole for each phase.

megohmmeter indicates infinity (∞). As indicated in Figure 15.7, the fault is between adjacent junctions that have a zero resistance reading and an infinity resistance reading, respectively, on the megohmmeter. In most cases the "open" is a loose connection at the junction, where the end of one coil connects to the beginning of another coil.

To locate the phase that contains the open in a wye-connected stator, as shown in Figure 15.6a, the open is in the phase that indicates infinity when tested with the other two phases. Thus in Figure 15.6a, an infinity reading will be indicated when testing between T_2 and T_1, and between T_2 and T_3.

The actual location of the open coil can be determined by testing the faulty phase with a megohmmeter and needle-point probes as illustrated in Figure 15.7.

Multicircuit Stators

Large machines often have the coils of each phase connected in series–parallel arrangements, as illustrated in Figure 15.8. Such machines will operate with an open in one coil but will not deliver rated power. The fault can be determined by opening the line and phase connections at A, B, and C and following the procedure outlined for single-circuit machines. The leads should be marked so that proper reconnection is possible after the repair is made.

An emergency repair of an open stator may be made by following the repair procedure outlined for a shorted coil.

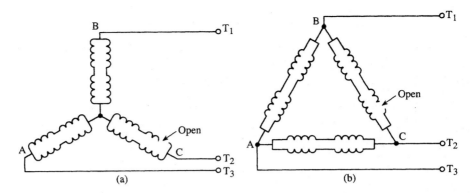

FIGURE 15.8
Two-circuit, three-phase stator connections, showing open coils: (a) wye connection;
(b) delta connection.

15.5 SQUIRREL-CAGE-ROTOR TROUBLES

Broken rotor bars cause a squirrel-cage motor to operate unsatisfactorily. The machine is noisy, has a low starting torque, and does not come up to speed, even though rated voltage, rated frequency, and rated load are applied. Open rotor bars that are not apparent by visual inspection can be detected by applying 10 to 20 percent of rated voltage to only two stator terminals, and the line current measured with a digital clamp-on ammeter as the rotor is turned slowly by hand. This is illustrated in Figure 15.9.

A broken rotor bar causes variations in the stator current as the rotor is turned. The number of variations in one revolution of the rotor is equal to the number of poles in the stator.[2] If the percent difference between high and low stator current (as the rotor is slowly turned) is 3 percent or greater, it indicates a broken bar [1]. Breaks usually occur at the rotor end ring. An emergency repair of bar-type rotors may be made by rebrazing the defective bars to the end ring. Cast-aluminum rotors can seldom be repaired.

---

[2]See Section 11–4, Chapter 11, for determining the number of poles for a given speed and frequency.

FIGURE 15.9
Ammeter test for broken rotor bars.

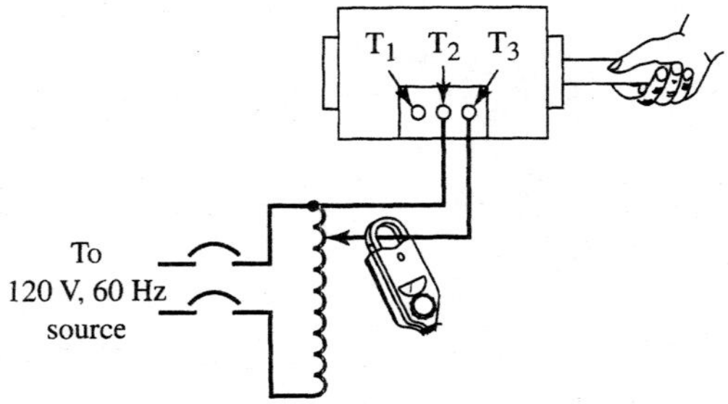

T_1 T_2 T_3

To
120 V, 60 Hz
source

EXAMPLE 15.1 Assuming stator current fluctuations, when testing, a certain squirrel-cage rotor is 22 A and 21 A. Does the rotor have a broken bar?

Solution

$$\text{Current fluctuation} = \frac{22 - 21}{21} \times 100 = 5\%$$

The rotor has a broken bar that should be repaired or the rotor should be replaced.

15.6 WOUND-ROTOR TROUBLES

Defective Rheostat

When a wound-rotor motor fails to start or operates at reduced speed, its rheostat should be checked for an open resistor or bad contact at the rheostat slider. The brushes should be checked for freedom of movement and for good contact with the slip rings. If the rheostat is at fault, short-circuiting the three slip rings will cause the motor to run. If shorting the slip rings does not start the machine, there is an open in either the stator or the rotor circuit. Note that the starting torque with all resistance out is relatively low. Hence, the motor should have no load or be only lightly loaded when making this test.

Shorted Rotor Coils

A shorted coil within the rotor causes the rotor to draw excessive current at the instant of starting. This is usually evidenced by arcing, smoking, or sparks at the slip rings (collector rings). For a quick test, raise the brushes from the slip rings and apply rated three-phase voltage to the stator. If the rotor rotates, one or more rotor coils are shorted. Shorted coils in a wound rotor may be located with the use of a growler as discussed for shorted stator coils in Section 15.1. Note that it is a good possibility that the short is at the slip rings or leads connecting to the slip rings.

15.7 LOCATING A SHORTED FIELD COIL IN A SYNCHRONOUS MOTOR

To check for short circuits in the field coils of a salient-pole generator or synchronous motor, a 120-V or 240-V, 60-Hz source should be applied to the slip rings via the brush pigtails, and the voltage drop measured across each field coil, as illustrated in Figure 15.10. If the voltage drop across a coil is less than 90 percent of the average, one or more turns in that coil may be shorted.

FIGURE 15.10
Locating a shorted field coil.

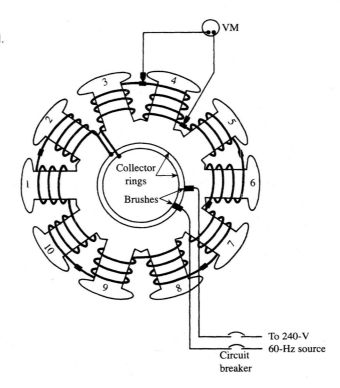

Making this test with direct current does not produce significant results; one shorted turn has very little effect on the voltage drop of a coil. For example, if each coil has 200 turns, one shorted turn will result in a resistance change of only one two-hundredth of the coil resistance. This very small change in resistance will have no noticeable effect on the voltage measured across the coil.

However, with alternating current applied, the transformer action of only one shorted turn appreciably lowers the impedance of the affected coil and the two adjacent coils, resulting in noticeably lower voltages in those coils.

EXAMPLE 15.2 A short-circuit test of the field winding of a certain 10-pole synchronous motor was made using a 240-V, 60-Hz source. The voltage drops across the field coils are listed in Table 15.1. Which field coils have shorted turns?

Solution

The average voltage drop is 240 ÷ 10 = 24 V
90% voltage = 24 × 0.90 = 21.6 V
Since coil 4 has a voltage drop of 17 V, it has shorted turns.

TABLE 15.1
Voltage Drop across Field Coils

Coil	Voltage Drop	Coil	Voltage Drop
1	25	6	25
2	25	7	25
3	24	8	25
4	17	9	25
5	24	10	25

Faulty field coils should be removed and the coils unwound to note the general condition of the insulation and to locate and reinsulate the fault. If the insulation is very brittle and cracked, the adjacent coils may be similarly deteriorated and, though not shorted, should be replaced.

15.8 LOCATING AN OPEN FIELD COIL IN A SYNCHRONOUS MOTOR

An open coil may be located by connecting the field circuit to its rated DC voltage or less and testing with a voltmeter from one line terminal to successive coil leads, as illustrated in Figure 15.11. The voltmeter will indicate zero until it passes the coil with the open; then full line voltage will be indicated. An open coil may also be located with a megohmmeter test across each one; a closed coil will indicate zero, and an open coil will indicate infinity. An emergency repair of an open field coil can be made by unwinding the coil to expose the break, splicing the broken ends together, and reinsulating.

15.9 LOCATING A GROUNDED FIELD COIL IN A SYNCHRONOUS MOTOR

A grounded coil may be located by connecting the field circuit to its rated DC voltage or less and measuring the voltage drop between each coil and ground, as shown in Figure 15.12. A grounded coil will indicate very low or no voltage when tested from its terminals to ground. A grounded coil may be located with a megohmmeter, but each coil must be disconnected from the others and tested to ground individually; a grounded coil will indicate zero ohms. The megohmmeter test is recommended for shipboard use, because the voltage test requires the field frame to be insulated from the ship's hull or floor plates, and this is often difficult to do.

FIGURE 15.11
Locating an open field coil.

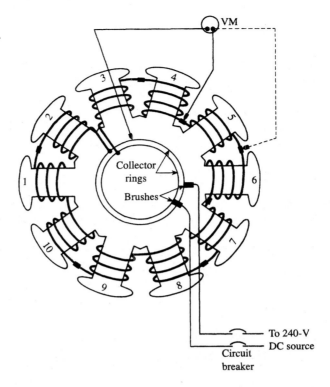

A grounded field coil should be corrected as soon as possible, because a second ground may result in a short circuit of serious magnitude; the resultant magnitude of magnetic imbalance may cause severe vibration and may damage the machine.

Grounds caused by the buildup of dirt and carbon dust can generally be cleared by cleaning and revarnishing. However, a coil that is grounded by chafing of insulation against the pole iron should be reinsulated or replaced if the copper shows any signs of having been worn away.

15.10 TESTING FOR A REVERSED-FIELD POLE IN A SYNCHRONOUS MOTOR

A reversed-field pole, caused by a reversed coil connection during disassembly and reassembly in a salient-pole generator or motor, may be detected by bridging the gap between poles with two identical steel bolts, as shown in Figure 15.13. Correct polarity will be indicated by a strong attraction between the ends of the bolts, but a reversed coil will cause the ends to repel. When making the test, the field winding should be connected to a DC source of approximately 50 percent of rated field

FIGURE 15.12
Locating a grounded field coil.

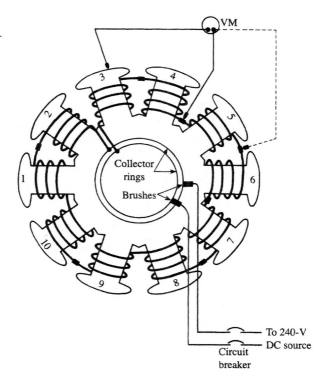

voltage. Because of the very strong magnetic field at the pole faces, testing with a single bolt may give misleading results. Two bolts should be used, as shown in Figure 15.13.

15.11 TROUBLESHOOTING CHART

The troubleshooting chart given in Table 15.2 lists the most common troubles and their possible causes and references the pertinent book section.

FIGURE 15.13
Testing for reversed polarity at field poles.

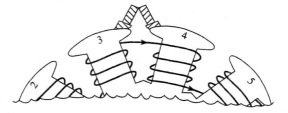

TABLE 15.2
Troubleshooting Chart for AC Motors

Trouble	Possible Causes	Book Section
Three-Phase Squirrel-Cage Induction Motors		
Fails to start	1. Overload	———
	2. Blown fuses	29–1, 2
	3. Defect in control panel	22–2, 9, 10
	4. Open in one phase	13–10
	5. Low voltage	13–1
Runs hot	1. Overload	———
	2. Clogged ventilating ducts	26–1 through 5
	3. Dirty air filters	21–3
	4. Shorted stator coils	15–1, 3
	5. Low voltage	13–1, 3
	6. High voltage	13–1, 3
	7. Low frequency	13–3, 4
	8. Open stator coils	15–4
	9. One phase open	13–10
	10. Grounded stator coils	15–2
	11. Bearing problems	21–4
	12. Unbalanced voltages	13–5
Runs slow	1. Overload	———
	2. Low voltage	13–1, 3
	3. Low frequency	13–1, 3
	4. Broken rotor bars	15–5
	5. Shorted stator coils	15–1
	6. Open stator coils	15–4
	7. One phase open	13–10
Vibration	1. Misalignment	21–5
Wound-Rotor Induction Motors		
Fails to start	1. Overload	———
	2. Blown fuses	29–1, 2
	3. Defect in control circuit	22–2, 9, 10
	4. Open in one phase of stator	13–10
	5. Open in rheostat	15–6
	6. Inadequate brush tension	19–7, 14
	7. Brushes do not touch slip rings	19–14
	8. Open in rotor circuit	15–6
	9. Low voltage	13–1, 3
Runs hot	1. Overload	———
	2. Clogged ventilating ducts	26–1 through 5
	3. Dirty air filters	21–3
	4. Low voltage	13–1
	5. High voltage	13–1
	6. Shorted stator coils	15–1
	7. Open stator coils	15–4
	8. One phase open	13–10
	9. Low frequency	13–1, 3

TABLE 15.2
Continued

Trouble	Possible Causes	Book Section
	10. Grounded stator	15–2
	11. Bearing problems	21–4
	12. Unbalanced voltages	13–5
Runs slow	1. Overload	——
	2. Low voltage	13–1, 3
	3. Low frequency	13–1, 3
	4. Too much resistance in rheostat	11–10, 15–6
	5. Shorted stator coils	15–1
	6. Open stator coil	15–4
	7. One phase open	13–10
	8. Open in rotor circuit	15–6
Vibration	1. Misalignment	21–5
Synchronous Motors		
Fails to start	1. Overload	——
	2. Blown fuses	29–1, 2
	3. Defect in control circuit	22–2, 9, 10
	4. Open in one phase	13–10
	5. Low voltage	13–1, 3
Runs hot	1. Overload	——
	2. Clogged ventilating ducts	26–1 through 5
	3. Shorted stator coils	15–1
	4. Open stator coils	15–4
	5. High voltage	13–1, 3
	6. Grounded stator	15–2
	7. Field current set too low	12–6
	8. Field current set too high	12–6
	9. Bearing problems	21–4
	10. Unbalanced voltages	13–2, 3
	11. One phase open	13–10
Runs fast	1. Frequency too high	11–3, 12–2
Runs slow	1. Frequency too low	11–3, 12–2
Pulls out of synchronism	1. Overload	——
	2. Open in field coils	15–8
	3. Open in field rheostat	12–4
	4. Rheostat resistance set too high	12–4
Will not synchronize	1. Field current set too low	12–2
	2. Open in field coils	15–8
	3. Open in field rheostat	12–3
	4. Brushes do not contact slip rings	19–14
Vibrates severely	1. Out of synchronism	12–4
	2. Open stator coil	15–4
	3. Open phase	13–10
	4. Shaft misaligned	21–5
	5. Reversed field pole	15–10
Shaft current	1. Defective bearing insulation	21–6

GENERAL REFERENCE

Rosenberg, R., and Hand, A., *Electric Motor Repair,* 3rd ed., Holt, Reinhart and Winston, New York, 1987.

SPECIFIC REFERENCE KEYED TO TEXT

[1] *IEEE Recommended Practice for the Repair and Rewinding of Motors for the Petroleum and Chemical Industry,* IEEE Std 1068–1990, Institute of Electrical and Electronic Engineers, New York, 1990.

REVIEW QUESTIONS

1. What is the first thing that should be checked when a motor fails to start?
2. What type of fault (ground, short, or open) or combination of faults is identified by burning, overheating, or discoloration? Explain.
3. Explain the principle of growler operation.
4. Describe a method for locating a grounded stator coil. Using suitable sketches, state how an emergency repair can be made.
5. Describe a test for locating an open stator coil in a single-circuit stator. Using suitable sketches, state how an emergency repair can be made.
6. How would you locate an open coil in a multicircuit stator?
7. What effect do broken rotor bars have on the performance of a squirrel-cage induction motor?
8. Describe a test for detecting the presence of broken rotor bars in a squirrel-cage rotor. Can broken rotor bars be repaired? Explain.
9. What faults other than blown fuses or voltage failure can prevent a wound-rotor motor from starting?
10. Describe a simple test to determine whether or not the starting rheostat for a wound-rotor motor is defective.
11. Describe a test for locating shorted field coils in the rotating-field member of a salient-pole alternator.
12. Explain why alternating current is better than direct current when testing for a shorted field coil in the rotating field of a synchronous motor.
13. Describe a test for locating an open field coil; state the procedure for making an emergency repair.
14. Describe a test for locating a grounded field coil; state the procedure for making an emergency repair.
15. Describe a test procedure for checking the polarity of the field poles of a synchronous motor.

PROBLEMS

15–1/5. Assuming stator current fluctuations, when testing, a certain squirrel-cage rotor is 48 A and 45 A. Does the rotor have a broken bar?

15–2/5. Assuming stator current fluctuations, when testing, a certain squirrel-cage rotor is 62.4 A and 60 A. Does the rotor have a broken bar?

15–3/5. Assuming stator current fluctuations, when testing, a certain squirrel-cage rotor is 88 A and 85.6 A. Does the rotor have a broken bar?

16

Single-Phase Induction Motors

16.0 INTRODUCTION

Single-phase (split-phase) induction motors are used extensively in industrial, commercial, and domestic applications. They are used in clocks, refrigerators, freezers, fans, air conditioners, blowers, pumps, washing machines, and machine tools and range in size from a fraction of a horsepower to 15 horsepower. The principle of induction-motor action in the single-phase motor is based on the principle of induction-motor action developed in Chapter 11 for three-phase motors. However, to obtain a rotating magnetic field from a single-phase system, the motor current is split into two separate windings whose currents are out of phase with respect to each other. Hence the name *split-phase motor*.[1]

16.1 RESISTANCE-START SPLIT-PHASE MOTORS

Figure 16.1 illustrates the winding layout for a four-pole, split-phase motor. The rotating magnetic field, necessary to produce induction-motor action, is caused by the current in the main winding and that in the auxiliary winding, with each attaining its respective maximum value at different instants of time. The main winding is also called the running winding, and the auxiliary winding is called the starting winding.

Resistance-start split-phase motors use additional resistance in the auxiliary winding (it is wound with a smaller diameter copper wire) so that its current is less lagging than the current in the main winding. The circuit for a resistance-start split-phase motor is shown in Figure 16.2. The flux in the auxiliary winding will reach its maximum value first and the flux in the main winding will attain its maximum value later. Thus referring to Figure 16.1, the net result is a counterclockwise (CCW)

---

[1] The reader is urged to review Section 11-2, Chapter 11, before starting this chapter.

FIGURE 16.1
Winding layout for a single-phase motor.

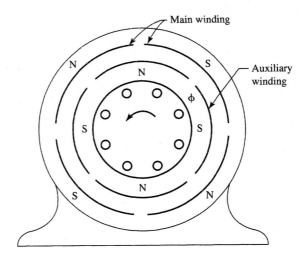

rotation of flux. The CCW rotating flux sweeping the squirrel-cage rotor produces counterclockwise rotation of the rotor.

Once the motor starts, the stator develops its own rotating magnetic field, and the auxiliary winding is no longer needed. When the rotor attains approximately 75 percent of rated speed, the auxiliary winding is disconnected by a centrifugally operated switch, magnetic relay or solid-state switch.

The resistance-start split-phase motor is adaptable to loads such as centrifugal pumps, oil burners, blowers, home laundry machines, and other loads of similar characteristics that require moderate torques and constant speed.

In some installations, such as small refrigerator units or water coolers, the switch that cuts the starting winding in or out may be external to the motor and operated by a coil connected in series with the machine. The high starting current drawn by the running winding actuates the relay, causing it to close and excite the starting winding. When the machine attains its rated speed, the countervoltage produced by motor

FIGURE 16.2
Resistance-start split-phase motor.

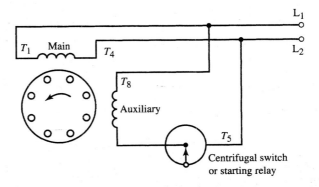

action reduces the current in the windings; the relay coil releases the starting-circuit contacts, and the starting circuit opens.

16.2 SPLIT-PHASE CAPACITOR MOTORS

The three basic types of capacitor motors are capacitor-start motors, permanent-split capacitor motors, and two-value capacitor motors.

Capacitor-Start Motor

The capacitor-start motor, shown in Figure 16.3a, uses a capacitor in series with the auxiliary winding to cause the current in the auxiliary winding to lead the current in the main winding. Thus, the flux in the auxiliary winding will reach its maximum value first and the flux in the main winding will attain its maximum value later. Thus, referring to Figure 16.1, the net result is a CCW rotation of flux, resulting in a counterclockwise rotation of the rotor.

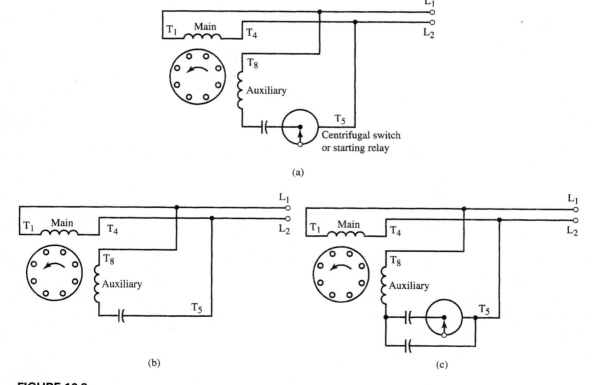

FIGURE 16.3
(a) Capacitor-start split-phase motor; (b) permanent-split capacitor motor; (c) two-value capacitor motor.

The significant difference between the capacitor-start motor and the resistance-start motor is the starting torque, which is about 130 percent rated for the resistance-start split-phase motor, and 300 percent rated for the capacitance-start motor. The high starting torque and good speed regulation of the capacitor-start motor make it well suited for applications to stokers, compressors, reciprocating pumps, and other loads of similar characteristics.

Permanent-Split Capacitor Motor

A permanent-split capacitor motor, shown in Figure 16.3b, has no switch in the auxiliary circuit, and its operation is smoother and quieter than a capacitor-start or resistance-start motor of the same power rating. The value of capacitance for this type of motor is smaller than the one used in the capacitor-start motor and results in a compromise between the best starting and best running performances. The primary field of application for a permanent-split capacitor motor is for shaft-mounted fans used in unit heaters and in ventilating fans.

Two-Value Capacitor Motor

The two-value capacitor motor, shown in Figure 16.3c, uses two capacitors. The use of two capacitors provides a greater amount of capacitance for starting than for running. This enables a greater locked-rotor torque than is obtainable with the permanent-split capacitor motor, and the reduced capacitance when running results in improved power factor, improved efficiency, and a higher breakdown torque.

16.3 SPLIT-PHASE MOTOR CONSTRUCTION

A cutaway view of a split-phase motor that uses a centrifugally operated switch is shown in Figure 16.4. The rotor-mounted centrifugal mechanism opens a switch, mounted in the end shield, at approximately 75 percent rated speed, opening the auxiliary winding. The auxiliary windings for both the resistance-start and capacitor-start motors are designed for starting duty only. If the centrifugal switch or relay fails to open, the auxiliary winding will burn out in a very short time.

Thermal-Overload Protector

A thermal protector, shown in Figure 16.5, is used in some motors to protect the machine against the overheating caused by overload or excessive repetitive starts. It accomplishes this by automatically opening when the temperature exceeds a predetermined value.

After tripping, the protector continues to sense the heat from the motor windings and does not allow the motor to operate until the motor has cooled. If the motor has a manual reset protector, pushing the red flag to reset the protector will restart the

FIGURE 16.4
Cutaway view of a split-phase motor (courtesy Magnetek Century Electric Company).

motor. However, if the motor has not cooled sufficiently, the reset will not lock in. If an automatic protector is used, it will reset automatically when the motor cools to a safe operating temperature. Never short-circuit the protector to prevent tripping; such action will result in a burned-out motor. If the protector trips, search for the fault and correct it.

Terminal Markings Identified by Color [1]

Single-phase motors, shown in Figures 16.2, 16.3, 16.5, and 16.6, are sometimes manufactured with NEMA recommended color-coded leads instead of letter/number

FIGURE 16.5
Circuit for a single-phase motor with a thermal protector.

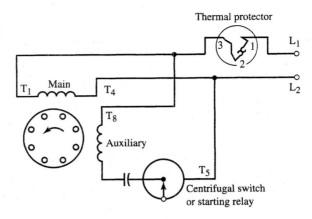

terminal markings. The color code for the above-mentioned motors is related to the letter/number markings in the following manner:

T_1—Blue T_2—White T_3—Orange T_4—Yellow T_5—Black
T_8—Red P_2—Brown P_1—No color assignment

Reversing a Split-Phase Induction Motor

The diagrams for single-phase induction motors shown in Figures 16.2, 16.3, 16.5, and 16.6 are standard NEMA diagrams. Connections shown are for CCW rotation, as observed when facing the end opposite the drive shaft. For clockwise (CW) rotation, interchange leads T_5 and T_8 of the auxiliary winding.

Single-Phase Double-Voltage Induction Motors

The standard connections for low and high nameplate voltages of double-voltage machines are illustrated in Figures 16.6a and 16.6b, respectively.

16.4 SHADED-POLE MOTOR

The shaded-pole motor, illustrated in Figure 16.7, utilizes a short-circuited coil or copper ring, called a shading coil, to provide starting torque. The shading coil is not connected in the circuit; it is wound around a part of the pole face and acts as the short-circuited secondary of a transformer. In accordance with Lenz's law, the current induced in the shading coil sets up a magnetomotive force (mmf) in a direction to oppose the mmf of the field that produced it. This causes a time delay in the buildup of flux in the shaded portion of the pole face. The flux reaches its maximum in the unshaded portion of the pole face before it reaches its maximum in the shaded portion. This causes a shifting flux in the direction from the unshaded part to the shaded part. The shifting flux "sweeping" the squirrel-cage rotor develops induction-motor action.

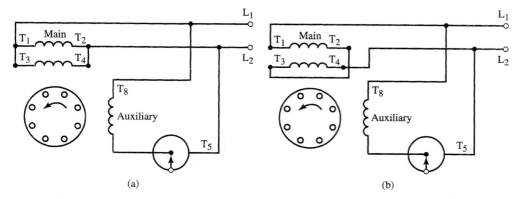

FIGURE 16.6
Circuits for single-phase double-voltage motors: (a) lower voltage; (b) higher voltage.

A shaded-pole motor cannot be reversed unless it has two sets of shading coils, each with a separate switch.

The principal field of application for shaded-pole motors is in clocks, record players, small fans, and other loads with similar characteristics.

16.5 SPEED CONTROL OF SINGLE-PHASE INDUCTION MOTORS

Speed control using pole-changing methods is used for resistance-start or capacitance-start motors. Depending on the application, consequent-pole windings or multiple windings may be used.[2]

Speed control for non-pole-changing permanent-split capacitor motors or shaded-pole motors can be accomplished by a tapped or slide-wire autotransformer in the main line, as shown in Figure 16.8; by using an external resistor or reactor in series

---

[2] See Reference [2].

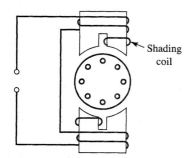

FIGURE 16.7
Shaded-pole motor.

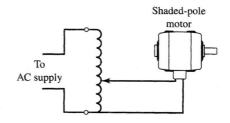

FIGURE 16.8
Speed control of shaded-pole motor using an autotransformer.

with the main winding or in series with both windings; or by adjusting the number of turns in the main winding through the use of taps and a selector switch, or by solid-state control.

Representative connection diagrams for multispeed, nonreversible, permanent-split capacitor motors and shaded-pole motors that use tapped windings are shown in Figures 16.9a and 16.9b, respectively. If the motor is designed for one speed, only black and (purple or white) leads will be present; for two speeds, only black, red, and (purple or white) leads will be present; for three speeds, only black, blue, red, and (purple or white) leads will be present; for four speeds, only black, yellow, orange, red, and (purple or white) leads will be present, etc. [1]. These small motors often have colored leads instead of numbered terminals. White is ground, purple ungrounded.

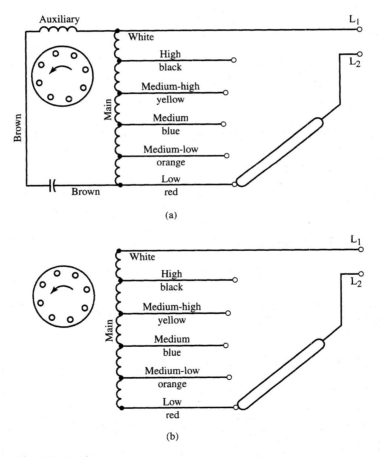

FIGURE 16.9
Representative connection diagrams for motors that use tapped windings: (a) multispeed, nonreversible, permanent-split capacitor motors; (b) shaded-pole motors.

16.6 SINGLE-PHASE INDUCTION-MOTOR PROBLEMS [3]

When a resistance-start or capacitor-start motor fails to start, the trouble may be in the starting winding, running winding, centrifugal mechanism, centrifugal switch, or capacitor if one is used.

To diagnose the trouble, the rotor should be spun rapidly by hand, and then the line switch should be closed. If the motor accelerates to rated speed and then continues to run, the trouble is in the starting circuit. However, if the motor accelerates to about one-half or three-fourths speed then slows down to a very slow speed, and then starts to accelerate again, etc., the trouble is in the running winding.

If the trouble is in the starting circuit, the machine should be disassembled, and the starting winding tested for continuity. The centrifugal switch should be checked for burned or worn contacts, and the capacitor should be tested if one is used. Substituting a replacement capacitor that is known to be good will resolve any doubts as to whether or not the capacitor is at fault. Causes of capacitor failure are excessive voltage, excessive temperature, overload, sticky contacts in the centrifugal switch or magnetic relay, and excessive duty cycle.

Capacitors used for motor starting are electrolytic capacitors, and are rated for starting duty only; too many starts without adequate cooling periods will overheat and burn out the capacitor. Capacitors may be tested for shorts and grounds with a megohmmeter as outlined in Section 5.7, Chapter 5. **Warning:** because of shock hazards, capacitors should be discharged before and after testing.

Measuring Capacitance

The capacitance of a motor starting capacitor can be determined by using the voltmeter-ammeter method shown in Figure 16.10. The applied voltage should be rated capacitor voltage or lower, and a 10-A fuse or 10-A breaker should be used to protect against the possibility of a shorted capacitor. The capacitor should be at room temperature when conducting the test. To prevent overheating, the circuit should not be energized for more than 10 seconds.

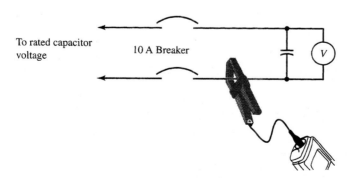

FIGURE 16.10

Voltmeter-ammeter method for determining the capacitance of a capacitor.

The capacitance is determined by substitution into the following equations:

$$X_C = \frac{V_{cap}}{I_{cap}} \qquad C = \frac{1}{2\pi f X_C} \times 10^6$$

where: V_{cap} = voltage across the capacitor (V)
I_{cap} = current in capacitor (A)
X_C = capacitive reactance (Ω)
C = capacitance (μF)
f = frequency (Hz).

TABLE 16.1
Trouble-shooting Chart for Single-Phase Induction Motors

Trouble	Possible Cause	Book Section
Fails to start	1. Overload	——
	2. Defective starting mechanism	16–6
	3. Blown fuses	29–1, 2
	4. Open in auxiliary winding	13–10
	5. Open in main winding	13–10
	6. Shorted capacitor	16–6
	7. Open capacitor	16–6
Runs hot	1. Overload	——
	2. Defective starting mechanism	16–6
	3. Low voltage	13–1, 3
	4. High voltage	13–1, 3
	5. Clogged ventilating ducts	26–1 through 5
	6. Shorted stator coils	15–1
	7. Worn bearings	21–4
	8. Low frequency	13–3, 6
	9. Rotor rubbing on stator	——
Runs slow	1. Overload	
	2. Low voltage	13–1, 3
	3. Low frequency	13–3, 6
	4. Broken rotor bars	15–5
	5. Shorted stator coils	15–1

EXAMPLE 16.1 For the circuit in Figure 16.10, the ammeter and voltmeter read 1.9 A and 117 V, respectively, and the frequency is 60 Hz. Determine the capacitance of the capacitor.

Solution

$$X_C = \frac{V_{cap}}{I_{cap}} = \frac{117}{1.9} = 61.58 \ \Omega$$

$$C = \frac{1}{2\pi f X_C} \times 10^6 = \frac{1}{2\pi \times 60 \times 61.58} \times 10^6 = 43.1 \ \mu F$$

16.7 TROUBLESHOOTING CHART

The Table 16.1 troubleshooting chart lists the most common troubles and their possible causes and references the pertinent book section.

GENERAL REFERENCE

Hubert, Charles I., *Electric Machines: Theory, Operation, Applications, Adjustment, and Control*, 2nd ed. Prentice-Hall, Upper Saddle River, NJ, 2001.

SPECIFIC REFERENCES KEYED TO THE TEXT

[1] *Motors and Generators*, NEMA Standards Publication No. MG 1–1998.
[2] Veinott, C. G., and Martin, J. E., *Fractional and Subfractional Horsepower Electric Motors*, 4th ed. McGraw-Hill, New York, 1986.
[3] Rosenberg, R., and Hand, A., *Electric Motor Repair*, 3rd ed. Holt, Reinhart and Winston, New York, 1987.

REVIEW QUESTIONS

1. What is the difference in construction details between a split-phase squirrel-cage motor and a three-phase squirrel-cage motor? What is the purpose of the rotor-mounted centrifugal mechanism?
2. Explain how a rotating magnetic field is produced in the stator of a split-phase motor.
3. Explain how a shaded-pole motor develops a rotating magnetic field.
4. How is a resistance-split-phase motor reversed?
5. Can a shaded-pole motor be reversed? Explain.
6. State two applications for each of the following single-phase motors: (a) resistance-start; (b) capacitor-start; (c) permanent-split capacitor; (d) two-value capacitor; (e) shaded-pole.
7. How can the speed of resistance-start and capacitor-start split-phase motors be changed?

8. How can the speed of permanent-split capacitor motors and shaded-pole motors be changed?

9. What is the purpose of a thermal overload relay, where is it located, and how does it work?

10. A grindstone operated by a resistance-start split-phase motor will start only when spun by hand; it will accelerate to near rated speed and then go into a recycling state: slow down, speed up, slow down, speed up, etc. Determine the cause and explain the behavior.

11. A capacitor-start motor does not run when the breaker is closed; it hums loudly and starts to smoke. However, if the breaker is closed while the shaft is spun by hand, the machine comes up to speed and operates properly. What are the three probable faults?

12. State three operating conditions that may damage a starting capacitor.

13. Describe megohmmeter tests that can be used to test a capacitor for shorts and grounds.

PROBLEMS

16–1/6. The voltage and current readings recorded when testing a 60-Hz, motor-starting capacitor are 125 V and 8.2 A, respectively. Determine the capacitance.

16–2/6. The voltage and current readings recorded when testing a 60-Hz, motor-starting capacitor are 230 V and 25 A, respectively. Determine the capacitance.

17

Direct-Current Generators: Principles and Operational Problems

17.0 INTRODUCTION

This chapter illustrates the general construction details of direct-current (DC) generators and explains how an alternating voltage, generated in the armature coils, is converted to direct voltage for use in the external circuit. The function of interpoles and compensating windings in reducing sparking and preventing flashover at the commutator is explained.

The buildup of voltage in self-excited and separately excited generators is presented, along with procedures for safe and efficient parallel operation. Load-sharing problems, reversed polarity, voltage failure, and the necessary corrective measures are discussed.

17.1 ELEMENTARY DC GENERATOR

Figure 17.1a shows the armature coil of an elementary DC generator rotating counterclockwise (CCW) within a magnetic field. A three-dimensional view of the coil is shown in Figure 17.1b. At zero degrees all of the flux that crosses the air gap passes through the window of the armature coil. As the coil rotates from the 0° position, the amount of flux passing through the window decreases linearly with angular position until, at 90°, the flux through the window is zero. When past the 90° position, the flux passes through the opposite face of the coil. The flux increases linearly in the opposite direction reaching its maximum value at 180°, then decreases, to zero at 270° and so forth. Thus, the flux through the coil window of this elementary two-pole generator changes direction twice in every revolution.

A graph showing the magnitude and direction of the pole flux (Φ) passing through the coil window as the coil rotates within the magnetic field is shown in Figure 17.1c. Note that the graph of flux is triangular in shape.

337

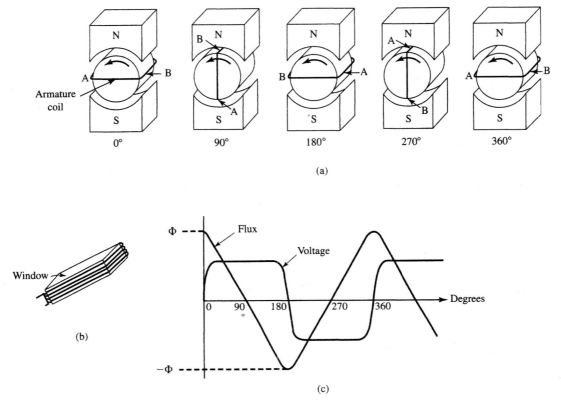

(a)

(b)

(c)

FIGURE 17.1
(a) Armature coil of an elementary DC generator rotating CCW within a magnetic field; (b) three-dimensional view of coil; (c) graph showing the magnitude and direction of the flux passing through the armature coil and the resultant voltage curve.

Generated Voltage

The changing flux through the window of the coil, as it rotates within the magnetic field, generates a voltage within the coil. In accordance with Faraday's law for magnetically induced voltages, the magnitude of the voltage generated in the armature coil is dependent on the *rate of change of flux* through the coil window. Figure 17.1c shows the *rate of change of flux* to be constant and downward from 0° to 180°, and constant and upward from 180° to 360°, assuming a constant speed and a constant pole flux.

In accordance with Lenz's law, a negative slope of flux through the window will generate a positive voltage, and a positive slope of flux through the window will generate a negative voltage. Thus, referring to Figure 17.1c, the voltage generated in the coil is positive from 0° to 180° and negative from 180° to 360°. Note that the voltage generated in the armature coil of a DC generator has an alternating but rectangular shape; this alternating voltage is converted to DC by means of a mechanical rectifier called a *commutator* for use in the external circuit.

17.2 COMMUTATION

Commutation in a direct-current generator is a mechanical rectifying process whereby the voltage and current that alternate in the armature coils are made unidirectional (DC) in the load circuit. The commutation process is illustrated with a one-turn armature coil in Figure 17.2.

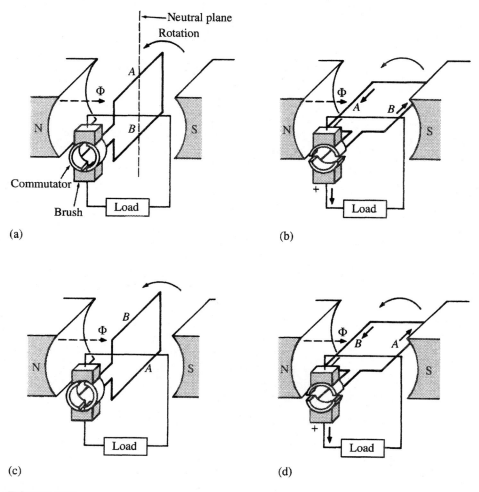

FIGURE 17.2
Commutation process.

FIGURE 17.3

Output voltage of an elementary two-pole DC generator.

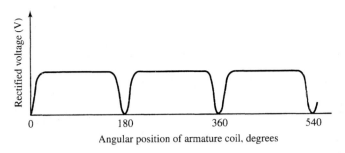

The coil consists of two conductors, A and B, joined at the back, and connected at the front to a two-bar commutator; conductor A to bar A, and conductor B to bar B. Connections to the external circuit are made via small stationary blocks of graphite, called brushes, that are pressed against the commutator by springs. The coil and its commutator are driven in a CCW direction by a prime mover (not shown).

As the armature coil rotates within the magnetic field, the voltage generated in the conductors changes direction every 180°. At zero degrees (Figure 17.2a), conductors A and B have moved into the magnetic-neutral plane, and no voltage is generated. The magnetic-neutral plane is located midway between adjacent poles of opposite polarity. At 90° (Figure 17.2b) the voltage generated in conductor A is directed toward the reader, and the voltage in conductor B is directed away from the reader. At 180° (Figure 17.2c), conductors A and B are again in the neutral plane, and no voltage is generated. At 270° (Figure 17.2d) the direction of the voltages in the conductors are reversed; the voltage generated in conductor A is directed away from the reader and the voltage in conductor B is directed toward the reader.

Although the voltage and current generated within armature conductors A and B change direction every 180 electrical degrees of rotation, the voltage and current in the external circuit do not reverse direction. The rotating commutator and stationary brushes constitute a rotary switch that provides a switching action, called *commutation,* that switches the internal alternating voltage and current within the coil conductors to direct voltage and direct current in the external circuit. The output voltage of this elementary generator is shown in Figure 17.3.

When the coil is rotating through the magnetic-neutral plane, as shown in Figures 17.2a and 17.2c, it is shorted by the brushes. However, since no armature voltage is generated, no short-circuit current occurs.

A practical machine has many coils distributed around the armature, and the coils pass through the neutral plane one at a time. As one coil moves into the neutral plane another moves out, producing an essentially constant voltage.

17.3 DC GENERATOR CONSTRUCTION

A cutaway view of a DC machine (generator or motor) is shown in Figure 17.4. The shunt-field coils and series-field coils (discussed later in Section 17.12) are wound around the same pole iron, and they provide specific machine characteristics. The

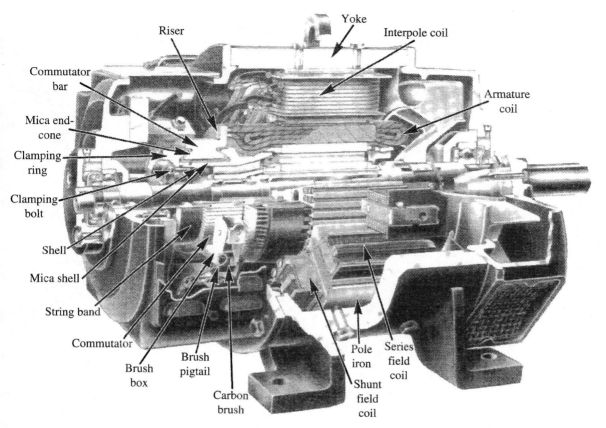

FIGURE 17.4
Cutaway view of a DC machine (courtesy Reliance Electric).

shunt-field coils are wound with many turns of small diameter copper wire, whereas the series-field coils are wound with only a few turns of large cross-section copper conductor.

The iron poles are bolted to a yoke or frame of cast steel. Graphite or metal-graphite brushes provide the connection between the rotating commutator and the external load. These brushes are designed to slide freely in metallic brush boxes, called brush holders. Each brush is connected to its box by a short flexible copper conductor called a pigtail or shunt. The commutator is composed of alternate sections of copper bars and mica separators, clamped together with mica-insulated vee-rings. The number of commutator bars is determined by the number of coils in the armature, the number of poles, and the type of winding.

A partially assembled armature is shown in Figure 17.5. Form-wound coils are set in slots and then connected to the commutator.

FIGURE 17.5
Partially wound armature of a direct current generator (courtesy Elliott Co.).

17.4 LAYOUT OF A SIMPLE ARMATURE WINDING

Figure 17.6 illustrates the layout of one type of armature winding (called a lap winding) for an eight-slot, eight-coil armature designed for operation with a two-pole field.[1] Figure 17.6a shows the layout of the armature winding and its connections to the commutator. Note that each slot contains two coil-sides; one side from each of two coils. For example, as seen in Figure 17.6a, the left side of coil 1 is in the top of slot 1, and the right side of coil 1 is in the bottom of slot 5; the letters T and B refer to the top and bottom conductors in a single slot. Figure 17.6b is an end view of the armature, showing the physical layout of top and bottom conductors in the armature slots. Figure 17.6c is an elementary diagram that shows how the brushes divide the armature winding into two parallel paths.

17.5 GENERATOR OUTPUT VOLTAGE

The magnitude of the generated voltage (also called electromotive force or emf) depends on the magnitude of the pole flux, and the rotational speed of the armature. Expressed mathematically,

---

[1] See Reference [1] for other types of machine windings.

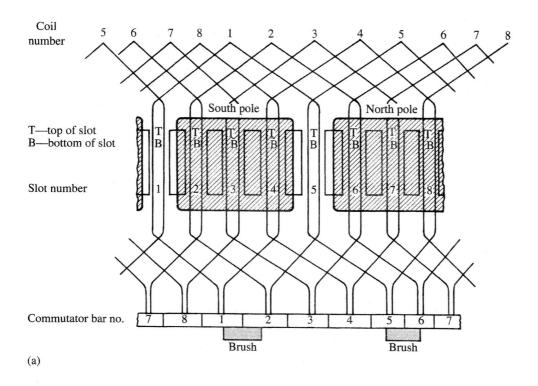

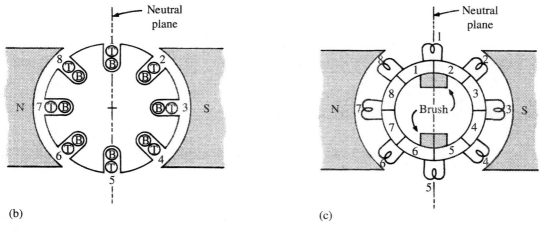

FIGURE 17.6

(a) Layout of a simple armature winding for an eight-slot, eight-coil armature with a two-pole field; (b) end view of armature and physical layout of conductors; (c) simplified view of the armature winding and its external connections.

$$E_A = n\Phi_P k_G \qquad\qquad (17\text{--}1)$$

where: E_A = generated voltage (V)
   n = armature speed (rpm)
   Φ_P = pole flux (Wb)
   k_G = generator constant.

The generator constant includes such design factors as the number of armature coils, arrangement of the armature windings, and the number of poles. If the generated voltage for a given speed and flux are known, the voltage for some other speed and pole flux may be determined by substituting into the following equation derived from Eq. (17–1):

$$\frac{E_2}{E_1} = \frac{n_2\Phi_2}{n_1\Phi_1} \qquad\qquad (17\text{--}2)$$

where: E_1, n_1, and Φ_1 are the original conditions
   E_2, n_2, and Φ_2 are the new conditions.

EXAMPLE 17.1

A DC generator running at 1800 rpm generates 256 V. If the speed is dropped to 1700 rpm, and the pole flux is increased to 1.16 times its original value, determine the new voltage.

Solution
From Eq. (17–2),

$$E_2 = E_1 \times \frac{n_2\Phi_2}{n_1\Phi_1} = 256 \times \frac{1700 \times 1.16\,\Phi_1}{1800 \times \Phi_1} = 280.5 \text{ V}$$

17.6   EFFECT OF ARMATURE INDUCTANCE ON COMMUTATION

The effect of armature inductance on the commutation process is explained with the aid of Figure 17.7. Figure 17.7a shows coils 2 and 3 feeding current (I) to a load via the commutator and brush. In Figure 17.7b, the armature has rotated to a position in which coil 2 is short-circuited by the brush. At this instant the coil lies in the magnetic-neutral plane, the magnetic field no longer induces an emf in the coil, and the current in the coil starts to decrease. As the process continues, the decreasing current causes an emf of self-induction in the coil, which, in accordance with Lenz's law, delays the change. Hence, coil 2 rotates into the position shown in Figure 17.7c without having had sufficient time for its current to drop to zero and reverse direction. As commutator bar 3 slides off the brush, the current in coil 2 rapidly decreases, generating a high emf of self-induction that causes an arc to jump between commutator bar 3 and the brush. This will occur with every coil, in turn, as their corresponding commutator bars slide

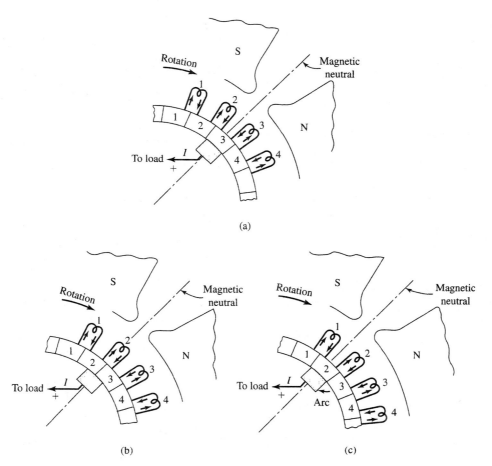

FIGURE 17.7
Effect of armature inductance on commutation.

off the brush. If allowed to continue, such arcing will severely damage both the commutator and the brushes. To prevent this condition, *commutating poles* are incorporated within the machine.

17.7 COMMUTATING POLES (INTERPOLES)

Narrow poles, called interpoles or commutating poles, are placed in the magnetic-neutral plane to reduce or eliminate sparking caused by the inductance of the armature coils. The strength and polarity of the commutating poles are designed to induce a voltage in the armature coil that is equal and opposite to its emf of self-induction.

The field frame of a DC machine, driven as a generator or as a motor, with two field poles and two commutating poles is shown in Figure 17.8. The narrow width of the commutating pole influences only those coils in the neutral plane. The polarity of

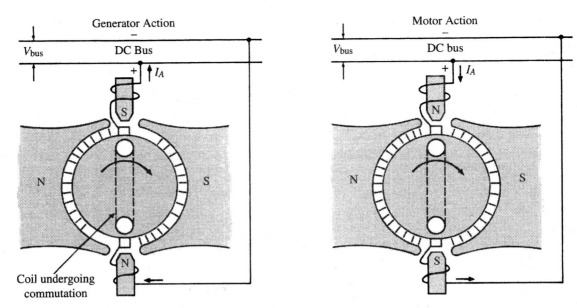

FIGURE 17.8
Location of interpoles, their connections, and their polarity for generators and motors.

the commutating poles relative to that of the field poles (for a generator) is shown in Figure 17.8 for clockwise (CW) rotation.[2] Thus, when the armature coil of a generator rotates into the neutral plane, its respective conductors come under the influence of an interpole whose polarity is opposite that of the main field pole that it just left. When driven as a motor the polarity of the interpole is the same as the main field pole it just left. The interpole windings are connected in series with the armature so that their strength will be proportional to the armature current. The directions of current when driven as a generator and when driven as a motor are shown in Figure 17.8.

Except for repairs, never break the connection between the interpole and the armature; the interpole windings must be treated as though they are a part of the armature. The resultant arcing, between brush and commutator, will be much worse if the interpoles are improperly connected than if no interpoles are used.

When operating at rated load or below rated load, the interpole flux remains in proportion to the armature current, and, as a result, the neutralizing voltage will always be equal and opposite to the emf of self-induction. However, if the generator is operating well above its rated load, the interpole iron saturates, and its flux is no longer proportional to the armature current. The interpole can no longer generate a neutralizing voltage in proportion to the emf of self-induction, and a relatively large current will appear in the coils short-circuited by the brushes, causing sparking and burning at the commutator–brush interface.

---

[2] The standard direction of shaft rotation for DC generators is CW facing the end opposite the drive [2].

17.8 ARMATURE REACTION AND COMPENSATING WINDINGS

When a generator is loaded, the current in the armature coils develops a magnetomotive force (mmf) of its own that interacts with the mmf of the field poles, disturbing the uniform flux distribution in the air gap. This behavior is called *armature reaction*. The effect of armature reaction on the flux distribution in a four-pole, clockwise-rotating, DC generator is shown in Figure 17.9.

Figure 17.9a illustrates the flux distribution when operating at no load; the flux is uniformly distributed about the pole face, and the magnetic neutral plane is equidistant from both poles.

Figure 17.9b illustrates the respective directions of the armature mmf and the field-pole mmf when the generator is operating under load. The mmf caused by armature conductors reacts with the mmf of the field to produce an additive effect on the trailing edge of each pole and a subtractive effect on the leading edge.[3] The result is a distorted flux pattern, as shown in Figure 17.9c; there is a concentration of flux on the trailing edge of the poles and a weakened flux on the leading edge. This causes a shift

---

[3] When an armature conductor enters the interpolar region, it departs from the trailing edge of one pole and approaches the leading edge of the next pole.

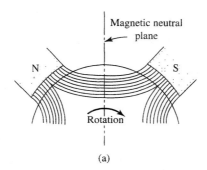

(a)

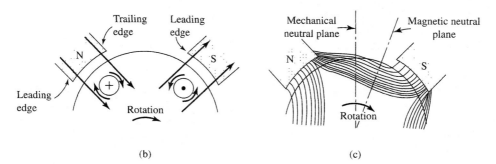

(b)   (c)

FIGURE 17.9

Effect of armature reaction on the flux distribution of a DC generator.

in the magnetic-neutral plane, with the magnitude of the shift dependent on the magnitude of the load current.

If the load current is great enough, the concentration of flux in the trailing edge may be sufficient to saturate the iron at the trailing edge. When saturation occurs, some of the flux will be "spilled" into the region between the poles (interpolar region) as shown in Figure 17.9c, resulting in sparking at the commutator–brush interface. Furthermore, with less flux at the pole face, the generator voltage will be reduced.

If the commutating poles are made sufficiently strong by the addition of more turns, additional opposing mmf will be produced to nullify the effect of the armature flux in the neutral plane and yet remain equally effective in bucking down the emf of self-induction in the coils undergoing commutation. However, although they eliminate or reduce sparking, commutating poles do not correct the field distortion brought about by armature reaction; they merely prevent the shift of flux into the neutral plane. The total field flux is still reduced by saturation at the trailing edge of the poles.

The effect of armature reaction is particularly severe in generator applications where large and rapid load changes occur. The sudden application of heavy loads will cause large and rapid swings of flux across the pole face, inducing high transient voltages in all of the armature coils. If the transient voltages are high enough to cause arcing between commutator bars, a brush-to-brush flashover may occur.

To prevent flashovers, DC machines designed for large and rapid load swings, such as in steel mills and ship propulsion, are equipped with pole-face windings. The pole-face winding, called a *compensating winding,* is shown in Figure 17.10. The compensating winding is connected in series with the armature and interpoles, and it essentially eliminates armature reaction by setting up a mmf that is always equal and opposite to the armature mmf. However, the machine still requires interpoles to buck down the emf of self-induction in the coils undergoing commutation.

17.9 SELF-EXCITED DC GENERATOR

The two basic types of DC generators are self-excited generators and separately excited generators. Self-excited generators use residual magnetism in the pole iron to build up voltage. Separately excited generators use a separate source of current to establish the flux in the pole iron. An elementary circuit diagram for a self-excited DC generator is shown in Figure 17.11a. The terminal markings and polarity correspond to a clockwise (CW) shaft rotation as viewed from the end opposite the drive shaft [2].

Assuming that the prime mover is operating at rated speed and there is sufficient residual magnetism in the pole iron, the armature coils sweeping the residual magnetic field generate a low voltage. This low voltage causes a small current in the shunt-field windings. As shown in Figure 17.11b, the flux contribution of the shunt-field winding adds to the residual flux. The slightly greater flux produces a somewhat higher voltage, which in turn causes a still higher shunt-field current and a further increase in flux. The process of voltage buildup continues until the combined effect of magnetic saturation and field-circuit resistance prevents further increases in flux. The rheostat is used to adjust the field current, which adjusts the amount of flux and, thus, changes the output voltage.

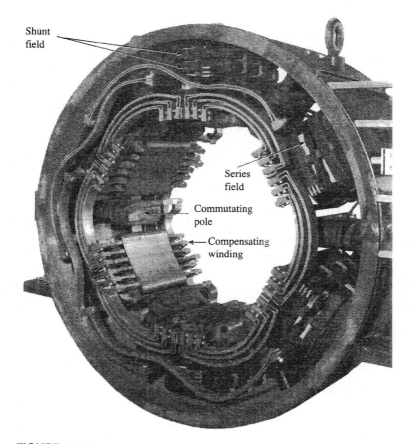

FIGURE 17.10
Stator of a DC machine equipped with interpoles and compensating windings (courtesy Marathon Electric Mfg. Corp.).

The iron pole pieces of a self-excited generator are designed to have relatively high magnetic retentivity. This ensures adequate residual magnetism for the development of sufficient initial voltage. However, high retentivity results in a slow response of the magnetic flux to a change in the shunt-field current. Thus a change in shunt-field current is not accompanied by an instantaneous change in the generated voltage.

17.10 SEPARATELY EXCITED GENERATOR

The circuit for a separately excited generator is shown in Figure 17.12. The separately excited generator utilizes an external source of energy to create the magnetic field and therefore is not dependent on residual magnetism for the buildup of voltage. The mate-

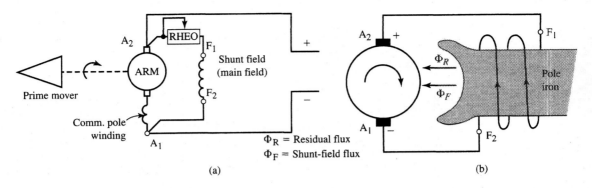

FIGURE 17.11
Self-excited DC generator: (a) elementary circuit; (b) sketch showing the buildup of flux.

rial used in the pole iron of separately excited generators is designed for low retentivity to enable a quick response to changes in field current. The source of excitation for the separately excited machine is most often a self-excited generator, called an exciter, operating at a fixed voltage setting.

17.11 GENERATOR VOLTAGE REGULATION

To prevent the dimming of lights and the malfunctioning of motors and other electrical equipment, the output voltage of a DC generator must be held relatively constant as electrical loads are switched on and off. Unfortunately, when load is connected to a generator, the following undesirable reactions cause a net reduction in output voltage:

1. Armature reaction, if uncompensated for, causes a net reduction in pole flux and, thus, a reduction in generated voltage.

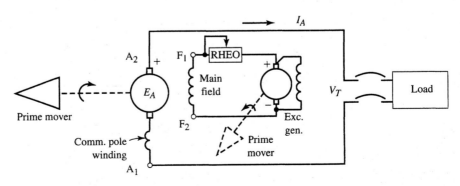

FIGURE 17.12
Separately excited DC generator.

2. Current in the armature conductors sets up a magnetic field that interacts with the magnetic field of the magnets, resulting in a counter torque that slows down the prime mover, causing a reduction in generated voltage.
3. The resistances of the armature winding and interpole windings cause an additional drop in the output voltage.

The relationship between the generated voltage, the resistance of the armature circuit, and the voltage at the armature terminals is obtained by applying Kirchhoff's voltage law to the armature circuit in Figure 17.12:

$$E_A = V_T + I_A R_{acir}$$

$$R_{acir} = R_A + R_{IP} + R_{CW}$$

(17–3)

where: V_T = voltage at the armature terminals (V)
E_A = generated voltage (V)
I_A = armature current (A)
R_A = resistance of armature (Ω)
R_{IP} = resistance of interpole (Ω)
R_{CW} = resistance of compensating winding (Ω)
R_{acir} = resistance of armature circuit (Ω).

The resistance of the armature circuit is the resistance of the armature plus everything within the generator housing that is in series with the armature, such as the commutating winding and compensating winding.

The percent change in generator voltage when going from no load to rated load is called the *voltage regulation*. Expressed mathematically,

$$VR = \frac{V_{nl} - V_{rated}}{V_{rated}} \times 100$$

(17–4)

where: $V_{nl} = E_A$ = voltage at no load (V)
V_{rated} = voltage at rated load (V)
VR = voltage regulation (%).

EXAMPLE 17.2

A certain, separately excited DC generator, with interpoles and a compensating winding, is operating at its rated load of 40 kW and 250 V. The armature resistance is 0.045 Ω, and the combined resistance of the interpole and compensating windings is 0.019 Ω. Determine (a) armature current; (b) no-load voltage; (c) voltage regulation.

Solution
The circuit is similar to that of Figure 17.12.

a. $P = V_T I_T \Rightarrow I_T = \dfrac{P}{V_T} = \dfrac{40 \times 1000}{250} = 160 \text{ A}$

b. $R_{\text{acir}} = R_A + R_{\text{IP}} + R_{\text{CW}} = 0.045 + 0.019 = 0.064$
$E_A = V_T + I_A R_{\text{acir}} = 250 + 160 \times 0.064 = 260.2 \text{ V}$

c. Voltage regulation $= \dfrac{V_{\text{nl}} - V_{\text{rated}}}{V_{\text{rated}}} \times 100 = \dfrac{260.2 - 250}{250} \times 100 = 4.1\%$

17.12 COMPOUND GENERATOR

Modification of the basic DC generator to obtain certain desirable voltage-current characteristics is generally accomplished by the addition of an auxiliary field winding called the series field. The series-field coils are wound with heavy copper conductors on the same pole iron as the shunt-field coils as shown in Figure 17.13a. These coils are connected in series with the armature, as shown in Figure 17.13b.[4] Such machines are called short-shunt compound generators. However, if shunt-field terminal F_2 in Figure 17.13b is connected to terminal S_1 instead of to A_1, the machine is called a long-shunt compound generator. Although either connection may be used, generators are usually connected short-shunt; a short-shunt connection results in less voltage variation with changes in load.

If the current in the series-field coils is in the same direction as the current in the respective shunt-field coils (cumulative compound), the flux contribution due to the

---

[4] The terminal markings and polarity correspond to a CW rotation as viewed from the end opposite the drive [2].

FIGURE 17.13
Compound generator: (a) series- and shunt-field windings; (b) circuit diagram; (c) typical current-voltage characteristics.

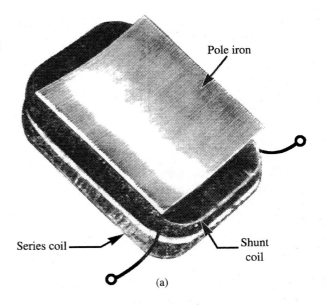

Pole iron

Series coil

Shunt coil

(a)

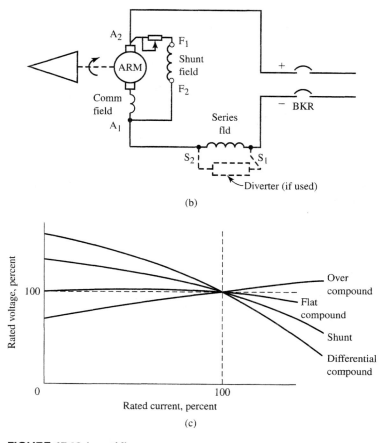

(b)

Over
compound

Flat
compound

Shunt

Differential
compound

(c)

FIGURE 17.13 (cont'd)

series field adds to the shunt-field flux; if the current in the series-field coils is in the opposite direction with respect to the current in the corresponding shunt-field coils (differential compound), the series-field flux subtracts from the shunt-field flux. Figure 17.13c illustrates the typical voltage-current characteristics of compound and shunt-connected generators.

The amount of compounding may be increased by adding more turns of copper conductor to each series-field coil. Thus, with the same armature and shunt-field winding, the overcompound generator has more turns of copper conductor in its series-field coils than the flat compound generator, and the flat compound machine has more turns of copper conductor in its series-field coils than the undercompound generator. The compounding effect of the series-field coils can be reduced by connecting a low-resistance path, called a *diverter,* in parallel with the series-field windings to bypass some of the armature current that would otherwise pass through the series-field coils. The connections for the diverter are shown with broken lines in Figure 17.13b.

Overcompound generators have limited applications and are used to offset voltage drops in long-distance transmission lines. Flat compound generators, undercompound generators, and shunt generators with automatic voltage regulators are most commonly used to supply DC distribution systems. The differential compound generator has applications in electric hoist systems where it is desirable for the voltage to decrease as the load on the hoist increases.

17.13 NEMA STANDARD TERMINAL MARKINGS AND CONNECTIONS FOR DC GENERATORS

The NEMA standard terminal markings and connections for DC generators rotating clockwise or counterclockwise, as viewed from the end opposite the drive, are shown in Figure 17.14. Although different directions of rotation may be used, the standard direction of rotation for DC generators is clockwise.

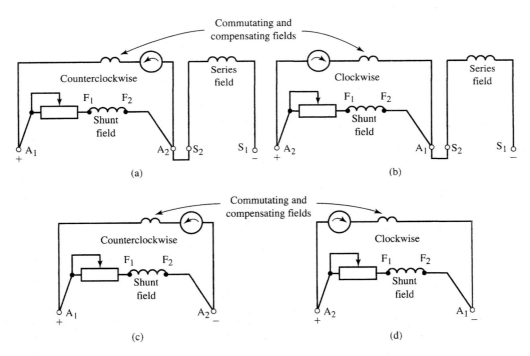

FIGURE 17.14
NEMA standard terminal markings and connections for DC generators: (a, b) compound generators; (c, d) shunt generators.

17.14 THREE-WIRE DC GENERATOR

A three-wire DC generator is used to provide two voltages for a three-wire DC distribution system; 120 volts for lighting circuits and 240 volts for power circuits. The construction details of the three-wire DC armature differ from that of the standard armature by the addition of two slip rings, as shown in Figure 17.15a. The slip rings, also called collector rings, are connected internally to tap the alternating voltage at appropriate points in the armature winding that are 180 electrical degrees apart.

Carbon brushes riding on the slip rings connect to an autotransformer, called a balance coil. The connection diagram for the three-wire DC generator and connections to representative loads are shown in Figure 17.15b; the internal connections to the generator are not shown.

The center or common line is called the neutral line. The current in the neutral line (unbalanced current) is the difference between the currents in the positive and negative lines. Alternating current appears in the balance coil, but because of the rectifying action of the commutator, the current in the positive, negative, and neutral lines is direct current. The indicated polarity is for clockwise rotation facing the end opposite the drive. The commutating winding and series-field winding are split in half, one-half connected to the positive side and the other half connected to the negative side; this is necessary to ensure that good commutation and series-field compounding will occur with an unbalanced load.

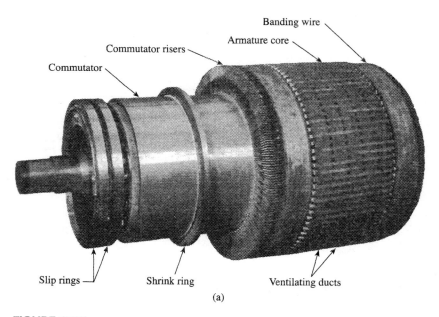

(a)

FIGURE 17.15

Three-wire direct current generator: (a) armature (courtesy Marathon Electric Mfg. Corp.); (b) three-wire generator and distribution system.

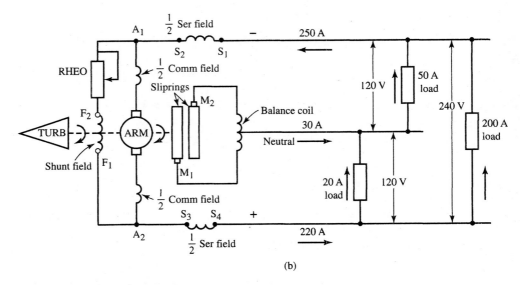

(b)

FIGURE 17.15 (cont'd)

17.15 DC GENERATOR AND DISTRIBUTION SYSTEM

An elementary circuit diagram and its one-line diagram counterpart for a self-excited DC generator and distribution system are shown in Figure 17.16. The generator feeds a distribution bus, which in turn supplies various connected loads. The millivoltmeter, calibrated in amperes, is used in conjunction with an ammeter shunt[5] to measure high values of direct current.

Changes in prime-mover speed caused by the application of load are corrected by an automatic speed governor that uses electronic or mechanical sensors to detect a change in speed and then automatically increases or decreases the energy input to the prime mover. Similarly, an automatic voltage regulator is used to detect a change in voltage, and then automatically adjusts the shunt-field current to raise or lower the voltage, as required.

17.16 INSTRUMENTATION AND CONTROL OF DC GENERATORS

Figure 17.17 is a functional one-line diagram for a two-generator system, showing the minimum instruments, switches, and adjustable controls that an operating engineer must be acquainted with when operating DC generators singly or in parallel. The connecting lines indicate what each control device operates and where each meter gets its signal.

---

[5] See Section 2.2, Chapter 2, for additional information on ammeter shunts.

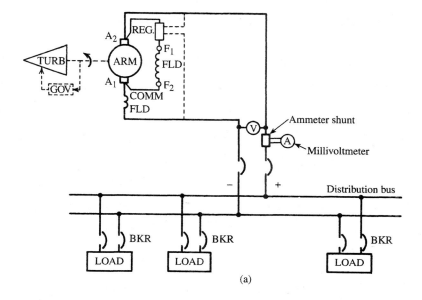

(a)

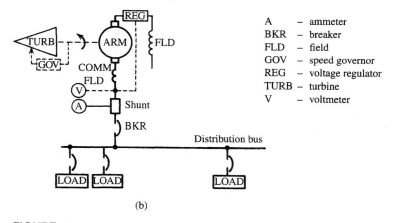

(b)

A	–	ammeter
BKR	–	breaker
FLD	–	field
GOV	–	speed governor
REG	–	voltage regulator
TURB	–	turbine
V	–	voltmeter

FIGURE 17.16
Elementary circuit diagram and its one-line diagram counterpart for a self-excited
DC generator and distribution system.

17.17 PROCEDURE FOR SINGLE-GENERATOR OPERATION

Before starting the prime mover, be sure that the generator breaker is open. Then
close the disconnect links located in the rear of the switchboard.[6] The purpose of the
links is to allow the complete isolation of the breaker from the bus for test and main-

---

[6]On old installations a disconnect switch is mounted on the front of the switchboard. Newer installations use
draw-out breakers for isolation purposes.

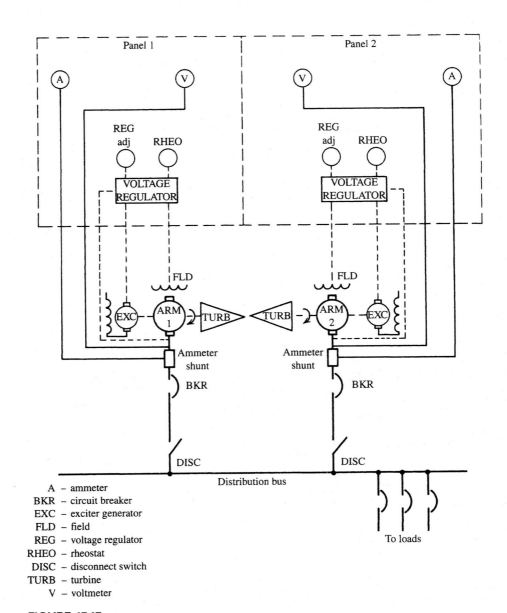

A – ammeter
BKR – circuit breaker
EXC – exciter generator
FLD – field
REG – voltage regulator
RHEO – rheostat
DISC – disconnect switch
TURB – turbine
V – voltmeter

FIGURE 17.17
Functional diagram for a two-wire DC generator and distribution system.

tenance purposes. These links must be bolted to the connection bars with an insulated wrench.

Warning: The circuit breaker must always be opened before opening or closing the disconnect links.

The following procedure may be used as a guide for single-generator operation:

1. Be sure the generator breaker is open.
2. Bolt the disconnect links, if used, to the connection bars. (Use an insulated wrench.)
3. Switch the voltage regulator to "manual."[7]
4. Start the turbine or engine and bring it up to speed.
5. Adjust the manual voltage control to obtain rated voltage.
6. Switch the regulator to automatic and adjust to rated voltage.
7. Close the circuit breaker manually or by operating the breaker-closing-switch.

17.18 PROCEDURE FOR PARALLEL OPERATION OF DC GENERATORS

When the load on a DC system exceeds the amount that can be supplied by a single generator, additional machines must be connected (paralleled) to the system to supply the required energy. The incoming machine must be paralleled in a manner that enables each machine to supply its proper share of power to the common load.

1. Be sure that the generator breaker of the incoming machine is open.
2. Bolt the disconnect links to the connection bars. (Use an insulated wrench.)
3. Switch the voltage regulator to "manual."
4. Start the turbine or engine and bring it up to speed.
5. Adjust the manual voltage control so that the voltage of the incoming machine is equal to or slightly higher than the bus voltage. If the voltage of the incoming machine is less than that of the bus, the incoming machine will be motorized at the instant it is paralleled. The energy to motorize the incoming machine comes from the other generators on the bus. If these machines are heavily loaded, the additional current required to motorize the incoming generator may be sufficient to trip them both off the bus and black out the plant. Hence, when DC generators are to be paralleled, it is necessary that the voltage of the incoming machine be at least equal to the bus voltage.

The generator circuit breaker provides protection against overcurrents, and through a reverse-current trip provides protection against sustained motorization. When a generator is motorized, it continues to run in the same direction as before, but its ammeter deflects in the negative direction.

---

[7]Raising the voltage by manual control, before switching the regulator to automatic mode, is a precautionary measure to avoid excessive current in the regulator and field coils during start-up.

Although light motorization does not harm the machine, it does result in a wasteful expenditure of energy by acting as a load connected to the other generators.

6. Switch the voltage regulator to automatic and adjust to rated voltage.
7. Close the circuit breaker.
8. Adjust the automatic voltage regulator of the incoming machine in the "raise voltage" direction, and that of the other machine in the "lower voltage" direction, until the ammeters of both machines indicate the same. If on manual control, use the field rheostats instead of the automatic voltage regulators.

The generator that is to accept more load should have its field excitation increased until one-half of the load that is to be transferred has been shifted, and the machine that is to lose some of its load should have its field excitation decreased until the remaining half has been shifted. If convenient, a simultaneous adjustment may be made on both machines.

In factories, hotels, or large ships that require two or three DC generators in parallel to supply the connected load, it is good operating practice to balance the current between machines in proportion to their ampere ratings. By doing so, transient overcurrents and short circuits in the distribution system are shared by all generators in proportion to their ratings, reducing the probability of one generator tripping. The tripping of one generator due to overcurrent will throw the entire load on the remaining unit or units, which may cause them to trip, blacking out the plant.

Removing a DC Generator from the Bus

To remove a DC generator from the bus, transfer its load to the other machines by adjustment of the field rheostats or automatic voltage regulators. Turn the field rheostat of the machine that is to give up load in the "lower voltage" direction and the field rheostats of the other machine(s) in the "raise voltage" direction, observing the ammeters until the transfer is essentially completed. Then trip the circuit breaker. If the generator is separately excited, its field switch should be opened last (after the generator breaker is opened); opening the field circuit while still in parallel with the bus causes the generator voltage to drop to a very low value, and if the machine is still on the bus, the other machines will feed into it. Severe damage to the commutator and brushes may result.

From a safety standpoint, if a generator is to be out of service for an extended period of time, the disconnect switch or links should be opened (after generator shutdown) and kept open until the generator is ready to go back on the line. This prevents damage to the machine in the event that the breaker is inadvertently closed.

17.19 EFFECT OF LOAD-VOLTAGE CHARACTERISTICS ON DC GENERATORS IN PARALLEL

For optimum performance when operating in parallel, DC generators should have the same or very similar load-voltage characteristics. That is, they should have the same voltage regulation. Generators with identical characteristics share the load in propor-

FIGURE 17.18

Typical voltage droop characteristic of a
DC generator.

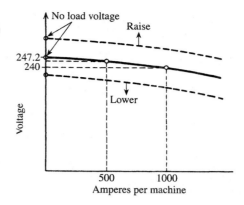

tion to their ampere rating. Generators with different characteristics, even with the same ampere rating, do not proportionately share any increase or decrease in bus load. The generator with the smallest percentage droop takes the greater portion of the increasing load. Figure 17.18 illustrates a typical droop characteristic for a DC generator. A voltage droop (voltage regulation) of 1 to 3 percent is essential for stable operation when in parallel. The desired voltage droop is obtained by adjusting the droop of the speed governor or by using a series field diverter.

The point of intersection of the droop characteristic with the voltage axis (vertical axis) is the no-load voltage of the generator for the given setting of the field rheostat. Turning the field rheostat in the "raise voltage" or in the "lower voltage" direction raises or lowers the no-load voltage setting of the generator but does not change the droop. This is indicated by the broken lines in Figure 17.18; adjustment of the rheostat raises or lowers the entire characteristic. The droop remains unchanged.

For a given no-load voltage setting of the field rheostat, a fixed relationship exists between the generator voltage and the ampere load. Increasing the ampere load causes the output voltage to decrease; decreasing the ampere load causes the output voltage to increase.

EXAMPLE 17.3

Assuming that the DC generator referred to in Figure 17.18 is operating at its rated values of 1000 A and 240 V, with a 3 percent voltage droop, determine the voltage when the load is removed.

Solution

From Eq. (17–4),

$$\text{VR} = \frac{V_{nl} - V_{rated}}{V_{rated}} \times 100 \Rightarrow V_{nl} = \frac{\text{VR} \times V_{rated}}{100} + V_{rated}$$

$$= \frac{3 \times 240}{100} + 240 = 247.2 \text{ V}$$

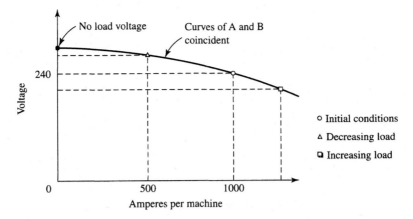

FIGURE 17.19
Two identical generators with identical droops and the same no-load voltage settings operating in parallel.

Figure 17.19 shows the characteristics of two DC generators (A and B) with identical droops and the same no-load voltage settings operating in parallel. The increased bus load is divided equally between the two generators.

Figure 17.20 shows the effect of radically different voltage droop characteristics on the division of load current. Starting with an initial balanced condition of 500 A per generator and a bus voltage of 240 V, the lowered voltage caused by the increased bus load causes generator B to take more of the additional load than generator A. If, as loading continues, the ampere load on generator B causes its breaker to trip, all the bus load will fall on generator A and it too will trip, blacking out the plant. Referring again to Figure 17.20, if the bus load is decreased, the system voltage will increase. For this

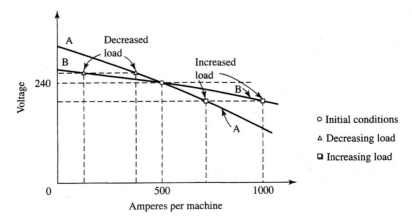

FIGURE 17.20
Effect of radically different voltage droop characteristics on the division of load current.

condition generator B loses a greater amount of the overall reduction in bus load than generator A. A large reduction in bus load can cause generator B to lose all its load and then be motorized by generator A. If this occurs, a reverse-current relay will trip generator B off the bus.

If, after balancing the bus amperes between two DC generators with supposedly identical voltage droops, the switching of load on or off the bus causes an unbalance condition, it is an indication that the governor droops of the two prime movers are not the same. One or both require droop adjustment, which should be done by experienced personnel.

17.20 LOAD-SHARING PROBLEMS OF COMPOUND DC GENERATORS IN PARALLEL

Stability of operation of two or more compound generators in parallel requires a third connection, called an equalizer connection, that parallels the series-field windings of all generators. This is illustrated for the two paralleled compound generators in Figure 17.21. The equalizer serves two functions. It ensures the division of load in accordance with the generator's ampere rating, and it prevents the reversal of current in the series field in the event that one generator becomes motorized.

With the series fields in parallel, the current through each series field is in the inverse ratio of its respective resistance. Hence, all incoming load cannot be "hogged" by one generator. If the generators are identical, the equalizer ensures an equal current through each series field, regardless of the division of load by the generator armatures.

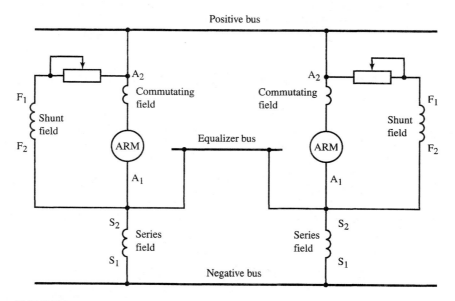

FIGURE 17.21
Compound generators in parallel with the series fields connected to an equalizer bus.

The equalizer connections of all generators are made to common bus, called the equalizer bus, using heavy copper cable so as not to add appreciably to the resistance of the circuit.

Paralleling without an Equalizer

If two identical compound generators are connected in parallel without an equalizer (this must not be done), any attempt to transfer load from one to the other will cause the generator with the greater excitation to take the entire load and drive the other as a motor. The generator that has its excitation increased takes load from the other and, in doing so, causes an increase in its series-field current. The generator that loses some of its load has its series-field current reduced and thus suffers a net reduction in its series-field strength. The result is an even greater internal voltage in the generator that gains some load and a further reduced internal voltage in the generator that loses some load; a rapid buildup of voltage occurs in one generator, accompanied by a simultaneous drop in the other. The net result is the absorption of all load by one generator and motorization of the other.

When a compound generator *without an equalizer* is motorized, the direction of current in its series field is reversed. If the series flux is greater than the shunt-field flux, the residual magnetism of the pole iron will be reversed, causing the polarity of the motorized generator to be reversed. This will be apparent after the circuit breaker trips.

17.21 FACTORS AFFECTING VOLTAGE BUILDUP IN DC GENERATORS

Factors other than excessive field-circuit resistance or low speed that affect the buildup of voltage in a self-excited generator are reversed shunt-field connections, reversed rotation, and reversed residual magnetism. These adverse effects can be visualized by studying the circuits in Figure 17.22 and using the right-hand rule to establish the direction of coil flux. For simplicity, only one field pole is shown.

Figure 17.22a represents normal operation; the prime-mover rotation is clockwise, and both the residual flux and the field-coil flux are directed to the left. In Figure 17.22b, reversed connections of the field circuit cause Φ_F to oppose Φ_R and the voltage builds down from its original residual value.

In Figure 17.22c, reversed rotation causes the armature voltage to reverse. This reverses the field current, causing Φ_F to oppose Φ_R, and the voltage builds down from its original residual value. In Figure 17.22d, reversed residual magnetism causes the armature voltage to reverse. This reverses the field current. In this case, both Φ_F and Φ_R are reversed. The result is voltage buildup in the reverse direction. The generator will operate at rated voltage with reversed polarity.

Reversed Polarity in a Self-Excited Generator

Reversed polarity in a self-excited shunt generator is caused by reversal of the residual magnetism in the pole iron. It can be caused by a heavy fault current, such as that produced by a short circuit at the bus or the main feeders, or by paralleling compound generators without an equalizer.

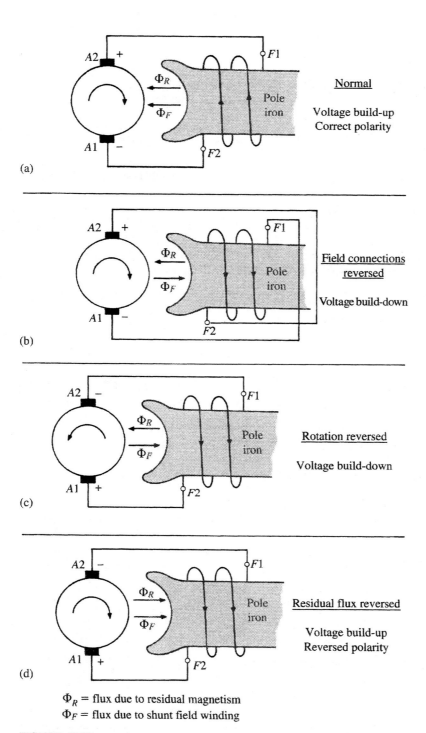

Φ_R = flux due to residual magnetism
Φ_F = flux due to shunt field winding

FIGURE 17.22
Factors affecting voltage buildup in a self-excited generator.

Reversed polarity is indicated by a reverse reading of the voltmeter on the generator panel. Correction of reversed polarity is accomplished by using an external DC source to remagnetize the iron in the correct direction. The procedure is called *field flashing*.

Under severe fault conditions, such as might occur when a self-excited shunt generator is accidentally short-circuited, the very high armature current causes complete saturation of the interpole iron, making the interpole ineffective (its field strength is no longer proportional to the armature current). The resultant high current in the commutated coil produces a strong demagnetizing mmf. Furthermore, short-circuiting the generator also short circuits the shunt-field winding. This results in a very low shunt-field current and, hence, a very weak shunt-field mmf. The combination of a weak shunt-field mmf and a relatively high opposing mmf produced by the commutated coil reverses the residual magnetism in the main-pole iron. Thus, after the circuit breaker trips, the reversed residual flux will cause the generator to build up voltage in the reverse direction.

Motorization of a shunt generator in parallel with the bus will not cause reverse polarity. Likewise, motorization of an equalizer-connected self-excited compound generator, in parallel with other equalizer-connected self-excited generators, will not result in reverse polarity. However, motorization of a self-excited compound generator without an equalizer connection can result in reverse polarity.

Correcting Reversed Polarity in a Self-Excited Generator

In order for the direction of the residual magnetism of a self-excited generator to be corrected, the current from another source must be fed (in the correct direction) into the shunt-field winding of the affected unit.

Batteries or another generator can be connected to the shunt-field winding to establish the correct magnetic flux. To do this, stop the prime mover, mark the location of the F_1 and F_2 leads, and disconnect them from the affected generator. Connect the shunt-field leads to another DC generator or to a 12-V storage battery with the F_1 lead connected to the positive terminal, as shown in Figure 17.23. If the shunt-field winding is then connected in accordance with NEMA standards for the given direction of rotation (Figure 17.14), the voltage will build up in the correct direction.

FIGURE 17.23
Using a 12-volt storage battery to restore residual magnetism.

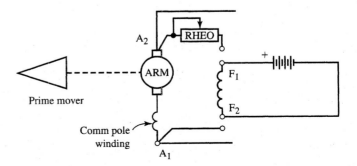

If the affected generator is self-excited, equipped for parallel operation, and another generator is on the bus, the correction is easily made. The prime mover of the affected generator should be stopped, its circuit breaker opened, and the brushes insulated from the commutator with pieces of manila file folder or thin cardboard. Do not change the brush-spring tension adjustment, and be careful not to scratch or smudge the surface film on the commutator. Double-check to make sure that all brushes are insulated from the commutator.

Then, with the prime mover still stopped, the field rheostat of the affected generator should be turned to its maximum resistance position, and the generator circuit breaker closed. The field rheostat should be turned to its minimum resistance position (raise voltage direction), then back to its maximum resistance position (lower voltage direction), and the circuit breaker tripped. The insulation separating the brushes from the commutator should then be removed, the prime mover started, and the generator rheostat adjusted to obtain the desired voltage.

If the brushes of the affected generator are not insulated from the commutator, the generator on the bus will feed into the armature of the affected generator, cause severe damage to the commutator and brushes, and black out the plant.

Reversed Polarity in a Separately Excited Generator

Reversed polarity in a separately excited generator may be caused by the wrong direction of rotation, the reversal of the field connections, or reversal of the exciter voltage. Correcting reversed polarity in a separately excited generator is relatively simple; it requires reversing the exciter connections or changing the direction of rotation, as applicable.

17.22 VOLTAGE FAILURE OF DC GENERATORS

If the generator is revolving at rated speed, failure to build up voltage may be attributed to a failure of the magnetic field.

Separately Excited Generator

In a separately excited generator, this can be caused by an open somewhere in the shunt-field circuit or by a failure of the exciter generator. A voltage check of the exciter generator will determine if it is at fault. To test the rheostat, short-circuit its terminals and observe the generator voltage. A buildup of voltage indicates an open rheostat. If the generator is operating at its rated speed, the exciter is operating at its rated voltage, and the generator fails to build up when the rheostat is shorted, the trouble is in the field circuit or the wires connecting the exciter to the field.

Self-Excited Generator

If a self-excited generator fails to build up voltage, the trouble may be caused by low speed, the rheostat set at too high a resistance, an open in the field rheostat, a poor contact between the brushes and the commutator, inadequate brush-spring pressure at the commutator, an open somewhere in the shunt-field circuit, or too low a value of residual magnetism. If the generator voltmeter indicates 10 volts or higher, the trouble is not a lack of residual magnetism but one or more of the other factors.

SUMMARY OF EQUATIONS FOR PROBLEM SOLVING

$$E_A = n\Phi_P k_G \tag{17-1}$$

$$\frac{E_2}{E_1} = \frac{n_2\Phi_2}{n_1\Phi_1} \tag{17-2}$$

$$E_A = V_T + I_A R_{acir} \tag{17-3}$$

$$VR = \frac{V_{nl} - V_{rated}}{V_{rated}} \times 100 \tag{17-4}$$

GENERAL REFERENCE

Hubert, Charles I., *Electric Machines: Theory, Operation, Applications, Adjustment, and Control,* 2nd ed. Prentice Hall, Upper Saddle River, NJ, 2001.

SPECIFIC REFERENCES KEYED TO THE TEXT

[1] Rosenberg, R., and Hand, A., *Electric Motor Repair,* 3rd ed. Holt, Reinhart and Winston, New York, 1986.
[2] *Motors and Generators,* NEMA Publication No. MG 1–1998, National Electrical Manufacturers Association.

REVIEW QUESTIONS

1. What is the function of the commutator and brushes in a DC generator?
2. What is the magnetic-neutral plane?
3. What is the function of interpoles? Where are they located?
4. What is armature reaction, what are its adverse effects, and how can it be eliminated?
5. Explain in detail how a self-excited generator builds up its voltage from a low value, initiated by residual magnetism, to its rated value.
6. When a relatively heavy load is connected to a DC generator, the speed and voltage output of the generator are reduced. Explain why this happens.
7. In what way does a compound generator differ from a shunt generator? Sketch the voltage vs. current characteristics of these generators, and explain why they are different.

8. Explain how an overcompound generator may be given a flat compound characteristic without changing the number of turns in the field windings and without changing the speed.
9. How does a three-wire DC generator differ from the standard DC generator? What is the function of the balance coil?
10. Outline the steps required for paralleling a DC generator with another already on the bus. Assume that the generator on the bus is operating at 250 volts and has a bus load of 1200 amps. Include the correct procedure for dividing the bus current equally between generators.
11. Why is it good practice to balance the load currents in proportion to the generator kW ratings?
12. Assume that two identical DC generators are operating in parallel and sharing a small bus load. Outline the correct procedure for removing one of the generators from the bus.
13. What effect do different load-voltage characteristics have on the division of oncoming load between generators? Explain with the aid of a sketch.
14. For stability of operation in parallel, compound generators require an equalizer connection. Explain how the equalizer provides this stability.
15. What is motorization? Is it harmful? How can it be detected and corrected?
16. What damage could occur if a DC generator, in parallel with other generators, loses its field excitation?
17. What fault condition can cause a generator to reverse its polarity? Assume that the direction of rotation is not changed.
18. Describe a method for correcting reversed polarity of a DC generator. Assume that the only source of energy available is that of another generator on the bus.
19. Describe a method that can be used to strengthen the residual magnetism of a self-excited DC generator.
20. What three fundamental fault conditions can prevent a self-excited generator from building up to rated voltage?
21. Describe a simple test that does not require instruments and that can determine whether or not a field rheostat is at fault when a generator fails to build up voltage.

PROBLEMS

17–1/5. A 500-V, 200-kW, 1150-rpm DC generator running at no load has a no-load voltage of 560 V. If the speed is raised to 1560 rpm, and the pole flux is increased to 1.1 times its original value, determine the new no-load voltage.

17–2/5. A certain DC generator develops 230 V when operating at no load and 1800 rpm. Determine (a) voltage if the speed is increased to 2000 rpm; (b) voltage for a speed of 1800 rpm and a flux that is reduced to 75% of its original value.

17–3/5. A shunt generator generates 200 V when running at 1800 rpm. Determine the speed required to maintain 200 V if the pole flux is reduced to 95% of its original value.

17–4/5. The speed of a certain DC generator operating at 600 V is 3600 rpm. If a fault in the prime-mover governing system causes the machine to operate at 3200 rpm, determine the percent increase in pole flux required to maintain 600 volts.

17–5/11. A certain separately excited generator, with interpoles and a compensating winding, is operating at its rated load current of 140 A at 240 V. The armature circuit resistance is 0.157 Ω. Determine (a) no-load voltage; (b) voltage regulation.

17–6/11. A 200-kW, 250-V, 850-rpm, separately excited generator, with interpoles and a compensating winding, is operating at its rated conditions. The armature resistance is 0.0057 Ω; the resistance of the interpole and compensating windings are 0.0023 Ω and 0.0011 Ω, respectively. Determine (a) armature current; (b) no-load voltage; (c) voltage regulation.

17–7/18. A DC generator with a 3% voltage droop is operating at its rated 800 A and 600 V. Determine the voltage if the circuit breaker is tripped.

17–8/18. A DC generator with a 2% voltage droop is operating at its rated 150 A at 240 V. Determine the voltage when the entire load is removed.

18

Direct-Current Motors: Principles and Operational Problems

18.0 INTRODUCTION

This chapter builds on the principles and construction details of DC generators presented in Chapter 17 and explains how motor torque is developed. The torque-speed characteristics of shunt, series, and compound motors are discussed, along with strong warnings about the dangers of the differential connection. Methods of speed control for above and below nameplate speed are presented. Standard connections for clockwise and counterclockwise motor rotation and test procedures to identify unmarked terminals are included.

Direct current motors and their control circuits are of particular concern for operating and maintenance personnel in that loose or high resistance connections in the shunt field circuit may cause the armature to accelerate to destruction.

18.1 PRINCIPLE OF MOTOR ACTION

Figure 18.1a illustrates a *very elementary* two-pole permanent-magnet DC motor with a one-coil armature. The armature of a DC motor is always the rotating member. The ends of the armature coil are connected to the insulated segments of an elementary two-bar commutator. Graphite brushes provide the connection to the power source. The brushes in this elementary motor are very narrow so as not to touch both commutator bars at the same time.

Figure 18.1b is the commutator end view of Figure 18.1a; it shows the direction of mechanical force on the conductors resulting from the interaction of the magnetic field of the magnets with the magnetic field caused by the current in the conductors. The force, as determined from flux bunching, is upward on conductor A and downward on conductor B. The net result is a counterclockwise (CCW) movement of the armature coil. When the armature coil rotates to the position shown in Figure 18.1c,

371

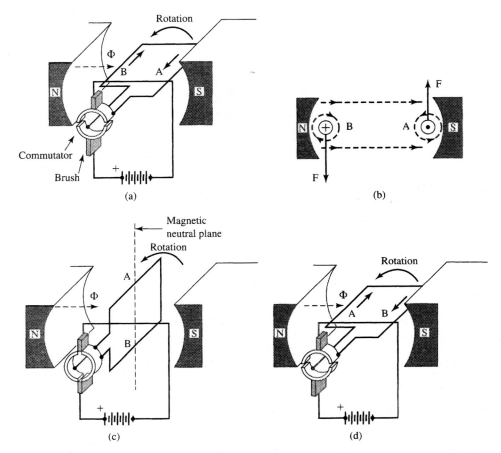

FIGURE 18.1
Elementary two-pole, permanent-magnet DC motor with a one-coil armature.

called the magnetic-neutral plane, the brushes no longer make contact with the commutator bars. The current is switched off and the coil coasts through the magnetic-neutral plane. The magnetic-neutral plane is located midway between adjacent poles of opposite polarity.

When the coil coasts past the neutral plane, the commutator switches the current on again as shown in Figure 18.1d; this switching action is called *commutation*. Note that although conductor A and conductor B have changed places, the commutation process ensures that the directions of current in each conductor are such that the motor torque remains in the same direction.

Developed Torque

The torque developed in a DC motor is proportional to the current in the armature and the density of the magnetic flux provided by the magnets. Expressed mathematically,

$$T_D \propto B_F I_A \tag{18-1}$$

where: T_D = shaft torque (lb-ft)
B_F = flux density in tesla (T)
I_A = Armature current (A).

Increasing the armature current or increasing the density of the pole flux will result in an increase in the developed torque. If the developed torque for one set of conditions is known, the torque for another set of conditions can be determined by converting Eq. (18–1) to the following ratio, where subscript 1 represents the initial conditions, and subscript 2 represents the new conditions.

$$T_{D2} = T_{D1}\frac{B_{F2}I_{A2}}{B_{F1}I_{A1}} \tag{18-2}$$

EXAMPLE 18.1   When the switch is closed to a certain DC motor, the starting torque and armature current are 165 lb-ft and 112 A, respectively. Determine the starting torque if the armature current is limited to 85 A. Assume the field flux is not changed.

Solution
From Eq. (18–2), and noting that the flux density did not change, $B_{F1} = B_{F2}$.

$$T_{D2} = T_{D1}\frac{B_{F2}\,I_{A2}}{B_{F1}\,I_{A1}} = T_{D1}\frac{B_{F1}\,I_{A2}}{B_{F1}\,I_{A1}} = 165 \times \frac{85}{112} = 125.2 \text{ lb–ft}$$

18.2 REVERSING A DC MOTOR

To reverse the direction of rotation of a DC motor, we merely reverse the polarity of the field poles or reverse the direction of the current to the armature, but not both. This is illustrated in Figure 18.2 for a representative armature coil and field poles. Figure 18.2a represents the original conditions as developed in Figure 18.1b. Figure 18.2b shows the effect of reversing the armature current on the direction of rotation, and Figure 18.2c shows the effect of reversing the polarity of the magnets. Flux bunching determines the direction of the developed force. Reversing both the polarity of the field poles and the direction of the current supplied to the armature does not reverse the direction of rotation of the armature.

The circuit connections for a DC motor with an eight-coil armature and a two-pole field is shown in Figure 18.3. The field winding and armature are connected in parallel and supplied by a DC source.[1] To reverse the direction of rotation, interchange armature leads A_1 and A_2, or interchange field leads F_1 and F_2. Interchanging power lines L_1 and L_2 changes the direction of the current in both the armature and the shunt field and, hence, will not reverse the rotation of the motor.

---

[1] The motor shown in Figure 18.3 is the same machine shown connected and operated as a generator in Figure 17.6c, Chapter 17.

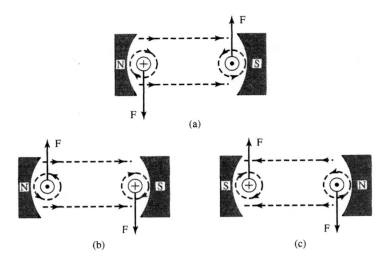

FIGURE 18.2
(a) Directions of pole flux and armature current for the given direction of rotation;
(b) reversing the direction of rotation by reversing the armature current; (c) reversing the
direction of rotation by reversing the polarity of the poles.

FIGURE 18.3
Circuit connections for a DC motor with an
eight-coil armature and a two-pole field.

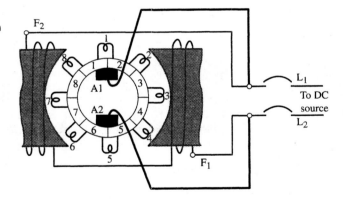

18.3 DC MOTOR CONSTRUCTION

The construction details of DC motors are essentially the same as that for the DC generators discussed in Chapter 17; both generators and motors have an armature, magnetic poles, a commutator, and carbon brushes. A cutaway view of a DC motor for outdoor use in severe weather conditions is shown in Figure 18.4. The shunt-field coils and series-field coils are wound around the same pole iron and provide specific machine characteristics.

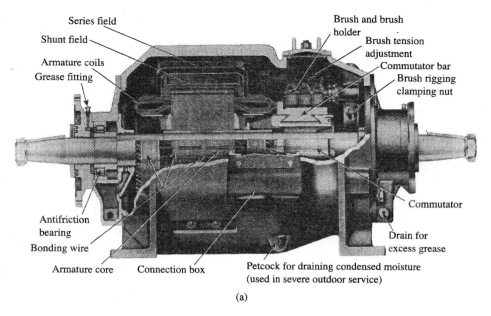

Series field
Shunt field
Armature coils
Grease fitting
Brush and brush holder
Brush tension adjustment
Commutator bar
Brush rigging clamping nut
Commutator
Antifriction bearing
Bonding wire
Drain for excess grease
Armature core   Connection box
Petcock for draining condensed moisture (used in severe outdoor service)

(a)

FIGURE 18.4
Construction details of a DC motor (courtesy TECO Westinghouse).

Interpoles

The interpole windings (also called commutating windings) and the compensating windings serve the same purpose in a DC motor as they do in a DC generator.[2] However, the polarity of the interpoles relative to the main field poles for a motor is opposite that of a generator for the same direction of rotation. Thus, when the armature coil of a motor rotates into the neutral plane, the armature conductors of a motor come under the influence of interpoles whose polarity is the same as the polarity of the main pole that it just left. A comparison of interpole polarity for motors and generators, for the same direction of rotation is shown in Figure 17.8, Chapter 17.

Compensating Windings

The effect of armature reaction is particularly severe in motor applications where quick reversals or large and rapid load changes occur. The sudden application of heavy loads, or quick reversal of a motor, will cause large and rapid swings of flux across the pole face, inducing high transient voltages in all of the armature coils. If the transient voltages are high enough to cause arcing between commutator bars, a brush-to-brush flashover may occur. Such machines require compensating windings.

---

[2] See Sections 17.6 and 17.7 of Chapter 17 for a discussion of armature inductance and interpoles, and Section 17.8 for a discussion of armature reaction and compensating windings.

Except for repairs, never break the connections between armature, interpole windings, and compensating windings. They must be treated as a unit. *If the motor is to be reversed by reversing the direction of armature current, the series branch consisting of the armature, interpole windings, and compensating windings must be reversed as a unit.*

18.4 COUNTER-EMF OF A DIRECT CURRENT MOTOR

The application of a driving voltage to a DC motor causes the armature to rotate within the magnetic flux set up by the field poles; this generates a voltage within the armature coils, which, in accordance with Lenz's law, is always in opposition to the driving voltage. The magnitude of this opposing voltage (also called counter-electromotive force, counter-emf, or cemf) depends on the magnitude of both the pole flux and the rotational speed of the armature. Expressed mathematically,

$$E_A = n\Phi_P k_M \tag{18-3}$$

where: E_A = cemf (V)
  n = rotational speed of armature (rpm)
  Φ_P = pole flux (Wb)
  k_M = motor constant.

Note: Equation (18–3) for the cemf of a DC motor is identical to Eq. (17–1), which is the electromotive force (emf) in a DC generator. Motors can be used as generators, and generators can be used as motors; the countervoltage generated in a DC motor becomes a source voltage when driven as a generator by a prime mover; and the torque developed in a DC motor becomes the countertorque when driven as a generator.

The relative directions of applied voltage, cemf, and armature current in a shunt motor are shown in Figure 18.5. It is called a shunt motor because the field is in shunt (parallel) with the armature. Applying Kirchhoff's law to the armature circuit in Figure 18.5,

$$V_T - E_A = I_A R_{\text{acir}} \tag{18-4}$$

FIGURE 18.5
Elementary circuit diagram of a shunt motor.

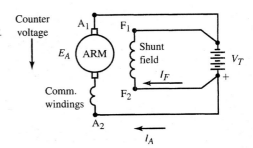

Solving for I_A,

$$I_A = \frac{V_T - E_A}{R_{\text{acir}}} \tag{18-5}$$

where: I_A = armature current (A)
V_T = applied voltage (V)
E_A = cemf (V)
R_{acir} = resistance of armature circuit (Ω).

The resistance of the armature circuit is the resistance of the armature plus everything within the generator housing that is in series with the armature, such as commutating winding, etc.

EXAMPLE 18.2

The armature of a 230-V, 1200-rpm shunt motor draws 100 A when operating at full load and rated speed. If the armature circuit has a resistance of 0.04 Ω, calculate the cemf.

Solution
From Eq. (18-4),

$$E_A = V_T - I_A R_{\text{acir}}$$
$$E_A = 230 - 100 \times 0.04 = 226 \text{ V}$$

At rated armature speed the cemf is nearly equivalent to the applied terminal voltage. The difference between the cemf and the terminal voltage is very slight, only 4 volts in the foregoing problem. As expressed in Eq. (18-3) the cemf developed in a motor armature is a function of the magnitude of the field flux and the speed of the armature.

The cemf is zero when the motor is not running but increases in magnitude as the machine accelerates. Since the speed of a motor is affected by the magnitude of the load on its shaft and the cemf, the armature current is different for different loads. The application of load to a motor causes the shaft power output to become greater than the electrical power input, and the motor slows down. This decreases the cemf and more current enters the armature to carry the load at the new lower speed.

The cemf functions to admit the correct amount of current to a motor required by its load conditions, just as the governor of an engine responds to admit the necessary energy demanded by its load. Both the cemf of a motor and the governor of an engine are modified by changes in speed to increase or decrease the input energy.

18.5 STARTING A DC MOTOR

With the exception of fractional horsepower motors, a DC motor requires the addition of an external resistor to limit the current in the armature circuit when starting. Excessively high starting current will cause severe arcing and sparking at the commutator–brush interface.

The starting resistance must be left in the armature circuit until the cemf builds up sufficiently to reduce the armature current to a safe value. The value of starting resistance is generally selected to limit the current from 150 to 250 percent of the rated value, depending on the starting torque required.

The shunt field is always connected across full-line voltage when starting, so that less armature current is required to develop the required torque. The resistance of the shunt field is many times that of the armature. The shunt-field current varies from less than 1 percent rated motor current for machines larger than 100 hp to as high as 5 percent rated motor current for machines below 5 hp.

Figure 18.6 illustrates a manually operated rheostat-type starter connected to a shunt motor. The circuit is closed when the lever is moved to the first contact, at which position all the rheostat resistance is in series with the armature. As the machine accelerates, the lever should be moved slowly to the run position. The lever should not be left too long in an intermediate position or the starter will be damaged by overheating.

The holding coil serves to hold the rheostat lever in the run position after all the resistance is cut out. The holding coil is connected in series with the shunt field of the motor, so that an open in the field circuit would de-energize the coil and permit a spring (not shown) to pull the lever back to the off position. The spring also returns the lever to the off position if a voltage failure occurs or if the lever is left in some intermediate position. Attempts to start integral horsepower motors (machines above 1 hp) by connecting them across the line may result in serious damage to the commutator, brushes, and armature windings.

A motor starting circuit using a fixed resistor (R_X) and a contactor for shorting out the resistor when the motor comes up to speed is shown in Figure 18.7. The total input current to the motor is

$$I_T = I_A + I_F \tag{18-6}$$

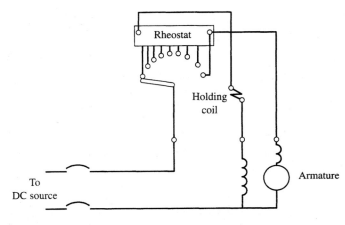

FIGURE 18.6
Connection diagram for a manually operated DC motor starter.

FIGURE 18.7
Motor starting circuit.

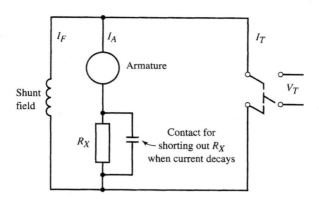

where: I_T = total current to motor (A)
I_A = armature current (A)
I_F = field current (A).

Shunt-Field Current

The shunt-field coils are wound with many turns (several hundred to several thousand turns) of relatively small-diameter wire, so that the field-circuit resistance is fairly high and the field current low. The many turns of wire around each field pole produce a relatively high flux with only a small current. Applying Ohm's law to the shunt-field circuit in Figure 18.7,

$$I_F = \frac{V_T}{R_F} \tag{18-7}$$

where: R_F = resistance of the shunt-field circuit (Ω).

The resistance of the shunt-field circuit includes any resistors or rheostats that may be in series with the shunt-field windings.

EXAMPLE 18.3   A 15-hp, 230-V, 1800-rpm shunt motor draws 56 A when operating at rated load. The resistance of the armature and shunt field are 0.28 Ω and 120.1 Ω, respectively. Determine (a) the starting current that the armature would draw if no starting resistance were used; (b) the additional resistance required in series with the armature that would limit the armature current on starting to 150% rated current; (c) shunt-field current.

Solution
a. From Eq. (18–5), and noting that the cemf is zero,

$$I_A = \frac{V_T - E_A}{R_A} = \frac{230 - 0}{0.28} = 821.4 \text{ A}$$

This magnitude of current would severely damage the armature, commutator, and brushes. Destructive arcing and burning will occur at the commutator–brush interface. Furthermore, the sudden impact to the driven equipment by the associated very high starting torque may cause mechanical damage to the driven equipment.

b. The motor circuit with the starting resistance in series with the armature is shown in Figure 18.7.

$$I_A = \frac{V_T - E_A}{R_A + R_X} \Rightarrow R_X = \frac{V_T - E_A}{I_A} - R_A$$

$$R_X = \frac{230 - 0}{1.5 \times 56} - 0.28 = 2.458 \ \Omega$$

c. Applying Ohm's law to the shunt-field circuit,

$$I_F = \frac{V_T}{R_F} = \frac{230}{120.1} = 1.91 \text{ A}$$

As the armature accelerates from standstill, the armature cemf E_A builds, causing the armature current to decrease from its high starting value to the normal value for the particular shaft load and starting resistance. The starting resistance is then shorted, and the motor accelerates to its final steady-state speed for the given load. In most DC motor applications, the starting resistance is removed in steps. Direct-current motors rated at less than 1/2 horsepower seldom require current-limiting starters.

18.6 SPEED ADJUSTMENT OF DC MOTORS

The relationship between motor speed, armature current, and pole flux is obtained by substituting Eq. (18–3) into Eq. (18–4) and solving for the speed. The result is the following DC motor speed equation:

$$n = \left. \frac{V_T - I_A R_{\text{acir}}}{\Phi_P k_M} \right|_{\Phi_P \neq 0} \tag{18–8}$$

$$R_{\text{acir}} = R_A + R_{\text{IP}} + R_S + R_{\text{CW}} + R_X \tag{18–9}$$

where: R_A = resistance of armature windings (Ω)
 R_{IP} = resistance of interpole windings (Ω)
 R_S = resistance of series field winding (Ω)
 R_{CW} = resistance of compensating windings (Ω)
 R_X = resistance of rheostat in series with the armature (Ω)
 Φ_P = pole flux (Wb)
 k_M = motor constant.

Equation (18–8) indicates the general trend of motor speed when the various factors are changed, *providing the pole flux is not zero,* and sufficient torque is developed to produce the necessary acceleration. If the pole flux is zero, there is no torque, no matter how high the armature current.

The *base speed* (nameplate speed or rated speed) of a motor is its speed when operating with rated shaft load, at rated temperature, with rated voltage applied, and no external resistors or rheostats connected in series with the armature or field circuits.

Adding external resistance in series with the armature, as shown in Figure 18.8a, reduces the armature current; this in turn reduces the developed torque, and the motor slows down.

Adding external resistance in series with the shunt field, as shown in Figure 18.8b, reduces the pole flux. With less pole flux, the cemf is reduced. This results in a higher armature current. As long as the percent increase in armature current is greater than the percent decrease in pole flux, the developed torque will increase and the motor will accelerate.

A large and rapid reduction in the shunt-field current of a motor may result in so high a magnitude of armature current that the circuit breaker or other overcurrent protective device will operate and remove the machine from the line; the high armature current is caused by the large reduction in cemf. If protective devices are absent or inoperative, the commutator and brushes may be severely damaged. Hence, speed adjustments using a rheostat in the shunt-field circuit should be made gradually. Large motors of high inertia, or motors driving high-inertia loads, are more susceptible to damage than are small machines. Small machines can accelerate rapidly; the associated rapid buildup of cemf in a low-inertia motor quickly reduces the high armature current.

The speed of a DC motor is extremely sensitive to changes in the field flux. If the shunt field is overly weakened by too much field rheostat resistance, or by an open or loose connection in the shunt-field circuit, and the load on the motor is light, the motor may accelerate to destruction. If the shunt-field circuit is opened, the flux does not instantaneously drop to zero; the hysteresis of the pole iron provides sufficient residual

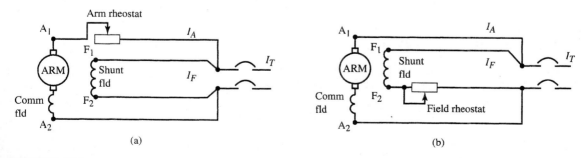

FIGURE 18.8

Speed adjustment by (a) armature rheostat control; (b) field rheostat control.

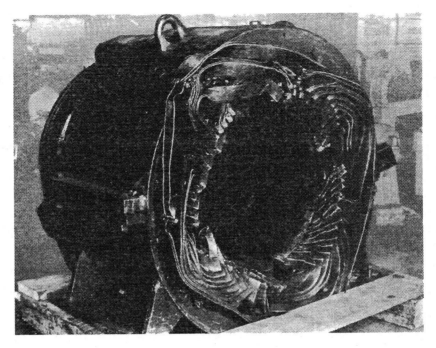

FIGURE 18.9
Damage to a DC motor caused by excessive speed (courtesy TECO Westinghouse).

magnetism to permit acceleration to speeds that may result in destruction of the machine. Figure 18.9 shows the damage caused by overspeed when the shunt-field circuit of a DC motor running at light load was opened.

Field rheostat adjustments should always be made slowly; a sudden increase in field-rheostat resistance or a sudden decrease in armature-rheostat resistance causes excessive armature current that may damage the commutator–brush interface.

18.7 MECHANICAL POWER DEVELOPED

The mechanical power developed in a direct current motor is equal to the total power input to the armature minus the I^2R losses in the armature windings. Expressed mathematically,

$$P_{\text{mech}} = V_T I_A - I_A{}^2 R_{\text{acir}} \tag{18–10}$$

From Eq. (18–4),

$$V_T - E_A = I_A R_{\text{acir}} \tag{18–11}$$

Multiplying both sides of Eq. (18–11) by I_A and rearranging terms, we obtain

$$E_A I_A = V_T I_A - I_A^2 R_{acir} \qquad (18\text{–}12)$$

A comparison of Eq. (18–10) with Eq. (18–12) shows that the mechanical power developed (P_{mech}) is numerically equal to the product of the cemf (E_A) and the armature current (I_A):

$$P_{mech} = E_A I_A \qquad (18\text{–}13)$$

All power drawn by the field windings is expended in the form of heat:

$$P_F = I_F^2 R_F \qquad (18\text{–}14)$$

where: P_F = heat energy expended (W)
             I_F = field current (A)
             R_F = resistance of field windings (Ω).

The field of a motor does no useful work. It merely provides the necessary medium for the armature conductors to push against when developing rotary motion. It is similar to the road bed against which the wheels of an automobile push. The road bed does no work, but without it the car could not move. The energy supplied to the field is expended as I^2R losses in the field windings. This may be proven easily by connecting a wattmeter to measure the power drawn by the shunt field. Observation of the wattmeter will indicate no change in the power input to the field as the machine is loaded. However, a wattmeter placed in the armature circuit will indicate proportional increases in power with increased shaft load.

EXAMPLE 18.4

A certain 240-V shunt motor operating at rated voltage and 1750 rpm draws a current of 70.2 A. The resistances of armature, interpole, and shunt-field windings are 0.0912 Ω, 0.045 Ω, and 105 Ω, respectively. Determine (a) field current; (b) armature current; (c) cemf; (d) mechanical power developed; (e) heat power losses in armature; (f) heat power losses in field winding.

Solution

a. $I_F = \dfrac{V_F}{R_F} = \dfrac{240}{105} = 2.286 \Rightarrow 2.29$ A

b. $I_A = I_T - I_F = 70.2 - 2.286 = 67.914 \Rightarrow 67.9$ A

c. $R_{acir} = R_A + R_{IP} = 0.0912 + 0.045 = 0.1362 \;\Omega$
    $E_A = V_T - I_A R_{acir} = 240 - 67.914 \times 0.1362 = 230.75$ V

d. $P_{mech} = E_A I_A = 230.75 \times 67.914 = 15{,}671.2$ W

e. $P_{A.heat} = I_A^2 R_A = 67.914^2 \times 0.0912 = 420.646 \Rightarrow 420.6$ W

f. $P_{F.heat} = I_F^2 R_F = 2.286^2 \times 105 = 548.57 \Rightarrow 548.6$ W

18.8 SHAFT POWER, SHAFT TORQUE, AND EFFICIENCY

The relationship between shaft power out, shaft torque out, and motor speed is given in the following equation:

$$P_{shaft} = \frac{T_{shaft} \times n_{shaft}}{5252} \qquad (18–15)$$

where: P_{shaft} = shaft power out (hp)
T_{shaft} = shaft torque out (lb-ft)
n_{shaft} = shaft speed (rpm).

The total input current and total electric power in to the motor are

$$I_T = I_A + I_F \qquad (18–16)$$

$$P_{in} = V_T I_T \qquad (18–17)$$

where: V_T = input voltage to motor (V)
I_T = total input current to motor (A).

Motor efficiency is the ratio of shaft power out (with shaft horsepower converted to watts) to electric power in, multiplied by 100 to express it as a percent:

$$eff = \frac{P_{shaft} \times 746}{P_{in}} \times 100 \qquad (18–18)$$

where: P_{in} = input power (W)
eff = efficiency (%).

EXAMPLE 18.5

A certain 20-hp, 1750-rpm, 230-V shunt motor, operating at rated conditions, draws a line current of 85.8 A. Determine (a) shaft torque; (b) input power; (c) efficiency.

Solution
From Eq. (18–14),

a. $T_{shaft} = \dfrac{5252 \times P_{shaft}}{n_{shaft}} = \dfrac{5252 \times 20}{1750} = 60$ lb–ft

b. $P_{in} = V_T I_T = 230 \times 85.8 = 19{,}734$ W

c. $eff = \dfrac{P_{shaft} \times 746}{P_{in}} \times 100 = \dfrac{20 \times 746}{19{,}734} \times 100 = 75.6\%$

18.9 COMPOUND AND SERIES MOTORS

Modification of the basic DC motor to obtain certain desirable torque-speed characteristics is generally accomplished by the addition of an auxiliary winding called the series field. The series-field coils are wound with heavy copper conductors on the

FIGURE 18.10
Circuit for a cumulative compound motor.

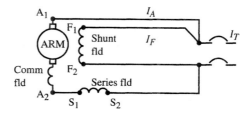

same pole iron as are the shunt-field coils and are connected in series with the armature, as shown in Figure 18.10. Such machines are called compound motors.

Cumulative Compound Motors

If the current in each series-field coil is in the same direction as the current in its corresponding shunt-field coil, the flux contribution due to the series-field coils adds to the flux of the shunt-field coils, and the machine is called a *cumulative compound motor*. If the series field has only one or two turns or sometimes a half turn, it is called a *stabilized-shunt motor;* the series-field winding of a stabilized-shunt motor is used to counteract the demagnetizing effect of armature reaction in shunt motors that do not have a compensating winding.

Reversing the Rotation of a Cumulative Compound Motor

Reversing the direction of rotation of a cumulative compound motor is always accomplished by reversing the armature-interpole-compensating-winding branch as a unit; the series-field winding and shunt-field winding are not reversed.

Differential Compound Motors

If the current in the series-field coils is in the opposite direction with respect to the current in the corresponding shunt-field coils, the series-field flux subtracts from the shunt-field flux, and the motor is called a *differential compound motor*. However, because of its instability, differential compound motors are not used in power applications. A differential connection for a motor poses a hazard to life and property and must be avoided. The behavior of a differential motor is both peculiar and frightening, especially in the higher horsepower machines; if connected to the power supply, it may start in the wrong direction, then quickly reverse and accelerate to destruction.

Series Motor

Motors with only a heavy series-field winding and no shunt-field winding, called series motors, have extremely high starting torques. Series motors must be directly coupled or geared to the load, because loss of load could result in destruction by overspeed. The circuit for a series motor is shown in Figure 18.11.

FIGURE 18.11
Circuit for a series motor.

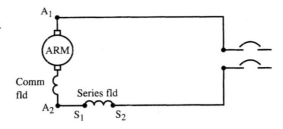

The current in the series-field winding is the armature current, as shown in Figure 18.11. Thus, the flux density of the series field will be proportional to the armature current. Expressed mathematically,

$$B_F \propto I_A$$

From Eq. (18–1),

$$T_D \propto B_F I_A$$

Hence,

$$T_D \propto I_A{}^2 \tag{18–19}$$

Thus, the developed torque in a series motor is proportional to the square of the armature current.

18.10 OPERATING CHARACTERISTICS OF DC MOTORS

The operating characteristics of shunt, cumulative compound, and series motors for the same rated values of speed and torque are shown in Figure 18.12. The characteristics of a stabilized-shunt motor lie somewhere between the shunt and compound characteristics.

Shunt Motor

As indicated by its operating characteristics in Figure 18.12, the speed of a shunt motor is relatively constant from no load to rated load and is attributed to the constant flux provided by the shunt field. The internal torque developed by a DC motor is dependent on the flux density and the armature current. Since the field current of a shunt motor is constant, the developed torque will be proportional to the armature current. The shunt motor has its greatest field of application with loads that require essentially constant speed and where high starting torques are not required. Shunt motors and stabilized-shunt motors are used to drive centrifugal pumps, fans, winding reels, conveyors, machine tools, and other loads of similar characteristics.

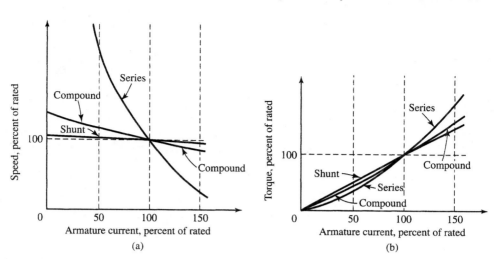

FIGURE 18.12
Typical operating characteristics of shunt, compound, and series motors of equal rated horsepower and equal rated speed.

Cumulative Compound Motor

The torque that the cumulative compound motor develops for given values of armature current above the rated value is much higher than that of a shunt machine with the same rated horsepower and rated speed. However, the higher torque values of the compound machine are accompanied by lower speeds. The compound motor has its greatest field of application with loads that require high starting torques or have pulsating loads. They are used to drive reciprocating pumps, hoists, compressors, and other loads of similar characteristics.

Compound motors smooth out the energy demand required by a pulsating load. The motor slows down during the work stroke, giving up the energy stored in its moving parts, and at the same time develops a greater internal torque. During the recovery stroke the motor accelerates and restores energy to the moving parts. Thus, the pulsating demand from the electrical system is less for a compound motor than for a shunt motor. A shunt motor would tend to maintain the same speed during the work stroke, causing excessive current demands from the electrical system.

Series Motor

The high ratio of light-load speed to full-load speed, coupled with its very high starting torque, makes the series motor very adaptable to driving some types of hoists, railroad cars, and other loads of similar characteristics.

As indicated by its speed curve in Figure 18.12a, removing the load from a series motor will cause it to "run away." At the instant the load is removed from a series motor, its speed increases. This is followed by an increase in the cemf and a resultant decrease in armature current. However, although decreased from its load value, the armature current and field flux are still high enough to cause sustained acceleration. Unlike shunt and compound machines, the internal torque developed by a series motor at no load is great enough to accelerate the series motor to speeds that may cause destruction by centrifugal force.

The behavior of the series motor may be likened to that of a shunt machine that has its field flux weakened by adjustment of its field rheostat every time the load is decreased. The result is a very high speed at no load. Hence, series motors must be directly connected to the load by gears or by a solid coupling; it should not be applied to loads that can be removed or reduced sufficiently to cause overspeed. The maximum safe speed for a series motor must be stamped on the motor nameplate [1].

Fractional horsepower series motors used to drive small power tools, such as drills, hammers, vacuum cleaners, and mixers, are usually designed for operation on both alternating and direct current. Such machines are called *universal motors*.

The application of load to a running motor causes the shaft output to become greater than the electrical input, and the motor slows down. This decreases the cemf and more current enters the armature to carry the load.

Speed Regulation

The percent change in speed from no-load speed to rated speed with respect to rated speed is called the motor speed regulation. Expressed mathematically,

$$\text{SR} = \frac{n_{nl} - n_{rated}}{n_{rated}} \times 100 \tag{18–20}$$

where: n_{nl} = no-load speed (rpm)
n_{rated} = rated speed (rpm)
SR = speed regulation (%).

EXAMPLE 18.6   A 230-V, 75-hp shunt motor has a rated speed of 1760 rpm, and a no-load speed of 1800 rpm. Determine the speed regulation.

Solution

$$\text{SR} = \frac{n_{nl} - n_{rated}}{n_{rated}} \times 100 = \frac{1800 - 1760}{1760} \times 100 = 2.3\%$$

18.11 STANDARD TERMINAL MARKINGS AND CONNECTIONS FOR DC MOTORS

Direct-current motors may be operated in either direction of rotation. However, the standard direction of rotation is counterclockwise facing the end opposite the drive shaft.

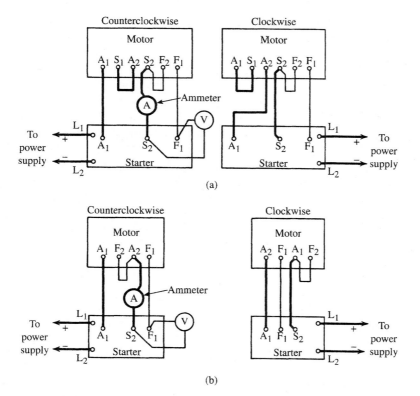

FIGURE 18.13
Connections to motor starters for different directions of rotation of DC motors:
(a) compound motor; (b) shunt motor.

The standard terminal markings for DC motors, associated starters, and the proper connections for different directions of rotation are shown in Figure 18.13 [1]. The same starter can be used for both shunt and compound motors of equal voltage and power ratings.[3] The ammeter and voltmeter should be inserted at the indicated points when checking current and voltage. If the motor overheats and voltmeter and ammeter readings indicate rated voltage and rated or below-rated current, respectively, the ventilating ducts may be clogged and the machine requires cleaning or the ambient temperature may be above normal.

Identifying Unmarked Terminals

Unmarked terminals should be identified by test before the machine is connected to the starter; otherwise blown fuses or damage to the starter and motor may result. The

---

[3] Circuit diagrams for motor starters are illustrated and explained in Chapter 22.

shunt-field leads are generally smaller in diameter than the leads of the armature and series field, and are therefore easily identified. However, a continuity test with an ohmmeter set on the high-resistance scale will resolve any doubts. This is illustrated in Figure 18.14a.

One ohmmeter lead should be connected to one motor lead, and the other ohmmeter lead should be used to probe the remaining motor leads until continuity is indicated (low reading on ohmmeter). The pair of motor leads that indicate continuity should be pushed to the side, and the remaining four motor leads tested in the same manner until the other two pairs are determined. The pair of motor leads that indicate the highest resistance is the shunt-field lead pair, and should be so marked.

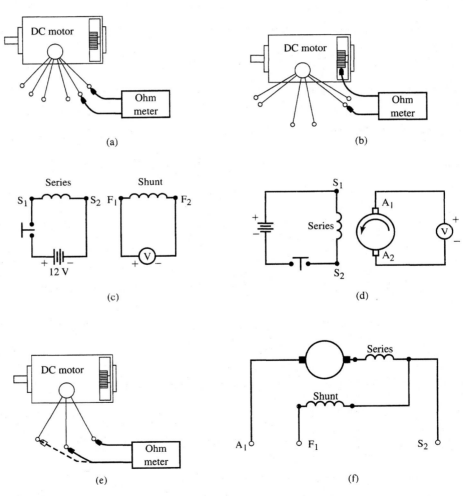

FIGURE 18.14
Procedure for identifying unmarked terminals.

The two armature leads can be identified by testing from the brush holders to each of the other four leads, as shown in Figure 18.14b; the armature leads will indicate approximately 0 Ω. The remaining two leads are those of the series-field winding.

Because it is extremely dangerous to operate a compound motor that is differentially connected, the relative polarity of the series and shunt fields must be established. This can be determined in a safe manner with the simple circuit shown in Figure 18.14c. Connect a 12-volt auto battery in series with a pushbutton and the series field, and connect a 10/50 volt DC analog-type voltmeter[4] across the shunt field. Make a momentary contact by depressing and releasing the button quickly. If the voltmeter reads upscale when the button is pushed and downscale when released, mark the shunt-field lead that connects to the + terminal of the voltmeter F_1, and mark the series-field lead that connects to the + terminal of the battery S_1. If the correct deflection is not indicated, interchange the two shunt field leads and try again.

The circuit shown in Figure 18.14d can be used to determine the terminal markings for the armature. Be sure to connect the positive terminal of the battery to the S_1 lead, and connect the voltmeter across the armature. Manually, or otherwise, rotate the armature in the CCW direction, as viewed from the end opposite the drive shaft. While rotating CCW, hold down the pushbutton, and note the deflection on an analog-type voltmeter. If the pointer deflects upscale, the armature lead that connects to the + terminal of the voltmeter should be marked A_1; if the correct deflection is not indicated, interchange the two armature leads and try again. Although the presence of residual magnetism in the iron pole pieces may cause a small voltage to be generated, it is the direction of the additional voltage generated when the button is depressed and held that is significant. After establishing the relative terminal markings of all the motor leads, the machine can be connected in accordance with the circuits shown in Figure 18.13a, with the assurance that it is not differentially connected.

To identify the leads of a DC motor that has only three external leads, one ohmmeter test point should be connected to one of the three leads, and the remaining two leads should be probed with the other point, as shown in Figure 18.14e. The F_1 lead will indicate the higher resistance when tested separately with each of the other two leads. To identify the S_2 lead, lift the brushes from the commutator and test from the F_1 lead to the other two leads. Lead A_1 will indicate infinity (∞). The remaining lead is S_2. Figure 18.14f shows the in-housing connections of a three-terminal DC motor.

18.12 CHECKING MOTOR PERFORMANCE

Overloading of DC motors results in the generation of excessive heat; the higher temperatures shorten the life of insulation, cause sparking at the brushes, loosen soldered banding wire, and loosen soldered connections to the commutator. If emergency conditions require a machine to supply more load than it can safely handle without excessive heating, the machine can be cooled with portable blowers and fans. This should be done only for a short time.

---

[4] An analog-type multimeter that can be used for this purpose is shown in Figure 1.3, Chapter 1.

SUMMARY OF EQUATIONS FOR PROBLEM SOLVING

$$T_{D2} = T_{D1} \frac{B_{F2} I_{A2}}{B_{F1} I_{A1}} \qquad (18\text{--}2)$$

$$E_A = n\Phi_P k_M \qquad (18\text{--}3)$$

$$I_A = \frac{V_T - E_A}{R_{acir}} \qquad (18\text{--}5)$$

$$I_T = I_A + I_F \qquad (18\text{--}6)$$

$$I_F = \frac{V_T}{R_F} \qquad (18\text{--}7)$$

$$n = \frac{V_T - I_A R_{acir}}{\Phi_P k_M}\bigg|_{\Phi_P \neq 0} \qquad (18\text{--}8)$$

$$R_{acir} = R_A + R_{IP} + R_S + R_{CW} + R_X \qquad (18\text{--}9)$$

$$P_{mech} = E_A I_A \qquad (18\text{--}13)$$

$$P_F = I_F^2 R_F \qquad (18\text{--}14)$$

$$P_{shaft} = \frac{T_{shaft} \times n_{shaft}}{5252} \qquad (18\text{--}15)$$

$$P_{in} = V_T I_T \qquad (18\text{--}17)$$

$$\text{eff} = \frac{P_{shaft} \times 746}{P_{in}} \times 100 \qquad (18\text{--}18)$$

$$T_D \propto I_A^2 \quad \text{(series motor)} \qquad (18\text{--}19)$$

$$\text{SR} = \frac{n_{nl} - n_{rated}}{n_{rated}} \times 100 \qquad (18\text{--}20)$$

GENERAL REFERENCE

Hubert, Charles, I., *Electric Machines: Theory, Operation, Applications, Adjustment, and Control*, 2nd ed. Prentice Hall, Upper Saddle River, NJ, 2001.

SPECIFIC REFERENCE KEYED TO THE TEXT

[1] *Motors and Generators*, NEMA Publication No. MG 1–1998, National Electrical Manufacturers Association.

REVIEW QUESTIONS

1. Explain how motor torque is developed.
2. What is the function of the commutator in a DC motor?
3. How can a DC motor be reversed?
4. What is the function of interpoles and compensating windings?
5. Explain why a running motor develops a countervoltage. What is the relative magnitude of this voltage compared with the driving voltage?
6. Why is additional resistance needed to limit the current when starting a DC motor?
7. What is meant by the base speed of a DC motor? What methods are used to adjust the speed above and below the base speed?
8. What precautions should be observed when increasing the resistance of a rheostat in series with the shunt-field circuit? Explain.
9. What effect does an open in the shunt-field circuit have on the performance of a lightly loaded motor?
10. What is a compound motor?
11. Explain why a differentially connected compound motor poses a hazard to life and property.
12. What is a stabilized-shunt motor?
13. Why must a series motor be geared directly to the load?
14. State the correct procedure for starting a motor with a manually operated starter. Can improper operation of the control handle damage the motor and starter? Explain.
15. Describe a simple polarity test for determining the relative polarity of the series and shunt fields when the terminals are unmarked.
16. State how you would identify the unmarked terminals of a compound motor that has six external leads.
17. State how you would identify the unmarked terminals of a compound motor that has three external leads.
18. State how you would identify the unmarked terminals of a shunt motor that has four external leads.
19. What adverse effect does continued overloading have on a DC motor? What protective action should be taken if emergency conditions require prolonged operation above rated load?

PROBLEMS

18–1/5. A 10-hp, 230-V, 3500-rpm DC shunt motor has a combined armature and interpole resistance of 0.393 Ω. The shunt field resistance is 208 Ω. Sketch the circuit and (a) neglecting the resistance of the cables, determine the starting current drawn by the armature if no starting resistance is used; (b) determine the shunt-field current.

18–2/5. A 30-hp, 230-V DC shunt motor has an armature resistance of 0.118 Ω, an interpole winding resistance of 0.029 Ω, and a shunt-field resistance of 58.9 Ω. A starting resistance must be connected in series with the armature to limit the armature current to 250 A. Sketch the circuit and, assuming that the resistance of the connecting cables is negligible, determine the resistance of the starting resistor.

18–3/5. The armature of a 50-hp, 230-V, 850-rpm, uncompensated shunt motor draws 172 A when operating at full load and rated speed. If the armature and interpole windings have a combined resistance of 0.070 Ω, calculate the cemf.

18–4/5. A 60-hp, 230-V, 400-rpm shunt motor operating at rated conditions draws an armature current of 204 A and a shunt-field current of 7.62 A. The armature resistance and interpole resistance are 0.039 Ω, and 0.012 Ω, respectively. Determine (a) resistance of shunt field; (b) cemf; (c) resistor that must be placed in series with the armature to limit the armature current to 200% rated current when starting.

18–5/7. A 40-hp, 240-V shunt motor operating at rated voltage and 650 rpm draws a current of 144 A. The resistances of armature, interpole, and shunt-field windings are 0.0552 Ω, 0.019 Ω, and 62 Ω, respectively. Determine (a) field current; (b) armature current; (c) cemf; (d) mechanical power developed; (e) heat power losses in armature; (f) heat power losses in field winding.

18–6/7. A 100-hp, 240-V shunt motor operating at rated voltage and 1750 rpm draws a current of 339 A. The resistances of armature, interpole, and shunt-field windings are 0.0116 Ω, 0.00444 Ω, and 61.2 Ω, respectively. Determine (a) field current; (b) armature current; (c) cemf; (d) mechanical power developed; (e) heat power losses in armature; (f) heat power losses in field winding.

18–7/8. A 150-hp, 500-V, 1180-rpm compound motor, running at rated conditions, draws a line current of 246 A. Determine (a) shaft torque; (b) input power to motor; (c) efficiency.

18–8/8. A 50-hp, 240-V, 1150-rpm compound motor, running at rated conditions, draws a line current of 174 A. Determine (a) shaft torque; (b) input power to motor; (c) efficiency.

18–9/10. A 240-V, 150-hp compound motor has a base speed of 800 rpm and a no-load speed of 850 rpm. Determine the speed regulation.

18–10/10. A 30-hp shunt motor has a base speed of 1150 rpm and a speed regulation of 2.3%. Determine the no-load speed.

19

Commutator, Slip-Ring, and Brush Maintenance

19.0 INTRODUCTION

One of the biggest maintenance considerations—and one of the least understood—is the cleaning and servicing of commutators, slip rings, and brushes. Yet, the service life of many electrical machines is dependent to a good extent on the care given to these components.

This chapter includes detailed procedures for the cleaning and servicing of commutators and slip rings, brush adjustment, brush seating, and methods for locating the correct neutral setting for the brushes. A detailed troubleshooting chart for commutators and brushes is included at the end of the chapter.

19.1 COMMUTATOR AND SLIP-RING SURFACE FILM

Commutators are constructed of copper, whereas slip rings (also called collectors) may be constructed of heat-treated steel, bronze, or other alloys of copper. During normal operation, commutators and slip rings acquire a shiny protective gloss that reduces wear and thereby lengthens service life. This surface film generally contains metallic oxide and graphite in varying proportions and is formed after several days or weeks of operation. Water vapor adsorbed by this film plays an important role in providing the lubrication required for reduced brush, commutator, and slip-ring wear. In fact, in areas of extremely low humidity, for instance, at high altitudes or during extended dry periods of cold winter months, the surface film disappears, and brushes, commutators, and slip rings wear severely.

Operating conditions, atmospheric conditions, and brush grade affect the makeup and color of this film, which, for copper, may range from a light straw color to jet black; the color most common to a good surface film for copper is chocolate brown.

A good surface film has the added advantage of higher contact resistance between the brush face and the commutator, or brush face and slip ring, than would be obtained with a raw metallic surface. In addition, this film is beneficial in reducing circulating currents that occur in DC machines, each time the brush short-circuits an armature coil. Although under ideal conditions the short-circuit current is zero, under very heavy loads saturation of the interpole iron permits a short-circuit current. A raw metallic surface may be caused by the wrong brush grade, some mechanical or electrical fault, or low current density at the brush surface brought on by consistently low loads. In the latter case, a current density of about 40 A/in.[2] is often necessary to maintain a good surface film. This may be accomplished by removing one or more brushes from each brush arm. The increased current in each of the remaining brushes raises the brush temperature, thereby reducing the friction between brush and commutator.

Figure 19.1 illustrates the effect of brush-face temperature on the coefficient of friction for electrographitic brushes. The greatest friction and, hence, the greatest rate of brush wear occurs at about 70°C; at temperatures between 85°C and 115°C, the friction is minimal, increasing again for temperatures above 115°C. The low brush current of lightly loaded machines does not generate sufficient heat for efficient brush operation, and excessive brush wear occurs. A similar relationship exists with graphite and metal-graphite brushes.

Atmospheric contamination, like that produced by oil vapors, salt air, hydrogen sulfide gas, or silicone vapors, is also very destructive to the surface film. The silicone vapors from only a very small amount of silicone material, such as silicone tape, silicone varnish, silicone-rubber-insulated leads, or protective hand creams containing silicones, cause very rapid brush wear. Furthermore, a considerable amount of carbon dust, caused by rapid brush wear, is deposited on the windings, which reduces the creepage distance and causes a lowering of the insulation resistance to ground. This adverse characteristic of silicone insulation is particularly bad in enclosed machines. Preventive maintenance personnel must not use hand creams containing silicones when working on any part of a machine that contains a commutator or slip rings.

Oil contamination is generally indicated by a relatively thick black film that causes poor brush contact and excessive heating at the conducting spots. If the temperature of the commutator at the relatively few conducting spots becomes high

FIGURE 19.1

Brush friction vs. temperature for electrographitic brushes (courtesy Dow Chemical Company).

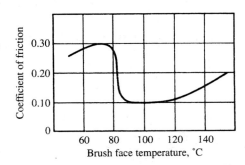

enough, very small globules of copper will break away from the commutator surface. Some of these tiny globules embed themselves in the brush and cut threadlike lines in the commutator.

19.2 CARBON BRUSHES

The carbon brushes used on commutators and slip rings may be composed of lampblack, coke, or graphite, to which various other elements can be added to obtain special characteristics. For example, copper and silver are added to some brushes to increase their current-carrying capacity, and silica is added to others to provide a cleaning action. Very soft brushes deposit graphite on the commutator, and hard abrasive brushes scour the commutator. Each brush type has its field of application, and no single brush grade serves well in all machines. The softer film-forming brushes are generally preferred for service at light loads and the harder scouring grade for operation with heavy loads in an oily atmosphere.

Brushes of different grades should never be mixed. When replacement brushes are ordered, the specifications of the manufacturer should be followed. Except for mechanically damaged brushes, all brushes on a given machine must be replaced at the same time.

19.3 BRUSH SPARKING

Brush sparking is a symptom of a fault that may be of a mechanical, electrical, or operating nature and can be caused by a great variety of disorders. Hence, the cause of sparking is not always easily discerned. More often than not, the sparking is instigated by a dirty commutator, dirty slip rings, or some mechanical fault.

A mechanical imbalance, machine misalignment, commutator and slip-ring eccentricity, incorrect positioning of brushes, wrong brush tension, incorrect brush grade, etc., are all detrimental to good operation and may cause sparking.

19.4 CLEANING COMMUTATORS AND SLIP RINGS

Commutators and slip rings can be cleaned with a wiper made of 16 layers of 6- or 8-oz. hard-woven duck canvas fastened to the end of a flexible wooden stick, as shown in Figure 19.2a. The rivets or screws used to fasten the canvas should be countersunk to avoid scratching the commutator.

The frequency of application is determined by the circumstances surrounding the machine, and should be often enough to prevent the formation of a smutty surface. In the case of open machines, an application once every 24 hours of operation will prolong the life of the commutator and extend the periods between shutdowns. Canvas is not abrasive and will not damage the surface film that is so necessary for good commutation. To be effective, the wiper should be applied with considerable pressure. It

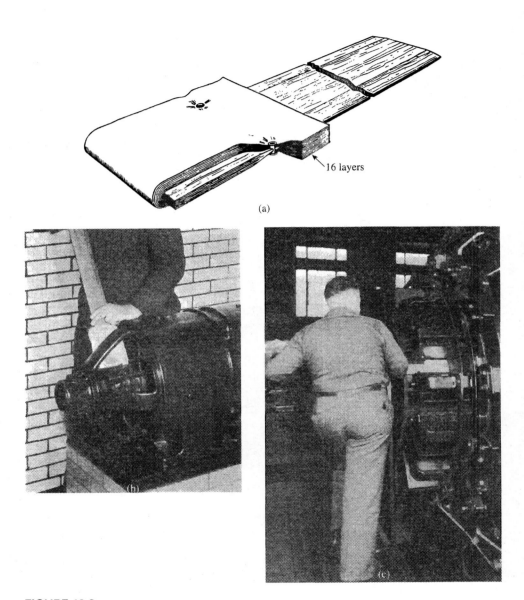

16 layers

(a)

FIGURE 19.2
(a) Canvas wiper for cleaning commutators and slip rings; (b) applied edgewise to the commutator of a small machine; (c) applied sidewise to the commutator of a large machine (courtesy Martindale Electric Co.).

may be used edgewise on small machines or sideways, as shown in Figures 19.2b and 19.2c, respectively. The use of solvents on the commutator should be avoided, because they can damage the surface film.

19.5 COMMUTATOR AND SLIP-RING RESURFACING

Resurfacing should not be done if there is little or no sparking and brush wear is normal. The best maintenance procedure for slip rings and large commutators that have a good operating history is to leave them alone. If there is little to no sparking and brush wear is normal, resurfacing of the commutator should not be performed.

Before a commutator is resurfaced, it should be checked for loose bars. Loose commutators should be tightened with a calibrated torque wrench, in accordance with the manufacturer's specifications of bolt torque at a specified temperature. Indiscriminate tightening of commutator nuts, without regard to torque limits, can cause buckled bars, and cause the steel V-ring to cut through the mica insulation, grounding the commutator. A cross section of a large commutator is shown in Figure 19.3.

Never disturb the commutator clamping bolts on properly operating commutators. They should only be adjusted if there is definite evidence of loose bars. The clamping bolts or nuts on any commutator can be turned if sufficient torque is applied. However, it should not be done unless there is reason to believe that they have loosened somewhat due to vibration.

If the commutator clamping ring is ever removed, a new mica V-ring should be installed, and the commutator should then be cured by spinning under heat. Check with the manufacturer before removing a clamping ring.

Commutator risers, shown in Figure 19.4, are used in large machines to connect the commutator bars to the armature coils. These should be checked for cracks or breakage, which may occur where the riser joins the commutator.

Commutators or slip rings that are out of round, grooved, burned, have flat spots, or are otherwise in bad condition should be resurfaced by grinding or turning

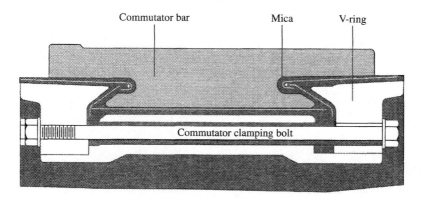

FIGURE 19.3
Cross section of a large commutator (courtesy Ideal Industries Inc.).

Risers

FIGURE 19.4
Grinding rig in use on a commutator (courtesy Martindale Electric Co.).

with a lathe tool. However, excessive grinding of a commutator should be avoided, because a considerably reduced commutator diameter may cause the brushes to contact the commutator at an off-neutral point. The preferred method is grinding with a grinding rig, as shown in Figure 19.4. The abrasive stone (also called a commutator stone) is moved slowly back and forth across the commutator or slip ring, with the machine revolving at rated speed. Operating the machine at rated speed allows centrifugal force to position any loose commutator bars that may have escaped detection.

Several light cuts are preferred, because excessive pressure on the stone will make the grinding rig deflect and cause an inferior job. Some commutators and slip rings have a very hard glasslike surface that is difficult to cut without considerable pressure, particularly on slow-speed machines. However, once the glaze is removed, the cutting proceeds with ease. Badly worn commutators may be ground with a coarse stone and then finished with a fine one. If only a light dressing of the commutator is necessary, a fine-grit stone is still preferred; sandpaper polishes the commutator but does not remove any flat spots.

Any grease, oil, or dirt that is present must be wiped off the commutator or rings before the stone is applied, because it will interfere with the cutting process. The armature windings and field coils should be protected against the entry of copper dust by using a vacuum cleaner and by shielding the windings with fiber or stiff cardboard.

FIGURE 19.5
Assorted hand stones used for hand grinding commutators (courtesy Ideal Industries Inc.).

Hand stones, shown in Figure 19.5, can be used in the absence of a grinding rig. Hand stones should be wide enough to cover the distance between adjacent sets of brushes and its length should be less than the length of the commutator to allow for a side-to-side motion while grinding. However, the stones should be longer than the largest flat spot if the flatness is to be removed.

Figure 19.6 illustrates the application of a hand stone to a commutator. Although hand stoning will remove flat spots, it will not remedy an eccentric commutator. Such correction must be made by a grinding rig with an abrasive stone or a diamond-shaped lathe tool.

Commutator stones are nonconductive and easy to apply. Hence, if necessary, they can be used while the machine is operating at reduced load and reduced voltage, provided that the voltage does not exceed 300 volts. However, for safety reasons, whenever practicable the commutator should be resurfaced with the machine de-energized. This is easily done with generators, because the prime mover does the driving. In the case of a motor, it is safer to accelerate the machine under its own power, then de-energize it and grind the surface while it is coasting, repeating as often as necessary.

If an exceptionally smooth finish is desired, the commutator can be burnished with a hardwood block applied manually with considerable pressure. Burnishing serves a twofold purpose: It removes the microscopic edges that may remain from the grinding process, and the heat generated helps to develop a thin oxide film, which is necessary to good commutation. The wood block should be applied with the end grain in contact with the commutator surface.

Freshly ground commutators are sometimes air-cured with dry compressed air at 30 to 40 lb/in.^{2}. With the machine assembled and operated at no load, the compressed air is applied to the commutator at the leading edge of the brush, and the load is gradually applied. The compressed air compacts the residual dirt, composed of copper and carbon dust, and burns it out; the controlled burning action assists in forming a thin oxide film on the commutator.

FIGURE 19.6
Application of a hand stone to a commutator (courtesy Martindale Industries).

Do not touch the freshly ground surface of a commutator or slip ring; the oily film deposited on the fresh surface will prevent or delay the development of a normal surface film.

19.6 MICA UNDERCUTTING AND COMMUTATOR REPAIR

Mica undercutting is essential to good commutation and should not be neglected. Mica is harder than copper and, if not undercut, will form ridges as the copper wears away. The mica ridge will cause sparking and chipping of brushes. Mica should be undercut to a depth approximating its thickness. This is easily done with a slotting file as shown in Figure 19.7. Portable undercutting machines using miniature circular saws are used in most armature repair facilities. Hacksaw blades can also be used for this purpose, provided that the sides of the blade are ground to remove the set (the protruding sides of the teeth).

During the process of undercutting, some or all of the copper bars may acquire a featheredge, or mica fins, as shown in Figure 19.8. To remedy this, the corners of the copper bars should be side cut or chamfered to ensure that the mica is well below the surface.

This chamfering process can be accomplished with a tool fashioned from a power hacksaw blade as illustrated in Figure 19.9. This is a very important part of the undercutting process and must not be neglected. If allowed to remain, the featheredges of mica and copper will flake off, embed in the brush face, and score the commutator.

FIGURE 19.7
Slotting file being used to smooth the rough edges after the undercutting operation (courtesy Ideal Industries Inc.).

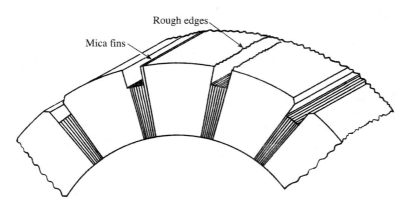

Rough edges

Mica fins

FIGURE 19.8
Rough edges and mica fins that must be removed by side cutting.

FIGURE 19.9
Side-cutting tool fashioned from a power
hacksaw blade.

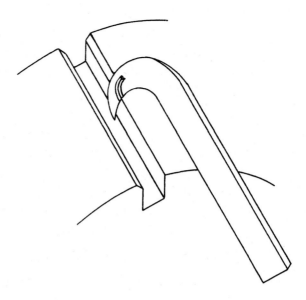

Carbonized mica between commutator bars should be dug out and the holes filled with commutator cement. Burned or carbonized mica, caused by the absorption of oil, moisture, and carbon dust, form conducting paths between the bars and between the bars and ground.

After undercutting and side cutting, the slots between the commutator bars should be cleaned out with a toothbrush and insulated with a light coating of air-drying varnish. This will prevent the absorption of oil by the mica and thus reduce the possibility of shorts by accumulated carbon dust.

19.7 BRUSH ADJUSTMENT, CLEANING, AND REPLACEMENT

A good surface film on commutators or slip rings is dependent to a large extent on correct brush adjustment. Figure 19.10 shows the effects of dirt and incorrect brush adjustment on the brush face.

When a machine is scheduled for overhaul, the brushes should be removed and tagged to ensure their replacement in the same brush holders; otherwise the unavoidable small variations in brush-holder angle and brush clearance may cause faulty commutation and require reseating of all brushes. The brushes and brush holders should be thoroughly cleaned and inspected for sticky brushes or excessive movement in the holders. Excessive play permits unnecessary forward and side motion, resulting in partial seating of the brush in several positions. Partial seating increases the current density at the contact surface, causing overheating and sparking. On the other hand, insufficient play interferes with the free movement of the brush, causing it to stick in the holder and arc at the commutator. Brushes should be replaced when they have worn down to one-quarter of their useful length.

FIGURE 19.10
Comparison of the effects of dirt and incorrect brush adjustment on the brushes.
(a) Appearance of brushes requiring cleaning; (b) brushes damaged by sparking caused by incorrect and nonuniform brush-spring pressure; (c) effect of high mica and rough commutator surface on brushes; (d) brush appearance resulting from excessive clearance in holder (courtesy GE Industrial Systems).

Cleaning of brushes and holders should be done with a lintless rag slightly moistened with an approved safety solvent. If the brush boxes are to be cleaned in place, the commutator or slip rings should be covered with several layers of heavy wrapping paper to prevent dirt and solvent from damaging the surface film.

When replacing brushes, the whole set should be replaced at the same time. If this is not practicable, all the brushes on a single arm should be replaced; replacing only a few brushes on a multiple-brush arm may cause unequal current distributions and lead to commutation or slip-ring difficulties.

Short brushes draw higher current because of lower resistance and eventually cause burning of the commutator and brush pigtails, as shown in Figure 19.11.

Under no circumstances should brushes be replaced while the machine is running. One slip or wrong move can cause injury or death to the mechanic, as well as damage to the machines and plant.

FIGURE 19.11
Damage to motor caused by short brushes. Arcing under the brushes caused a flashover that burned the brush pigtails and the commutator (courtesy GE Industrial Systems).

19.8 BRUSH ALIGNMENT AND CLEARANCE

All brushes of a given machine should be accurately aligned with the commutator bars. Misalignment of brushes results in some brushes being off neutral while others on the same stud are in the neutral position. Such conditions may give rise to serious sparking.

The clearance between the brush holders and the commutator varies with different machines and may be 1/16 inch in some machines and 3/16 inch in others. All brush holders of a given machine should be set the same distance from the commutator. A gauge made from several thicknesses of fiber can be used on commutators to obtain the desired clearance. Each brush-holder clamp is loosened, and the box is pressed firmly down on the fiber and then reclamped.

19.9 BRUSH STAGGERING ON COMMUTATORS

Staggering of the brush holders on commutators should be done in positive and negative pairs, as shown in Figure 19.12a. Each set of positive brushes must be in line with a negative set. If brushes of the same polarity track each other, the commutator will wear unevenly as shown in Figure 19.12b. The alternate bands of bright copper and dull brown or black are characteristic of incorrect staggering.

This phenomenon, called electrolysis, may be explained with the aid of Figure 19.13, where it is shown for the commutator of a DC generator. The moisture at the

(a)

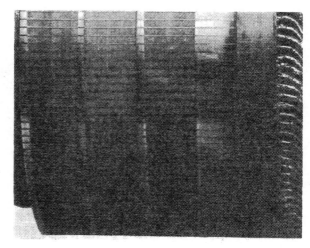

(b)

FIGURE 19.12

(a) Correct method of staggering brushes; (b) characteristic light and dark bands caused by incorrect brush staggering (courtesy GE Industrial Systems).

FIGURE 19.13
Anodes and cathodes formed at contact surfaces between the brushes and commutator of a DC generator.

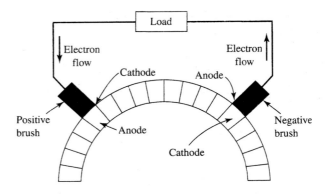

commutator–brush interface is decomposed by electrolysis; oxygen is formed at the anodes and hydrogen at the cathodes.

At the negative brush, the oxygen formed at the anode causes the carbon surface to oxidize, forming carbon dioxide; the hydrogen formed at the cathode has a reducing action, thereby maintaining a bright copper surface. Hence, both the negative brush and the commutator surface below it will wear relatively fast.

At the positive brush, the oxygen formed at the anode causes a protective copper oxide film to form on the commutator; the hydrogen formed at the cathode acts as a reducing agent at the brush and lessens the brush wear. Hence, in a generator the brush and the commutator surface on the positive side will wear less than the brush and commutator surface at the negative side. If the machine is operated as a motor, the cathodes become anodes and the anodes become cathodes. This will cause the positive brush and the commutator surface below it to wear relatively fast.

Correct staggering of brushes, as shown in Figure 19.12a, prevents the bad effect of electrolysis on commutators and brushes. Similarly, and for the same reason, slip rings that carry direct current should have their connections reversed every 6 months to provide for better brush and ring wear.

19.10 BRUSH ANGLE

The angle that the brush makes from the radial depends on the type of brush holder used. However, all brush holders on a given machine must have the same angle. Brushes set at the wrong angle do not make contact with the commutator at the correct place and cause severe sparking. The two most common types of brush holders are the box type and the reaction type, shown in Figure 19.14.

Box-Type Brush Holder

The box-type brush holder, shown in Figures 19.14a, 19.14b, and 19.14c, can be operated in the radial, trailing, or leading position. Box-type brush holders should be mounted in the radial position for motors in reversing service. If the brush angle devi-

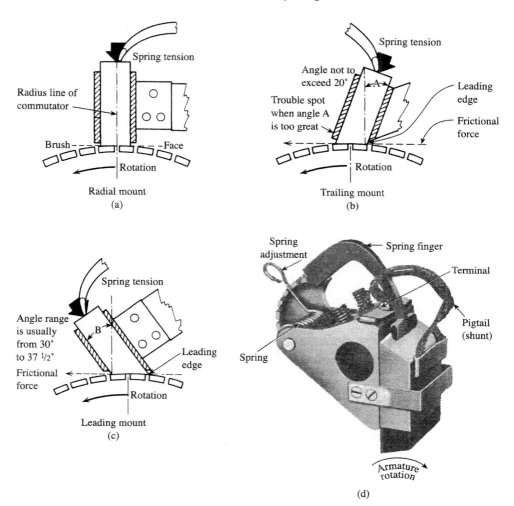

FIGURE 19.14
Brush holders: (a–c) box type; (d) reaction type (courtesy Dow Chemical Company).

ates from the radial by more than 4°, brush chatter, chipped brushes, and sparking may occur when reversing.

For nonreversal service, box-type brushes may be used in the trailing position, and the angle may be any value up to 20°; but when used in the leading position, the angle should be between a minimum of 30° and a maximum of 37.5°. An angle of less than 30° in the leading position causes severe chattering.

Box-type-brush holders are designed with very little clearance to prevent excess side motion that would cause wear on both the brush and holder. Unfortunately, this small clearance makes it subject to clogging with dirt, and it should therefore be checked periodically for sticky brushes; each brush should be checked by moving it up and down in the holder.

Reaction-Type Brush Holder

The reaction type of holder, shown in Figure 19.14d, is positioned to hold its brush at an angle of 30° to 37.5° from the radial and must be operated in the leading direction. Hence, machines that have this type of holder should not be reversed. If set properly, the reaction type of holder offers the advantage of smooth performance with no brush chatter, and its open construction eliminates the possibility of sticky brushes.

19.11 BRUSH SPACING ON COMMUTATORS

The spacing between positive and negative sets of brushes can be checked by wrapping a strip of paper around the commutator, as shown in Figure 19.15. The paper is marked at the full circumference, then removed and carefully marked off in as many equal parts as there are brush arms. The paper is then replaced, and one of the marks is placed coincident with the edge of a brush and held in position with masking tape. If all the brush arms are properly spaced, the brush edge of each will coincide with the respective markings on the paper strip. A deviation greater than 1/64 in. should be corrected through adjustment of the brush arms.

FIGURE 19.15
Checking brush spacing (courtesy United States Merchant Marine Academy).

19.12 CONNECTING BRUSH PIGTAILS

When installing brushes, the brush pigtail, also called the brush shunt, must be properly bolted to the brush box, as shown in Figure 19.14d. A loose shunt terminal causes some of the current to be conducted through the brush box rather than through the pigtail, causing burning or rough spots on the inside of the box.

If two or more brushes are in parallel, as shown in Figures 19.15 and 19.16, and the pigtail of one is not tightly bolted to its brush box, it will not carry its share of the current. The properly connected brush will carry a greater percentage of current and get hotter. Unfortunately, the resistance of carbon decreases with increasing temperature, causing the properly connected brush to take an even greater percentage of the total current. The high current density at this one brush may cause sparking at the commutator.

19.13 BRUSH SEATING

The seating of all brushes to the exact curvature of the commutator and collector rings is a prerequisite to good brush performance. Improper seating of some brushes causes the others to carry a greater portion of the load. The net result is overheating and sparking, destructive to both the commutator surface film and the brushes.

Before new brushes are seated, it is essential that the holders be inspected, and any roughness filed smooth. The pigtails should be checked for frayed wires and loose connections. Brushes with defective pigtails should not be used, because there will be current through the sides of the brush, damaging the holder. The brushes should be checked for freedom of movement in the holders without excess side play.

The actual fitting of brushes to the commutator or collector rings should be done with sandpaper when the machine is not in operation, or with a brush-seating stone if the seating is to be done while the machine is running. Emery cloth or emery paper should never be used for seating brushes or polishing the commutator or slip rings. Emery is very abrasive, and any particles that become embedded in the brush face will score the commutator and rings.

When sandpaper is used for seating, it should be pulled back and forth under the brush with the sand side facing the brush. The brush tension should be adjusted for maximum pressure during the sanding operation, and the ends of the sandpaper should be pulled along the curvature of the commutator or slip rings to prevent rounding the edges of the brushes. If the machine is nonreversible, the last few strokes should be made in the known direction of rotation. To facilitate the sanding operation, coarse sandpaper, grade 1½, can be used for the initial cutting, followed by a fine grade, such as grade 0, for the final cut. Figure 19.16 shows the proper method of sanding brushes.

Long sheets of sandpaper, which are purchased on a roll and are available to fit the entire circumference of the commutator or slip rings, can be used for the simultaneous sandpapering of all brushes. The sandpaper should be wrapped around the

FIGURE 19.16
Proper method of sanding brushes (courtesy TECO Westinghouse).

commutator or slip rings and held in position with masking tape; the brushes should be placed in the holders, and the rotor should be revolved slowly until the brushes are shaped to the correct curvature. After the sanding operation is completed, the carbon dust should be removed with a vacuum cleaner or blown out with dry compressed air. You should avoid blowing the dust into the windings.

Seating of brushes to a commutator may be done with a special brush-seating stone as illustrated in Figure 19.17. The stone is of fine-grain composition and is applied at the commutator close to the brush holder while the machine is in motion. The fine grains worn off the stone are carried under the brush, grinding it to the curvature of the commutator. The stone should be applied a little at a time, and the brush should be checked after each application until complete seating is obtained.

In addition to ease of operation, a brush-seating stone offers the advantage of grinding the brushes to the exact curvature of the commutator and, if not used to extremes, will not noticeably affect the commutator surface film.

Sandpaper has the disadvantage of grinding the brush surface to a radius equal to that of the commutator plus the thickness of the sandpaper; thus the true curvature will not be obtained until the machine has been in operation for some time. It is not advisable to operate a machine at full load until the brushes are worn to the true curvature. Machines should not be placed under load until at least 75 percent of the brush face is properly seated.

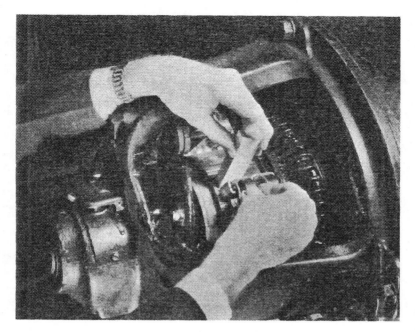

FIGURE 19.17
Brush-seating stone applied to a commutator (courtesy Martindale Electric Co.).

19.14 BRUSH PRESSURE

A curve showing the relationship between the rate of brush wear and brush pressure for electrographitic brushes is shown in Figure 19.18a; the best brush wear is obtained with a brush pressure of between 2 and 3 lb/in.^{2}.

Insufficient pressure causes poor electrical contact and hence chattering and arcing between the commutator and the brush, rapidly wearing the brush. On the other hand, too much pressure causes excessive friction, also resulting in excessive brush wear. However, as indicated by the curve, pressures below the optimum value cause much greater brush wear than the correspondingly higher pressures. Similar effects occur with other grades of brushes, but the range of optimum performance may be different. The optimum brush pressure for other grades of brushes is given in Table 19.1.

The total force on the brush may be measured with a spring balance and a strip of paper, as shown in Figure 19.18b. The leather loop must be placed under the end of the spring finger, where the finger touches the brush. The scale is read when the strip of paper placed between the commutator and the brush can be drawn out with very little effort. The amount of spring force necessary to provide the correct brush pressure may be obtained by multiplying the recommended pressure in pounds per square inch by the cross-sectional area of the brush face and then adjusting this figure, in the case of horizontal machines, by the weight of the brush. Thus, brushes that ride the top of

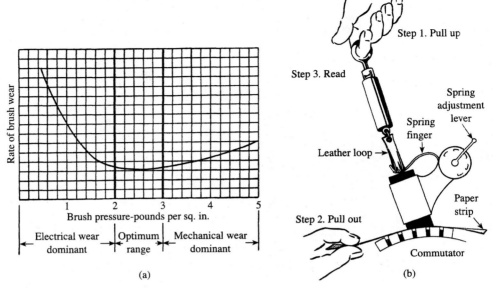

FIGURE 19.18

(a) Relationship between brush pressure and rate of brush wear for lampblack base electrographitic brushes; (b) measuring total brush force with a spring balance (courtesy Dow Chemical Company).

the commutator should have a lower spring force, and those that ride the bottom of the commutator should have a greater spring force than those that ride on the side.

Note: Special-purpose machines, such as electric drilling motors and mud pumps for off-shore drilling rigs, can be designed for brush pressures as high as 7 lb/in.^{2}.

TABLE 19.1
Recommended Brush Pressure

Brush Grade	Pressure (lb/in.^{2})
Carbon	1¾–2½
Carbon-graphite[a]	1¾–2½
Graphite-carbon[a]	1¾–2½
Electrographitic	2–3
Graphite	1¼–2
Metal graphite	2½–3½
Fractional-horsepower motors	4–5

[a] A carbon-graphite brush has more coke type of carbon than graphite. A graphite-carbon brush has more graphite than coke type of carbon (Union Carbide Co.).

EXAMPLE 19.1   The manufacturer's brush-pressure specification for a certain electrographitic brush is 2.5 lb/in.^{2}. The width and thickness of the brush are 1.5 and 0.5 in., respectively. Neglecting the weight of the brush, determine the required spring force on the brush.

Solution

$$F_s = P_b \times A_b$$

where:

F_s = spring force (lb)
P_b = brush pressure (lb/in.^{2})
A_b = brush surface area (in.^{2})
$F_s = P_b \times A_b = 2.5 \times (1.5 \times 0.5) = 1.9$ lb

19.15   BRUSH NEUTRAL SETTING AND PHYSICAL INSPECTION TEST

The neutral setting of the brush holders should not be adjusted unless there is reasonable evidence that the original installation was incorrect, or that the neutral was shifted during overhaul, or because of commutator wear. A commutator whose diameter has been reduced by excessive wear or excessive grinding may cause trailing or leading brushes to make contact at an off-neutral position.

A chisel mark, covered with white paint for easy location, is often used to mark the neutral position. The mark is placed on the bracket that supports the brush rigging and on the end bell to which it is clamped. If off neutral, the brush-rigging clamp is loosened and the rigging rotated until its chisel mark lines up with the one on the end bell. Some manufacturers of large machines mark two armature slots and three commutator bars to assist the operator in locating the proper brush position. To obtain the correct setting, the armature should be rotated so that the marked slots are directly under the center of the interpoles. With the armature held in this position, the brush rigging should be adjusted so that one set of brushes is centered on the three marked commutator bars, as shown in Figure 19.19.

If no markings are present, and the neutral position is in question, tests should be made to determine the correct position.

19.16   INDUCTIVE-KICK TEST FOR APPROXIMATE LOCATION OF NO-LOAD NEUTRAL

The inductive-kick test is a fairly accurate method for locating the no-load neutral and may be used when the brushes must be positioned prior to operation of the machine. It is equally effective with interpole and non-interpole machines. To make the test, the shunt field should be energized by a source of direct current in series with a rheostat and switch, as shown in Figure 19.20a.

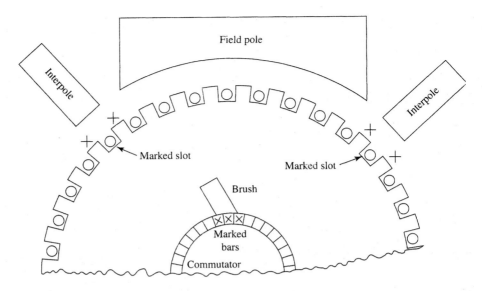

FIGURE 19.19
Factory marks on armature slots and commutator bars to locate the correct neutral setting of the brushes.

All brushes should be removed, and two beveled brushes installed, one in a positive holder and the other in a negative holder. The beveled brushes must not span more than one commutator bar, as shown in Figure 19.20b. A DC voltmeter with an adjustable voltage range from 0.5 to 15 volts should have its 15-volt range connected across the two brushes of opposite polarity. The interpoles, series field, and compensating winding must not be in the voltmeter circuit.[1]

With the rheostat adjusted to obtain a value of field current equal to 20 percent of its normal value, the switch is quickly opened and the momentary deflection of the voltmeter pointer is observed. A knife switch with a quick-break blade, or a snap switch, provides the same decay rate for the field current each time the switch is opened. The brush rigging should then be shifted a few degrees and the test repeated. If the magnitude of the voltmeter deflections increases, the brushes must be shifted in the opposite direction. The 15-volt connection should be used until the deflections fall within the range of the lower scales. The neutral position for the brush rigging is located when there is no deflection on the 0.5-volt range of the voltmeter when the field circuit is opened.

The principle of the inductive-kick test is explained with the aid of Figures 19.20c and 19.20d. Because the armature is stationary, the only voltages generated in the armature coils are those resulting from the "inductive kick" caused by transformer action when making and breaking the field circuit. When the field circuit is closed, the poles establish north and south polarity as shown; when the field circuit is opened, the

---

[1] For additional information on this and other methods for locating the neutral setting of the brushes, see Reference [1].

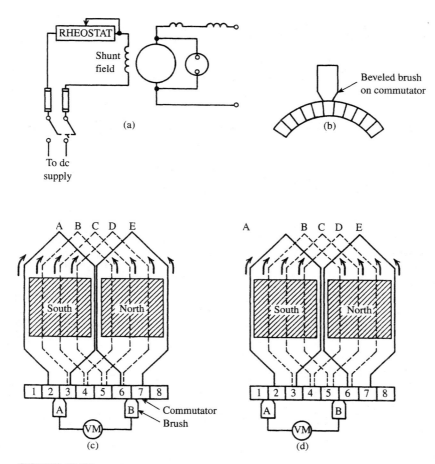

FIGURE 19.20
Inductive-kick test for locating the brush neutral location: (a) circuit diagram; (b) beveled brush resting on the commutator; (c) brushes in neutral position; (d) brushes off neutral.

collapse of this flux through the windows of the armature coils generates voltages in the coils in a direction to oppose the collapse of flux (Lenz's law).

Referring to Figure 19.20c, flux from only the south pole passes through the window of coil A; hence, the voltage induced in coil A, because of the collapse of flux, will be in the clockwise direction. Similarly, flux from only the north pole passes through the window of coil E; hence the voltage induced in coil E will be in the counterclockwise direction. However, flux from both poles passes through the windows of coils B, C, and D, causing voltages to be induced in each of these coils in both the clockwise and the counterclockwise directions. The direction of the induced voltages in each coil, as a result of opening the field circuit, is shown with arrows in Figures 19.20c and 19.20d.

The relative magnitude of the voltage generated by each pole in each one of the five coils depends on the percentage of the total pole flux that passes through the

TABLE 19.2
Magnitude and Direction of Voltage in Armature Coils Due to Buildup
of Flux in Main Field Poles

Armature Coil	South Pole	North Pole	Net Voltage in Each Coil
A	+4 (CW)	0	+4
B	+3 (CW)	−1 (CCW)	+2
C	+2 (CW)	−2 (CCW)	0
D	+1 (CW)	−3 (CCW)	−2
E	0	−4 (CCW)	−4

respective coil window. For example, all of the south pole flux passes through the window of coil A, three-quarters through coil B, one-half through coil C, one-quarter through coil D, and none through coil E.

Assuming that the collapse of south pole flux causes the voltage in coil A to attain a peak value of 4 volts, then the corresponding voltages in the other coils, because of the action of this same pole, will be 3 volts in coil B, 2 volts in coil C, 1 volt in coil D, and 0 volts in coil E.

Similarly, the collapse of north pole flux will cause the generated voltages in coils E, D, C, B, and A to have peak values of 4, 3, 2, 1, and 0 volts, respectively. The magnitude and direction of the voltages generated in each coil by the collapse of flux in the north and south poles are listed in Table 19.2, along with the net voltage. Counterclockwise voltages are assumed negative ($-$), and clockwise voltages are assumed positive ($+$).

With the brushes in the neutral position (Figure 19.20c), the voltmeter reads the voltage between commutator bars 3 and 6; coil A and coil E are each shorted by a brush and, therefore, do not contribute to the total voltage. Thus, the voltmeter reads the summation of the net voltages in coil B, coil C, and coil D, which adds up to zero:

$$E_B + E_C + E_D = +2 + 0 - 2 = 0$$

With the brushes off neutral, a distance of only one commutator bar (Figure 19.20d), the voltmeter reads the summation of voltages between commutator bars 2 and 5; discounting the coils shorted by the brushes, the voltmeter indicates the summation of the net voltages in coil A, coil B, and coil C. This voltage adds up to

$$E_A + E_B + E_C = +4 + 2 + 0 = +6 \text{ V}$$

19.17 CAUSES OF UNSATISFACTORY BRUSH AND COMMUTATOR PERFORMANCE

Tables 19.3 and 19.4, along with Figure 19.21 (p. 423), provide an effective means for determining the primary source of most brush and commutator troubles; the numbers listed in the third column of Table 19.3 are keyed to Table 19.4 (pp. 419–422).

TABLE 19.3

Troubleshooting Chart for Commutators and Brushes

Indication Appearing at Brushes	Immediate Cause	Key to Primary Sources of Poor Brush Performance[a]
Sparking	Commutator surface condition	1, 2, 3, 43, 44, 45, 46, 49, 59, 60
	Overcommutation	7, 12, 31, 33
	Undercommutation	7, 12, 30, 32
	Too rapid reversal of current	7, 12, 30, 32
	Faulty machine adjustment	8, 9, 11
	Mechanical fault in machine	6, 14, 15, 16, 17, 18, 19, 20, 21, 22
	Electrical fault in machine	25, 27, 28, 29
	Bad load condition	38, 39, 40, 41, 42
	Poorly equalized parallel operation	7, 13, 23, 34
	Vibration	51, 52
	Chattering of brushes	See "Chattering or noisy brushes"
	Wrong brush grade	55, 57, 59
	Fluctuating contact drop	50
Etched or burned bands on brush face	Overcommutation	7, 12, 31, 33
	Undercommutation	7, 12, 30, 32
	Too rapid reversal of current	7, 12, 30, 32
Pitting of brush face	Glowing	See "Glowing at brush face"
	Embedded copper	See "Copper in brush face"
Rapid brush wear	Commutator surface condition	See specific surface fault in evidence; also 50
	Severe sparking	See "Sparking"
	Imperfect contact with commutator	11, 14, 15, 16, 51, 52
	Wrong brush grade	54, 58
Glowing at brush face	Embedded copper	See "Copper in brush face"
	Faulty machine adjustment	7, 12
	Severe load condition	38, 39, 41, 42
	Bad service condition	46, 47
	Wrong brush grade	57, 61, 62
Copper in brush face	Commutator surface condition	2, 3
	Bad service condition	43, 46, 47, 48, 49
	Wrong brush grade	59, 61
Flashover at brushes	Machine condition	14, 35
	Bad load condition	38, 39, 41, 53
	Lack of attention	5, 11
Chattering or noisy brushes	Commutator surface condition	See specific surface fault in evidence
	Looseness in machine	15, 16, 17
	Faulty machine adjustment	10, 11
	High friction	6, 43, 45, 49, 52, 58, 59
	Wrong brush grade	55, 58, 59

continued

TABLE 19.3
(Continued)

Indications Appearing at Commutator Surface	Immediate Cause	Key to Primary Source of Poor Commutator Performance[a]
Brush chipping or breakage	Commutator surface condition	See specific surface fault in evidence
	Looseness in machine	15, 16, 17
	Vibration	52
	Chattering	See "Chattering or noisy brushes"
	Sluggish brush movement	14
Rough or uneven surface		1, 2, 3, 4, 17
Dull or dirty surface		5, 44, 60
Eccentric surface		1, 19, 22, 52
High commutator bar	Sparking	17
Low commutator bar	Sparking	2, 25
Streaking or threading of surface (Figs. 19.21d and 19.21e)	Sparking	43, 44, 45, 46, 49, 59
	Copper or foreign material in brush face	2, 3, 46, 47, 48, 61
	Glowing	See "Glowing at brush face"
Bar etching or burning	Sparking	2, 3, 7, 12, 30, 31, 32, 33
	Flashover	5, 11, 14, 35, 38, 39, 41, 53
Bar marking at pole pitch spacing (Fig. 19.21b)[b]	Sparking	25, 37
Bar marking at slot pitch spacing (Fig. 19.21c)[c]	Sparking	7, 12, 30, 57, 60
Flat spot	Sparking	19, 23, 25, 41, 42, 53
	Flashover	5, 11, 14, 35, 38, 39, 41, 53
	Lack of attention	1, 5, 11
Discoloration of surface	High temperature	See "Heating at commutator"
	Atmospheric condition	44, 46
	Wrong brush grade	60
Raw copper surface	Embedded copper	See "Copper in brush face"
	Bad service condition	43, 45, 47, 49
	Wrong brush grade	59, 61
Rapid commutator wear with blackened surface	Burning	2, 3, 11, 14
	Severe sparking	See "Sparking"

TABLE 19.3
(Continued)

Indications Appearing at Commutator Surface	Immediate Cause	Key to Primary Sources of Poor Commutator Performance[a]
Rapid commutator wear with bright surface (Fig. 19.21f)	Foreign material in brush face	43, 45, 47, 49
	Wrong brush grade	61
	Brush vibration	39, 52, 58, 59
Copper dragging (Fig. 19.21a)		

Indication Appearing as Heating	Immediate Cause	Key to Primary Sources of Overheating[a]
Heating in windings	Severe load condition	38, 41, 42, 53
	Unbalanced magnetic field	18, 19, 20, 21, 27, 28, 29
	Unbalanced armature currents	8, 19, 22, 25, 27, 28, 29, 37
	Poorly equalized parallel operation	7, 13, 23, 34
	Lack of ventilation	6
Heating at commutator	Severe load condition	38, 41, 42
	Severe sparking	7, 8, 9, 12, 20, 33, 45, 57
	High friction	10, 11, 36, 43, 45, 49, 58, 59
	Poor commutator surface	See specific surface fault in evidence
	High contact resistance	56
Heating at brushes	Severe load condition	38, 41, 42
	Faulty machine adjustment	7, 10, 11, 12, 26
	Severe sparking	See "Sparking"
	Raw streaks on commutator surface	See "Streaking or threading of surface"
	Embedded copper	See "Copper in brush face"
	Wrong brush grade	57, 58, 59, 61, 62

Source: Union Carbide Co.

[a] See Table 19.4 for key to sources.

[b] Bar marking at pole pitch spacing is indicated by discoloration or etching of the commutator in groups of bars equally spaced around the commutator. The number of discolored areas is related to one-half the number of main field poles. For example, a six-pole machine may have three discolored areas equally spaced around the commutator (see Fig. 19.21b).

[c] Bar marking at slot pitch spacing is indicated by a film discoloration or mark every second, third, or fourth commutator bar. This is related to the number of coils contained in one set of armature slots (see Fig. 19.21c).

TABLE 19.4
Primary Sources of Poor Brush and Commutator Performance

Numbers are keyed to Table 19.3

Preparation and Care of Machine
1. Poor preparation of commutator surface
2. High mica
3. Featheredge mica
4. Bar edges not chamfered after undercutting
5. Need for periodic cleaning
6. Clogged ventilating ducts

Machine Adjustment
7. Brushes in wrong position
8. Unequal brush spacing
9. Poor alignment of brush holders
10. Incorrect brush angle
11. Incorrect spring tension
12. Interpoles improperly adjusted
13. Series field improperly adjusted

Mechanical Fault in Machine
14. Brushes tight in holders
15. Brushes too loose in holders
16. Brush holders loose at mounting
17. Commutator loose
18. Loose pole pieces or pole-face shoes
19. Loose or worn bearings
20. Unequal air gaps
21. Unequal pole spacing
22. Dynamic unbalance
23. Variable angular velocity
24. Commutator too small

Electrical Fault in Machine
25. Open or high resistance connection at commutator
26. Poor connection at pigtail
27. Short circuit in field or armature winding
28. Ground in field or armature winding
29. Reversed polarity on main pole or interpole

Machine Design
30. Commutating zone too narrow[a]
31. Commutating zone too wide[a]
32. Brushes too thin

33. Brushes too thick
34. Magnetic saturation of interpoles
35. High bar-to-bar voltage
36. High ratio of brush contact to commutator surface area
37. Insufficient cross connection of armature coils

Load or Service Condition
38. Overload
39. Rapid change of load
40. Reversing operation of non-interpole machine
41. Plugging
42. Dynamic braking
43. Low average current density in brushes
44. Contaminated atmosphere
45. "Contact poisons"
46. Oil on commutator or oil mist in air
47. Abrasive dust in air
48. Humidity too high
49. Humidity too low
50. Silicone contamination

Disturbing External Condition
51. Loose or unstable foundation
52. External source of vibration
53. External short circuit or very heavy load surge

Wrong Brush Grade
54. "Commutation factor" too high[b]
55. "Commutation factor" too low[b]
56. Contact drop of brushes too high
57. Contact drop of brushes too low
58. Coefficient of friction too high
59. Lack of film forming properties in brush
60. Lack of polishing action in brush
61. Brushes too abrasive
62. Lack of carrying capacity

[a] The commutating zone is the neutral region between adjacent main field poles. Any flux in this region is caused by the commutating poles (interpoles). The size of the commutating zone depends on the number of commutator bars spanned by a brush.
[b] Commutation factor is a measure of the shock-absorbing characteristic of the carbon brush. The factor depends on the brush grade.

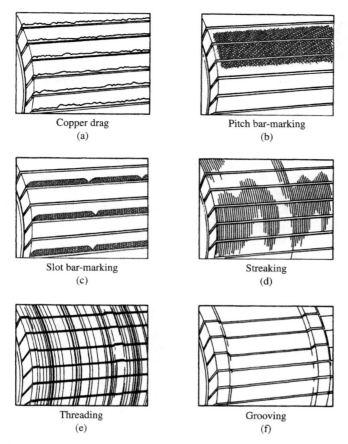

Copper drag
(a)

Pitch bar-marking
(b)

Slot bar-marking
(c)

Streaking
(d)

Threading
(e)

Grooving
(f)

FIGURE 19.21
Indications of unsatisfactory brush and commutator performance.

GENERAL REFERENCE

Kalb, W. C., and Lutz, F. K., *Carbon Brushes for Electrical Equipment,* Union Carbide Corp., Carbon Products Division, 1966.

SPECIFIC REFERENCE

[1] Institute of Electrical and Electronic Engineers Guide: Test Procedures for Direct Current Machines. IEEE Std. 113–1985.

REVIEW QUESTIONS

1. What is the composition of the surface film that forms on commutators and slip rings during normal operation?
2. What is the advantage of maintaining a good surface film on a commutator?
3. What causes a raw copper surface?
4. How does the brush temperature affect brush wear?
5. What adverse effect does silicone vapor have on the performance of DC machines?
6. What causes threading on commutators and slip rings?
7. State the operating conditions that require (a) very soft brushes and (b) very hard brushes.
8. What are some of the more common causes of brush sparking?
9. How should a commutator be cleaned?
10. How can a loose commutator bar be detected? What is the correct procedure for tightening a commutator?
11. What is the best method for resurfacing a commutator? Describe it in detail.
12. Why is sandpaper undesirable for resurfacing commutators, even if only a light dressing is needed?
13. What method should be used to remedy an eccentric commutator?
14. Explain why excessive grinding of a commutator may result in sparking.
15. What is the one commutator problem that hand stoning cannot correct?
16. Is it safe to use a commutator stone while the machine is carrying a load? Explain.
17. What is the advantage of burnishing a commutator, and how is it done?
18. What is air curing as applied to a commutator?
19. Why should one avoid touching a freshly ground commutator?
20. Describe several methods for undercutting mica. When should it be done, and why is it necessary?
21. What can be done to prevent the mica from absorbing oil and moisture?
22. If the brushes are to be removed during overhaul, why is it important that they be tagged for return to the same holder?
23. Make a sketch showing the correct way to stagger the brushes of a six-pole machine.
24. What bad effect does excessive brush play in a brush holder have on the operation of the machine?
25. When should brushes be replaced? Why should all brushes on a multiple-brush arm be replaced at the same time?
26. Explain why slip rings that carry direct current should have their connections reversed every 6 months.
27. What is the essential difference between the reaction type and the box type of brush holder? What are the advantages and disadvantages of each?
28. Describe the entire procedure, in proper sequence, for installing and seating new brushes.
29. What is a brush-seating stone? How is it used?

30. How much of the brush-contact area should be seated before full load can be applied?
31. What adverse reactions will be instigated by brush-spring pressure values that are too high or too low?
32. Should the neutral setting of the brushes be adjusted every 6 months? Explain.
33. Describe the physical inspection test for locating the brush neutral setting.
34. Describe the inductive-kick test for locating the brush neutral setting of a DC machine.

20

Troubleshooting and Emergency Repair of DC Machines

20.0 INTRODUCTION

This chapter is devoted to logical methods for determining some of the more common problems that occur in DC machinery. Suggestions for emergency repairs are also provided to allow equipment to operate safely until a major overhaul can be scheduled. Difficulties of a mechanical nature, such as brush position, brush tension, commutator irregularities, bearing wear, and vibration, are covered in other chapters.

Electrical apparatus (by itself) does not reverse the connections to its field, armature, interpole, etc. Hence, unless there is reasonable suspicion that the connections were changed, we will assume that they are correct. The nameplate data of the machine should always be checked against the actual operating values of voltage, current, temperature, speed, etc. Poor performance may often be traced to operating above or below the rated nameplate values.

Figure 20.1 illustrates some of the more visible effects that overload can inflict on electrical machinery. High temperature caused the armature banding wire to throw solder, grounding the field coils and the armature. A major overhaul was necessary to put the machine back in service.

20.1 TROUBLESHOOTING A DC MOTOR

When a motor fails to start, the branch-circuit fuses (or molded-case breaker), thermal overload relay, and fuses on the motor control panel should be the first things checked. If the fuses are good and the breaker and relays are not tripped, the control panel should be checked.[1]

---

[1] See Chapter 22 for operation and maintenance of motor controllers.

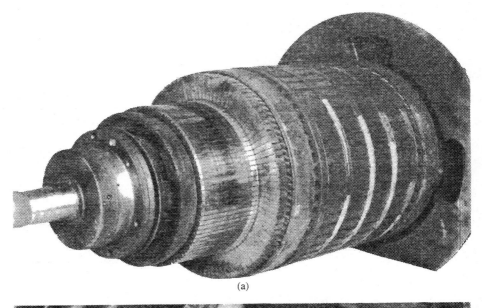

(a)

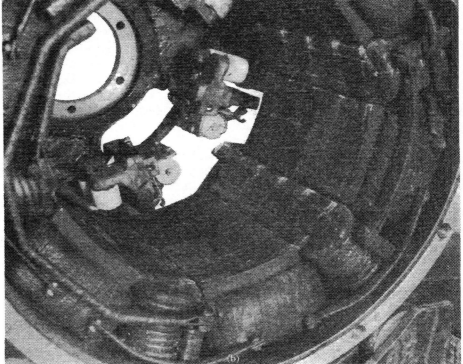

FIGURE 20.1
(a) Overheating causes commutator and armature bands to throw solder; (b) solder thrown from armature bands deposited on field coils and interpole coils (courtesy GE Industrial Systems).

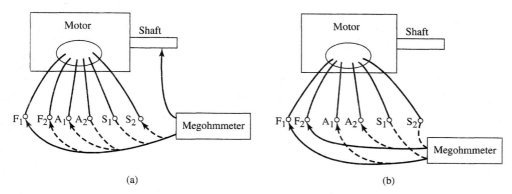

FIGURE 20.2
(a) Ground test of motor circuits; (b) continuity test of motor circuits.

Preliminary Tests

Open the branch-circuit breaker, lock it in the open position, and attach a "PERSON WORKING ON LINE" sign. Disconnect the armature and field leads from the controller, and make continuity and ground tests of the shunt-field circuit, the series-field circuit, and the armature circuit.

The ground test is made by connecting one terminal of a megohmmeter to the shaft or to an unpainted part of the motor frame and connecting the other terminal of the megohmmeter to each of the motor terminals in turn, as shown in Figure 20.2a. A zero reading indicates a grounded circuit.

The continuity test (open-circuit test) is made by connecting the two megohmmeter terminals to the corresponding circuit leads F_1 and F_2, A_1 and A_2, and S_1 and S_2, in turn, as shown in Figure 20.2b. A zero reading indicates a complete circuit. A high reading on the megohmmeter indicates an open circuit.

A listing of some of the common problems experienced with DC motors and their probable causes is given in Section 20–14.

20.2 LOCATING A SHORTED SHUNT-FIELD COIL

To check for short circuits in the shunt-field coils of a DC motor or generator, a 120-V or 240-V, 60-Hz source should be applied to the shunt-field circuit, and the voltage drop measured across each field coil, as illustrated in Figure 20.3. Because each field coil has the same number of turns, the resistance and inductance of each coil are the same, and thus the voltage across each coil should be the same. Slight differences may be due to a few turns (more or less) or stretching of the wire during the winding process.

If the voltages across the coils are equal, we can assume that all six coils are free of shorted turns. A shorted turn in one coil causes its voltage to be noticeably lower than the average value. Also, because of transformer action, the voltage across coils adjacent to the shorted coils will also be slightly lower than the average value. If the voltage drop across any coil is less than 90 percent of the average value, one or more turns may be shorted.

FIGURE 20.3
Testing for a shorted shunt-field coil.

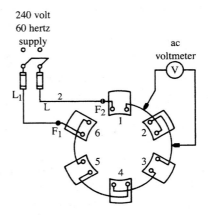

Short-circuit tests of field coils should be performed when the windings are hot, such as immediately after shutdown. Otherwise, shorts caused by defective insulation between adjacent turns (caused by thermal expansion of the copper) might not be detected.

EXAMPLE 20.1

The results of a short-circuit test of shunt-field coils for the six-pole DC motor shown in Figure 20.3 are given in Table 20.1. The voltage applied to the shunt field is 240 volts, 60 Hz. Are there any shorted coils?

TABLE 20.1
Voltage Drops for Shunt-Field Test

Coil	Voltage
1	42
2	42
3	41
4	32
5	41
6	42

Solution
The average voltage is determined by dividing the impressed voltage by the number of field poles. Thus,

$$V_{avg} = \frac{V_{applied}}{\text{Number of poles}} = \frac{240}{6} = 40 \text{ V}$$

Ninety percent of the average value is $40 \times 0.90 = 36$ V. Hence, coil 4 is shorted.

If an AC test voltage is not available, the test may be performed with direct current. However, the DC test is not as sensitive as the AC test, and will not detect a short involving one or two turns. Nevertheless, the DC test will detect a shorted coil if 10 percent or more of its turns are shorted.

If each coil in Example 20.1 has 200 turns, one shorted turn will result in a resistance change of only one two-hundredth of the coil resistance. This very small change in resistance will have no noticeable effect on the DC voltage drop measured across the coil. However, with AC applied, the transformer action of the shorted turn will lower appreciably the impedance of the affected coil, and to some extent lower the impedance of the adjacent coils, resulting in noticeably lower voltages.

Short-circuit tests of coils wound with strap copper, such as series coils and interpole coils, are best done with low resistance ohmmeters capable of measuring resistance as low as 1×10^{-6} ohms.

Faulty field coils should be removed, and the coils unwound to note the general condition of the insulation and to locate the fault. If the insulation is very brittle and cracked, there is a good probability that the other coils are similarly deteriorated. Even if there is no indication of a short in the remaining coils they should be replaced.

20.3 LOCATING AN OPEN SHUNT-FIELD COIL

An open shunt-field coil may be located by connecting the field circuit to its rated DC voltage, or less, and testing with a voltmeter from one line terminal to successive coil leads, as shown in Figure 20.4. The voltmeter will indicate zero until it passes the coil with the open; then full-line voltage will be indicated. The test voltage may be either AC or DC.

An open coil may also be located with an ohmmeter or megohmmeter test across each one; an open coil will indicate infinity. An emergency repair of an open field coil can be made by unwinding the coil to expose the break, soldering the broken ends together, and then reinsulating the coil.

FIGURE 20.4
Testing for an open field coil.

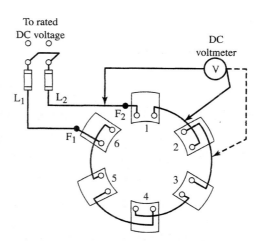

20.4 LOCATING A GROUNDED SHUNT-FIELD COIL

A grounded shunt-field coil may be located by connecting the field circuit to its rated DC voltage, or less, and measuring the voltage drop between each coil and ground, as shown in Figure 20.5. A grounded coil will indicate very low or no voltage when tested from its terminals to ground. A voltmeter test to locate grounded field coils should be made only if a megohmmeter test indicates that the field circuit is grounded. A voltmeter test of an ungrounded field circuit will have no significance; it will indicate 0 V for every coil.

A grounded coil may also be located with a megohmmeter, but each coil must be disconnected from the others and tested to ground individually. A grounded coil will indicate zero.

A ground caused by dirt and carbon dust can generally be cleared by cleaning and revarnishing. However, a coil that is grounded by chafing of insulation against the pole iron should be replaced if the copper shows signs of having worn away.

20.5 TESTING FOR A REVERSED SHUNT-FIELD COIL

A reversed shunt-field coil in a generator or motor may be detected by testing with a magnetic compass. To make the test, the field should be connected to a source of direct current no higher than the rated voltage, and the polarity of each pole should be tested with a compass. The poles should test alternately north and south around the frame. The compass should not be held too close to the field, or its own polarity may be reversed. An alternate method for testing magnetic polarity makes use of two short steel bolts bridged across the gap between adjacent field poles; correct polarity will cause a strong attraction between the bolts, but a reversed coil will cause the ends to repel. These tests are illustrated in Figure 20.6. Because of the very strong magnetic field at the pole faces, testing with a single bolt may give misleading results. Two bolts should be used.

FIGURE 20.5
Locating a grounded shunt-field coil.

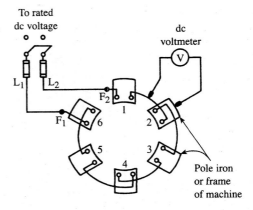

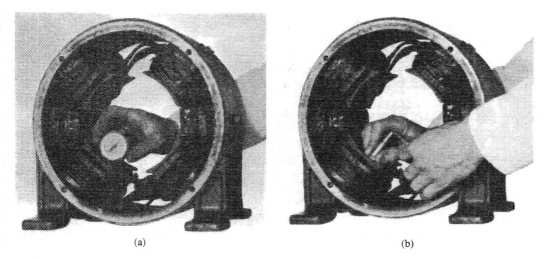

(a)   (b)

FIGURE 20.6
Polarity test for shunt-field poles: (a) using a magnetic compass; (b) using two steel bolts (courtesy United States Merchant Marine Academy).

20.6 TESTING FOR REVERSED COMMUTATING WINDINGS (INTERPOLES)

The function of a commutating winding is to reduce or eliminate sparking. Hence, its reversal will result in worse sparking than if no commutating winding were used. Thus, if a rebuilt or repaired machine sparks severely when under load and when starting, the commutating poles may have been reversed. Shorting the interpole connections with a heavy copper jumper (in series with a switch) while the machine is under load will confirm or deny the suspicion. If the sparking is reduced when the winding is shorted, then the commutating winding is connected in reverse and should be corrected.

20.7 INSPECTION AND REMOVAL OF FIELD COILS

Shunt-field coils, series-field coils, and interpole coils should be inspected for discolorations, brittleness, cracks in the insulation, defective leads, loose connections, loose mounting bolts, and evidence of solder on the pole faces.

Discoloration of the field coil insulation indicates excessive field current. Brittle insulation (caused by age or overheating) can be determined by tapping the insulation gently with the handle of a screwdriver or a rubber or rawhide mallet. Discolored leads generally indicate overheating due to poor lead connections.

Solder on the pole faces indicates overheating of the armature with the resultant melting and throwing of solder from the banding wire.

When field poles or commutating poles are removed for inspection, repair, or replacement, the poles and any metallic spacers (called shims) must be marked so that they can be returned to their original positions. Shims are placed in back of the pole iron to obtain the desired distance (air gap) between the pole face and the armature. Failure to correctly install the shims may cause sparking at the brushes.

20.8 LOCATING AN OPEN ARMATURE COIL

An open in the armature of a DC machine can generally be detected by a visual inspection of the commutator. The bars that connect to the defective coils will be partly burned or discolored by excessive heating and arcing, as shown in Figure 20.7. Most opens originate from overheating due to overload or inadequate ventilation, which causes the solder to soften and be thrown from the coil connections at the commutator or commutator risers.[2] Inspection of the risers will reveal this condition.

The risers and coil connections should be thoroughly cleaned and resoldered with a high-temperature solder, and the machine should be thoroughly cleaned to clear any clogged ventilating ducts. A bar-to-bar test of the armature will reveal any

---

[2] See Figures 19.4 and 20.1a.

FIGURE 20.7
Open armature coils indicated by burned commutator bars (courtesy GE Industrial Systems).

4-pole machine   6-pole machine   4-pole machine   6-pole machine

Parallel-wound armatures   Series-wound armatures

(a)   (b)

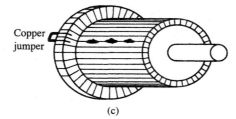

Copper jumper

(c)

FIGURE 20.8

(a,b) Relationship between burn spots and type of armature winding; (c) emergency bypass of an open armature coil.

opens or high-resistance connections not detected by visual inspection. When the machine is put back in service, the load current should be checked and compared with the rated nameplate value. A higher than normal current is an indication of overload.

If the open is not in the risers, identification of the type of winding and the required emergency repair can be determined by the number of burned spots and their geometric location on the commutator, as shown in Figure 20.8.

20.9 EMERGENCY REPAIR OF AN OPEN ARMATURE COIL

The number and location of burned spots on the commutator, for an open in only one armature coil, are determined by the type of winding.

Lap Winding

A lap winding, also called a parallel winding, will have only one burned spot on the commutator for each open, regardless of the number of poles. This is shown in Figure 20.8a. An emergency repair can be made by bridging over the affected bars. This is shown in Figure 20.8c. The jumper should be of the same size of armature wire and must be soldered in place. If the open is in a fractional-horsepower machine, the repair may be made with a lump of solder bridging the affected risers.

Wave Winding

A wave winding, also called a series winding, will have the number of burned spots equal to the number of poles divided by 2, as shown in Figure 20.8b. However. only one of the burned spots should be bridged over (any one will do). Bridging more than one will cause circulating currents in two or more coils, overheating the armature.

20.10 LOCATING SHORTED ARMATURE COILS

A shorted armature coil can usually be detected by sparking at the brushes, as well as discoloration, and burning of its insulation. If a shorted coil is allowed to operate, the heat it generates may loosen the soldered connections to the commutator. A growler test or a bar-to-bar test can be used to locate a shorted coil that is not discernible by visual inspection; the very low coil resistance of integral-horsepower armatures makes it quite difficult, if not impossible, to check for shorted coils with a conventional ohmmeter.

Short-circuit tests of armature coils should be made when the armature is hot, such as immediately after shutdown. Otherwise, defective insulation between turns, caused by thermal expansion, may not be detected.

Growler Test

The growler test, shown in Figure 20.9, represents a quick and reliable method for locating short-circuited armature coils. The growler is energized from the 120-volt, 60-hertz supply voltage, and a feeler fashioned from a hacksaw blade with its teeth ground off is "wiped" across the armature from slot to slot, as shown. The feeler will vibrate with a loud growling noise when directly over a slot containing the shorted

FIGURE 20.9
Testing an armature for short-circuited coils with a portable growler (courtesy United States Merchant Marine Academy).

coil. All slots should be feeler tested with the growler set in different positions around the armature. The growler should be shifted two or three slots at a time.[3]

The changing flux through the window of the armature coil generates an alternating voltage in the armature coil. If the coil is shorted, the circuit is complete, and an alternating current will appear. The resultant magnetic field encircling the armature conductors will attract the hacksaw blade, causing it to vibrate in synchronism with the alternating current. A vibrating noise (growling) indicates that the coil is shorted. The size of the armature that can be tested in this manner depends on the size of the growler used. The portable growler shown in Figure 20.9 is most effective in the range of 0.1 hp to 15 hp. However, with careful observations and a sensitive feeler (thinner than a hacksaw blade), it can be extended to 50 hp.

Bar-to-Bar Test

A bar-to-bar test is made by passing about 10 percent of the rated current through the armature and then measuring the voltage from bar to bar with an analog-type millivoltmeter. The current may be supplied by a battery or generator connected in series with a rheostat or bank of lamps, as shown in Figure 20.10a for a four-pole machine. The current may be conducted to the armature through two strips of copper secured to the commutator with cord or electrician's tape at points that are one pole apart. Thus the copper strips for a two-pole machine would be 180° apart, a four-pole machine 90° apart, a six-pole machine 60° apart, etc. A millivoltmeter may be improvised by using an external shunt ammeter without its shunt.

If the millivoltmeter deflection for a given bar-to-bar measurement is less than the average bar-to-bar value, a shorted coil is indicated. A zero reading indicates a dead short, and a very high or off-scale deflection indicates a high-resistance connection of the soldered joints at the risers or an open in the coil.

Warning

1. Before starting this test, the rheostat or the number of lamps in parallel should be adjusted to obtain no more than a half-scale deflection on the millivoltmeter for a few sample bar-to-bar readings. This will prevent overvoltage, caused by high-resistance armature connections, from damaging the millivoltmeter.
2. When making the bar-to-bar test, do not span more than one mica segment at a time; spanning two or three mica segments doubles or triples the voltage across the millivoltmeter and may damage it.

The principle of the bar-to-bar test is explained with the aid of Figure 20.10b, which shows the armature test connections for a two-pole machine. For the representative

---

[3]The principle of growler action is explained in Section 15–1, Chapter 15.

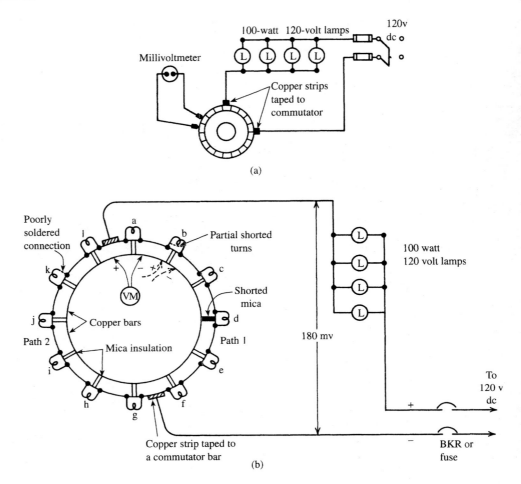

FIGURE 20.10

(a) Circuit connections for a bar-to bar test of a four-pole machine; (b) circuit connections for a two-pole machine (three different types of faults are indicated).

example, it is assumed that the lamp bank limits the voltage across the armature to 180 millivolts (0.180 volt).

The armature current divides into two parallel paths consisting of six coils in each path. Path 1 has coils a, b, c, d, e, and f; path 2 has coils g, h, i, j, k, and l. If there are no defective connections or defective coils, the voltage across each coil will be the same and equal to 180/6 = 30 mV.

However, for the faults shown in Figure 20.10b, the voltages across the different coils will not be the same. Shorted coils will offer less opposition to the current and high-resistance connections more opposition to the current, causing the voltage drops to be, respectively, lower and higher than the common voltage in the path being tested. A tabulation of representative voltage drops for the fault conditions shown in Figure

TABLE 20.2

Voltage Drops for Bar-to-Bar Test of Figure 20.10b

Path 1	Millivolt Drop	Path 2	Millivolt Drop
a	39	g	27
b	24[a]	h	27
c	39	i	27
d	0[b]	j	27
e	39	k	45[c]
f	39	l	27

[a] Partially shorted turns.
[b] Shorted mica.
[c] Poorly soldered connection.

20.10b is provided in Table 20.2. The common voltage in path 1 is 39 volts, and the common voltage in path 2 is 27 volts.

Note: If there is an open in an armature coil, or a completely unsoldered connection at the commutator, there will be no current in that section of the armature circuit that contains the open coil. Thus, when testing bar to bar in the armature section containing an open coil, the millivoltmeter will indicate zero until the test points bridges the open coil, at which time it will indicate the total millivolts across the armature. This could damage the millivoltmeter if it is not properly fused.

Lamp-Test for Fractional-Horsepower Armatures

A lamp test can be used to check for shorted coils in the armatures of fractional-horse-power machines, such as drills, vacuum cleaners, mixers, etc. The circuit hookup shown in Figure 20.11 is very simple. The choice of lamp size (watts) is by trial and error; the correct size lamp will burn at reduced brilliance when connected across a normal coil, but will burn brightly when connected across a shorted coil. Lamps with ratings of 25 W, 40 W, 60 W, and 100 W (all 120 V) should be kept handy for such tests.

FIGURE 20.11

Lamp test of fractional-horsepower armatures.

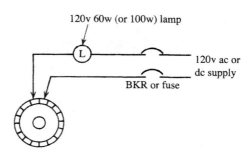

20.11 EMERGENCY REPAIR OF A SHORTED ARMATURE COIL

An emergency repair can be made by cutting the shorted coil in half at the back end of the armature. This is done with a sharp chisel, hacksaw, or diagonal-cutting pliers, as shown in Figure 20.12. Care must be taken to prevent damage to the adjacent coils. The open ends should be taped and secured to the coils. After the coil is opened, its commutator bars must be bridged over; use the procedure outlined for the emergency repair of an open armature coil, as illustrated in Figure 20.8c.

Shorts caused by the absorption of carbon dust, oil, dirt, and moisture in the mica insulation between the commutator bars can be repaired by digging out and filling with commutator cement.

20.12 LOCATING GROUNDED ARMATURE COILS

A grounded armature coil can be located with a bar-to-ground test. To make this test, about 10 percent of the rated armature current should be passed through the armature, and the voltage should be measured from bar to ground with a millivolt-meter, as shown in Figure 20.13a for a four-pole machine; the current can be supplied by a battery or by a generator connected in series with a rheostat or a bank of lamps.

If this test is to be made aboard ship, the shaft and armature core must be insulated from the steel deck. The current can be conducted to the armature through two

FIGURE 20.12
Cutting an armature coil as part of an emergency repair (courtesy United States Merchant Marine Academy).

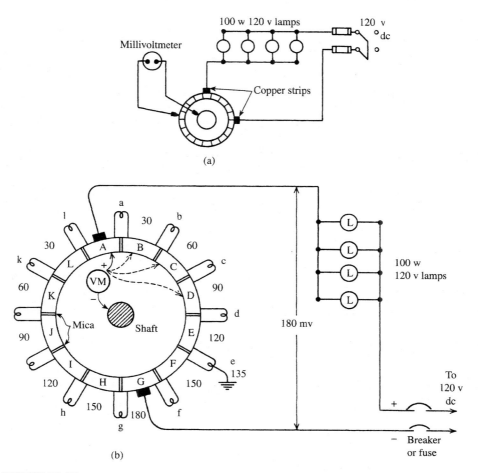

FIGURE 20.13

Bar-to-ground test: (a) circuit test for a four-pole machine; (b) circuit test for a two-pole machine (a ground is indicated).

strips of copper conductor secured to the commutator with cord or electrician's tape at points that are one pole apart. One terminal of the millivoltmeter should be connected to the shaft, and the other should be moved from bar to bar along the commutator surface. Every bar should be tested and those that indicate zero volts should be marked with chalk.[4] To complete the test, both copper strips must be shifted several bars in the same direction, and the bar-to-ground test must be reapplied to the chalk-marked bars

---

[4]An ungrounded armature will indicate zero volts between every bar and ground. Hence, the bar-to-ground test should be applied only to armatures that have been determined to be grounded in the preliminary tests. See Section 20–1.

only. Those that indicate zero on the second test are the true grounds. The bars that indicate zero on the first test, but not on the second, are called phantom or sympathetic grounds.

Phantom grounds are indicated at points in the armature that have the same voltage as the grounded coil. Because all armature windings have at least two parallel paths, there will always be twice as many ground indications as actual grounds. Moving the power connections several bars shifts the phantom grounds, but the true grounds remain unchanged. The principle of the bar-to-ground test is explained with the aid of Figure 20.13b, where coil e is shown with a ground at its center; the armature is from a two-pole machine. Assuming no opens, shorts, or high-resistance connections, the voltage across each coil will be 30 mV.

To locate the grounded coil, connect one terminal of the millivoltmeter to the armature shaft. This connects it electrically to the center of coil e, which for this example is at a potential of 135 mV. Then probe the commutator with the other test point, bar by bar, watching for the lowest meter readings. Because a voltmeter (or millivoltmeter) indicates the difference in potential (voltage) across its terminals, a zero or minimum reading in the bar-to-ground test indicates that the "roving" test point is electrically connected to the grounded coil. A tabulation of the millivoltmeter indications for the bar-to-ground test of Figure 20.13b is provided in Table 20.3; the lowest readings occur at the actual ground and at the phantom ground.

TABLE 20.3
Millivoltmeter Readings for Bar-to-Ground Test of Figure 20.13b

Bar	Millivolts
A	135
B	105
C	75
D	45
E	15[a]
F	15[a]
G	45
H	15[b]
I	15[b]
J	45
K	75
L	105

[a] Grounded coil.
[b] Phantom ground.

20.13 EMERGENCY REPAIR OF A GROUNDED ARMATURE

The coil connections to the marked bars should be unsoldered, and each wire should be individually tested for grounds with an insulation-resistance tester, as shown in Figure 20.14. The wires that indicate zero are grounded and should not be reconnected; they should be insulated from the others and secured to the armature with stout cord to prevent shifting by centrifugal force. The remaining wires can then be reconnected and an emergency repair effected for the now open armature circuit, using the procedure described for open armature coils in Section 20.9.

However, if the coil connections test clear of grounds, the fault lies in the commutator bars. A megohmmeter ground test of the marked bars will identify the faulty one. The most likely place for the ground to occur is at the string band that holds the mica in position; see Figure 20.1a. Absorption of carbon dust, oil, dirt, and moisture may in time cause carbonized paths to occur in the string band between the commutator bars and the steel clamping ring, resulting in a ground. The ground can be cleared by digging out the "carbonized" mica, filling with commutator cement, and replacing the string band.

FIGURE 20.14
Ohmmeter connections for the ground test
of individual coils after disconnecting from
the commutator.

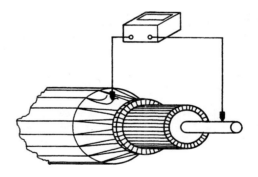

20.14 TROUBLESHOOTING CHART

The DC motor troubleshooting chart in Table 20.4 lists the most common DC motor problems and their possible causes. Problems of DC generators are discussed in Chapter 17.

TABLE 20.4
DC Motor Troubleshooting Chart

Trouble	Possible Cause	Book Section
Fails to start	1. Blown fuses	29–2
	2. Frozen shaft	21–4
	3. Brushes not contacting commutator	19–14
	4. Open in shunt field	20–1, 20–3
	5. Open in armature circuit	20–1
Runs hot	1. Overload	18–12
	2. Shorted armature coils	20–10
	3. Clogged ventilating ducts	26–3, 26–4, 26–5
	4. Brushes off neutral	19–7, 19–8, 19–15, 19–16
	5. Shorted field coils	20–2
	6. High ambient temperature	———
Sparks at brushes	1. Commutator and brushes dirty	19–4
	2. Wrong brush grade	19–3
	3. Wrong brush adjustment	19–7, 19–8
	4. Brushes off neutral	19–7, 19–8, 19–15, 19–16
	5. Open armature coil	20–8, 20–9
	6. Commutator eccentric	19–5
	7. High mica or high commutator bars	19–6
	8. Brush chattering	19–14
	9. Shorted or reversed interpole	20–5, 20–6
Runs fast	1. Series field bucking shunt field	18–9, 18–11
	2. Open in shunt field of a lightly loaded compound motor	18–6
	3. Shunt field rheostat resistance set too high	18–6
	4. Shunt field coil shorted	20–1, 20–2
Runs slow	1. Overload	18–12
	2. Short circuit in armature	20–10
	3. Brushes off neutral	19–7, 19–8, 19–15, 19–16
	4. Starting resistance not cut out	18–5

GENERAL REFERENCES

Rosenberg, R., and Hand, A., *Electric Motor Repair,* Holt, Rinehart and Winston, New York, 1987.

Veinott, C. G., and Martin, J. E., *Fractional Horsepower Electric Motors,* McGraw-Hill, New York, 1986.

REVIEW QUESTIONS

1. State the routine procedure that should be followed when a motor fails to start.
2. Describe a continuity test applicable to the shunt field of a motor.

3. Explain why alternating current is better than direct current when testing the shunt-field coils for short circuits. How should this test be made?
4. Describe a test for locating an open shunt-field coil. State the procedure for making an emergency repair.
5. Describe a test for locating a grounded shunt-field coil. Can a grounded coil be repaired? Explain.
6. Describe a test procedure that does not require a compass for checking the sequence of polarity of the shunt-field poles.
7. What test procedure can be used to determine whether or not the interpole circuit is reversed?
8. What commutator bar markings are indicative of an open armature coil?
9. State the correct procedure for making an emergency repair of (a) a wave-wound armature; (b) a lap-wound armature.
10. Describe a growler test for locating a shorted armature coil.
11. Sketch the circuit and state the correct procedure for making a bar-to-bar test of an armature. What precautions should be observed? What are the test indications for a high-resistance connection at the commutator?
12. Describe the lamp test for fractional-horsepower armatures.
13. State the correct procedure for making an emergency repair of a shorted armature coil.
14. Describe a test for locating a grounded armature coil.
15. State the correct emergency procedure for repairing a grounded armature coil.
16. What causes commutators to become grounded? Can grounded commutators be repaired?

21

Mechanical Maintenance of Electrical Machinery

21.0 INTRODUCTION

Mechanical maintenance of electrical machinery is a very important part of an overall maintenance program. A high percentage of burned-out motor windings are caused by mechanical failures. A worn bearing could cause the rotor to rub the stator, ripping the windings apart. The high locked-rotor current caused by a frozen bearing will cause the windings to burn out and possibly cause an electrical fire if the circuit is not tripped by an overcurrent device. A loose cooling fan can damage the winding insulation, causing a ground or short circuit. Severe vibration could cause electrical connections and windings to loosen, resulting in overheating and damaged insulation. Misalignment of shafts, blockages in the ventilating system, etc., all take their toll.

A good mechanical maintenance program includes proper lubrication of bearings, measurement of bearing temperature, measurement of vibration, and checking the tightness of bolted parts. Sealed bearings are not discussed; they are lubricated for life and hence are essentially maintenance free.

21.1 DISASSEMBLY OF ELECTRIC MOTORS

The disassembly of machines is an important phase of repair work that must be done without damage to the component parts. Before starting the disassembly process, the circuit breaker and disconnect switch feeding the machine must be opened and a sign "DO NOT ENERGIZE; PERSON WORKING ON LINE" must be fastened to the breaker or switch. A voltmeter should be connected across the machine terminals to confirm that the correct breaker is open and the circuit is dead.

The outside of the machine should be cleaned to prevent outside dirt from entering the machine. The stator end shields, bearing housing, clamping plate, and other removable parts should be carefully match-marked with respect to one another to

447

facilitate proper positioning when they are reassembled. The match marks should be made with a punch set or a chisel.

All cables and wires connecting to the motor should be clearly marked to facilitate reconnection, and the incoming power lines must be insulated to prevent shock or fire in the event that the breaker is accidentally closed. To prevent loss, all small parts should be stored in a box and tagged for identification; bearings should be wrapped in clean lint-free cloth.

All leads to the brush rigging should be disconnected and marked, brushes should be lifted or removed, and thin fiber or several layers of heavy paper should be tied or taped in place around the commutator to prevent scratching of the surface film. If the air gap between the rotor and stator is large, several thicknesses of sheet fiber placed there will relieve the strain on the remaining bearing when one end shield is removed. If necessary, remove keys and file any burrs on the keyway and shaft.

Machines equipped with bearing housings that slide through the end shield (cartridge type) should not have the bearing cap removed unless the bearing is to be changed. Bearings that are locked to both the shaft and the end shield must have the locknut removed before the machine can be disassembled. Bearings that contain oil should be drained before disassembly.

The end shields are generally closely fitted to the frame of the machine with a rabbet type of joint. To remove the end shield, take out the tie bolts, and then tap the lip or knockoff lugs with a mallet, or a hammer on hardwood blocks, as shown in Figure 21.1. Some ball-bearing and roller-bearing machines have very tight bearing fits and require special tools, available from the manufacturer, to assist in disassembly.

The disassembly of explosion-proof machines requires exceptional care and attention to the manufacturer's instructions. Hammers and prying tools must not be

FIGURE 21.1
Removing the end shield by tapping knockoff lugs with a mallet.

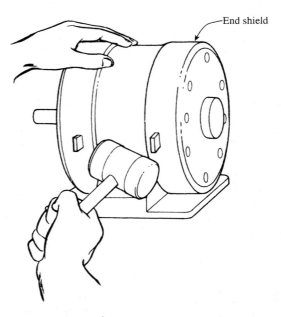

FIGURE 21.2
Spreader bar used to remove rotor from machine: (a) position prior to lifting; (b) lowering to a platform.

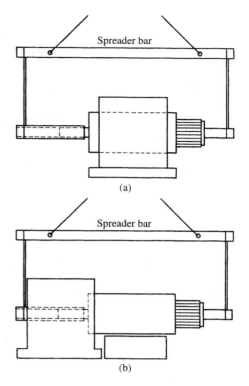

used on the machined surfaces that form the flame-path surfaces of the enclosure. The tie bolts used to secure the end shields to the motor body are made of high-tensile-strength material in order to withstand the high pressures that may occur during an explosion. They must never be replaced with conventional low-tensile-strength bolts. These special bolts should be placed in a separate bag and marked with the motor serial number and the words "high-strength bolts."

The rotors of heavy machines should be lifted and removed in the manner shown in Figure 21.2. A spreader bar is used to prevent the cable from cutting into the commutator or windings, and a length of pipe is used as an extension for short shafts. The rotor should never be lifted by the commutator, slip rings, or windings or allowed to rest on them. The armature should never be rolled on the floor, because the coils and banding wire are easily injured.

21.2 REASSEMBLY OF ELECTRIC MOTORS

Before reassembling a machine, inspect all the mating surfaces for burrs and corrosion. Clean any corroded surfaces, file the burrs smooth, and apply an extremely light coating of oil or grease to the mating surfaces to aid in assembly.

The bolts and nuts used for connecting the field coils to one another should be positioned so as to have maximum clearance between the electrical wiring and the frame, and between points of opposite polarity. Sometimes reversing the respective positions of the nut and bolt will result in a smaller distance between it and the frame, or between it and a conductor of opposite polarity. Too small a distance sets the stage for a possible damaging flashover.

When reassembling, line up the holes in the end shields with the corresponding holes in the stator; square the mating parts by supporting the weight of the end shield; insert the bolts and then gently tap the end shield in position. It is generally easier to insert the side bolts first and the top and bottom bolts last. Do not force the parts together. Tap gently on the top, bottom and sides, checking to see that the end shield is not cocked. The bolts should be tightened evenly and alternately, until the end shield is properly secured. If an inner bearing cap is used, an unusually long stud screwed into one of its mounting holes will facilitate alignment with the corresponding holes in the end shield.

When reassembling a machine with sleeve bearings, lift the oil ring and hold it in place with a piece of wire while the shaft is inserted. Do not use your fingers to hold the ring. Your finger could be severed by the shaft as it slides into the bearing.

When driving pulleys, couplings, or fans onto the motor shaft, the opposite end of the shaft should be supported against a stop or backed up with a heavy piece of brass or steel to absorb the blow and prevent damage to the bearing. Heating the fan or coupling in an oven will expand the hole and permit easy mounting on the shaft. The additional clearance required for a particularly tight fit may be obtained by using dry ice to cool and thereby shrink the shaft. Check the bolts or set-screws that hold the fan or coupling in position. If the fan loosens during operation, it could destroy the windings.

21.3 MAINTENANCE OF AIR FILTERS

Many large motors used in dirty environments, such as in cement plants, use filtering systems to provide clean cooling air to the motor. Clogged or blocked filters will cause overheating of the motor, thus shortening insulation life.

The many different types of air filters used on electric machines include fiberglass throwaway filters, electrostatic filters, oil-impregnated washable filters and water-spray filters. Specific instructions for filter maintenance are provided by the manufacturer and must be adhered to; different equipment requires different procedures.

21.4 MAINTENANCE OF BEARINGS

The proper lubrication of electrical machinery extends the useful life of the bearings and prevents burned-out motors caused by a frozen shaft. Always use the correct type and amount of lubrication recommended by the manufacturer. Excessive oiling or greasing forces the lubricant into the machine, where it may come into contact with the windings and damage the insulation. On the other hand, underlubrication causes excessive bearing wear; the rotor may rub on the stator, or the shaft may wipe the bearing and freeze [1]. The wrong lubricant can also lead to trouble. If lubricants with the wrong viscosity, pressure rating, or composition are used, premature bearing wear may result.

When lubricating bearings, it is wise to check the operating conditions; belts that are too loose cause excessive pounding, and those that are too tight cause wear by overloading. Hot bearings indicate trouble. The safe operating temperature for most bearings, operating under rated load, is 40°C rise above the ambient temperature as measured by a thermometer, or with a thermocouple surface probe that plugs into a digital multimeter as shown in Figure 21.3. The surface probe is ideal for quick temperature checks of accessible bearing surfaces.[1]

Sleeve Bearings with Removable Liners

The bearing housing, which contains the liner, is usually an integral part of the end shield. The bearing liner, shown in Figure 21.4a, is pressed into the bearing housing. A slot cut in the liner allows an oversized brass ring, called an oil ring, to ride on the steel shaft. The oil ring dips into a reservoir of oil. When the motor is running, the revolving oil ring carries oil to the shaft and liner. A cutaway view of an older type AC motor equipped with sleeve bearings, oil reservoirs, and settling chambers for collecting accumulated dirt is shown in Figure 21.5.

To clean this type of bearing, drain the old oil, flush with light mineral oil heated to about 75°C (an approved safety solvent may be used for flushing), and then refill with oil specified by the manufacturer. To drain the oil from machines not equipped with drain plugs, remove the end-shield bolts and rotate the end shield so that the filler cap is upside down. If the manufacturer's specification for oil is not available, use a good grade of mineral oil; SAE 20 or SAE 10 turbine oil is generally recommended, depending on the ambient temperatures.

Bearing liners are removed and installed with an arbor press. Figure 21.4b illustrates a split arbor for removing sleeve-bearing liners after the end shield is removed.

---

[1] The digital multimeter shown in Figure 1.3b, Chapter 1, accepts a thermocouple surface probe.

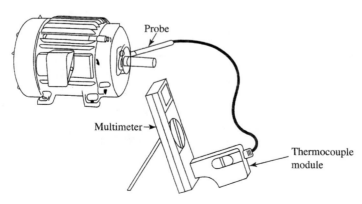

FIGURE 21.3
Surface probe connected to digital multimeter for measuring bearing temperature.

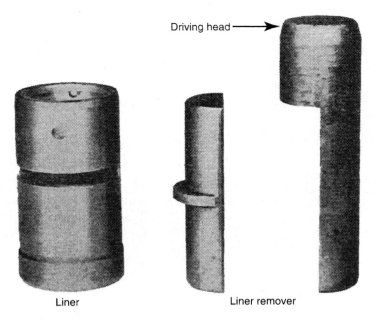

Driving head

Liner   Liner remover

FIGURE 21.4
(a) Liner; (b) split-ring bearing-liner remover (courtesy GE Industrial Systems).

Oil ring

Oil reservoir

Stator windings   Rotor

FIGURE 21.5
Cutaway view of an older AC motor equipped with sleeve bearings, oil rings, and an oil reservoir (courtesy TECO Westinghouse).

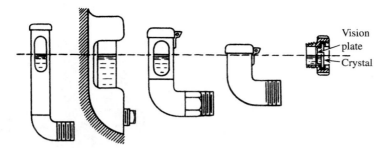

FIGURE 21.6
Correct oil level for different types of oil gauges (courtesy GE Industrial Systems).

The section with the projection is inserted in the liner, and the projection is fitted into the oil-ring slot. The other section, or driving head, is then placed in the liner and an arbor press is used to remove the bearing. The oil ring should be held clear with a piece of wire while the liner is driven out.

Felt washers that are used to seal the ends of the bearing housing should be replaced whenever new bearings are installed. The felt washer prevents airborne dust from contaminating the oil and prevents oil vapors from being drawn into the machine enclosure, where it can damage the insulation.

The correct oil level for different types of oil gauges is illustrated in Figure 21.6. The oiling of sleeve bearings while the motor is in operation should be avoided, because it may result in overfilling.

Bearing wear can be determined by measuring the air gap between the rotor and stator and comparing with previous measurements. Measuring on both sides and the bottom with a tapered, long-blade, feeler gauge is the accepted method. If bearings are removed, wear can be determined by measuring the inside bearing dimensions with vernier calipers and micrometers.

Ball Bearings Equipped with a Pressure Relief System

Bearing housings equipped with pressure fittings and relief plugs can be cleaned and relubricated without ever removing the bearings from the housings. When this method of cleaning is used, both the bearing and housing are simultaneously purged of old grease. The cleaning agent can be either a light mineral oil heated to about 75°C or an approved safety solvent. If the latter is used, the bearing should be rinsed with a light mineral oil to remove all traces of solvent before regreasing. *The motor must be unloaded when performing this operation.*

Figures 21.7a through 21.7d illustrate the recommended method for cleaning the ball bearings of horizontal motors that are equipped with pressure fittings and relief plugs.

The bearing housings and relief fittings should be wiped clean to prevent the entry of dirt. The grease fittings and pressure relief plugs, located at the top and bottom, respectively, of the bearing housings, should be removed, and a screwdriver should be

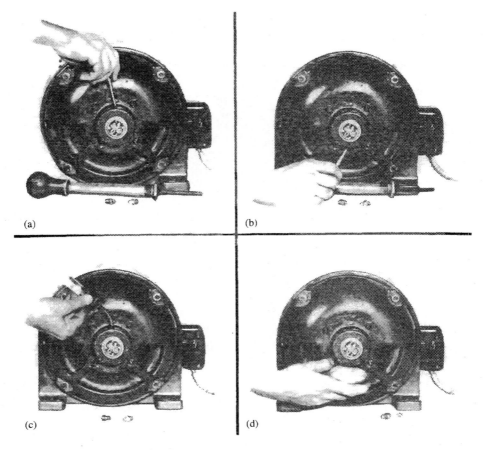

(a)　　　　　　　　　　　　　(b)

(c)　　　　　　　　　　　　　(d)

FIGURE 21.7
Cleaning ball-bearing motors that are equipped with a pressure relief system (courtesy GE Industrial Systems).

used to free the openings of hardened grease. Then with the motor running unloaded, an approved type of grease solvent should be injected into the bearing housing. This is easily done with a syringe, injecting the solvent into the bearing through the top hole. As the grease becomes thinned by the solvent, it drains out through the relief hole.

Solvent should be added in small quantities until it drains out reasonably clear. The relief plug should then be replaced, and a small amount of solvent should be added and allowed to churn for a few minutes. The relief plug may then be removed, and the solvent allowed to drain. If the drainage is not clear, replace the relief plug, add more solvent, and allow it to churn a few minutes more. This process should be repeated until the solvent drains clear. The bearing must then be flushed out with light mineral oil to remove all traces of solvent. The cleaning of vertical machines or machines not equipped with pressure relief systems must be done by disassembling the bearing housing or completely disassembling the machine.

The recommended procedure for regreasing machines that are equipped with pressure fittings and relief plugs is illustrated in Figures 21.8a through 21.8d. The bearing housings, pressure plugs, relief fittings, and grease gun should be wiped clean and the relief plugs removed. Removal of the grease plugs permits the expulsion of old grease and prevents the buildup of excessive pressure in the bearing housing that might rupture the bearing seals. Then with the motor running, grease should be added with a hand-operated grease gun until all old grease is expelled and new grease begins to flow from the relief hole. Use only grease specified by the manufacturer.

The motor should be allowed to run long enough to permit the bearing to expel all excess grease from the housing. The machine should then be stopped, the relief plug replaced, and the housing wiped clean. If it proves dangerous to solvent-clean and lubricate the motor bearings while it is running, follow the same procedure with the motor at standstill. Then start the machine, and allow it to run until all excess grease is expelled.

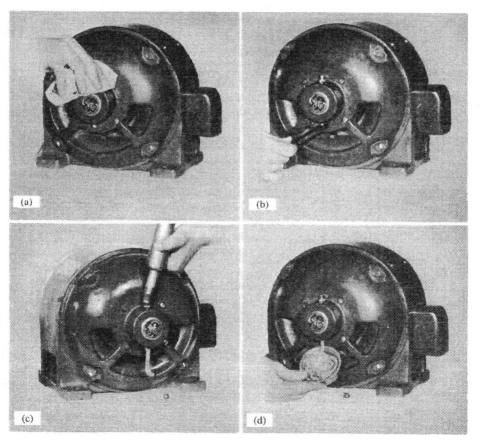

FIGURE 21.8
Greasing ball-bearing motors that are equipped with a pressure relief system (courtesy GE Industrial Systems).

Ball Bearings That Are Not Equipped with a Pressure Relief System

The greasing of machines that are not equipped with pressure relief systems can be done by disassembling the bearing housings. The bearing and housing should be washed with an approved safety solvent, rinsed with light mineral oil, and then packed with grease. Only the lower half of the bearing housing and the space between the balls should be packed with grease. Do not fill the entire housing with grease; it may overheat and build up excessive pressure, thus forcing the grease into the motor housing.

Do not arbitrarily regrease motors; a regreasing schedule, including the type and quantity of grease, should be obtained from the manufacturer of the motors.

Ball Bearing Replacement

Ball bearings are precision made and are adversely affected by dirt. Hence, new bearings should not be removed from their original wrapper until ready for immediate installation. A defective bearing should be replaced with the same size and type as specified by the manufacturer. Before removing a defective bearing, the outside of the bearing housing should be cleaned with an approved safety solvent. The defective bearing can be removed from the shaft with a bearing puller or arbor press by applying pressure against the inner race. Specially designed hot plates and adjustable induction heaters expand the bearing, making removal easier. If the bearing is frozen in place, it can be removed with a cutoff wheel on a high-speed grinder. However, care must be taken to prevent damage to the shaft.

A replacement bearing should be preheated for 30 to 60 minutes in a temperature-controlled oven preheated to approximately 203°F (95°C). This will allow the bearing to expand sufficiently to slip on the shaft with little or no driving. Using clean thermo-insulated gloves, the bearing should be removed from the oven, slipped onto the shaft, and held in place until it cools and locks in place. If the bearing has a locknut, it must be torqued down immediately to the manufacturer's specifications. To compensate for contraction, the locknut should be retorqued after the bearing has cooled. Note that heating in excess of 230°F (110°C) should be avoided because the bearing may lose its hardness.

Some replacement bearings can be pressed on the shaft using an arbor press, or tapped on the shaft using a hammer and a clean brass tube or pipe that fits evenly against the inner race. Do not tap the outer race, because it may damage the bearing.[2]

21.5   MECHANICAL VIBRATION

Periodic vibration measurements of vital machinery can provide useful preventive maintenance data on its mechanical condition. Although a sudden and significant increase in the amplitude of vibration is a very apparent indicator that something is wrong, a gradual increase in vibration may not be noticed until serious damage has

---

[2] Tapping or pressing on the outer race is done only when the bearing is being installed into a bearing housing.

occurred. In addition to damaging the machine, excessive vibration can cause damage to nearby equipment and to the buildings and ships in which it is installed; it also increases worker fatigue and reduces efficiency. Excessive vibration can be caused by misalignment between motor and driven equipment, loose mounting bolts, badly worn bearings, a mechanically unbalanced rotor, a bent or cracked shaft, or an excessively pulsating load.

A small battery-operated handheld vibration meter for checking vibration on all types of rotating machinery is shown in Figure 21.9. The instrument can measure displacement in mils (thousandths of an inch), velocity in ips (inches per second), and acceleration in g's (one $g \approx 32.2$ ft/s^2).

When troubleshooting excessive vibration in rotating machinery, first check the mounting bolts, couplings, and foundations of the motor and the driven equipment. A loose mounting bolt can sometimes be detected by placing a finger at the parting line where the motor feet are bolted to the bedplate or deck. If this fails to indicate the trouble, and the bearings do not have excess play, uncouple the motor from the driven equipment and run a vibration test on the motor alone; if the motor vibrates it is proba-

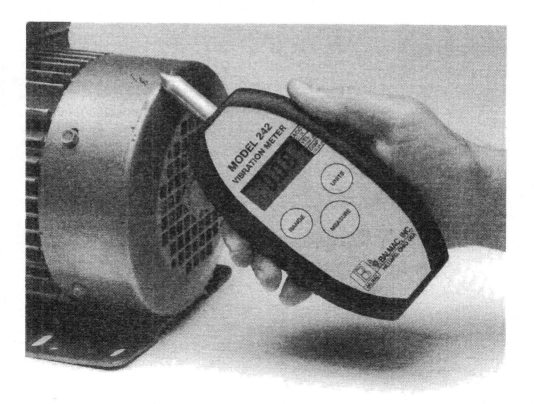

FIGURE 21.9
Pocket vibration meter (courtesy Balmac Inc.).

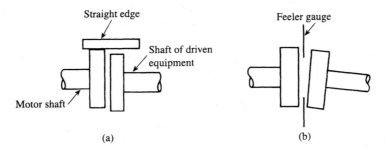

FIGURE 21.10
(a) Checking horizontal and vertical alignment; (b) checking axial alignment.

bly out of balance.[3] If the motor does not vibrate when running alone, the vibration may be in the driven equipment or caused by shaft misalignment. Figure 21.10 illustrates an emergency method for checking shaft alignment using only a straightedge and a feeler gauge.

21.6 SHAFT CURRENTS AND BEARING INSULATION [2]

Machines with very strong magnetic fields often generate a voltage in the shaft that causes current through the bearings. The resultant electric arcs and electrolytic action in the bearing will in time cause bearing failure and possible sludging of the lubricating oil. To eliminate shaft currents, either the outboard bearing shell is insulated from the housing, or one outboard bearing pedestal is insulated from the bedplate as shown in Figure 21.11.

---

[3] Electrical faults in the windings of motors or generators can also cause vibration. See Reference [2].

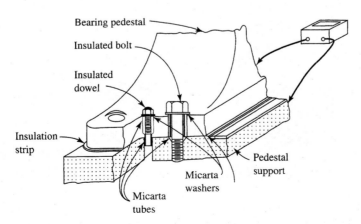

FIGURE 21.11
Insulated bearing pedestal (courtesy TECO Westinghouse).

Only the outboard bearing shell or outboard bearing pedestal is insulated. Insulating both bearings would permit the buildup of electrical charges (static electricity); the resultant high voltage would be hazardous to operating personnel and could cause a breakdown of the machine insulation. All lube-oil piping connections, exciter bearings, etc., that make connection to the shaft or bearing in question must also be insulated. The shell or pedestal insulation should be kept clean and should never be painted with metallic or other conducting paints.

The pedestal insulation of a *disassembled* machine can be tested with a 500-V megohmmeter. An insulation resistance of 500,000 Ω or greater, measured between the pedestal support (bedplate) and the bearing pedestal, as shown in Figure 21.11, is an indication of satisfactory insulation. The pedestal-insulation resistance of an assembled machine cannot be checked with an ohmmeter or megohmmeter, because the parallel path provided by the uninsulated pedestal will result in an erroneous indication.

Grounding Brushes

Electrostatic charges built up within the steam turbine of a turbine generator can be discharged with grounding brushes placed anywhere on the shaft except at the outboard bearing of the generator; the outboard end of a generator is the insulated end and must not be grounded.

SPECIFIC REFERENCES KEYED TO THE TEXT

[1] Bonnett, A. H., Cause and Analysis of Anti-Friction Bearing Failures in AC Induction Motors, *IEEE Industry Applications Society Newsletter,* September/October 1993.

[2] Institute of Electrical and Electronic Engineers, *Guide for Operation and Maintenance of Turbine Generators,* ANSI/IEEE Std 67–1995, IEEE, New York, 1995.

REVIEW QUESTIONS

1. Describe the correct procedure for disassembling a DC machine equipped with the cartridge type of bearing housings.
2. What special care should be taken when disassembling explosion-proof machines?
3. What is the purpose of a spreader bar?
4. State the correct procedure for reassembling an overhauled induction motor.
5. Outline the correct procedure for cleaning and lubricating ball bearings on motors equipped with pressure relief systems.
6. State the correct procedure for cleaning and lubricating sleeve bearings.
7. How can sleeve-bearing wear be detected without removing the bearing?
8. State the correct procedure for replacing worn sleeve bearings. What is the purpose of the felt washer?

9. State the correct procedure for removing a defective ball bearing and replacing it with a new one.

10. What are the principal causes of excessive vibration in electrical machinery? State the procedure that should be followed when troubleshooting a vibration problem.

11. What causes shaft currents, and why are they harmful?

12. Outline a procedure for checking the condition of bearing insulation: (a) prior to assembly; (b) during operation.

13. Under what conditions are both a grounding brush and pedestal insulation required?

22

Operation and Maintenance of Motor Controllers

22.0 INTRODUCTION

At the push of a button or the closing of a master switch, a motor controller will automatically start a machine, limit its current, provide uniform acceleration, and stop it. In addition to these basic functions, motor controllers can be used to limit torque, limit speed, change speed, change direction of rotation, and even apply dynamic braking. Motor controllers literally take the guesswork out of motor starting.

Motor controllers fall into two general categories: magnetic motor controllers (which use electromechanical devices with many moving parts) and solid-state controllers (which use electronic devices with very few moving parts).

This chapter provides a detailed look into the general construction, operation, and maintenance of motor controllers. In addition, the chapter provides a section on how to read and interpret control diagrams, which can be a very useful aid during troubleshooting. A comprehensive troubleshooting chart for magnetic starters and multispeed controllers is incorporated at the end of the chapter.

Although many new motor controllers are solid state, the majority of controllers in use today are magnetic; they will not be replaced for many years, and most will be in need of some form of maintenance. Maintenance requirements for solid-state control are primarily those of preventing overheating by removal of dirt, changing filters, checking cooling systems, inspecting for evidence of component deterioration, and replacement of defective components.

22.1 MAGNETIC CONTROLLER COMPONENTS

Magnetic controllers are composed of magnetically operated switches (called contactors), magnetic and thermal relays, and electrical and mechanical interlocks. These are used in conjunction with pushbuttons, selector switches, master switches, float switches, pressure switches, etc., as required by design considerations.

Magnetic Contactors

Figure 22.1a illustrates a cutaway view of a single-pole magnetic contactor used in a DC controller. It is a magnetically operated switch that serves to close or open an electric circuit. The assembly consists of stationary and movable contacts, operating coil,

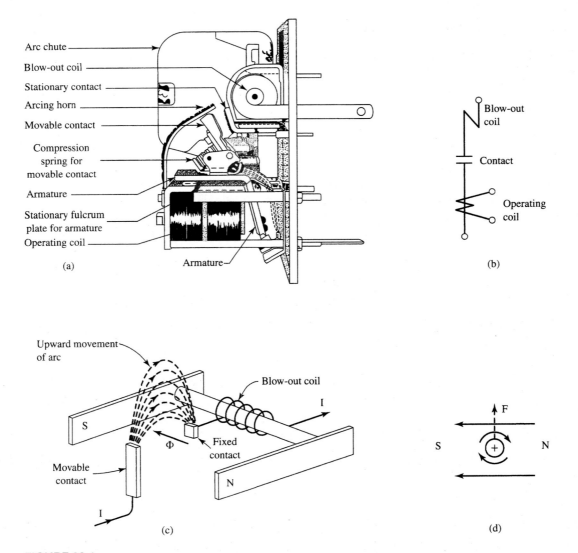

FIGURE 22.1
Magnetic contactor: (a) cutaway view of DC contactor (courtesy GE Industrial Systems); (b) elementary diagram; (c) behavior of the magnetic blowout; (d) interaction of magnetic fields.

armature, springs, mounting arms, blowout coil, arcing horn, and brackets. Energizing the operating coil sets up a magnetic field that attracts the armature, causing the movable contact to move up and make contact with the corresponding stationary contact.

The blow-out coil is connected in series with the stationary contact and provides a magnetic flux to "blow" the arc up the chute, where it is extinguished by elongation and cooling when the contacts are opened. Wear on the stationary and movable contacts is reduced by the arcing horn, which takes the brunt of the burning. The blow-out coil shifts the arc to the arcing horn and the upper curved part of the stationary contact, where it is elongated and extinguished. Figure 22.1b is an elementary diagram showing the contacts, operating coil, and blow-out coil. Note that the blow-out coil is not shown in all controller diagrams.

Figure 22.1c illustrates the behavior of the magnetic blow-out. The current through the blow-out coil produces north and south poles in steel plates that are mounted on the exterior sides of the arc chute. When the contacts separate, an arc is established in a direction perpendicular to the pole flux. The interaction of the magnetic flux produced by the arc with the pole flux produced by the blow-out coil pushes the arc upwards, stretching it until it breaks; this is shown in Figure 22.1d, where flux bunching under the arc causes an upward force on the arc.[1]

A three-pole alternating-current contactor, shown in Figure 22.2a, is also equipped with arcing horns and blow-out coils. However, the magnetic material of the armature and the magnetic core of the coil are laminated to reduce eddy currents, and a pole shader, shown in Figure 22.2b, is used to prevent the magnetic pull from dropping to zero each time the current wave goes through zero.

The pole shader or shading coil consists of a copper ring that surrounds a section of the iron core and has no external connections. The shading coil behaves as the short-circuited secondary of a transformer whose primary is the operating coil. In accordance with Lenz's law, the shading coil causes the flux in the shaded part of the pole face to lag behind the flux in the nonshaded part. This prevents the flux in the armature from falling to zero and thus reduces armature chatter; the flux in the shaded part of the pole piece holds the armature in the sealed position until the flux in the unshaded part builds up again in the opposite direction.

Thermal-Overload Relays

Excessive current drawn by an overloaded motor will, if allowed to continue for sufficient time, result in dangerous overheating of both motor and control. Correctly sized motor-overload relays with the proper time-current characteristics provide protection against sustained overloads, and yet permit the short-duration, high locked-rotor current necessary for motor starting. They also protect against overheating due to overcurrents caused by low-voltage, low-frequency, unbalanced voltages, and some other types of motor faults. Although overload currents are of relatively low magnitude compared with short-circuit currents, if sustained they will shorten the life of the insulation. Motor-overload relays do not protect against short circuits or

---

[1] See Sections 3–2 through 3–6 in Chapter 3 for interaction of magnetic fields.

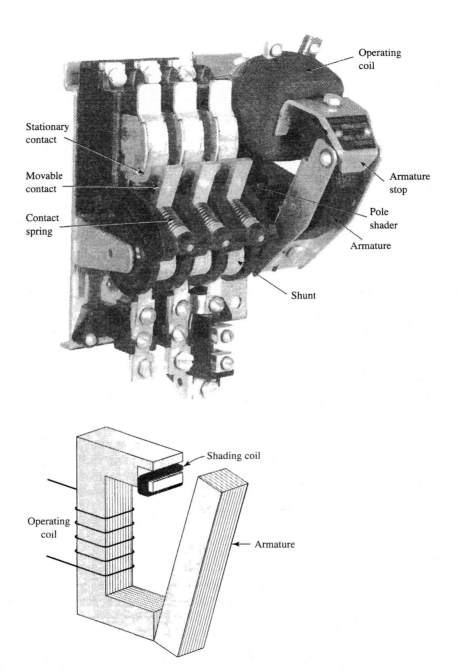

Operating
coil

Stationary
contact

Movable
contact

Contact
spring

Armature
stop

Pole
shader

Armature

Shunt

Shading coil

Operating
coil

Armature

FIGURE 22.2
(a) Alternating current contactor; (b) pole shader surrounding a section of the iron core
(courtesy GE Industrial Systems).

grounds. Protection against short circuits and grounds must be provided by branch-circuit protection devices such as fuses or circuit breakers.

A thermal overload relay (also called overcurrent relay) is shown in Figure 22.3(a). It consists of a bimetallic element, a heater element, normally closed contacts, normally open contacts, and a reset arm or button.

The bimetallic element is formed from two strips of dissimilar metals with different coefficients of linear expansion that are bonded together to form a single element. The heater is connected in series with the motor and simulates the I^2R heating of the motor windings. Should an overload occur, the increase in heater temperature caused by the greater-than-normal motor current will (in time) cause the bimetallic element to deflect. This opens the normally closed contacts, which de-energizes the motor, and closes the normally open contacts, which energizes an alarm circuit.

The reset arm is used to manually reset the relay contacts after the bimetallic element has cooled. Figure 22.3b shows how the heater and associated normally closed contacts are connected into a controller circuit. Controllers for large motors use current transformers (CTs) to provide proportional but lower values of current to a proportionally lower rated heater; this is shown with broken lines in Figure 22.3b. An overload of sufficient duration causes normally closed contact OL to open. This de-energizes coil M, causing the three normally open M contacts to fall to the open position.

The approximate current-time characteristics for tripping and resetting the bimetallic relay are shown in Figure 22.3c. Note that the tripping time decreases with increasing current (inverse-time characteristic). However, the reset time increases with increasing current. For the characteristics shown, an overload of 400 percent rated heater current causes tripping somewhere between 20 and 30 seconds. The resetting time for the specific family of relays, however, is roughly between 75 and 140 seconds, during which time the motor is at rest and cooling.

The overcurrent relay shown in Figure 22.3a is provided with an adjustable setting to permit tripping at approximately 15 percent above or 15 percent below the heater rating.

22.2 BASIC STARTING CIRCUITS

The two basic types of starting circuits are undervoltage release and undervoltage protection. Undervoltage-release circuits stop the motor when a power failure occurs and automatically resequence the control for automatic restarting when power is restored. Undervoltage-protection circuits protect against automatic restart, and must be manually restarted when power is restored.

Undervoltage-release circuits are used in vital applications such as sump pumps and fire pumps where it is imperative that the restart be automatic. Circuits with undervoltage protection are used in those applications where the unexpected start of a machine or drive (upon restoration of voltage) would present a hazard to operating personnel.

If all motors in a plant used undervoltage release, their simultaneous restarting when voltage is restored may overload the generator or distribution system, blacking out the plant. Hence, whenever possible, motors that are not vital to the operation of the plant should have their control circuits arranged for undervoltage protection.

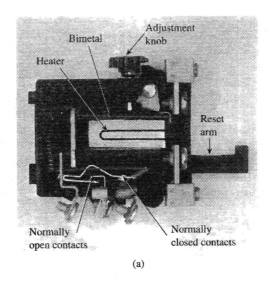

(a)

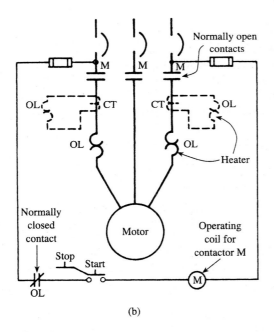

(b)

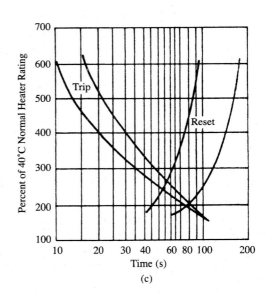

(c)

FIGURE 22.3

Thermal overload relay: (a) cutaway view; (b) circuit diagram; (c) current-time characteristic for tripping and resetting (courtesy GE Industrial Systems).

Undervoltage Release (UVR)

The circuit in Figure 22.4a provides undervoltage release. The "start" button that is used in undervoltage release circuits is a maintaining type of pushbutton that remains depressed when pushed. Depressing the "stop" button causes the "start" button to come back to its original position and vice versa.

Pushing the "start" button in Figure 22.4a energizes operating coil M, causing the normally open contacts M to close and the motor to start. If a power failure occurs, coil M is de-energized, the M contacts open, and the motor stops. When power is restored, the motor will automatically restart; it is not necessary to repush the "start" button, because it remains in the start position after pushing.

Undervoltage Protection (UVP)

The circuit shown in Figure 22.4b has the same power circuit as in Figure 22.4a, but its control circuit provides undervoltage protection instead of undervoltage release. The basic difference is in the type of pushbutton used. The UVP circuit uses momentary-contact buttons with spring-return to the normal position. The "start" button is held in the open position by a spring, and the "stop" button is held in the closed position by a spring. Thus the "start" button is *normally open*, and the "stop" button is *normally closed*.

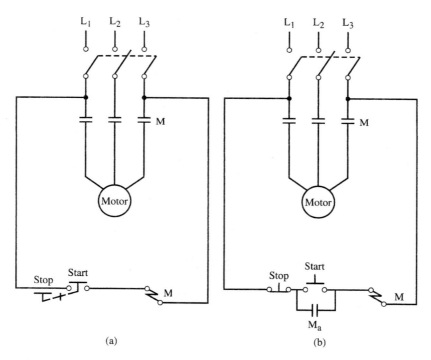

(a)   (b)

FIGURE 22.4
Basic starting circuits: (a) undervoltage release; (b) undervoltage protection.

Referring to Figure 22.4b, pushing the "start" button energizes coil M, which closes normally open contacts M in the power circuit and closes normally open auxiliary contact M_a in the control circuit. Closing of contact M_a (called a sealing contact) provides a bypass around the "start" button, so that releasing the "start" button does not stop the motor. If a power failure occurs, coil M is de-energized, causing contacts M and contact M_a to open, and the motor stops. When power is restored, the motor does not automatically restart; it is necessary to push the "start" button again to restart the motor. A starter with undervoltage protection protects against automatic restarting.

Standardized graphic symbols most frequently used in motor control diagrams are shown in Figure 22.5. Switch, pushbutton, and contact symbols show the condition of the contacts when in the de-energized position, called the *normal position.*

22.3   READING AND INTERPRETING CONTROL DIAGRAMS

Control diagrams are divided into two general classifications, the connection diagram and the elementary diagram. The connection diagram, shown in Figure 22.6a, shows the actual physical location of the component devices, their relative size, and the position and location of each wire as viewed from the back or front of the panel. This diagram facilitates the location of the components if repairs or adjustments are necessary. The numbering of the different terminals in a controller diagram serves as a guide for determining points of equal potential and is of considerable assistance when troubleshooting; terminals with the same number should have like potentials irrespective of contactor position.

The power circuit, drawn with heavy lines in Figure 22.6a, shows the path of motor current via contacts, overload heaters, disconnect switches, etc. The control circuit, drawn with light lines, includes pushbuttons, operating coils, relays, etc., that provide the logic and commands for the operation of the power circuit.

The elementary diagram, shown in Figure 22.6b, is a simplified diagram that separates the power and control components. However, all components with the same letter designation are part of the same control device. Thus, the three M contacts in the power circuit, the M_a contact in the control circuit, and the M coil are all components of the magnetic contactor M shown enclosed by the broken lines in the connection diagram (Figure 22.6a). The elementary diagram is easier to read than the connection diagram, aids understanding of the operation of the controller, and is used extensively when troubleshooting the control system.

22.4   REVERSING CONTROLLERS FOR AC MOTORS

The elementary circuit diagram for a simple reversing controller using undervoltage protection is shown in Figure 22.7a. The "Rev" button has a set of normally closed contacts (2–3), and directly below it a set of normally open contacts (5–4). Pushing the "Rev" button opens normally closed contacts (2–3) and closes normally open contacts (5–4). Do not confuse pushbutton contacts (5–4) with auxiliary contact R_a (5–4); pushbutton contacts are always shown as small circles. Contact R_a (5–4) closes only when coil R is energized.

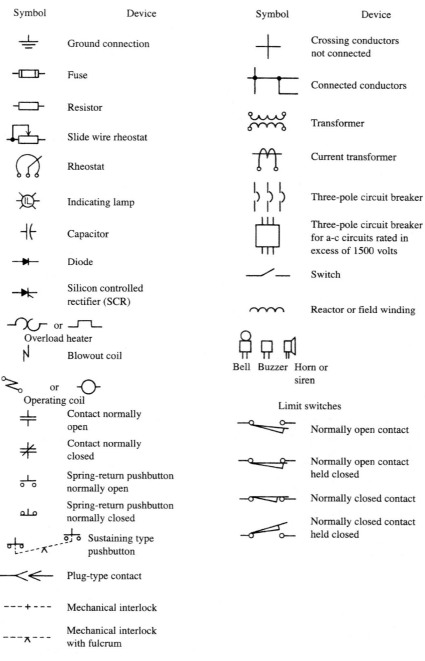

Symbol	Device	Symbol	Device
	Ground connection		Crossing conductors not connected
	Fuse		Connected conductors
	Resistor		Transformer
	Slide wire rheostat		Current transformer
	Rheostat		Three-pole circuit breaker
	Indicating lamp		Three-pole circuit breaker for a-c circuits rated in excess of 1500 volts
	Capacitor		Switch
	Diode		Reactor or field winding
	Silicon controlled rectifier (SCR)		Bell Buzzer Horn or siren
or	Overload heater		Limit switches
	Blowout coil		Normally open contact
or	Operating coil		Normally open contact held closed
	Contact normally open		Normally closed contact
	Contact normally closed		Normally closed contact held closed
	Spring-return pushbutton normally open		
	Spring-return pushbutton normally closed		
	Sustaining type pushbutton		
	Plug-type contact		
	Mechanical interlock		
	Mechanical interlock with fulcrum		

FIGURE 22.5

Standard symbols for control components.

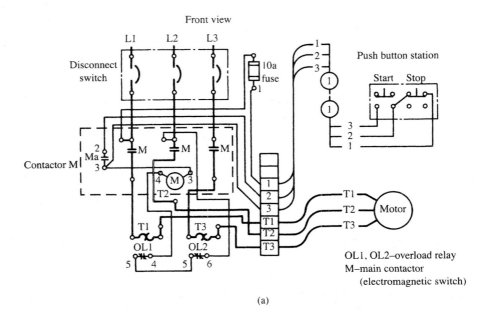

(a)

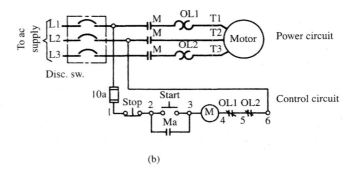

(b)

FIGURE 22.6
(a) Connection diagram; (b) elementary diagram.

The double pushbutton arrangement for each of the "Rev" and "Fwd" pushbuttons provides protective interlocking to prevent energizing both the R and F coils at the same time. Closing of the R and F contacts at the same time would short-circuit the connecting lines between the disconnect switch and the motor. In addition, a mechanical interlock (not shown) is generally provided to ensure that both contactors cannot be mechanically closed at the same time.

A reversing controller with undervoltage release is shown in Figure 22.7b. The three-position switch is snapped to positions "Fwd," "Stop," or "Rev" as desired. Switching the control to "Fwd" energizes the F coil, which closes the F contacts in the motor circuit starting the motor in the forward direction.

FIGURE 22.7

(a) Reversing controller using undervoltage protection; (b) reversing controller with undervoltage release.

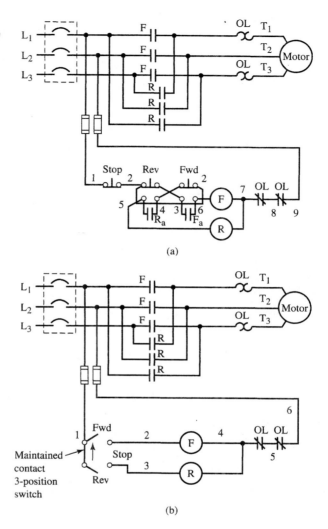

(a)

(b)

22.5 TWO-SPEED CONTROLLERS FOR AC MOTORS

The elementary circuit diagram for a two-speed *two-winding* induction-motor controller is shown Figure 22.8a. Note that each winding has its own overload protective device, indicating that the power and current ratings are different for the two speeds.

Sequence interlocks in series with each operating coil prevent the HS contactor from operating while the LS contactor is energized, and vice versa. The sequence interlocks are LS_{6-7} and HS_{10-9}. Pushing the high-speed button energizes the HS coil, which closes sealing contact HS_{5-6}, opens sequence interlock HS_{10-9}, and closes the three HS contacts in the power circuit.

Figure 22.8b shows the control circuit for a two-speed *single-winding* consequent-pole induction motor. Note that the control circuit is identical to that for the two-winding

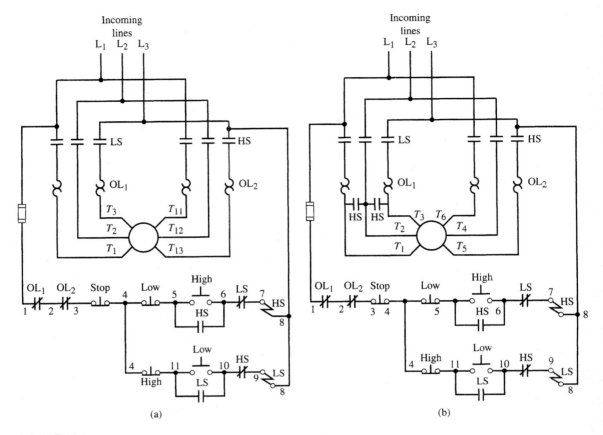

FIGURE 22.8
Two-speed controllers for AC motors: (a) two-speed two-winding motor; (b) two-speed single-winding motor.

motor shown in Figure 22.8a, but the power circuit connections are different. For the single-winding motor in Figure 22.8b, at high speed, contactor HS connects terminals T_1, T_2, and T_3.

22.6 AUTOTRANSFORMER STARTING

The circuit diagram for a typical three-phase, reduced-voltage starter using an autotransformer is shown in Figure 22.9. The voltage taps are selected on the basis of the starting requirements. They are permanently connected and are not adjustable by the operator.

Timing relay TR is used to permit starting at reduced voltage for a preset time, and then to connect the machine across full voltage. The timing relay has one normally open contact (TR_{4-5}), two normally closed, time-opening contacts (TR_{5-6} and TR_{5-10}), one normally open, time-closing contact (TR_{5-8}), a timing motor (TM), and an operating coil (TR_{5-7}). Note that the letters "TO" associated with a contact denote time-delay opening, and the letters "TC" denote time-delay closing.

FIGURE 22.9
Reduced-voltage starter using an auto-transformer (courtesy GE Industrial Systems).

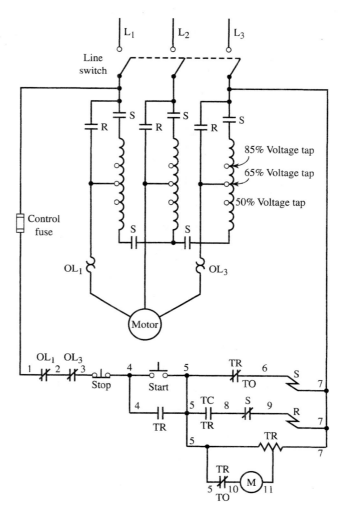

When the start button is pushed, the current in the control circuit takes three parallel paths: through coil S to energize the autotransformer, starting the motor; through the timing motor TM to time the starting operation; and through relay coil TR.

When the five S contacts in the power circuit close, they connect the autotransformer in wye, and the motor to the 65 percent tap. With TR energized, contact TR_{4-5} closes, sealing the starting circuit so that the start button may be released.

After a time delay of 1 to 10 seconds, adjustable, the timing motor causes the time-delay contacts to operate. Contact TR_{5-6} opens, contact TR_{5-8} closes, and contact TR_{5-10} opens. Coil S is de-energized, causing the five S contacts in the power circuit to open, and contact S_{8-9} in the control circuit to close, energizing coil R. This closes running contacts R, connecting the motor across full voltage. Opening of contact TR_{5-10} de-energizes the timing motor, which is no longer needed.

22.7 AC HOIST CONTROL

Figure 22.10 illustrates a reversing magnetic controller for a wound-rotor induction motor used to drive a hoist. A cam-operated master switch energizes relays and contactors in a prescribed manner to provide five hoisting and five lowering speeds. Changes in speed are accomplished by changing the amount of resistance in the rotor circuit (secondary resistors).

The X marks on the elementary circuit diagram for the master switch designate the cam-operated contacts that are closed for that particular position of the master switch, regardless of whether it was open or closed when in the off position. Thus, for any position of the master switch, an X mark associated with an MS contact indicates it to be closed, and the absence of an X mark indicates that the contact is open. For example, in position 2-Hoist, cam-operated contacts MS2, MS4, and MS5 are closed; in position 4-Lower, contacts MS2, MS3, MS5, MS6, and MS7 are closed. Cam-operated contact MS2 is used if an electromechanical brake is used.

With the master switch in the off position, there will be a path of current from power line L_3 to power line L_1, through the control switch, cam-operated contact MS1, overload-relay contact OL_{2-3}, and undervoltage relay coil UV; the energized UV coil causes contacts UV_{1-2} and UV_{4-8} to close.

Moving the master switch to position 1-Hoist establishes a path of current from line L_1 through contact UV_{1-2}, contact MS4, hoist limit switch LSH, normally closed contact L_{10-11}, and hoist coil H; the energized H coil closes contacts H_{10-12} in the control circuit and the H power contacts in the power circuit. Closing of the H power contacts causes the motor to run in the hoist direction, with all the resistance in the rotor circuit.

Relays 1T, 2T, and 3T are time-delay-closing relays; when the respective operating coils are energized, the relay contacts close after a predetermined time has elapsed. The time-delay relays protect the motor against such poor operating practices as rapidly throwing the master switch lever from off to 5-Hoist, or from 5-Lower to 5-Hoist. This is accomplished by delaying the operation of contactors 2A, 3A, and 4A.

Throwing the master switch rapidly from off to 5-Hoist closes cam-operated contacts MS2, MS4, MS5, MS6, MS7, and MS8. Cam contact MS4 energizes coil H, which establishes the direction of rotation, and also closes H_{10-12}, starting the following accelerating sequence:

1. Coil 1A is energized, closing the two 1A power contacts at the secondary resistor, causing the motor to attain a higher speed. Coil 1A also closes contact $1A_{8-14}$, energizing coil 1T.
2. After a time delay, contact $1T_{16-14}$ closes, energizing coil 2A. Coil 2A closes the two 2A power contacts at the secondary resistor, shorting out another block of resistance, causing additional acceleration. It also closes contact $2A_{14-19}$, energizing coil 2T.
3. After a second time delay, contact $2T_{18-19}$ closes, energizing coil 3A. Coil 3A closes the two 3A power contacts at the secondary resistors, causing additional acceleration. Coil 3A also closes contact $3A_{19-22}$, energizing coil 3T.

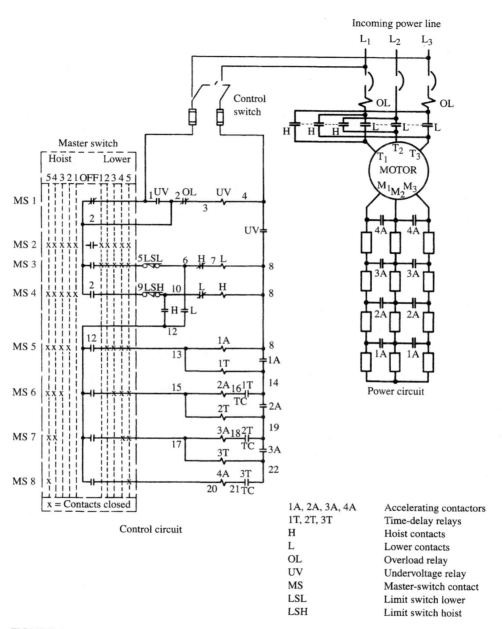

FIGURE 22.10
Reversing magnetic controller for a wound-rotor motor.

1A, 2A, 3A, 4A	Accelerating contactors
1T, 2T, 3T	Time-delay relays
H	Hoist contacts
L	Lower contacts
OL	Overload relay
UV	Undervoltage relay
MS	Master-switch contact
LSL	Limit switch lower
LSH	Limit switch hoist

4. After a third time delay, contact $3T_{21-22}$ closes, energizing coil 4A. Coil 4A closes the two 4A power contacts, shorting the remaining resistance, and the motor then accelerates to its full speed.

22.8 DC MOTOR CONTROLLER

The elementary diagram shown in Figure 22.11 is that of a controller for a DC motor driving a forced draft fan used in a power plant. The motor can be started from three different places and stopped from four places.

Pushing any one of three start buttons energizes coil AU, which closes sealing contact AU_{5-6}, closes contact AU_{6-8}, and opens contact AU_{6-12}. The closing of contact AU_{6-8} energizes coil LE, which closes contact LE, starting the motor with resistor R_1-R_4 in series with the armature. As long as the starting resistance is in series with the armature, a voltage will appear across coil FF, which closes contact FF, shorting the field rheostat; the resultant high shunt-field flux ensures a high starting torque. Opening of contact AU_{6-12} reduces the current in coil AU making it more sensitive to reduced voltage.

As the motor accelerates, the contacts on the accelerating unit, in time-delay sequence, short out the starting resistor, accelerating the motor to its base speed. When the last block of resistance is shorted, the voltage across coil FF is zero, and contact FF across the field rheostat opens, causing further acceleration.

Weakening the shunt-field flux by turning the field rheostat in a clockwise direction decreases the cemf in the armature, causing an increase in armature current, which accelerates the motor. As the motor increases in speed, the buildup of cemf in the armature limits the magnitude of the armature current.

Changes in field rheostat position should be made gradually. A quick turn of the field rheostat in the direction to increase speed will cause a large reduction in armature cemf, resulting in excessive armature current; damage to the commutator and brushes may occur, and the circuit breaker may trip. To protect the armature from excessive current brought about by rapid field weakening, the manufacturer provided a field accelerating relay (FA), shown in Figure 22.11. The FA coil is connected in series with the motor armature so that high values of armature current will actuate the relay, closing the FA contact and shorting out the field rheostat. When the high value of armature current subsides, the FA contact opens, and the shunt-field rheostat is back in the circuit; the pickup (PU) and dropout (DO) values of current are listed in the nomenclature table in Figure 22.11.

The control relay CR is used to energize circuits that light signal lamps and actuate an electrically operated fuel valve.

22.9 PROCEDURE FOR TROUBLESHOOTING MAGNETIC STARTERS AND MULTISPEED CONTROLLERS

Whenever a motor fails to operate, the main line fuses, circuit breaker, and control fuses should be tested for continuity. Opens do not generally occur in the wiring of a control panel. Most breaks in the control circuit can usually be traced to opens in the

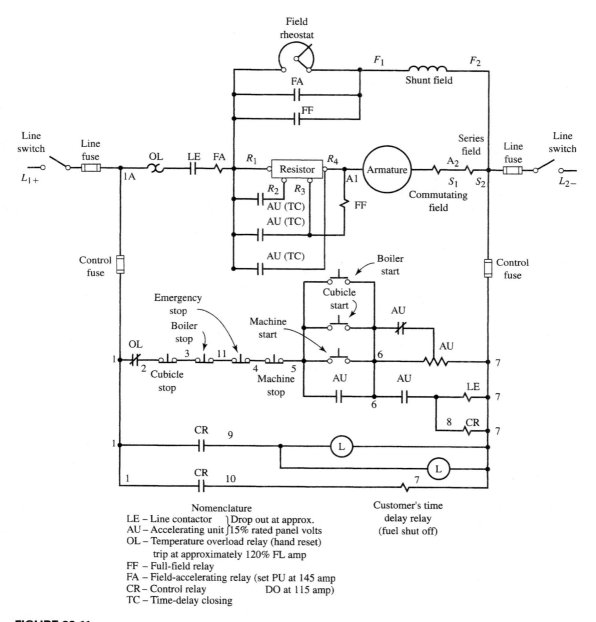

FIGURE 22.11
Elementary diagram of a DC controller for a forced draft fan in a power plant (courtesy GE Industrial Systems).

operating coils, to corroded dirty contacts (in the master switch or other contacts), or to pushbuttons that are in series with the operating coil.

To check the master switch and operating coils, the motor should be disconnected and the master switch operated in all positions.[2] If each contactor and relay operates properly for its respective master switch position, as indicated on the manufacturer's circuit diagram, the trouble is in the motor, transformer, or resistor bank. The resistor bank should be checked for signs of overheating, loose connections, or broken resistors. The motor should be checked for opens in the field and armature circuits.

If a contactor or relay does not close or open when the position of the master switch indicates that it should, the trouble may be with either the master switch contacts, an operating coil, an auxiliary contact in series with the coil, or a loose connection.

The voltage across the coil should be checked with a voltmeter. If it has rated voltage but no magnetic pull, the coil is open. However, if the voltmeter registers zero, it indicates that the master switch contacts or the auxiliary contacts in series with the coil are open or have a high-resistance connection.

Shorted coils are often indicated by discolored or charred insulation and by blown fuses in the control circuit.

Troubleshooting Multispeed Controllers

Regardless of the apparent complexity of schematic diagrams, learning how to read and interpret them is an essential part of speedy troubleshooting and repair. The following suggestions should prove helpful in troubleshooting the more complex problems posed by multispeed controllers that use a master switch along with contactors and relays, such as the controller shown in Figure 22.10:

1. Familiarize yourself with the elementary diagram by locating the contacts that are associated with each operating coil.
2. From the symptoms of motor behavior, determine which positions of the master switch are involved in the fault.
3. Using the elementary diagram as a guide, make a check list of the operating coils and contacts, including the master switch contacts, that should be energized for each master switch position in question.
4. Using the check list as a reference, operate the master switch and determine whether these contactors or relays are functioning properly. Correct or replace the faulty ones. When testing the control, the motor should be disconnected. This can be done by disconnecting the motor at the control terminals.
5. If all contactors and relays appear to function normally and the supply voltage is correct, the trouble must be in a resistor bank if DC, or autotransformer if AC, or in the motor.

---

[2] Some controllers have an easily removed test link that, when removed, disconnects the armature circuit.

Troubleshooting Chart

The troubleshooting chart given in Table 22.1 lists the most common troubles and their possible causes.

TABLE 22.1

Troubleshooting Chart for Magnetic Starters and Multispeed Controllers

Trouble	Possible Cause
Contact buzz (loud buzzing noise)	1. Broken shading coil (AC) 2. Misalignment of magnet faces (AC) 3. Dirt on magnet faces (AC) 4. Low voltage (AC)
Contact chatter (make-and-break action, staccato noise)	1. Poor contact in series with operating coil 2. Fluttering of a relay in the control circuit 3. Broken pole shader (AC)
Excessive burning of contacts	1. Excessive current (overload) 2. Weak spring pressure 3. Oxidized contacts 4. Poorly bolted connection 5. Arcing horn needs replacement
Welding of contacts	1. Excessive and rapid jogging 2. Low spring pressure 3. Excessive current (overload) 4. Low voltage on operating coil 5. Bouncing of contacts
Frequent tripping of thermal OL relay	1. Overload 2. Wrong size of trip coil or heater 3. Excessive ambient temperature (thermal relays)
Burned-out operating coil	1. Overvoltage 2. High ambient temperature 3. Shorted turns 4. Undervoltage (AC) 5. Dirt on magnet faces (AC) 6. Excessive jogging (AC)
Failure of contactor or relay to pick up	1. Low voltage 2. Open coil 3. Excessive magnet gap 4. Mechanical binding 5. Open circuit in series with operating coil
Failure of contactor or relay to drop out	1. Welded contacts 2. Improper adjustment 3. Accumulation of gum and other foreign matter 4. Misalignment 5. Voltage maintained at coil because of failure of contacts in series with operating coil

EXAMPLE 22.1 The motor in Figure 22.8a runs at high speed but does not operate at low speed. Furthermore, none of the contactors operate at low speed. What are the probable sources of trouble?

Solution
The most probable sources of trouble are:

1. Open in LS coil.
2. Open at contact HS(10–9).
3. The "High" (4–11) button is defective.
4. The "Low" (11–10) button is defective.

22.10 MAINTENANCE OF MAGNETIC CONTROLLERS

Control devices are essential to the successful control of electric motors, but, unfortunately, they are too often neglected. Failure of the control device can result in poor performance, failure to start, and in many instances destruction of the machine itself. Failure of an overload trip can burn out the motor; failure of a shunt-field contactor can accelerate a machine to destruction; failure of the main propulsion control of an electric-drive ship can place the ship at the mercy of the elements; etc.

Figure 22.12 illustrates the extensive damage that can occur when protective devices fail. The motor shown was a 500-hp, 600-volt, 600-rpm compound DC motor driving a strip mill in a steel plant. The circuit breaker shunt-trip coil (located within the circuit breaker and used for remote tripping) developed trouble and was disconnected to avoid production loss; it was to be replaced at the next scheduled shutdown. The control for this motor was equipped with an overspeed trip and a shunt-field failure relay, both of which were connected in series with the shunt-trip coil of the circuit breaker. Disconnecting the trip coil invalidated these protective features. Unfortunately, before the scheduled shutdown occurred, the shunt field lost its excitation and the machine accelerated to destruction. The motor had to be replaced and the mill was out of production for an extended period.

Protective devices are vital to equipment operation and safety and their function should be tested periodically. Examples of these protective devices include circuit breaker trips, overspeed trips, and loss-of-field relays. If found defective, they should be repaired or replaced and properly adjusted for the specific application. Under no circumstances should protective equipment be bypassed or made inoperative; to do so is to invite disaster.

To provide for the proper maintenance of control equipment, a schedule should be arranged for periodic inspections. The frequency of inspection should be dictated by usage. Controllers should be checked for loose or worn contacts, weak spring pressure, displaced or burned arc shields, defective coils, low insulation resistance to ground, corrosion, etc.

With the power off, check the mechanical operation of each contactor and relay by hand for freedom of motion. Check for loose pins, loose bolts, and loose wires. A

FIGURE 22.12
DC motor destroyed by overspeeding when a control circuit failed (courtesy Mutual Boiler and Insurance Co.).

gentle tug on a wire where it connects to a relay coil or contactor will indicate whether or not it is loose. Discolored insulation or discolored copper at connections is a sure indicator of loose connections.

An adequate supply of spare parts, as recommended by the manufacturer, should always be on hand for emergencies and preventive maintenance. Control equipment should be kept clean, dry, and in proper working condition. Electric strip heaters used to prevent condensation of moisture should always be connected to ensure against insulation failure and corrosion. Excessive vibration should be avoided.

Cleaning Contacts

Dirt and grease on contact surfaces increase the contact resistance and should be removed. The dust is easily removed with a vacuum cleaner, and the greasy film may be removed with a clean, dry cloth. Compressed air should be avoided, because it may force metallic dust, caused by contact wear, into the coil insulation. Clean delicate mechanical parts with a small stiff brush and an approved safety solvent.

Before any work is performed on control equipment, the circuit should be disconnected from the power lines and the fuses removed. A "PERSON WORKING ON LINE" sign should be fastened to the switch to prevent accidental closing.

Filing Contacts

Heavy copper contacts and cadmium-plated contacts should be inspected regularly and filed when they become badly pitted. A roughened butt contact can generally carry as much current as a smooth one and should be filed only when large projections caused by excessive arcing are evidenced. File carefully, with a smooth mill file, so that the contacts retain their original shape. Figure 22.13 illustrates badly worn butt contacts that require replacement. Contacts should never be allowed to wear to this extent.

Solid-silver contacts, used in relays and auxiliary control circuits, should not be filed unless sharp projections extend beyond the surface. The black silver oxide that forms on the surface should not be removed, because it is almost as good a conductor as the silver. When necessary, dress solid-silver contacts with a fine-cut file or grade 0000 sandpaper fastened to a flat stick. In some cases, fine sandpaper may be drawn between the contacts while they are held together with moderate pressure.

Silver-plated contacts should be dressed carefully with grade 0000 sandpaper and should be replaced if badly pitted. Filing or burnishing of silver-plated contacts should be avoided, because the plating may be damaged. Hard-alloy contacts that are pitted or corroded should be dressed with a burnishing tool instead of a file.

FIGURE 22.13
(a) The two contacts on the right are badly worn and should be replaced; (b) the two contacts on the left are still in good condition (courtesy United States Merchant Marine Academy).

After filing, burnishing, or sandpapering, the contacts should be wiped with a clean cloth moistened with cleaning fluid and then polished with a dry cloth.

Controller Lubrication

The unnecessary use of lubricant may impair the effective operation of the controller. The bearings of contactors and relays are generally designed for operation without a lubricant. Hence, they should not be lubricated unless specifically recommended by the manufacturer. Dust in the air adheres to the lubricant and forms a gum that may cause contactors or relays to stick. Butt-type contacts should never be lubricated. Sliding contacts, such as those used on rheostats and drum controllers, may be lubricated with a very thin film of electric-contact lubricant.

Burned-Out Coils

Operating coils used in magnetic contactors and relays operate satisfactorily between 85 and 110 percent of rated voltage. Excessive voltage (greater than 110 percent rated) will cause overheating and possible burnout of operating coils on both AC and DC controllers.

However, only the coils in AC controllers are affected by low voltage (less than 85 percent rated). When an AC contactor coil is energized, the magnetic field exerts an attractive force on the armature, pulling it to the magnet against the opposing force of a spring (see Figure 22.2). If the voltage is too low, the magnetic force may not be enough to seat the armature to the magnet surface. Inadequate seating will cause the coil to have a low inductive reactance, resulting in an excessively high coil current and a burned-out coil. Note that a burned-out coil can also occur at rated voltage if accumulated dirt prevents proper seating of the armature to the magnet.

Welding of Contacts

The welding of contacts is generally due to low contact pressure caused by low voltage or weak springs. Low operating voltage (less than 85 percent rated) causes a weak magnetic pull; the contacts do not seal properly, and a relatively high resistance occurs between the connecting surfaces. If the contact resistance is considerably greater than the resistance of the rest of the circuit, the heat generated at the contact surface may cause the contacts to weld together. Low spring force produces the same effect; although the magnetic pull of the coil may be very strong, if the spring is weak or improperly adjusted, the high contact resistance may weld the contacts. Welding of contacts may also be caused by excessively high in-rush currents at the instant the contacts close.

The spring force can be determined with the power off, using a spring balance and a thin strip of paper, as shown in Figure 22.14. Force is gradually applied, and the scale read at the instant the paper is released. The contactor should be operated mechanically by hand when the test is made.

FIGURE 22.14
Measuring contact spring pressure
(courtesy GE Industrial Systems).

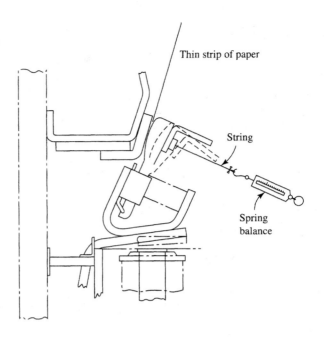

Thin strip of paper

String

Spring
balance

Resistors

Resistors failure can be caused by poor connections, excessive vibration, inadequate ventilation, overloading, defective insulation, and corrosion. Loose connections cause excessive heating and burning at junctions. Excessive vibration of cast grid types of resistors causes fractures to develop, and inadequate ventilation and overloads may result in burned-out units.

If some resistors in a bank of series-connected resistors have worn thin by corrosion or burning, it may be advisable to replace the entire bank rather than renew a few resistors. The reduction in cross-sectional area of worn resistors will cause the worn resistor to have a higher resistance than the new ones. Thus, placing new resistors in series with worn ones will cause the worn resistors to get hotter than the renewals, hastening their breakdown.

22.11 SOLID-STATE CONTROL OF MOTORS AND DRIVE SYSTEMS

Solid-state controllers use diodes, transistors, thyristors (silicon control rectifiers, SCRs), triacs, and other solid-state devices in different configurations to start, stop, reverse, brake, provide soft starting, and adjust the speed of electrical machinery. Because solid-state devices have no moving parts, they require considerably less maintenance than their magnetic counterparts. Furthermore, the absence of arcing or sparking makes them attractive for applications in explosive atmospheres. Solid-state devices can be used effectively and efficiently for the control of machinery from fractional horsepower units to tens of thousands of horsepower.

Programmable Logic Controllers (PLCs)

The controllers discussed in Sections 22–3 through 22–9 are hard-wired logic controllers programmed for a specific application. Any change in the programming requires physically adding or subtracting relays and rewiring as required for the new application. A programmable logic controller as defined by NEMA is a "digitally operating electronic apparatus that uses a programmable memory for internal storage of instructions for implementing specific functions such as logic, sequencing, timing, counting and arithmetic, to control machines or processes through digital or analog input and output modules."

Programmable controllers were developed in the 1970s to replace hard-wired control panels in the auto industry, and their success led to their gradual adoption in virtually all industries with large complex control systems that must be modified periodically to meet changes in production. The principal advantage of PLCs over hard-wired relay panels is the ability to program and reprogram relay logic through a computer terminal instead of manually wiring, rewiring, or replacing hard-wired panels. The programmable controller uses electronic components (solid-state relays) that simulate electromechanical relays, and it can have any number of electronically simulated normally open and normally closed contacts.

Programmable controllers are designed to withstand the high temperature, humidity, vibration, electrical noise, and power interruptions generally encountered in industrial environments. They are commercially available in small sizes ranging from 50 to 150 relays to 500 to 3000 relays in the larger sizes.

The principal parts of a programmable logic controller are shown as components of a block diagram in Figure 22.15. The central processing unit (CPU) contains logic memory, storage memory, power supply, and processor. The CPU receives data from input devices, performs logical decisions based on a stored program, and operates output devices to perform the desired functions.

The solid-state input/output (I/O) interface provides optical isolation to isolate the I/O voltages of field devices from the CPU and also helps reduce electrical noise. Input field devices include start and stop buttons, limit switches, sensors, and other command devices. Output field devices include such items as motors, magnetic contactors, electric heaters, lights, and electrically operated valves.

The programming keyboard and monitor ("TV" screen) are used to enter the desired relay logic. Programming a PLC does not require any knowledge of computers, computer language, nor any programming skills. The required control system for a particular application is drawn by the design engineer as a ladder diagram similar to the control diagram shown in Figure 22.10. The technician or plant engineer types the diagram into the CPU using a special keyboard.

The required number of relay contacts and their configuration, such as normally open, normally closed, series, parallel, etc., are typed into the computer memory. Then when changes are required, the computer logic can be easily reconfigured from the

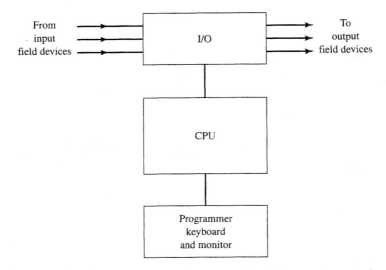

FIGURE 22.15
Basic components of a programmable logic controller.

keyboard; contacts can be added, erased, reconnected, etc. The number of contacts associated with a specific relay coil is limited only by the available memory.

For detailed descriptions, diagrams, and photographs of PLCs, their mode of operation, and programming techniques, see References [1] and [2].

Maintenance of Solid-State Control

The maintenance of solid-state controllers is relatively minor compared to the maintenance of magnetic controllers. However, certain housekeeping requirements are necessary to maintain satisfactory operation of static controllers; they must be kept clean, cool, and free from excessive vibration and mechanical shock. They must also be protected against voltage surges.

Overheating, which can cause breakdown of solid-state components, can be caused by inadequate ventilation, overloading, failure of fans, dirty filters, dirty heat sinks, or operating in an ambient temperature that is too high for the specific controller.

Severe vibration and mechanical stress can cause failure by loosening or breaking connections to the electronic components. The absorption of moisture from damp environments can cause corrosion of metallic components, and moisture in combination with dust deposits can cause short circuits.

Apparatus such as photoelectric devices, electronic speed controllers for motors, electronic battery chargers, electronic welding controls, and electronic heating, although different in circuitry, respond well to a carefully planned maintenance program. Electronic equipment should always be de-energized before servicing, because dangerously high voltages inherent in such apparatus may cause serious injury or death. All power capacitors should be discharged, and the circuits grounded to the chassis or metal housing.

Electronic equipment should be cleaned by vacuuming. This removes dust that might otherwise cause overheating by reducing the heat transfer characteristics of the components. Dust collected on high-voltage equipment can also result in short circuits. The screwed down or bolted wiring should be inspected for loose or broken connections; gently poking the wires with a clean, dry, wood stick assists in detecting such faults. Lenses, mirrors, and light sources on photoelectric apparatus should be cleaned as frequently as local conditions require.

Resistors and rheostats should be inspected for signs of overheating, and discolored ones replaced. Capacitors that start to drip wax or fluid or that are discolored by excess heat should be replaced. Transformers that overheat, smoke, or have melted sealing compound should be replaced. Relays and contactors should be checked for defective components, and the procedure outlined for the maintenance of magnetic controllers should be carefully adhered to.

The first thing to do when checking faulty electronic apparatus is to test all fuses and measure the line voltage. A volt-ohmmeter with a sensitivity of 1000 ohms per volt or a solid-state multimeter aids considerably in diagnosing troubles. The ohmmeter measurements of resistance should be checked against the values stipulated by the manufacturer in the wiring diagram or specification sheet. All questionable resistors should be replaced. The color code for small tubular resistors is tabulated in Appendix 4 and will be of assistance in the absence of manufacturer's specifications.

Although the preventive maintenance methods outlined above will reduce accidental shutdowns, they are not cure-alls. Electronic equipment is highly complex, and procedures for maintenance and troubleshooting should be performed as outlined by the manufacturer and performed by qualified maintenance personnel.

SPECIFIC REFERENCES KEYED TO THE TEXT

[1] Webb, J. W., *Programmable Controllers, Principles and Applications,* Merrill Publishing Co., Columbus, OH, 1988.
[2] Cox, R. A., *Technicians Guide to Programmable Controllers,* Delmar Publishers, New York, 1989.

REVIEW QUESTIONS

1. What is the function of a shading coil? How does it work?
2. What is the function of a magnetic blow-out? Explain how it works.
3. Explain the operation of a bimetallic type of motor-overload relay and state how it is connected in the control circuit.
4. Explain the basic difference in operation between UVR and UVP circuits. Sketch the circuits and state an application for each.

5. Sketch an autotransformer starting circuit, and state the sequence of operation when the start button is pushed.

6. Explain the sequence of operation of the controller shown in Figure 22.10 when the master switch lever is thrown rapidly from position 1 "HOIST" to position 4 "LOWER."

7. What do the X marks on the master switch indicate?

8. When inspecting motor controllers, what items in particular should be checked?

9. What are some of the disastrous consequences that can result from an inadequate controller maintenance program?

10. Outline the general approach you would follow when troubleshooting a motor controller.

11. State the correct procedure for cleaning contacts. What controller parts require lubrication?

12. State the correct procedure for dressing (a) heavy butt types of contacts; (b) solid-silver contacts; (c) silver-plated contacts; (d) hard-alloy contacts.

13. What conditions might cause contact surfaces to weld together?

14. Describe a method for determining the spring force on a normally closed butt contact.

15. What are some of the factors that can cause resistors to fail?

16. Can the renewal of one or more resistor units in a bank of resistors cause some of the remaining units in the same bank to burn out? Explain.

17. Referring to the control circuit shown in Figure 22.8a, what are the possible reasons for failure to start when the "Low" button is pressed?

18. Referring to the control circuit shown in Figure 22.6b, when the start button is pressed the motor runs, but when the button is released the motor stops. What are the possible faults?

19. What are the advantages of a solid-state starter over a magnetic starter?

20. Do solid-state starters require maintenance?

23

Classification, Characteristics, Aging, and Failure Mechanism of Electrical Insulation

23.0 INTRODUCTION

The insulation used on practically all electrical apparatus is partially composed of organic compounds that contain water as part of their chemical makeup. Excessive temperatures, which tend to dehydrate and oxidize the insulation, cause it to become brittle and disintegrate under vibration and shock. Thermoplastic materials, such as polyvinylchloride (PVC), soften and lose mechanical strength at elevated temperatures.

The insulation used in electrical apparatus does not last forever. It gradually deteriorates, slowly at low temperatures and more rapidly at higher temperatures. The greater the load, the higher the temperature and the shorter the life of the insulation. Thus, the question of how hot a machine may safely be operated can be answered only by how long a life is desired for the machine.

Economic factors, such as initial cost, replacement cost, obsolescence, and maintenance, are of prime importance when determining the years of useful service desired for the electrical insulation. The permissible horsepower and kilowatt rating of machines, as indicated on the nameplate, are determined and standardized by the allowable temperature rise dictated by economic reasons. Exceeding the load rating of the machine heats the insulation above the allowable limit and hastens its deterioration, whereas operating below the load rating prolongs its useful life.

23.1 MECHANISM OF INSULATION FAILURE

Deterioration of electrical insulation is directly dependent on age and usage. Machines operated under severe mechanical stress conditions such as frequent starts, reversals, pole slipping, overloads, voltage surges, or other poor operating techniques will have a shorter useful life than machines operating under light load with little mechanical or electrical stress.

Old insulation in an advanced state of deterioration will have lost almost all resiliency and is brittle and easily cracked. In such situations, transient electrical and mechanical forces associated with a fault can stress the insulation and winding ties to the breaking point.

Cleaning, drying, and revarnishing of insulation during its normal life span helps avoid early deterioration by improved cooling, by eliminating paths for surface currents, and by reducing current leakage through the insulation. However, the process of cleaning, drying, and varnishing does not constitute renewal of the insulation. The revarnishing of old lifeless insulation merely masks a deteriorated condition; clean but brittle insulation can still fail under stress.

Although failure of electrical insulation is generally associated with sparks, arcs, smoke, flames, tripped circuit breakers, and blown fuses, most insulation failures are predominantly mechanical in origin. The movement of conductors by mechanical vibration, electromagnetic forces, and thermal expansion and contraction causes cracked and flaked-off insulation, tape separation, mica separation, etc.

Leakage Paths

The development of surface cracks in insulation, with the accompanying absorption of contaminants, increases the number of leakage paths through the insulation, causing further deterioration and, thus, increasing the probability of ground faults or short circuits.

Tracking

Tracking across insulation is indicated by charred or discolored lines between the conductor and the framework of an apparatus, or between conductors of opposite polarity. Severe tracking across insulation surfaces is generally visible as glowing lines, a string of glowing pinpoints, or a string of tiny sparks, called *scintillation*. Although the current through such creepage paths may not be appreciable, the white-hot sparks, which are really tiny arcs, can eventually cause a short circuit or flashover across the insulation.

External Corona

An external corona is a luminous and sometimes audible electric discharge caused by ionization of the atmosphere surrounding high-voltage conductors. Its most harmful effect is the production of ozone and nitrogen oxides, which cause chemical deterioration of the organic materials used in insulation. Ozone, a form of oxygen (O_3), is an extremely powerful oxidizing agent and, in addition to damaging insulation, readily oxidizes such metals as copper and iron.

Another adverse effect of corona is severe local heating of insulation, resulting in carbonization, called *corona burning,* that produces leakage tracks (paths) in organic materials. The telltale marks of early corona discharge on the surface of varnished insulation are usually in the form of white, gray, or red dust spots. Each spot, which occurs at a point of high-voltage stress, has a matching spot on an opposite surface. Figure 23.1 shows damage from corona burning in the area where a 13.8-kV potential transformer lead passed through an insulation barrier. Figure 23.1a shows damage to the lead insulation, and Figure 23.1b shows the effect of corona burning inside the hole and on the surface of the insulating barrier. The corona discharge and surface leakage at the barrier were probably due to a buildup of conductive deposits on the cable sheath and on the insulating barrier. If such corona tracking is not found and corrected, the buildup of leakage paths can result in a line-to-line or line-to-ground flashover.

FIGURE 23.1
Damage from corona burning: (a) damage to lead insulation; (b) damage on the surface of an insulation barrier.

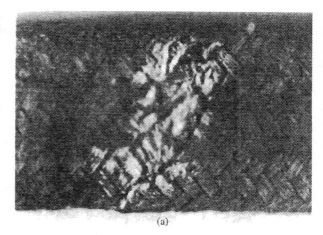

(a)

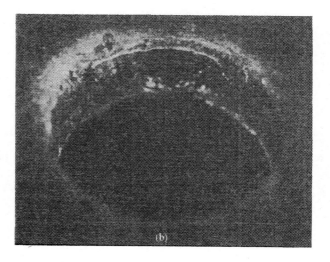

(b)

FIGURE 23.2
Tree-like channels formed in electrical insulation.

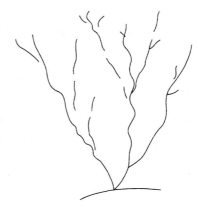

Internal Corona

The development of internal cracks and separations (voids) can cause internal corona deterioration. Although corona within the internal voids of insulation cannot be seen, the development and expansion of voids that allow corona to occur can be detected by insulation power-factor tests, partial discharge tests, and surge tests. Corona action in a void can develop tree-like channels within the volume of insulation. These *electrical trees,* also called *dendrites,* are thought to be caused by bombardment of the insulation material by high-speed electrons within an ionized void. Once initiated, the tree grows in a sporadic manner, developing more branches as the process continues. When the growth of tree branches advances sufficiently close to the opposite electrode, complete breakdown can occur. A sketch of the branch-like structure of a tree is shown in Figure 23.2. Trees that develop within cable insulation by the intrusion of moisture at faults in the cable shield are called *water-trees* [1].

23.2 ENVIRONMENTAL, MECHANICAL, THERMAL, AND VOLTAGE AGING OF INSULATION

Environmental Aging

Environmental aging includes the effects of humidity and of airborne pollutants that react chemically with the insulation and radiation (if in nuclear reactor plants).

Mechanical Aging

Mechanical aging refers to the changes that take place in electrical insulation due to forces that mechanically stress the insulation during normal and abnormal operation. Such stresses can cause cracks in insulation, tape separation, separation of insulation from the ground wall, etc. Examples of mechanical forces that take their toll in shortening the life of insulation are:

1. Electromagnetic vibration of conductors at twice the power frequency caused by alternating current

2. Vibration transmitted to the insulation from the driven equipment
3. Impact forces on insulation due to mechanical or electrical faults
4. Expansion and contraction of insulation due to temperature changes following changes in load
5. Compression forces on coil sides caused by developed motor torque and generator countertorque
6. Centrifugal forces on coils in generator and motor rotors
7. Erosion of insulation by cooling air (windage).

Voltage Aging

Voltage aging refers to the electric stresses and partial discharges within the insulation system that, over a period of time, cause changes in the properties of the insulating material. Partial discharges of stored capacitive energy occur through voids in the insulation volume.

Thermal Aging

Thermal aging refers to the changes that take place in electrical insulation during both normal and abnormal operating temperatures. The aging process is complex and behaves differently with different materials. Some materials soften and lose their mechanical strength, others lose resilience, become brittle, and then fail mechanically under stress.

23.3 SOFT VS. HARD INSULATION

Insulation containing asphalt binders, called soft insulation, is somewhat flexible and thus not grossly affected by vibration. However, it is hygroscopic and thus has a tendency to absorb moisture and other moisture-bearing contaminants. The failure mode of soft insulation is due primarily to thermal and electric aging.

Insulation impregnated with thermoplastic epoxy, called hard insulation, is not hygroscopic and thus is not adversely affected by moisture. However, because hard insulation is brittle, its failure mode is primarily due to mechanical aging. Mechanical stresses due to load changes and vibration cause the gradual development of cracks (voids) in the slot insulation and end turns, which can eventually result in insulation failure.

23.4 THERMAL CLASSIFICATION OF INSULATION SYSTEMS

An insulation system is an assembly of insulating materials and equipment parts. For example, the insulation system for a rotating machine includes coil insulation, ground-wall insulation, phase-to-phase insulation, binders, etc. The thermal classification of insulation systems for rotating machinery, as indicated on a motor nameplate, is listed by letter and temperature in Table 23.1. These are limiting internal hot-spot tempera-

TABLE 23.1
Temperature Classification of Insulation Systems

Thermal Class	Hot-Spot Limit (°C)
A	105
B	130
F	155
H	180
N	200
R	220
S	240
C	Over 240

tures for insulation systems selected on the basis of obtaining the same desired service life for all classes of insulating systems [2–4].

Motors with identical horsepower ratings, speed ratings, voltage ratings, and enclosures that are constructed with either Class A, Class B, Class F, or Class H insulation systems and operated continuously at their respective maximum hot-spot temperatures should have the same average life expectancy. Operating above these limiting hot-spot temperatures will shorten the insulation life.

23.5 TEMPERATURE RISE OF ELECTRICAL APPARATUS

During normal operation of electrical machinery, its temperature rises above that of the surrounding air. The ambient air temperature of machines operating in the United States seldom exceeds 40°C (104°F). Hence, this value is established as the reference temperature when the permissible temperature rise is determined.

Table 23.2 lists the maximum allowable temperature rise used by manufacturers for various windings and insulation classes, based on a 40°C ambient temperature, and an altitude of 1000 meters (3300 ft) or less [5]. The temperature rise for service factor motors is based on continuous operation at its service factor load. However, as indicated in Table 23.2, when operating continuously at its 1.15 service factor, the temperature rise will be 10°C higher than if running at 100 percent load.

To eliminate misinterpretations of temperature rise, the temperature rise has not been included on the nameplates of motors designed since 1969. Instead, temperature specifications on the nameplates of motors include only the ambient temperature and the insulation class. Motors are specified by horsepower, speed, voltage, frequency, service factor (SF), enclosure, insulation class, and ambient temperature within which it will operate. Temperature rise is taken care of by the manufacturer. As long as the machine is clean, has no defects, and is operated within its rating, the allowable temperature rise will not be exceeded.

TABLE 23.2

Maximum allowable temperature rise for medium single-phase and polyphase induction motors in °C, based on a maximum ambient temperature of 40°C[a]

Class of insulation system (see MG 1–1.65)	A	B	F[b]	H[b,c]
Time rating (shall be continuous or any short-time rating given in MG 1–10.36)				
Temperature rise (based on a maximum ambient temperature of 40°C), °C				
1. Winding temperature, by resistance method				
(a) Motors with 1.0 service factor other than those given in items 1(c) and 1(d)	60	80	105	125
(b) All motors with 1.15 or higher service factor	70	90	115	—
(c) Totally enclosed nonventilated motors with 1.0 service factor	65	85	110	135
(d) Motors with encapsulated windings and with 1.0 service factor, all enclosures	65	85	110	—
2. The temperatures attained by cores, squirrel-cage windings, commutators, collector rings, and miscellaneous parts (such as brush holders, brushes, pole tips, uninsulated shading coils) shall not injure the insulation or the machine in any respect.				

[a] Reproduced by permission of the National Electrical Manufacturers Association from *NEMA Standards Publication MG 1-1998, Motors & Generators.* Copyright 1999 by NEMA, Washington, DC.

[b] Where a Class F or H insulation system is used, special consideration should be given to bearing temperatures, lubrication, etc. (Footnote approved as Authorized Engineering Information.)

[c] This column applies to polyphase induction motors only.

Loading above the nameplate power rating to take advantage of a cooler ambient temperature is wrong; although it may not cause immediate failure, the adverse effects of increased internal temperature and increased mechanical stresses on the insulation caused by the overload are cumulative and will shorten the life of the insulation.

23.6 INSULATION HALF-LIFE RULE

Available statistics indicate the life expectancy of Class A insulation to be approximately halved with each 10°C rise in operating temperature. This is called the *ten-degree half-life rule.* For example, a machine with Class A insulation that is designed for continuous operation at 80°C will have its insulation life cut in half when operating continuously at 90°C. The 10-degree half-life rule was first demonstrated by V. M. Montsinger in 1930 [6].

Figure 23.3 illustrates the life expectancy curve for representative Class A insulation. The temperature of the conductor (at its hottest spot) is shown plotted against time. The sides of the shaded band give the maximum and minimum life expectancy for a given operating temperature, and the straight line through the center of the band

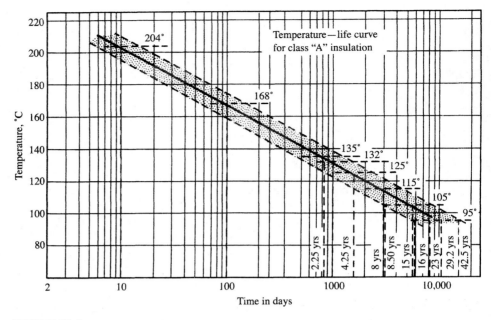

FIGURE 23.3
Life expectancy curve for representative Class A insulation (courtesy Allis Chalmers Mfg. Co.).

indicates the average value. The life curve of Figure 23.3 indicates that this machine has an average life expectancy of 16 years when operating at its internal hot-spot temperature of 105°C, but only 8 years when operating at 115°C.

23.7 TEMPERATURE DETERMINATION OF INSULATION BY THE CHANGE IN RESISTANCE METHOD

During steady-state operation, the temperature of the insulation surrounding and in contact with a conductor is equal to the temperature of the conductor.

The temperature of a winding can be determined by the change in resistance method. It is relatively easy and requires only two resistance measurements. The first resistance measurement (R_1) is generally made at the factory and is available from the manufacturer in ohms at a specified temperature. The second resistance measurement (R_2) is measured in the field. The temperature corresponding to the second measurement is calculated using the following equation:[1]

$$T_2 = \frac{R_2}{R_1} \times (T_1 + k) - k \tag{23-1}$$

---

[1] Justification for Eq. (23–1) is given in Section 1–8, Chapter 1. See also References [7] and [8].

where: R_1 = resistance of winding at temperature T_1 (Ω)
R_2 = resistance of winding at temperature T_2 (Ω)
k = 234.5, a constant for copper between 0°C and 125°C
k = 224.1, a constant for aluminum between 25°C and 125°C
T_1 = average winding temperature (°C)
T_2 = average winding temperature (°C).

The resistance may be determined by using a low-resistance ohmmeter or by the voltmeter-ammeter method as shown in Figure 23.4a. However, regardless of the method used, the temperature obtained is the average temperature of the winding.

The test should be made at the motor terminals, not at the starter. Furthermore, to minimize heating the test current should not exceed 15 percent of rated winding current, and meter readings should be taken as soon as the current reaches its steady-state value.

Effect of Winding Time Constants on Resistance Measurement

The windings of small machines have a relatively low time constant, and steady-state current is attained in a few seconds. However, large machines with highly inductive, low-resistance windings may require minutes or even hours for the current to attain steady state. The time can be reduced significantly by adding resistance in series with the winding and by applying a higher voltage to obtain a readable current, as shown in Figure 23.4b.

The time constant for the winding alone is

$$\tau_{wind} = \frac{L_{wind}}{R_{wind}} \tag{23-2}$$

The time-constant for the series circuit with the added resistance in Figure 23.4b is

$$\tau_{circ} = a \cdot \tau_{wind} = \frac{L_{wind}}{R_{wind} + R_x} \tag{23-3}$$

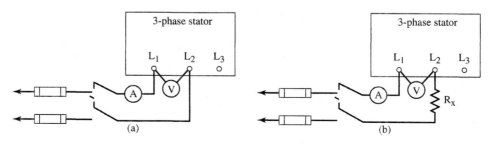

FIGURE 23.4
(a) Circuit for the voltmeter-ammeter method; (b) circuit with an external resistor to shorten the time constant.

where: a = desired time-constant reduction factor $(0 < a < 1.0)$
τ_{wind} = time constant of the winding (s)
τ_{circ} = time constant of the circuit (s)
R_{wind} = resistance of the winding (Ω)
L_{wind} = inductance of the winding, in henrys (H)
R_x = added resistance (Ω).

Dividing Eq. (23–2) by Eq. (23–3), and solving for R_x,

$$\frac{\tau_{wind}}{a \cdot \tau_{wind}} = \frac{L_{wind}/R_{wind}}{L_{wind}/(R_{wind} + R_x)} \Rightarrow \frac{1}{a} = \frac{R_{wind} + R_x}{R_{wind}} \tag{23–4}$$

$$R_x = R_{wind}\left(\frac{1 - a}{a}\right)$$

Ohms per Phase vs. Ohms between Terminals

Resistance data for three-phase stators, as provided by the manufacturer, can be expressed as ohms per phase, or ohms measured between two motor terminals. The relationship between ohms per phase and ohms between terminals depends on whether the winding is connected wye or delta.

If wye connected, as shown in Figure 23.5a,

$$R_{termY} = 2 \times R_{phase} \tag{23–5}$$

If delta connected, as shown in Figure 23.5b,

$$R_{term\Delta} = \frac{2}{3}R_{phase} \tag{23–6}$$

where: R_{term} (Y or Δ) = resistance measured between two terminals (Ω)
R_{phase} = resistance of one phase (Ω).

FIGURE 23.5
(a) Wye-connected winding; (b) delta-connected winding.

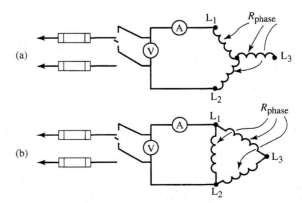

EXAMPLE 23.1 Data supplied by a motor manufacturer lists the stator resistance of a certain three-phase, 460-V, 60-Hz, 40-hp, 460-A, wye-connected induction motor to be 0.102 Ω per phase at 20°C. Stator windings are copper. (a) Determine the resistance between two motor terminals at 20°C. (b) After 8 hours of operation, a DC voltmeter-ammeter test made across two line terminals of the motor, as shown in Figure 23.4a, indicates 46.8 A at 12.0 V. Determine the temperature of the windings. (c) Determine the temperature of the windings if the manufacturer indicated that the stator is delta connected, and 6.2 V applied to two terminals results in a current of 72.5 A.

Solution

a. Referring to Figure 23.5a, the resistance between two terminals of a wye-connected stator is:

$$R_1 = R_{termY} = 2 \times R_{phase} = 2 \times 0.102 = 0.204 \ \Omega$$

b. $R_2 = \dfrac{V}{I} = \dfrac{12.0}{46.8} = 0.2564 \ \Omega$

Substituting into Eq. (23–1),

$$T_2 = \frac{R_2}{R_1} \times (T_1 + k) - k = \frac{0.2564}{0.204} \times (20 + 234.5) - 234.5 = 85.4°C$$

c. Referring to Figure 23.5b, the resistance between two terminals of a delta-connected stator is

$$R_1 = R_{termΔ} = \frac{2}{3} R_{phase} = \frac{2}{3} \times 0.102 = 0.068 \ \Omega$$

$$R_2 = \frac{V}{I} = \frac{6.2}{72.5} = 0.0855 \ \Omega$$

$$T_2 = \frac{R_2}{R_1} \times (T_1 + k) - k = \frac{0.0855}{0.068} \times (20 + 234.5) - 234.5 = 85.5°C$$

EXAMPLE 23.2 The resistance per phase of a certain wye-connected winding is 0.146 ohms. Determine (a) the resistance at two terminals; (b) the value of external resistance that should be added in series with the winding to reduce the time constant of the circuit to approximately one-tenth of its original value.

Solution

a. $R_1 = R_{termY} = 2 \times R_{phase} = 2 \times 0.146 = 0.292 \ \Omega$

b. $R_x = R_{wind} \left(\dfrac{1 - a}{a} \right) = 0.292 \left(\dfrac{1 - 0.1}{0.1} \right) = 2.63 \ \Omega$

23.8 RESISTANCE TEMPERATURE DETECTORS

Resistance temperature detector (RTDs), embedded in stator slots between the top and bottom coil sides, are often used to determine the temperature of motor or generator windings while the machine is in operation. Embedded temperature detectors are not affected by the cooling fan. A properly placed RTD will provide a temperature indication that very closely approximates the actual hot-spot temperature of the machine. The temperature-sensing element is a thin coil, usually wound with copper wire, whose length and diameter are sized to obtain a resistance of 10 ohms at 25°C.[2] A minimum of six RTDs are uniformly spaced around the stator and positioned in the center of the respective slots. Temperature measurements obtained from RTD resistance readings may be read directly on ohmmeters calibrated to read degrees Celsius instead of ohms. Other methods for determining winding temperature are presented in Reference [7].

Resistance temperature detectors have very useful applications in preventive maintenance programs. A temperature rise of more than 10°C to 15°C above the known normal operating temperature, when operating at normal loads and in a normal ambient temperature, at rated voltage and frequency, is an almost positive indication that the machine should be cleaned. The accumulation of dirt on the surface of the insulation and in the ventilating ducts reduces the dissipation of heat, raises the temperature, and causes thermal degrading of the insulation. Because the life expectancy of insulation is approximately halved with each 10°C increase in operating temperature, good preventive maintenance procedures require that periodic checks be made on the operating temperatures of machines, particularly those operating on a continuous basis.

Motors protected by RTDs in the control circuit will trip from overheating regardless of the cause; the RTDs actuate a relay in the control circuit, tripping the machine off the line. A system using RTDs takes into consideration heating caused by clogged ventilating ducts, accumulated dust on the windings, abnormal ambient temperatures, the thermal inertia of the mass of iron and copper, and the heat caused by an overload.

Temperature measurements for preventive maintenance purposes in old machines with skeleton frames can be made with an alcohol thermometer or thermocouple taped to the ends of the stator coils.[3] Although alcohol thermometers register a temperature about 15°C cooler than by RTD, a gradual rise in temperature over time (assuming the same load and same ambient temperature) can provide an early warning of lack of cooling because of dirt accumulation.

---

[2] Slot RTDs are also available in nickel (120 Ω at 0°C) and in platinum (100 Ω at 0°C).

[3] Mercury thermometers should not be used; breakage and the resultant release of mercury can cause a short circuit.

SUMMARY OF EQUATIONS FOR PROBLEM SOLVING

$$T_2 = \frac{R_2}{R_1} \times (T_1 + k) - k \qquad \textbf{(23–1)}$$

$$\tau_{wind} = \frac{L_{wind}}{R_{wind}} \tag{23-2}$$

$$\tau_{circ} = a \cdot \tau_{wind} = \frac{L_{wind}}{R_{wind} + R_x} \tag{23-3}$$

$$R_x = R_{wind}\left(\frac{1-a}{a}\right) \tag{23-4}$$

$$R_{termY} = 2 \times R_{phase} \tag{23-5}$$

$$R_{term\Delta} = \frac{2}{3}R_{phase} \tag{23-6}$$

SPECIFIC REFERENCES KEYED TO THE TEXT

[1] McMahon, E.J., A Tutorial on Treeing, *IEEE Trans. Electrical Insulation*, Vol. EI-13, No. 4, August 1978.

[2] *General Principles for Temperature Limits in the Rating of Electric Equipment and for the Evaluation of Electrical Insulation*, IEEE Std 1-2000.

[3] *Standard Test Procedure for Evaluation of Systems of Insulating Materials for Random-Wound AC Electric Machinery (Appendix)*, IEEE Std. 117-1991, IEEE, New York, 1991.

[4] *Standard for the Preparation of Test Procedures for the Thermal Evaluation of Solid Electrical Insulating Materials*, IEEE Std. 98-1993, IEEE, New York, 1993.

[5] *Motors and Generators*, ANSI/NEMA STD, MG1-12.42, 1998.

[6] Brancato, E. L., Insulation Aging, an Historical and Critical Review, *IEEE Trans. Electrical Insulation*, Vol. EI-13, No. 4, August 1978, pp. 308–317.

[7] *Recommended Practice for General Principles of Temperature Measurement as Applied to Electrical Apparatus*, IEEE Std. 119-1974, IEEE, New York, 1974.

[8] *IEEE Standard Test Code for Resistance Measurement*, IEEE Std. 118-1992, IEEE, New York, 1992.

REVIEW QUESTIONS

1. What effect does excessive temperature have on the life of electrical insulation?
2. What factors determine the useful life of electrical insulation?
3. Name three factors involved in environmental aging.
4. List six factors involved in mechanical aging.
5. Explain the process of thermal aging.
6. Explain the process of voltage aging.
7. What is the reference temperature used by manufacturers to determine the maximum allowable temperature rise for motors?
8. What is the ten-degree half-life rule?
9. How does cleaning electrical insulation help avoid early deterioration?

10. What is an electrical tree, and what causes it?

11. Define *corona*. How does it affect electrical insulation?

12. Define *tracking*. How is it detected, and what adverse effect does it have on electrical insulation?

13. What is an RTD, where are they used, and what is its function?

PROBLEMS

23–1/7. The manufacturer's data for a 60-hp, 440-V, 60-Hz, 1765-rpm, wye-connected induction motor specify a resistance between terminals of 0.136 Ω at 25°C. The motor was operating continuously for several days in an overheated environment. Upon shutdown, a 12-V battery was used to make an immediate voltmeter-ammeter test in order to determine the operating temperature. The windings are copper, and the meter readings were 11.2 V and 64.6 A. Determine the average temperature of the windings.

23–2/7. The rotor resistance for a 5400-kW, 3600-rpm, 2300-V, synchronous generator is 0.44 Ω measured at 25°C. Determine the average temperature of the rotor copper when operating at steady state with field current and field voltage at 327 A and 188 V, respectively.

23–3/7. A 25-hp, 60-Hz, 460-V induction motor has a winding resistance of 0.209 Ω/phase at 25°C. Determine the resistance per phase when operating at an average winding temperature of 115°C.

23–4/7. Data supplied by a motor manufacturer lists the stator resistance per phase of a certain three phase, 460-V, 60-Hz, 40-hp, delta-connected induction motor to be 0.204 Ω per phase at 20°C. Stator windings are copper. A DC voltmeter-ammeter test made across two terminals of the motor, as shown in Figure 23.4a, indicates 76.2 A at 11.8 V. Determine the average temperature of the windings.

23–5/7. The resistance per phase of a certain wye-connected winding is 0.01 ohms. Determine (a) the resistance between terminals; (b) the value of external resistance that should be added in series with the winding to reduce the time constant of the circuit to approximately two-tenths of its original value.

23–6/7. The resistance per phase of a certain wye-connected winding is 0.218 ohms. Determine (a) the resistance between terminals; (b) the value of external resistance that should be added in series with the winding to reduce the time constant of the circuit to approximately one-quarter of its original value.

24

Insulation Resistance:
Its Measurement
and Interpretation

24.0 INTRODUCTION

The first in-depth research on the characteristics of insulation, with particular emphasis on the effects of moisture and voltage on insulation resistance, was conducted by
Sydney Evershed in the early 1900s in England [1]. Since then, insulation-resistance
measurement has become an increasingly important part of an electrical maintenance
program.

Accumulated dirt on the insulation surface, as well as moisture and other contaminants absorbed in the insulation or condensed on the surface of the insulation,
causes a decrease in the measured values of insulation resistance. Thus, periodic
insulation-resistance tests, properly done with the same instrument under the same
operating conditions and plotted over many months or years, will indicate any downward trend in the "health" of the insulation.

Although a single insulation-resistance measurement cannot indicate any downward trend, it can provide some indication as to whether or not it is safe to operate the
machine at the time the test was made, and whether or not it is safe to undergo a high-
voltage stress test.

A more definitive test that clearly differentiates between clean-dry insulation
and dirty-moist insulation is the 10-minute dielectric absorption test.

Warning

The test procedures described in this chapter must be conducted by personnel
knowledgeable and skilled in the safe handling of high voltage.

24.1 INSULATION RESISTANCE

Insulation resistance can be measured by nondestructive tests applied between the conductors and the framework of the apparatus. The resistance value can be read directly from a megohmmeter, called a *megger,* or indirectly by calculation using the voltmeter-ammeter method. When properly made and evaluated, such tests assist in diagnosing impending trouble.

Moisture absorbed in the windings or condensed on the surface of insulation results in a decrease in the measured values of insulation resistance. Hence, for insulation measurements to be significant, the tests should be made immediately after shutdown. This avoids errors due to condensation of moisture on the windings. When the machine temperature is lower than the temperature of the surrounding air, moisture condenses on the windings and is gradually absorbed by the insulation. The insulation-resistance values of DC machines are generally more sensitive to changes in humidity than are those of AC windings; this is due to the greater number of leakage paths in the armature and fields of DC machines.

24.2 INSULATION CURRENT-TIME CHARACTERISTIC

The insulation current-time curve for relatively good insulation is shown in Figure 24.1a and the component currents that result in this behavior are illustrated separately in Figures 24.1b, 24.1c, and 24.1d. For clarity, the direction of the current is shown as actual electron flow ($-$ to $+$) rather than the conventional ($+$ to $-$).

The leakage component, shown in Figure 24.1b, passes through or across the surface of the insulation. The magnitude of this leakage current depends on the resistance of the leakage paths and the value of the driving voltage; it is an Ohm's law relationship. For good clean-dry insulation, only a very small amount of leakage current occurs.

The capacitive component, shown in Figure 24.1c, is caused by the capacitance between the wiring and the metal frame of the apparatus, and it is typical of the charging current to a capacitor. This component of test current starts high but drops rapidly, reaching almost zero in a very short time (five time constants). Because the duration of this current is very brief, it has very little effect on the indicated values of insulation resistance for motors, generators, and transformers.

The absorption component, shown in Figure 24.1d, converts electric energy to stored energy in the form of a molecular strain in the insulating material. Although each molecule is electrically neutral (the positive charge is equal to the negative charge), its positive and negative charges form an electric dipole. In the presence of an applied voltage, the positive end of the dipole is twisted toward the negative terminal, and the negative is twisted toward the positive terminal. Thus each electric dipole not already aligned in the direction of the applied voltage experiences a torque that tends to position it parallel to the line of action of the applied voltage. This behavior, called *dielectric absorption,* is a relatively slow process that may take many hours or days to complete. When the applied voltage is removed, and the wiring grounded, the molecules return slowly to their unstressed equilibrium position.

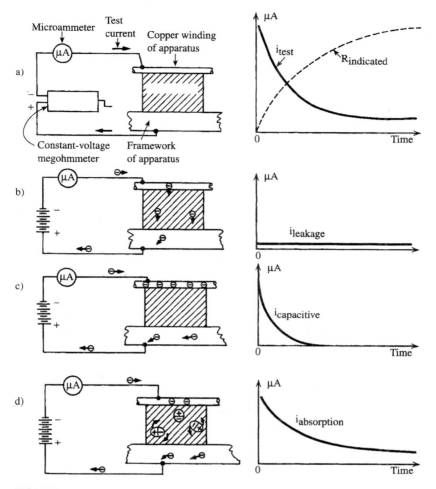

FIGURE 24.1
Current-time curves for relatively good insulation.

The dielectric-absorption characteristic of electrical insulation makes it both difficult and time consuming to obtain an absolute measurement of insulation resistance. The insulation resistance indicated by the megger, at any instant of time, is the ratio of the megger voltage to the megger test current (Ohm's law):

$$R_{indicated} = \frac{V_{megger}}{i_{megger}} \tag{24-1}$$

Assuming a constant megger voltage, the indicated insulation resistance depends solely on the test current. If the test current is high, the indicated resistance will be low; if the test current is low, the indicated resistance will be high. Because the test current in relatively good insulation starts high and gradually decreases with time, as shown in Figure 24.1a, the indicated resistance starts low and increases with the continued application of test voltage. Higher leakage currents through and across the surface of relatively poor insulation permit less accumulation of stored energy within the insulating material; this reduces the dielectric-absorption effect, causing both the current and resistance curves in Figure 24.1a to flatten faster.

24.3 TEMPERATURE CORRECTION OF MEASURED VALUES OF INSULATION RESISTANCE

Insulating materials have a negative temperature-resistance characteristic; that is, the resistance of the insulation decreases with increasing temperature. If insulation-resistance readings taken at different operating temperatures are to be compared, the readings should be corrected to a common reference temperature (usually 40°C).

The graph in Figure 24.2 is based on test data that show insulation-resistance approximately halving for each 10°C rise in winding temperature above 40°C, and doubling for each 10°C drop in temperature below 40°C. The graph is a plot of Eq. (24–2) with $H = 10$ [2]. Equation (24–2) is a mathematical expression of the half-life rule.

$$k_T = 0.5^{[(40 - T)/H]} \qquad (24\text{–}2)$$

where: k_T = correction factor
H = halving factor (°C)
T = actual temperature of winding (°C).

The halving factor (H) for most older types of insulation (late 1950s) is 10°C. The halving factor for some of the newer types of insulation may range from $H = 5$°C to 20°C. However, if the temperature-resistance characteristic of the insulation is not known, the curve in Figure 24.2 will provide a useful approximation.

To obtain the corrected value, multiply the observed insulation-resistance reading by the correction factor obtained from Figure 24.2 or from Eq. (24–2). That is,

$$R_{40} = k_T \times R_T \qquad (24\text{–}3)$$

where: R_{40} = insulation resistance corrected to 40°C (MΩ)
R_T = measured insulation resistance at T°C (MΩ).

EXAMPLE 24.1

The measured insulation resistance of a DC motor is 20 MΩ at a temperature of 60°C. Determine (a) correction factor; (b) insulation resistance corrected to 40°C.

Solution

Since no resistance-temperature data are available, use a halving factor of 10 or the curve in Figure 24.2.

a. The correction factor obtained from Figure 24.2 is 4.
b. $R_{40} = k_T \times R_T = 4 \times 20 = 80$ MΩ

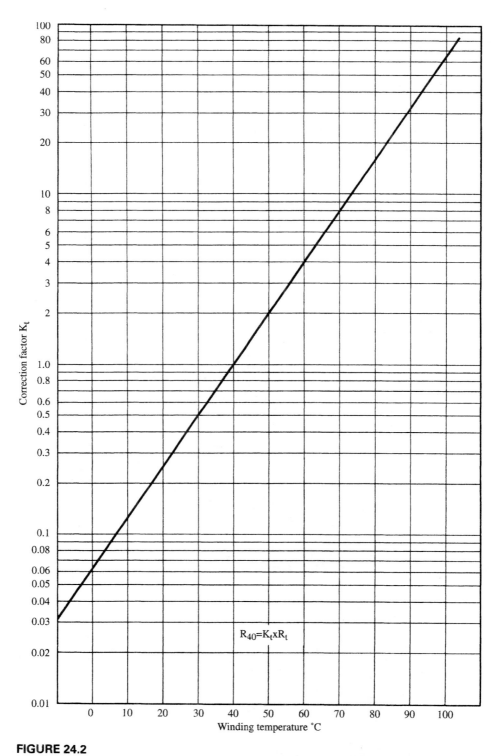

FIGURE 24.2

Temperature-correction factor for measured values of insulation resistance, with 10°C as the halving factor.

Figure 24.3 illustrates quite vividly the need for correcting insulation-resistance readings to a common reference temperature. Curve A is an uncorrected resistance curve, and curve B is a plot of the values adjusted to 40°C. The large fluctuations indicated by the unadjusted curve result in an erroneous picture of the actual insulation trend. The curve of the corrected values shows only slight changes in insulation resistance with time.

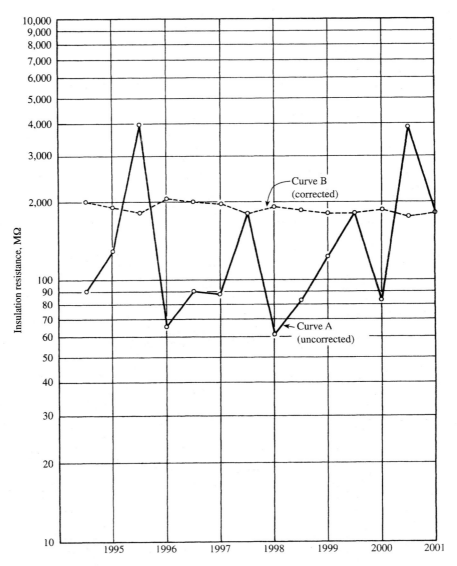

FIGURE 24.3
Representative plots of uncorrected and temperature-corrected insulation-resistance readings.

EXAMPLE 24.2

Assume that the insulation in Example 24.1 has a halving factor of 15°C. Determine (a) correction factor; (b) insulation resistance corrected to 40°C.

Solution

a. The correction factor obtained from Eq. (24–2) is

$$k_T = 0.5^{[(40 - T)/H]} = 0.5^{[(40-60)/15]} = 2.5$$

b. $R_{40} = k_T \times R_T = 2.5 \times 20 = 50 \text{ M}\Omega$

24.4 INSULATION-RESISTANCE METERS

Insulation resistance measurements can be made directly with an instrument called a megger that indicates the insulation resistance in megohms (MΩ). One megohm equals 1 million ohms. Meggers are available as a battery-operated instrument, a hand-cranked instrument using a built-in DC generator, a motor-driven instrument, and a rectifier instrument operated from an AC source. A hand-cranked megger is shown in Figure 24.4. Some meggers have two or more switchable test voltages. Insulation-resistance ranges are available up to 20,000 MΩ.

To avoid overstressing the insulation, the megger test voltage should be appropriate for the voltage rating of the motor, generator, transformer, etc. Voltage guidelines are listed in Table 24.1 [2].

Pretesting the Megger

With the test leads connected to the megger, and the free ends separated, energizing the megger should cause the instrument to read infinity (∞). With the free ends connected together, energizing the megger should cause the instrument to read zero.

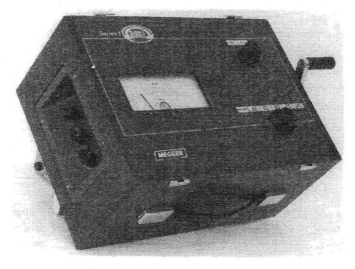

FIGURE 24.4
Hand-cranked megger (courtesy Biddle Instruments).

TABLE 24.1
Voltage Guidelines for Insulation-Resistance Testing

Rated Voltage of Apparatus[a] (V)	DC Test Voltage (V)
Up to 100	100–250
240 to <1000	500
1000 to 2500	500–1000
2501 to 5000	1000–2500
5001 to 12,000	2500–5000
>12,000	5000–10,000

[a] Rated line-to-line nameplate voltages for AC and DC machines.

Preparing Apparatus for Testing

Before making an insulation-resistance test, the apparatus must be disconnected from the power line and from all other equipment such as switches, breakers, cables, capacitors, and surge arrestors. Failure to do so is not only dangerous but will result in misleading interpretations. Ground-fault interrupting circuits must be disconnected or the electronic circuitry will be damaged. The circuit to be tested should then be discharged by grounding to the framework for no less than 5 min immediately prior to the test; the windings of large machines should be grounded for about 15 min immediately prior to the test.

Grounding will ensure removal of any accumulated electrostatic charge that would otherwise result in erroneous meter readings. Large machines with good insulation have a considerable capacitive effect between the conductors and the frame, and during the normal course of operation they may acquire a charge of static electricity and retain it for an extended time after shutdown.

The insulation on motor leads, or other apparatus being tested, should be cleaned before testing to minimize surface leakage.

Connecting and Operating the Megger

Connect the terminal marked "+," "Line," or "Hot" to the winding, and the earth terminal (E) to the frame as shown in Figure 24.5. Cranking the megger at its slip speed ensures a constant voltage and causes the pointer to move to a position on the scale corresponding to the value of insulation resistance under test.

Behavior of the Megger Pointer

When a megger is being operated, the behavior of the pointer should be carefully observed, because much can be learned from its movements. The leakage of current along the surface of dirty insulation is generally indicated by slight kicks downscale; whereas the response of the pointer when testing good insulation is a downward dip followed by a gradual climb to the true resistance value. The initial dip of the pointer

FIGURE 24.5
Connections for a megger test of a motor.

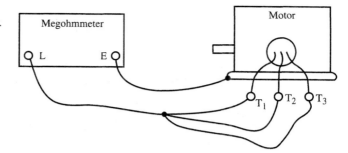

toward zero is caused by the capacitance of the windings and is especially noticeable in large machines, cables, and capacitors. However, the capacitive charging time is short, seldom more than several seconds. The gradual rise in pointer reading with continued cranking is caused by the dielectric-absorption effect of the insulation. It may take hours or days before the electrification is completed and the pointer ceases to rise.

Warning

All large machines that undergo a long-time insulation test should be discharged after the test is completed. This may be accomplished by grounding the wiring to the framework of the apparatus, and the discharge time should be at least as long as the charging time. When making the ground discharge connection, one end of the grounding wire must first be solidly connected to the framework of the apparatus (ground) and then the other end clipped to the wiring.

Special Cases

1. Insulation tests of hermetically sealed refrigerator compressor motors should be made after shutdown while still warm but at equalized pressure; testing the windings while under vacuum may damage the insulation.
2. If a rotor has brushless excitation, check the manufacturer's instructions before making an insulation resistance test. The high voltage may destroy the diodes. If the manufacturer's instructions are not available, the diodes should be bypassed with a jumper before making the test.

Using the Guard Terminal

Megohmmeters having ranges of 1000 MΩ and higher are equipped with a guard terminal to prevent leakage current, caused by dampness or dirt, from affecting the measurement. The megger shown in Figure 24.4 has a guard terminal. The difference in connections, with and without the guard terminal, is shown in Figure 24.6.

Figure 24.6a shows the connections for measuring the insulation resistance-to-ground of cable A, without using the guard terminal. Note that this connection does

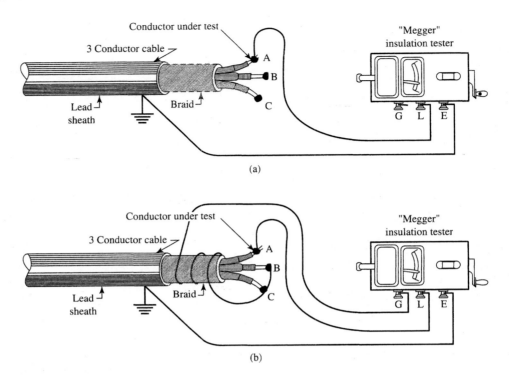

FIGURE 24.6
Measuring insulation resistance of a cable: (a) without the guard terminal; (b) with the guard terminal (courtesy Biddle Instruments).

not take into consideration the fact that there is a path of leakage current from conductor A to ground along the surface of the braid, and a path of leakage current from conductor A through the insulation to conductor B and to conductor C.

To prevent significant errors, when making insulation measurements in the 1000-MΩ range and higher, the guard terminal should be used. The higher megohm ranges on a megger use higher voltages to make the measurement. This causes greater leakage current and a greater error.

Figure 24.6b shows the connections for the same test, but with a connection to the guard terminal. Bare wire wrapped around the braided insulation is connected to conductor B, conductor C, and the guard terminal. The guard circuit offers a low-resistance path for the leakage current, going directly to the DC source without passing through the deflecting coil of the megger.

24.5 SIXTY-SECOND INSULATION-RESISTANCE TEST

The 60-second insulation-resistance test is recommended for comparison with previous records. In a 60-second test, sometimes called a *spot test*, the instrument is operated for 60 seconds and a reading recorded at the end of that time. If a hand-cranked

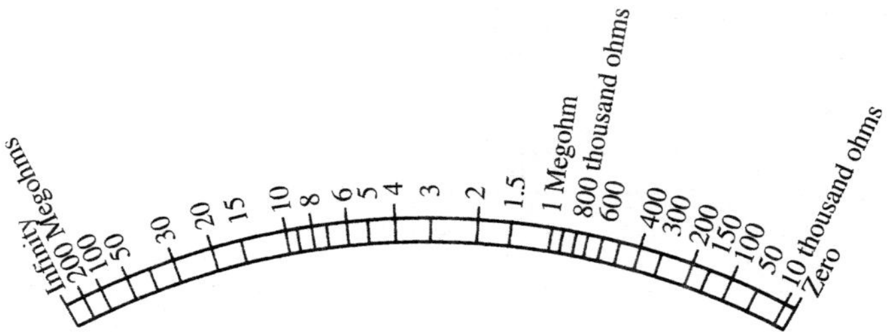

FIGURE 24.7

Typical scale for a megger insulation tester.

megger is used, the reading should be taken while still cranking at slip speed. Readings should be taken at the end of the 60-second period, even though the pointer is still climbing. All readings should be corrected to 40°C, using the graph in Figure 24.2 or Eq. (24–2) and (24–3) as appropriate.

If the meter pointer reaches the ∞ mark before the end of the 60-second period, as will happen when testing small machines with good insulation, the test can be terminated and the reading of ∞ recorded. A reading of ∞ is not the value of insulation resistance; it is an indication that the insulation resistance is too high to be measured on that particular test instrument. Figure 24.7 illustrates a typical scale for a megger insulation tester with a useful ohms range from 10,000 ohms to 200 megohms. A reading of ∞ on this scale indicates an insulation resistance in excess of 200 MΩ. Similarly, a reading of zero indicates a resistance of less than 10,000 Ω.

24.6 INSULATION-RESISTANCE RECORDS AND INTERPRETATION

A single insulation-resistance measurement, taken at random, without reference to previous records, is of little value. It is the trend of insulation-resistance values, measured with the same instrument, at the same test voltage, under the same conditions, and corrected to a common reference temperature, that provides an indication of insulation condition with respect to dirt, moisture, and other contaminants. Even though each individual insulation-resistance reading may be above the minimum recommended value, a consistent downward trend indicates the need for cleaning, drying, and revarnishing. An abrupt drop may indicate serious problems.

The keeping of accurate records is an essential first step to the logical analysis of insulation condition. Tests can be made on a monthly, semiannual, or annual basis as conditions demand. The insulation resistance, temperature, humidity, instrument used, and date should be recorded, along with any notes regarding unusual operating conditions such as excessive vibration, moisture, overheating, etc. The 60-second megger reading should be plotted on semilog paper for easy determination of the trend. Figure 24.8 illustrates one form of test-record card that facilitates recording and plotting of insulation-resistance data.

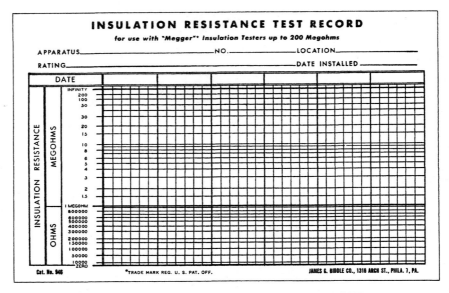

FIGURE 24.8
Test-record card to facilitate recording and plotting of insulation-resistance data (courtesy Biddle Instruments).

24.7 RECOMMENDED MINIMUM INSULATION RESISTANCE FOR ROTATING MACHINERY

There is no established rule or method for determining the minimum value of insulation resistance at which a machine may operate without breaking down. However, minimum values of insulation resistance have been recommended by the Institute of Electrical and Electronics Engineers and are defined as *the least value that a winding should have after cleaning, or if an appropriate overpotential test is to be applied.* Although it is possible to operate machines with much lower values of insulation resistance, it is not considered good practice.

The recommended minimum insulation resistance for AC and DC machine armature windings and for field windings of AC and DC machines, based on a 60-second test corrected to 40°C, can be determined from Table 24.2 [2].

TABLE 24.2
Recommended minimum insulation resistance for AC and DC armature
and field windings, corrected to 40°C after
a 60-second test

Winding under Test	Minimum Resistance (MΩ)
Windings manufactured before 1970, all field windings and others not described below	kV + 1
DC armature and AC windings manufactured after 1970 (form-wound coils)	100 MΩ
Machines with random-wound or form-wound stator coils rated below 1 kV	5 MΩ

Note 1: kV is the nameplate voltage rating of the machine in kilovolts.
Note 2: For three-phase windings, the test should be made on the entire winding. If
the test is made on only one phase, with the other two phases isolated and grounded,
the observed resistance should be equal to three times that indicated in this table.

Induction Motors

To avoid unnecessary maintenance costs, induction motors can be pretested at the
motor controller, providing the starter box, cable armor, and motor frame are
grounded; in addition, the starter cannot be solid state and no capacitors can be connected to the motor terminals. With the switch or breaker open, the three outgoing
leads are clipped together, and the combination connected to the "hot" terminal of the
megger as shown in Figure 24.9a. The earth terminal of the megger must be connected
to the steel frame of the control enclosure. Complete the 60-second test, correct the
reading to 40°C, and compare it to the recommended minimum value.

If the corrected reading is less than the recommended minimum, or if the
motor is not tested at the motor controller, the three stator leads must be disconnected at the motor terminal box, joined as illustrated in Figure 24.9b, and given a
60-second megger test. If the reading, corrected to 40°C, is still too low the motor
requires reconditioning.

If the motor insulation is satisfactory when tested alone, as in Figure 24.9b, leakage in the cables or control caused the low reading, and these pieces should be tested
separately.

Direct-Current Machines

Insulation-resistance tests on DC machines should be made separately on armature
and fields. The brushes must be lifted from the commutator when testing the
armature, and the insulation resistance of the brush rigging should be measured
separately.

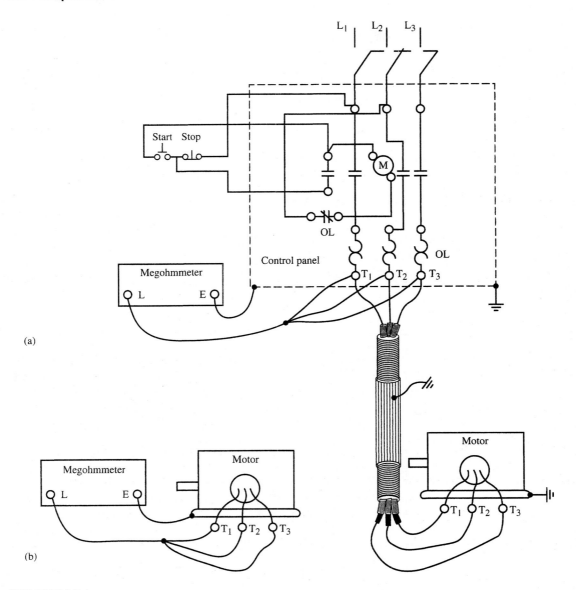

(a)

(b)

FIGURE 24.9
Insulation-resistance measurement of: (a) circuit plus cable and control; (b) motor alone.

EXAMPLE 24.3 A 200-hp, 2300-V, three-phase induction motor of 1968 vintage has an indicated insulation resistance to ground of 2 MΩ at a temperature of 60°C. The test was made with a 1000-V megger applied between stator conductors and motor frame (ground) for 1 min. (a) Determine the recommended minimum insulation resistance: (b) Does the insulation meet the recommended minimum resistance?

Solution

a. From Table 24.2,

$$R_{\text{min}} = \text{kV} + 1 = \frac{2300}{1000} + 1 = 3.3 \text{ M}\Omega$$

b. From Figure 24.2, $k_T = 4$

$$R_{40} = k_T \times R_T = 4 \times 2 = 8 \text{ M}\Omega$$

The insulation exceeds the recommended minimum.

24.8 MEASURING INSULATION RESISTANCE OF DRY-TYPE TRANSFORMERS

The minimum acceptable insulation resistance for a dry-type transformer should be as specified by the manufacturer or, in the absence of such data, should not be less than the minimum specified in Table 24.3 [3]. The insulation-resistance test should be a 60-second test at a voltage selected from Table 24.1. If the insulation resistance is equal to or greater than the recommended minimum for the transformer, it is safe to operate or to perform a high-voltage dielectric test.

To prepare a dry-type transformer for an insulation-resistance test, the circuit breaker or disconnect switch on both the high side and low side must be opened, and all windings short circuited. Each winding should be tested separately as illustrated in Figure 24.10a for the low-side winding. The winding not under test must be grounded. The insulation between windings should also be tested as shown in Figure 24.10b.

24.9 MEASURING INSULATION RESISTANCE OF CABLE

The capacitance charging current of long cables will take much longer than 60 seconds to decay, therefore a safe minimum cannot be determined. However, a record of 60-second tests with the same instrument, and under the same conditions, can be used as a guide to determine gradual decay. Both ends of the cable should be disconnected,

TABLE 24.3

Recommended Minimum Insulation Values for Dry-Type Distribution and Power Transformers Corrected to 40°C

kV Rating	Minimum Resistance (MΩ)
1.2	600
2.5	1000
5.0	1500
8.7	2000
15.0	3000

and the cable should be discharged to ground for at least 5 min before and after testing. Long cables should be tested using methods described in Chapter 25.

Effect of Cable Length on the Measured Values of Insulation Resistance

The effect that the length of metal armored cable has on its measured insulation resistance can be explained by reference to Figure 24.11. The cable insulation in Figure 24.11 is represented as the equivalent of many resistors in parallel along the length of the cable. Each resistor represents the insulation resistance between conductor and shield for one linear foot of insulation.

From Chapter 2, Section 2.5, the equivalent resistance of identical resistors in parallel is equal to the resistance of one resistor divided by the number of resistors.

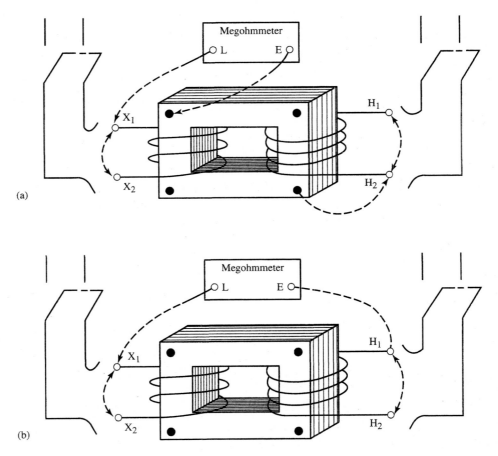

FIGURE 24.10
Measuring insulation resistance: (a) between windings and ground; (b) between windings.

FIGURE 24.11
Equivalent resistance circuit of a cable.

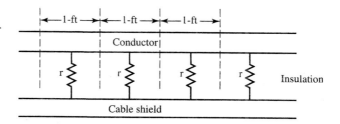

Thus, the equivalent insulation resistance of a cable is equal to the insulation resistance per unit length divided by the length of the cable. Expressed mathematically,

$$R_{\text{cable ins}} = \frac{r}{L} \qquad (24\text{--}4)$$

where: r = insulation resistance per linear foot of cable (MΩ)
$R_{\text{cable ins}}$ = insulation resistance of cable (MΩ)
L = length of conductor (ft).

Note: Each unit length of a cable is a leakage path. The longer the cable, the greater the number of leakage paths, and thus the lower the insulation resistance of the cable.

EXAMPLE 24.4

One thousand feet of steel-armored cable wound on a reel has a measured insulation resistance of 25 MΩ at 20°C. A 100-foot section was cut off the reel and is to be used to supply power to a motor. Determine (a) insulation resistance/foot of cable; (b) insulation resistance of the 100-foot length.

Solution

a. $R_{\text{cable ins}} = \dfrac{r}{L} \quad \Rightarrow \quad 25 = \dfrac{r}{1000}$

b. $r = 25{,}000$ MΩ/ft

c. $R_{\text{cable ins}} = \dfrac{r}{L} = \dfrac{25{,}000}{100} = 250$ MΩ

24.10 TWO-VOLTAGE INSULATION-RESISTANCE TEST

The two-voltage test is a step-voltage test in which only two different voltages are used. Sixty-second insulation resistance tests conducted at two different voltages, preferably in the ratio of 1:5 (for example, 500 volt and 2500 volt), can provide more conclusive indications of moisture or other contaminants than a single voltage test. Using a multivoltage megger, a 60-second test is made at the low-voltage setting and again at the high-voltage setting. The insulation should be discharged by grounding before each test. The megger voltages used should not exceed the voltage guidelines in Table 24.1.

Experience has indicated that a reduction of 25 percent or more in the measured insulation resistance at the higher voltage, with a 1:5 ratio of test voltages, is a sure indication of moisture or other contamination.[1] To determine the percent reduction, substitute into the following equation:

$$\text{Percent reduction in insulation resistance} = \frac{R_{5V} - R_V}{R_V} \times 100 \qquad (24\text{--}5)$$

where: R_V = insulation resistance at voltage V
R_{5V} = insulation resistance at voltage 5 × V.

Temperature corrections are not required, because it is the percent change in insulation resistance rather than the actual value of resistance that is significant.

Cracked, aged, or otherwise deteriorated insulation that is fairly clean and dry may pass a one-voltage test but generally fail to pass a two-voltage test; the insulation resistance of such defective insulation drops rapidly with higher voltages.

Insulation in good condition, and thoroughly clean and dry, will indicate substantially the same insulation resistance when measured with any test voltage up to the rated terminal voltage of the apparatus under test.

---

[1] AVO Biddle Instruments, *A Stitch in Time*, 1992.

EXAMPLE 24.5 Insulation-resistance measurements recorded during a two-voltage test of an induction motor stator are 8.4 MΩ at 500 V, and 5.6 MΩ at 2500 V. What is the condition of the insulation?

Solution

$$\text{Percent reduction} = \frac{R_{5V} - R_V}{R_V} \times 100 = \frac{5.6 - 8.4}{8.4} \times 100 = -33\%$$

The 33% reduction in insulation resistance with a 1:5 ratio of test voltages indicates that the stator winding has absorbed moisture or is otherwise contaminated and should be cleaned and dried.

24.11 DIELECTRIC ABSORPTION TEST

A dielectric absorption test is a 10-minute megger test made with a motor-driven, battery-operated, or rectifier-type instrument. Megohm readings are taken at 0.5, 0.75, 1.0, 1.5, and 2.0 min, and every minute thereafter up to 10 min. The data are plotted on log-log paper as shown in Figure 24.12. The slope of the curve provides information regarding the relative condition of the insulation. *A steadily rising curve indicates a clean-dry winding, but one that flattens out rapidly indicates dirt, moisture, or other contaminants* [2, 4].

The dielectric absorption test is particularly useful for diagnostic testing of machines with commutators or slip rings, machines with soft moisture-absorbing insu-

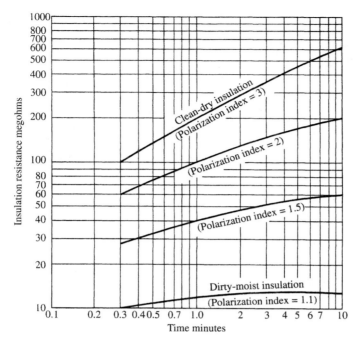

FIGURE 24.12
Typical curves showing the variation of resistance with time for Class B insulated armature windings.

lation, and machines with hard insulation operating in a badly contaminated environment. It is also useful for obtaining baseline data when no previous insulation-resistance records are available.[2]

Connections and precautions for the dielectric-absorption test are the same as for the insulation-resistance test. The apparatus to be tested should have its windings grounded for no less than 5 min prior to starting the dielectric-absorption test. If possible, the individual phases should be tested separately with the remaining phases grounded; this requires disconnecting both terminals of each phase.

Although the dielectric-absorption test is fairly independent of temperature, to prevent errors due to condensation of moisture on the insulation, it is generally best to conduct the test immediately after shutdown.

A dielectric-absorption test can also be conducted using a microammeter and voltmeter; the voltage is held constant throughout the 10-min test, and the indicated insulation resistance calculated for each reading, using Ohm's law:

$$R = \frac{V}{I}$$

[2] The dielectric absorption test has not proven to be a useful test for transformer windings or cables.

where: V = voltage (V)
I = current (μA)
R = insulation resistance (MΩ).

Warning

The dielectric-absorption characteristic of insulation enables it to accumulate considerable electric charge during the 10-min absorption test. When the test is completed the absorption charge gradually converts to a high-voltage capacitive-charge between winding and ground. To prevent electric shock, the windings must be grounded for at least 10 min after completion of a dielectric-absorption test.

24.12 POLARIZATION INDEX

The polarization index (PI) obtained from a dielectric-absorption test provides a quantitative appraisal of the condition of the insulation with respect to dirt, moisture, and other contaminants. The index, obtained by taking the ratio of the 10-min to 1-min insulation-resistance readings, is a measure of the slope of the absorption curve during the first 10 min of charging:

$$PI = \frac{R_{10}}{R_1} \qquad (24\text{-}6)$$

The polarization index can also be determined directly from a voltmeter-ammeter test. Substituting Ohm's law ($R = V/I$) into Eq. (24-6),

$$PI = \frac{(v/i)_{10}}{(v/i)_1} \qquad (24\text{-}7)$$

Since the test voltage is held constant Eq. (24-7) reduces to

$$PI = \frac{i_1}{i_{10}} \qquad (24\text{-}8)$$

Temperature correction of the measured or calculated values of PI is unnecessary unless there is an appreciable change in temperature between the 1-min and 10-min readings.

A record of the PI of an electrical apparatus is useful for comparing its value when new with subsequent tests on the same machine and with other machines of the same rating and manufacture. A steady downward trend in PI over a period of years indicates the gradual accumulation of dirt and the absorption of moisture or other contaminants.

The recommended minimum values of PI for clean-dry windings of AC and DC machines are listed in Table 24.4 [2]. Although no minimum PI values are recommended for other classes of insulation, a PI of less than 2.0 for classes of insulation not listed indi-

TABLE 24.4
Recommended Minimum Values of Polarization Index
(PI) for Clean-Dry Insulation of AC and DC Machines

Insulation Class	PI
A	1.5
B	2.0
F	2.0
H	2.0

cates the need for immediate cleaning and drying. If the PI is still less than 2.0 after cleaning and drying, the process should be repeated. If a second cleaning and drying does not bring the PI to the recommended minimums, the manufacturer should be consulted.

Note 1   Insulation surfaces treated with semiconducting material to eliminate corona will have a lower PI than identical untreated machines whose insulation is in the same general condition.

Note 2   A high polarization index indicates that the insulation is clean, dry, and free from contaminants, but it is not a positive indication of reliability. Clean, dry insulation that suffers from thermal and mechanical aging is brittle; it may have cracked insulation, tape separation, and other flaws that could cause it to fail catastrophically when subject to the high mechanical forces due to short circuit, out-of-synchronous operation, or plugging operations. For example, insulation that is 20 years old or older, and is clean and dry, may have a PI of 5.0 or more; such insulation has lost its resiliency, is brittle, and can fracture under severe mechanical stresses.

Note 3   A machine that has been out of service for an extended period of time has probably absorbed some moisture. However, if its PI is 1.0 or more, and the insulation resistance is above the minimum acceptable value as determined from Table 24.2, the machine may generally be returned to service; *this is a calculated risk.*

EXAMPLE 24.6   The 1-min and 10-min readings recorded during a dielectric absorption test of a 600-V DC armature with Class B insulation are 12 MΩ and 32 MΩ, respectively. The test voltage was 500 volts. (a) Determine the polarization index. (b) What does the PI indicate about the condition of the insulation?

Solution

a. $PI = \dfrac{R_{10}}{R_1} = \dfrac{32}{12} = 2.7$

b. From Table 24.4, the recommended minimum PI for Class B insulation is 2.0. The actual value is 2.7. It can therefore be concluded that the insulation is clean and dry.

SUMMARY OF EQUATIONS FOR PROBLEM SOLVING

$$R_{\text{indicated}} = \frac{V_{\text{megger}}}{i_{\text{megger}}} \tag{24-1}$$

$$k_T = 0.5^{[(40-T)/H]} \tag{24-2}$$

$$R_{40} = k_T \times R_T \tag{24-3}$$

$$R_{\text{cable ins}} = \frac{r}{L} \tag{24-4}$$

SPECIFIC REFERENCES KEYED TO THE TEXT

[1] Evershed, Sydney, *The Characteristics of Insulation Resistance*, The Institute of Electrical Engineers, London, September 1913.

[2] *Recommended Practice for Testing Insulation Resistance of Rotating Machinery*, IEEE 43–2000, IEEE, New York, 2000.

[3] *Recommended Practice for Installation, Application, Operation, and Maintenance of Dry Type General Purpose Distribution and Power Transformers*, USAS/IEEE C57.94–1987, IEEE, New York, 1987.

[4] *Guide for Insulation Maintenance for Rotating Electrical Machinery (5 hp to Less Than 10,000 hp)*, IEEE Std. 432–1998, IEEE, New York, 1998.

REVIEW QUESTIONS

1. What is insulation resistance? What instruments are used to measure it, and how are they used?
2. What three components of test current are present when meggering insulation?
3. What effect does capacitance have on the indicated values of insulation resistance? Explain.
4. What is dielectric absorption?
5. What effect does absorbed moisture have on the measured values of insulation resistance?
6. What effect does the temperature of a machine have on the measured values of insulation resistance?
7. What is a megger? How is it used?
8. Can electrical equipment be damaged as a result of a megger test? Explain.
9. An insulation-resistance test of an AC motor indicates ∞ on the megger scale. Does this mean that the insulation resistance is infinite? Explain.
10. Why should a large machine be grounded for several minutes before an insulation-resistance test is made?
11. When testing insulation resistance with a megger, what would cause the pointer to make a few downscale kicks while gradually climbing upscale?

12. What is the 60-second insulation-resistance test? Of what use is it in a preventive-maintenance program?
13. What is the recommended minimum value for insulation resistance of a 200-hp, 2300-V, three-phase induction motor?
14. What is the purpose of a dielectric-absorption test, and how is it done?
15. Define polarization index and state its significance.

PROBLEMS

24–1/3. The insulation resistance of a DC generator is 30 MΩ at a temperature of 70°C. Assuming a halving factor of 10, what is its value corrected to 40°C?

24–2/3. The insulation resistance of an AC generator is 15 MΩ at a temperature of 70°C. Assuming a halving factor of 20, what is its value corrected to 40°C?

24–3/3. A series of periodic insulation-resistance tests made over a 1-year period provide the tabulated data for a 100-hp, 2300-V induction motor as shown in Table 24.5. The motor was built in 1969 and has Class A insulation. Plot a curve of the observed and the corrected values of insulation resistance. What does the curve indicate?

TABLE 24.5
Tabulated Data for Problem 24–3/3

Month	Observed Resistance (MΩ)	Temperature (°C)
January	12	35
February	9	38
March	8	43
April	9	40
May	5	48
June	3	53
July	2	62
August	3	54
September	3.2	53
October	6	46
November	7	43
December	10	40

24–4/7. What is the recommended minimum insulation resistance for a 90-kW, 120-V DC armature manufactured in 1995?

24–5/7. What is the recommended minimum insulation resistance for a 2300-V, 500-kVA synchronous generator armature manufactured in 1955?

24–6/7. The armature of a 4160-V, 6000-kVA, 3600-rpm alternator manufactured in 1969 has a measured insulation resistance of 2 MΩ at a temperature of 80°C. The insulation is Class B, and the test was made with a 500-V megger applied for 1 min. Using a halving factor of 10, determine (a) resistance at 40°C; (b) recommended minimum resistance; (c) status of the insulation.

24–7/7. A 60-second insulation resistance reading on a 4000-V, 800-hp induction motor is 5 MΩ at 75°C. The motor was manufactured in June 2000. The insulation has a halving factor of 5. Determine (a) insulation resistance corrected to 40°C; (b) recommended minimum insulation resistance. (c) What is the status of the insulation?

24–8/10. A two-voltage test of a synchronous motor stator results in 25 MΩ when tested at 250 V, and 24 MΩ when tested at 2500 V. What is the status of the insulation?

24–9/10. Insulation-resistance measurements recorded during a two-voltage test of an induction motor stator are 85 MΩ at 1000 V, and 54 MΩ at 5000 V. What is the condition of the insulation?

24–10/12. A motor-driven megger is used to test the insulation resistance of a Class F-insulated AC armature. The readings observed after 1 min and 10 min are 12 megohms and 13 megohms, respectively. (a) What is the polarization index? (b) Would you recommend continued operation of the machine in its present condition? Explain.

24–11/12. A Class B-insulated armature winding has a measured insulation resistance of 200 megohms after 1 min of application of test voltage. After 10 min of continued application of test voltage the reading is 150 megohms. (a) Calculate the polarization index. (b) Would you recommend continued operation of the machine in its present condition? Explain.

24–12/12. The results of a 10-min high-potential test made on a large Class B-insulated induction motor show the leakage current to be 3.6 mA after 1 min and 1.5 mA after 10 min. (a) What is the polarization index? (b) Would you recommend continued operation of the machine in its present condition? Explain.

25

High-Potential Maintenance Testing of Electrical Insulation

25.0 INTRODUCTION

This chapter introduces the basics of high-potential maintenance testing. High-potential testing utilizes voltages in excess of rated voltage to stress the insulation between the winding and the frame (ground-wall insulation) to validate new or reconditioned apparatus or to search for signs of deterioration.

Preventive maintenance high-potential tests should be limited to those applications where failure of the insulation system during normal operation could endanger lives, cause extensive property damage, black out a community, compromise the safety of a ship, halt production lines, etc. Insulation that fails during a high-potential test, and is then rebuilt, prevents the disastrous consequences of failure under load. Testing of apparatus that is not vital to plant operation or safety should be limited to insulation resistance and dielectric absorption.

If the insulation passes a high-potential test, the probability of failure due to electrical stresses when returned to service is small. However, it must be understood that this is a voltage stress test, not a mechanical stress test. There is still a possibility that, when returned to service, the stresses associated with vibration and centrifugal force could cause failure.

Caution   Because of the possibility of breakdown, a high-potential test should not be made unless the following two conditions are met:

1. Management must be made aware of the consequences of failure. Insulation breakdown caused by the test voltage must be repaired or renewed before the apparatus can be returned to service. This may result in considerable downtime. In the case of shipboard apparatus, it should be done only when the ship is laid up for an overhaul or a survey or when major repairs have been made.

2. The polarization index (PI) obtained from a dielectric absorption test must indicate clean dry insulation. Insulation that would prove satisfactory on a high-potential test when clean and dry may break down if the insulation resistance is low because of dirt or moisture. If equipment for a dielectric absorption test is not available, an insulation resistance test using a megohmmeter with an appropriate voltage rating should be used. The measured value of insulation resistance should be above the minimum value set forth by the manufacturer.

Warning

High-voltage testing, whether AC or DC, represents a potential hazard to life, and every safety precaution recommended by the manufacturer or testing agency must be followed. The voltages used and the available current (although in milliamperes) are lethal. Electrician's rubber gloves, a long sleeve shirt, long pants, and safety glasses should be worn when making these tests. Instructions for operation of specific test equipment must be followed.

25.1 HIGH-POTENTIAL PROOF-TESTS (GO/NO-GO)

High-potential proof-tests, also called *hi-pot tests*, are go/no-go tests that demonstrate whether the insulation of a winding or other item is in good or bad condition. The test is made by applying a specified overvoltage between the terminals of the apparatus and the framework of the apparatus for a period of 1 minute.[1] If the insulation does not break down, the apparatus is safe to use. Breakdown is indicated by a neon light on the test equipment, by an arc or spark at the fault or weak spot, or by tripping of an over-current relay in the test circuit. Since sparking or arcing may precipitate a fire, fire extinguishers should be on hand nearby during the test. A power-frequency high-potential test set is shown in Figure 25.1.

Factory Proof-Test [2]

Armatures and Fields of DC Motors and DC Generators ≥1/2 hp   The factory proof-test voltage for newly manufactured components shall be:

$$V_{\text{factoryproof-test}} = (2V_{\text{rated}} + 1000) \text{ volts}$$

where: V_{rated} = rated line-to-line voltage of the machine (V).

---

[1] For an example of how one company uses the AC hi-pot test to estimate remaining service life of electrical apparatus, see Reference [1].

FIGURE 25.1
Power-frequency high-potential test set
(courtesy Associated Research, Inc.).

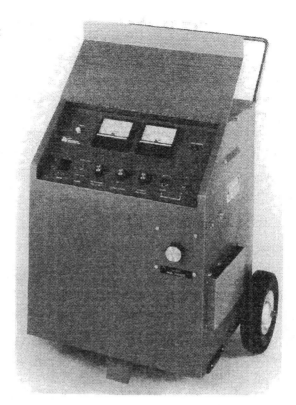

Armatures and Fields of DC Motors <1/2 hp and ≤240 V   The factory proof-test voltage for newly manufactured components shall be:

$$V_{factoryproof\text{-}test} = 1000 \text{ volts}$$

Armatures and Fields of DC Motors <1/2 hp and >240 V   The factory proof-test voltage for newly manufactured components shall be:

$$V_{factoryproof\text{-}test} = (2V_{rated} + 1000) \text{ volts}$$

Stators of Induction and Synchronous Machines ≥1/2 hp   The factory proof-test voltage for newly manufactured components shall be:

$$V_{factoryproof\text{-}test} = (2V_{rated} + 1000) \text{ volts}$$

Universal Motors ≥1/2 hp and ≤250 V   The factory proof-test voltage for newly manufactured components shall be:

$$V_{factoryproof\text{-}test} = (2V_{rated} + 1000) \text{ volts}$$

Universal Motors <1/2 hp   The factory proof-test voltage for newly manufactured components shall be:

$$V_{\text{factoryproof-test}} = 1000 \text{ volts}$$

Induction Motors Rated <1/2 hp and >250 V   The factory proof-test voltage for newly manufactured components shall be:

$$V_{\text{factoryproof-test}} = (2V_{\text{rated}} + 1000) \text{ volts}$$

Induction Motors Rated <1/2 hp and ≤250 V   The factory proof-test voltage for newly manufactured components shall be:

$$V_{\text{factoryproof-test}} = 1000 \text{ volts}$$

Slip-Ring-Type Field Windings [2]   Newly manufactured slip-ring-type generator-field windings rated ≤500 volts should be factory proof-tested with a 50- to 60-Hz voltage equal to 10 times the rated field voltage (excitation voltage), but in no case less than 1500 volts. Field windings rated >500 volts should be factory proof-tested with a 50- to 60-Hz voltage equal to two times the excitation voltage + 4000 volts.

Brushless-Exciter Field Windings   For proof-test voltages, see Reference [2].

Acceptance Proof-Tests [2]

New apparatus installed in factories, ships, etc., is given an acceptance proof-test before being put into service. This test should also be applied to a completely reinsulated armature. The acceptance test is performed at the power frequency, and is usually 75 percent of the factory-test voltage. Thus,

$$V_{\text{acceptanceproof-test}} = V_{\text{factoryproof-test}} \times 0.75 \qquad \textbf{(25–1)}$$

25.2   MAINTENANCE PROOF-TEST VOLTAGES [3–5]

High-potential maintenance proof-tests search for evidence of an insufficient factor of safety with respect to normal operating voltages, overvoltages, and deterioration while in service. Such tests should also be made on repaired equipment, and on equipment that has been subject to severe short circuits or ground faults. It also has very useful applications as a preventive maintenance test during periodic overhauls. The apparatus must be clean and dry and must pass an insulation-resistance test and a dielectric absorption test before being given a high-potential test.

Periodic maintenance proof-testing of equipment in service is made at a much lower voltage than that used for factory tests or acceptance tests. The recommended maintenance proof-test voltages for AC stators and DC armatures are given in Table 25.1. For maintenance testing of field windings, use the maximum available field voltage instead of rated terminal voltage.

TABLE 25.1

Recommended rms Test Voltages for High-Potential Maintenance Proof-Tests of AC Stator and DC Armatures

Testing Source	Test Voltage
Power frequency (50–60 Hz)	(1.25–1.50) × rated terminal voltage [5]
Very low frequency (0.1 Hz)	(1.25–1.50) × 1.63 × rated terminal voltage [6]
Direct current	(1.25–1.50) × 1.7 × rated terminal voltage [5]

Note: The specific multiplier selected (1.25–1.5) is usually based on experience in a particular environment or manufacturer's recommendations. If in doubt, use 1.25.

Maintenance Proof-Testing of AC Machines with AC

An AC high-potential maintenance proof-test on rotating machines is a test made on equipment that has been in service or has been repaired.

A careful visual inspection of the windings and laminations should be made just prior to making an AC high-potential proof-test. This may be accomplished by removing the end shields and using bright lights and mirrors to check for cracked insulation, dirt accumulation, abrasions, torn bindings, loose wedges, corona markings, etc. Note that the insulation must be clean and dry and pass an insulation-resistance test and dielectric absorption test prior to an AC hi-pot test (see Section 24.11, Chapter 24).

High-potential AC proof tests on three-phase machines should be made on each phase separately, with the remaining phases grounded as shown in Figure 25.2a. After each phase is given an AC high-potential proof-test, the insulation should be completely discharged by grounding for at least 15 minutes; dangerously high voltages could be built up by the release of stored energy if the ground is removed too soon. In those cases where only three or four leads are brought out, and the phases cannot be disconnected, all three phases must be tested at once as shown in Figure 25.2b.

Power-frequency proof-testing has been used since the beginning of the electric power industry. It has been a very effective tool in both maintenance and acceptance testing. The one drawback of power-frequency testing is the weight and space requirements of the test sets needed to supply the charging current (capacitive current) to large machines.

A very-low-frequency (VLF) test set has a frequency of 0.1 Hz, weighs considerably less than a power-frequency set, and provides a good approximation of the stresses produced by power-frequency testing [6].

Procedure

When conducting an AC maintenance-proof test, whether at the power frequency or at very low frequency, the voltage should be raised quickly to the specified test voltage and held there for 1 min. After the 1-min timing period, the voltage should be gradually reduced to 25 percent or less of the specified test voltage before de-energizing the test set; this gradual reduction in voltage prevents a possibly damaging surge voltage. The winding should be grounded for at least 15 minutes after de-energizing the test set.

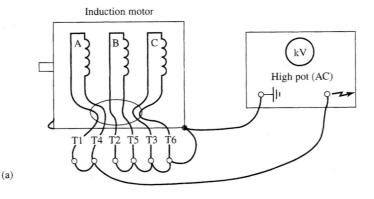

(a)

(b)

FIGURE 25.2
High-potential proof-testing of an induction motor: (a) testing one phase at a time;
(b) testing all three phases at once.

EXAMPLE 25.1

A 2300-V, 5000-kVA, three-phase, 60-Hz generator stator is to be given an AC hi-pot maintenance proof-test. What maximum rms test-voltage should be used for (a) a power-frequency test? (b) A VLF test?

Solution
From Table 25.1,

a. $V_{\text{line freq}} = 1.5 \times 2300 = 3450$ V

b. $V_{\text{VLF}} = 1.5 \times 1.63 \times 2300 = 5624$ V

Maintenance Proof-Testing of DC Machines with AC

Power-frequency hi-pot tests on DC machines should be made separately on the armature and on the fields, with the remaining circuits grounded as shown in Figure 25.3. The range of power-frequency test voltages is provided in Table 25.1.

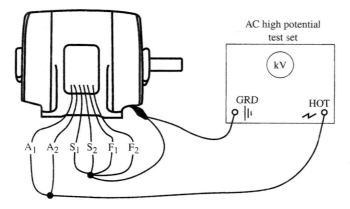

FIGURE 25.3
Maintenance proof-testing of DC machines with AC.

Caution The machine must be clean and dry, pass a visual inspection, and pass an insulation-resistance test and a dielectric absorption test before being subject to an AC hi-pot test.

EXAMPLE 25.2 A 240-V DC shunt motor resting in an ambient temperature of 30°C has been cleaned and dried and has passed a dielectric absorption test. It is to be given an AC maintenance proof-test. Determine the maximum rms test voltage that should be permitted.

Solution
From Table 25.1,

$$V_{test} = 1.5 \times 240 = 360 \text{ V}$$

25.3 MAINTENANCE PROOF-TESTING OF TRANSFORMERS

Maintenance proof-testing of dry-type transformers is not recommended. The severe stress that it places on the insulation can hasten breakdown [7]. Maintenance tests of insulation in dry-type transformers should be limited to periodic measurement of insulation resistance. Measurements should be taken immediately after shutdown, with the same instrument and same test voltage. Data should be recorded along with transformer temperature, relative humidity, and comments on the general conditions of the transformer and connections.

Maintenance testing of insulation of liquid-filled transformers should be primarily limited to liquid dielectric tests, power factor tests, and fluid filtering.[2]

---

[2] See Section 25.9.

25.4 CONTROLLED-DC OVERVOLTAGE TEST (STEP-VOLTAGE TEST) OF INSULATION FOR ROTATING MACHINES [3]

A controlled-DC overvoltage test (also called a step-voltage test, DC leakage test, or measured-current test) is a diagnostic test used to obtain an indication of expected service reliability. The test voltage is raised in equal voltage steps in 1-minute intervals until the preselected maximum test voltage is reached. The shape of the curve plotted from current and voltage test data is used to indicate weakness in the insulation that is not otherwise indicated in insulation resistance tests or dielectric absorption tests. The step-voltage test is made with a portable DC high-potential test set, illustrated in Figure 25.4.

Preparing Apparatus for Testing

Before making a step-voltage test, the apparatus must be disconnected from the power line and from all other equipment such as switches, breakers, cables, capacitors, and surge arrestors. Failure to do so is not only dangerous but will result in misleading interpretations. The circuit to be tested should then be discharged by ground-

FIGURE 25.4
Portable DC high-potential test set
(courtesy Associated Research Inc.).

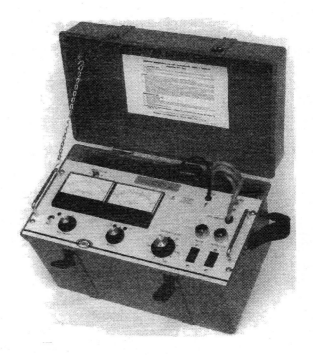

ing to the framework for no less than 5 minutes immediately prior to the test; the windings of large machines should be grounded for about 15 minutes immediately prior to the test.

Grounding will ensure removal of any accumulated electrostatic charge that would otherwise result in erroneous meter readings. Large machines with good insulation have a considerable capacitive effect between the conductors and the frame, and during the normal course of operation may acquire a charge of static electricity and retain it for an extended time after shutdown.

Since leakage-current due to condensed moisture will adversely affect the test results, the test should be performed immediately after shutdown. The insulation on motor leads, or other apparatus about to be tested, should be cleaned to minimize surface leakage, and the windings must pass an insulation-resistance test.[3]

Step-voltage tests should be made on each phase separately with the remaining phases grounded, as illustrated in Figure 25.5a. After each phase is tested, the insulation should be completely discharged by grounding before testing the next phase. In those cases where only three leads are brought out, and the phases cannot be disconnected, all three phases must be tested at once as illustrated in Figure 25.5b.

Dielectric Absorption Pretest

To avoid the possibility of unnecessary breakdown due to dirt, moisture, or other contaminants, a dielectric absorption test[4] to determine the polarization index should precede the step-voltage test and be an integral part of it.

An initial voltage step of approximately 1/3 of the recommended maximum allowable test voltage should be used for the dielectric absorption portion of the test. This is the absorption test voltage. The maximum allowable DC maintenance test voltage should be as recommended by the manufacturer or, in the absence of such information, the DC maintenance test voltage in Table 25.1 can be used as a guide.

Procedure

With the initial voltage step applied, and held constant, the current is recorded and plotted at the *end* of each minute for a total of 10 minutes, as shown in the left section of Figure 25.6. The current should decay as absorption proceeds. *Any rise in current during these first 10 minutes is a signal to stop the test and proceed with appropriate cleaning and drying methods.*

---

[3] See Section 24.7, Chapter 24.

[4] See Section 24.11, Chapter 24.

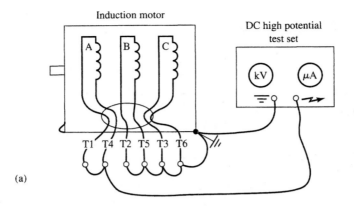

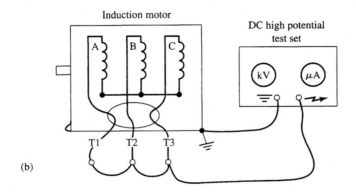

FIGURE 25.5
Connections for step-voltage test: (a) testing each phase separately; (b) testing all phases at once.

An advance determination of the maximum allowable test current at 10 minutes, based on the polarization index, that will permit immediate continuation from the dielectric absorption test into the step-voltage test, may be calculated from the following equation:

$$PI = \frac{i_1}{i_{10}} \Rightarrow i_{10} = \frac{i_1}{PI} \qquad (25\text{–}2)$$

where: PI = minimum polarization index from Table 25.2 [3]
i_1 = test current measured at 1 minute (μA)
i_{10} = maximum allowable test current at 10 minutes (μA).

If the actual test current at 10 minutes is greater than the calculated i_{10}, the test should be stopped to prevent possible damage to the insulation.

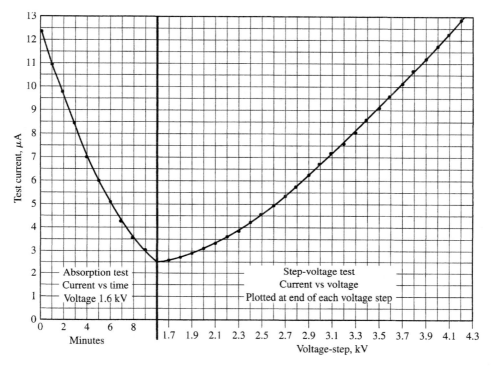

FIGURE 25.6
Dielectric absorption pretest followed by a step-voltage test.

Continuing into the Step-Voltage Test

If the actual test current is not more than the i_{10} value calculated using Eq. (25–2), continue into the step-voltage test. All voltage steps should be equal, each step not more than 3 percent of the recommended maximum allowable test voltage, and 1 minute apart, as shown in Figure 25.6. It is essential that each step increase in voltage be made quickly and not readjusted or errors will be introduced. If you overshoot, don't readjust; going back will complicate the charging process and may cause errors in interpretation [3].

With each increment increase in voltage, the current will increase abruptly (capacitive effect), and then gradually decrease with time as charging continues

TABLE 25.2
Minimum PI for Step-Voltage Test [8]

Class	PI
A	1.5
B	2.0
F	2.0
H	2.0

(absorption effect). At the *end* of each 1-minute period, record and plot current and voltage readings and then quickly raise the voltage to the next step. *However, if at any step the current continues to increase with time, the test must be stopped at once to prevent a current avalanche from destroying the insulation.* The test should also be stopped if extrapolation of the curve (with an exponential template) indicates that continuation of the test would cause it to be asymptotic to a voltage equal to or lower than the maximum DC test voltage, or if an abrupt decrease in the slope of the curve occurs when the test voltage is above the rated machine voltage.

As a safety precaution, when making a step-voltage test, the current relay should be adjusted to an initial setting approximately four times the steady current obtained after the 10-minute absorption part of the test. Because increased increments of test voltage cause the current to approach the trip setting of the current relay, the trip setting should be gradually inched upward. It is important that the relay setting not be set too high, or a sudden failure may cause arcing and extensive damage to the insulation. In any event, the test should be stopped when a sudden increase in current is observed.

Representative curves obtained from step-voltage tests for different insulation conditions are shown in Figure 25.7. Curves A and B indicate satisfactory insulation. Curve C shows a rapid, but not steep, rise in current with voltage; this is generally indicative of damp or contaminated insulation. A very steep slope (approaching the vertical), shown by curve D, indicates exceptionally dirty insulation or ionization of

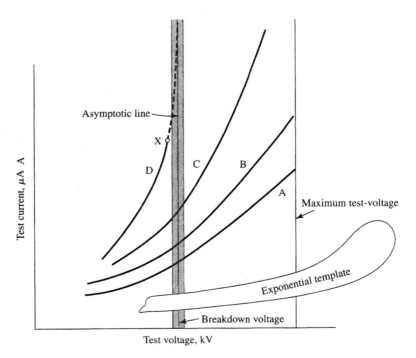

FIGURE 25.7
Representative step-voltage curves showing good and bad insulation.

voids (empty spaces or holes) within the insulation due to mechanical and thermal aging. The test should be stopped at point X if extrapolation of the curve with an exponential template shows it to be asymptotic to a test voltage equal to or lower than the prescribed maximum test value. The extrapolation is shown with a broken line. If testing is continued, insulation breakdown would probably occur within the narrow voltage band indicated by the shaded area.

The test curves, such as those shown in Figure 25.7, provide information that is useful in evaluating the present condition of insulation, and provides a permanent record for comparison with future tests. To compare future test results with previous tests, the same test voltages, time periods, and plotting scales should be used. The test should always be performed immediately after shutdown. Tests made at different conditions of temperature and humidity will have different results.

In most cases breakdown is preceded by an asymptotic warning, permitting the test to be stopped before failure occurs. If the knee of the curve occurs abruptly, and is almost a right-angle bend, there may be no time to stop the test before breakdown occurs. In those instances where breakdown occurs before the maximum test voltage is reached, and without an asymptotic warning, the insulation is obviously defective.

Repeated diagnostic DC high-potential tests on an apparatus during reconditioning, and during subsequent maintenance tests, should always be made with the same instrument, same connections, same voltage steps, and at the same temperature, preferably at shutdown to avoid condensation of moisture.

Grounding after a Step-Voltage Test

The apparatus must be discharged after testing by grounding for a period of time at least equal to four times the total test period, but in no case less than 1 hour [3]. Some high-voltage test sets have built-in discharge resistors that automatically complete the grounding circuit when the set is switched off.

A separate *grounding stick* is illustrated in Figure 25.8. It has a built-in resistor that should have a value of 10,000 Ω/kV of test voltage. When making the grounding-stick connection, the ground clamp must first be solidly connected to the framework of the apparatus, and then the stick end connected to the wiring. At very high test voltages (75,000 V, for example), appreciable voltage recovery may occur when the grounding wire is removed, even though the winding may have been grounded for

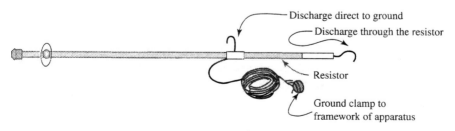

Discharge direct to ground
Discharge through the resistor
Resistor
Ground clamp to framework of apparatus

FIGURE 25.8
Grounding stick (courtesy Biddle Instruments).

over 30 minutes; this voltage is caused by the release of energy as the insulation dipoles return slowly to their unstressed equilibrium position.

25.5 MAINTENANCE TESTING OF POWER CABLES

Considerable differences in opinion are voiced regarding the need for high-voltage maintenance testing of cable. Whether or not to make such tests should be based on the degree of reliability required and on whether spare cables are in place or available for easy connection.

Weak cables may provide many years of service if not subject to the stress of high-potential tests. Thus, in low-priority applications, where a cable outage would not result in excessive financial loss or injury to personnel, it may be more economical to rely on insulation-resistance tests or dielectric absorption tests at rated or slightly higher than rated phase-to-phase voltage.

Testing at Line Frequency

High-voltage maintenance testing of cables at line frequency is not recommended; it is a go/no-go test that does not evaluate the condition of the insulation. A line-frequency test provides no information as to whether the cable insulation is in excellent condition or close to failing. Furthermore each high-potential test at line frequency causes further degradation of the insulation, and failure during testing can cause severe burning at the fault.

Testing with High Direct Voltage

Testing with high direct voltage is much less damaging to insulation than a line-frequency test and can detect gradual deterioration of insulation; failure during a DC test usually results in only a small pinhole at the fault. Periodic high direct-voltage testing of cable systems by *skilled personnel* can substantially reduce in-service failures by replacing weak cable before failure occurs, and thus minimize production loss and property damage, etc., as well as enhance in-service safety.

The maximum test voltage used in a high direct-voltage maintenance test of cables is approximately 75 percent of the acceptance (installation) test voltage. This reduction in test voltage provides an allowance for some deterioration that is normally expected in service, before repair or replacement is indicated. This normal deterioration with time is caused by voltage aging, thermal aging, and mechanical aging. Deterioration can be further aggravated by contact with oil, moisture, grime, and other contaminants. Suggested test voltages for power cable are provided in Table 25.3 [9]. See Section 10.14, Chapter 10, for the definition of basic impulse level (BIL).

A high direct-voltage test set for cable testing should have a voltage rating high enough for the test and be adjustable continuously or in small steps. The AC input to the test set should be supplied from a constant voltage transformer to prevent fluctuations in AC line voltage from causing fluctuation in both the DC voltmeter and DC microammeter readings.

TABLE 25.3

Field Test Voltages for Unshielded Power Cables from 5 kV to 500 kV System Voltage

System Voltage (kV rms) (phase-phase)	System BIL (kV) (crest)	Acceptance Test Voltage[a] (kV dc) (cond-gnd)	Maintenance Test Voltage[a] (kV dc) (cond-gnd)
5	75	28	23
8	36	36	29
15	110	56	46
25	150	75	61
28	170	85	68
35	200	100	75
46	250	125	95
69	350	175	130
115	350	175	130
	450	225	170
	550	275	205
138	450	225	170
	550	275	205
	650	325	245
161	550	275	205
	650	325	245
	750	375	280
230	850	425	320
	950	475	355
	1050	525	395
345	950	475	355
	1050	525	351
	1175	585	440
500	1300	650	490
	1425	710	535
	1550	775	381

Reprinted with permission from IEEE Std. 400–1991, *Table 1 Field Test Voltages for Unshielded Power Cables from 5 kV to 500 kV Systems*. Copyright 1991 IEEE. All rights reserved. The IEEE disclaims any responsibility resulting from the placement and use in the prescribed manner.

[a] Acceptance test voltage duration is normally 15 min. Maintenance test voltage duration is normally not less than 5 min or more than 15 min.

NOTES: (1) Voltages higher than those listed, up to 70% of system BIL, may be considered, but the age and operating environment of the system should be taken into account. The user is urged to consult the suppliers of the cable and any/all accessories before applying the high voltage.

(2) When older cables or other types/classes of cables or other equipment, such as transformers, switchgear, motors, etc., are connected to the cable to be tested, voltages lower than those shown in this table may be necessary to comply with the limitations imposed by such interconnected cables and equipment.

(3) The test voltage should not exceed 50% of system BIL unless surge protection against excessive overvoltages induced by flashovers at the terminations is provided in accordance with 5.1 and B2.

(4) There is relatively less experience with such tests on cables and cable accessories rated 115 kV and higher. It is strongly recommended that the user consult with the manufacture(s) of all components that will be subjected to such testing before performing any tests.

(5) It should be noted that this table and the test procedures suggested in this guide do not necessarily agree with the recommendations of other organizations, such as those of the Association of Edison Illuminating Companies (AEIC CS1–90 [1], AEIC CS2–90 [2], AEIC CS3–90 [3], AEIC CS4–79 [4], AEIC CS5–87 [5], AEIC CS6–87 [6], AEIC CS7–87 [7]).

(6) The dangerously high voltages used for testing power cables mandate that the test be performed only by or under the supervision of personnel experienced with high-voltage testing.

Before starting the test, the cable should be de-energized, and checked with a voltage tester of appropriate rating to ensure that the system is dead. The cable should then be disconnected from all other apparatus including surge and lightning arrestors, potential transformers, and breakers. All normal ground connections must be maintained, and the cable conductors grounded for at least 5 minutes to discharge any stored energy. If some apparatus cannot be disconnected from the cable, the test voltage must not exceed the maintenance test voltage for the apparatus with the lowest voltage rating.

Loose, dry dust on the terminal ends of the cable insulation should be wiped off with clean, dry, lintless cloths. Any encrusted dirt should be removed with a hardwood scraper. Failure to remove dirt from the terminating ends of cable insulation will cause surface leakage, resulting in false readings.

Cooling Cable to Ambient Temperature before Testing

The cable should be allowed to cool to ambient temperature before testing. If not allowed to cool, the temperature of the insulation at the conductor will be higher than the temperature of the insulation at the metallic shield. Since electrical insulation has a negative temperature coefficient (higher temperatures cause lower insulation resistance), if not allowed to cool to ambient temperature, the insulation resistance of the cable will be higher at the shield than at the conductor. This is illustrated in Figure 25.9a for homogeneous insulation; note the temperature distribution and insulation-resistance distribution.

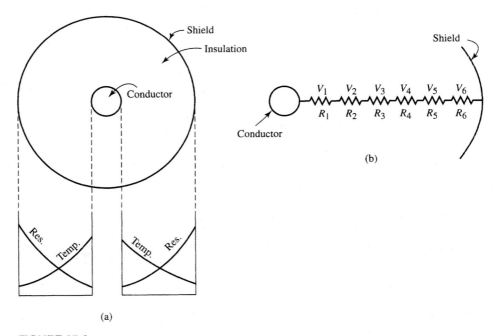

FIGURE 25.9
(a) Cross section of cable; (b) series equivalent of cable insulation resistance.

The actual insulation resistance between conductor and metallic shield is represented in Figure 25.9b as the equivalent of a number of series-connected resistors, each representing a radial section of insulation resistance.

Assuming that the cable is at ambient temperature, all resistors in Figure 25.9b will have the same value. Thus, a test current through the insulation would show an equal distribution of voltage (IR drop) across each resistor; $V_1 = V_2 = V_3 = V_4 = V_5 = V_6$. However, if the cable is tested while still hot, the temperature of R_6 will be less than that of R_1, causing $R_6 > R_5 > R_4 > R_3 > R_2 > R_1$. Hence a test current will cause $V_6 > V_5 > V_4 > V_3 > V_2 > V_1$. This means that a greater portion of the test voltage (IR drop) will occur across R_6, which represents the radial section of insulation nearest the shield. This could overstress the region near the shield, causing failure. Some cable terminations have a much higher negative temperature resistance coefficient than the cable insulation itself and this is one reason why many cables fail at the terminations when tested while hot.

Preparation for Cable Test

The "hot lead" of the test set should be checked for leakage before connecting it to the cable. This can be done in an open-air test by connecting a short piece of small-diameter bare wire to the test lead and raising the voltage until an audible corona occurs. This leakage current should be recorded and subtracted from the test current readings.

After all cables have been discharged and allowed to cool to ambient temperature, the cable conductors, cable shields, and ground terminal of the test set should be connected to a common ground as shown in Figure 25.10a. When ready to start the test, the ground connection should be removed from one conductor and replaced with the hot lead from the test set, as shown in Figure 25.10b. All cable conductors not under test should remain grounded.

Exposed surfaces of conductors under test should be wrapped with polyethylene film or enclosed in polyethylene bags to reduce or eliminate leakage current caused by the corona. Safety precautions for use with high-voltage equipment must be adhered to and manufacturer's instructions must be followed. Each insulated conductor of multi-conductor cable should be tested separately, and the remaining conductors grounded.

Conducting and Interpreting Power Cable Tests[5]

The test is started as a step-voltage test up to the maximum maintenance test voltage and is then held at that voltage for a 15-minute absorption stress test.[6] The step-voltage part of the test requires five or more steps of approximately equal voltage increments of 1-min duration. The first step should be no more than 1.8 times the rated line-to-line voltage. The last step should be equal to the corresponding maintenance test voltage given in Table 25.3 and held for 15 minutes [9].

---

[5] For a very detailed and informative discussion of maintenance testing of power cables, see Reference [10].

[6] When testing rotating machines, the absorption test precedes the step-voltage test. See Section 25.5.

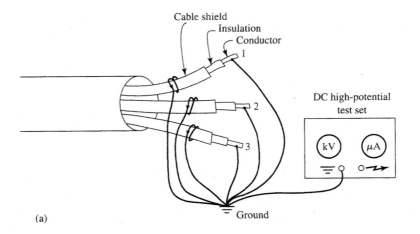

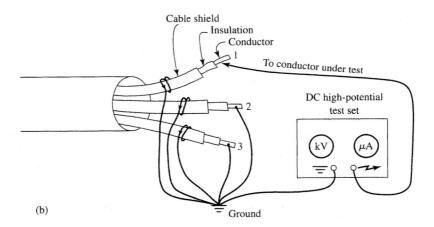

FIGURE 25.10

(a) Connections for grounding prior to high-potential test; (b) connections for high-potential test.

Current and voltage should be read and recorded 1 minute *after* the start of each voltage step, and *after* each minute during the 15-minute absorption test; the data should be plotted as the test proceeds.

If during the step-voltage part of the test, or in the absorption part of the test, the current starts to increase without any additional increase in test voltage, failure is imminent and the test voltage must be rapidly reduced if complete breakdown is to be avoided.

Representative curves for good and bad cable insulation are illustrated in Figure 25.11. If extrapolation of the *v-i* curve for bad insulation shows it to be asymptotic to a voltage lower than the maximum test voltage, the test should be stopped at point X, and the voltage reduced to zero. The test should also be stopped at point Y of the absorption-stress test.

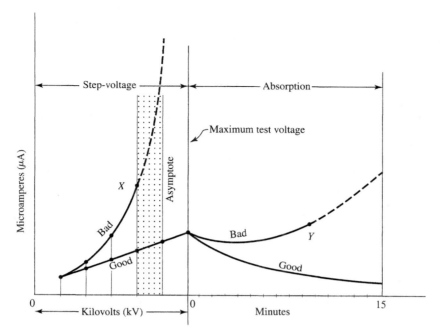

FIGURE 25.11
Representative curves for good and bad cable insulation.

After completing the test, the voltage should be reduced and the cable allowed to slowly discharge through the built-in resistance of the test set. The discharge time can be shortened by using an insulated grounding stick to connect a discharge resistor between ground and the conductor under test (refer to Figure 25.8). The discharge resistor should not be less than 10,000 Ω/kV of test voltage, and should be capable of withstanding the full voltage without overheating or flashing over. When the voltage is reduced to less than 40 percent of the maximum test value, the cable should be solidly grounded with the insulated grounding stick (by passing the resistor) for at least four times the duration of the test voltage. As a safety precaution, it is best to leave the cable grounded until it is to be put into service.

Some engineers prefer to use a plot of indicated insulation resistance vs. test voltage (v-r plot) instead of the v-i plot for the step-voltage part of the test. The indicated resistance is calculated from the step-voltage data using Ohm's law:

$$R = \frac{1000 \, V}{I}$$

where: R = insulation resistance (MΩ)
V = test voltage (kV)
I = test current (μA).

A substantial drop in insulation resistance with increasing test voltage indicates approaching failure.

25.6 INSULATION POWER-FACTOR AND DISSIPATION-FACTOR TESTS

Insulation power-factor tests and insulation dissipation-factor tests are used for trend testing of machine windings, transformers, bushings, lightning arrestors, and other power apparatus. The test applies mainly to large apparatus of 6000 volts or higher. The advantage of these tests is that they can detect the presence of deteriorated insulation even though it may be sandwiched between several layers of good insulation. Defective insulation may indicate infinity or a very high resistance value when measured with a megohmmeter, or have a high polarization index, and still be defective.

The power-factor test and dissipation-factor test measure the relative dielectric loss (energy loss in the insulation); ideal insulation would have a power factor of zero and a dissipation factor of zero. The power factor of electrical insulation should not be confused with the power factor of a circuit, motor, or system.[7]

Insulation voids, such as that caused by tape separation, cracked insulation, separation of conductors from the ground-wall insulation, and breakage of bonds between layers of mica, will develop corona and thus have greater dielectric losses under voltage stress as compared with healthy insulation. The development of such voids can be detected through periodic insulation power-factor tests or insulation dissipation-factor tests. A gradual or abrupt increase in the insulation power factor or dissipation factor over a period of years indicates a general deterioration of insulation, due to an increase in the size and number of voids [5, 11].

Since the power-factor and dissipation-factor measurements are significantly affected by moisture and other contaminants, the apparatus should be cleaned and dried and the PI determined immediately prior to making the tests.

A single insulation power-factor test or dissipation-factor test, taken at random without reference to previous records, is of little value. It is the trend of the power factor and dissipation factor, taken over a period of time, that provides an indication of the general condition of the insulation with respect to structural change, particularly the development of voids.

When making periodic tests, every effort should be made to apply the test under similar conditions of temperature and humidity and with the same test equipment. Higher temperatures and higher humidity result in greater dielectric losses and thus indicate higher power factors. Recorded data relating to insulation power-factor and insulation dissipation-factor tests should include the cleanliness and temperature of the windings, atmospheric temperature, humidity, PI, and insulation resistance. The serial number and manufacturer of the instrument should also be noted.

---

[7] The power factor of a circuit, motor, or system, as defined in Section 8.4, Chapter 8, is a measure of its effectiveness in utilizing the apparent power it draws from the generator. An ideal motor or system would have a power factor of 1.0 (unity).

The relationship between power factor and dissipation factor of electrical insulation can be developed with the aid of Figure 25.12a, which shows a representative block of insulation sandwiched between a copper conductor and the steel framework of the apparatus.

Resistance R_{eq} is the equivalent resistance of the insulation, and C is the equivalent capacitance, between the winding and the framework of the apparatus. The corresponding phasor diagram in Figure 25.12b shows the test current I_{test} and its two components. Component I_C represents the capacitive current, and component I_{loss} accounts for heat losses within the insulation due to leakage, corona, and dielectric-dipole[8] oscillation when an AC voltage is applied.

Referring to Figure 25.12b, angle θ is called the insulation *power-factor angle,* and angle δ is called the insulation *dissipation-factor angle* (also called the *loss angle*) of the insulation. Expressing the power-factor (PF) and the dissipation factor (DF) in terms of their respective angles,

$$PF = \cos \theta = \frac{I_{loss}}{I_{test}} \tag{25–3}$$

$$DF = \tan \delta = \frac{I_{loss}}{I_C} \tag{25–4}$$

The following relationship between DF and PF was derived from the geometry of the phasor diagram in Figure 25.12b.

---

[8] See Section 24–2, Chapter 2.

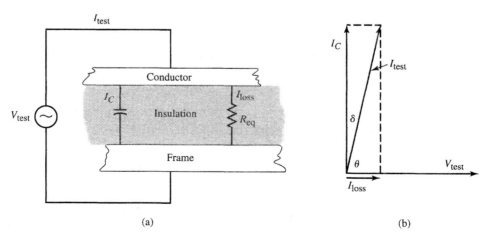

(a)   (b)

FIGURE 25.12
(a) Equivalent circuit of a block of cable insulation; (b) corresponding phasor diagram.

$$PF = \frac{DF}{\sqrt{1 + DF^2}} \qquad (25\text{--}5)$$

where: PF = insulation power factor, decimal form, called per-unit power
factor (pu)

DF = insulation dissipation factor, decimal form, called per-unit
dissipation factor (pu).

Note 1   If DF or PF is given in percent form it must be converted to pu by dividing by 100 before substituting into Eq. (25–5).

Note 2   For values of DF < 0.10 (less than 10%), cos θ ≈ tan δ, and PF ≈ DF.

EXAMPLE 25.3   The dissipation factor of a certain transformer, measured on a dissipation factor test set, is 8.05%. Calculate the percent power factor of the insulation.

Solution

$$DF = \frac{8.05}{100} = 0.0805$$

$$PF = \frac{DF}{\sqrt{1 + DF^2}} = \frac{0.0805}{\sqrt{1 + (0.0805)^2}} = 0.0802 \text{ or } 8.02\%$$

Note   The very small difference between DF and PF in this example.

The insulation power factor and dissipation factor of a complete winding of a transformer or machine represents an average value for all coils taken together. Therefore, it may not indicate a developing fault localized in one area. Splitting the phases, or if possible splitting the circuit into separate coils and testing each separately, will provide a more accurate picture of the condition of the insulation. Coils that are directly connected to the incoming power lines have the greatest voltage stresses during normal and adverse operation and are generally the first to indicate deterioration.

Power-Factor Tip-Up

The numerical difference between insulation power factors measured at two different voltages is called the *power-factor tip-up*. It is a far more significant diagnostic maintenance test than the actual power-factor values for determining void development in the stator winding insulation of rotating machines. Recommended test voltages for power-factor tip-up are 25 percent and 100 percent of the operating phase-to-ground voltage of the apparatus under test [12].

The sensitivity of the power-factor tip-up test depends on the number of coils involved in the test. Testing each stator coil individually provides the greatest sensitivity. If voids are not present, the test current will increase in proportion to the test volt-

FIGURE 25.13

Plot of power-factor tip-up representing two hypothetical coils of a synchronous generator.

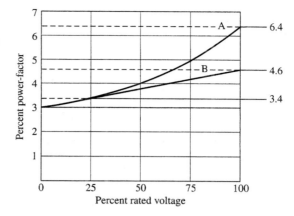

age, and phase angles θ and δ in Figure 25.12b will not change. Thus PF and DF will remain constant and the power-factor tip-up will be zero.

Figure 25.13 illustrates a plot of power-factor tip-up representing two hypothetical coils of a synchronous generator. The corresponding power-factor tip-up for each coil is:

$$\text{Tip-up (coil A)} = 6.4\% - 3.4\% = 3.0\%$$

$$\text{Tip-up (coil B)} = 4.6\% - 3.4\% = 1.2\%$$

The plot and calculated tip-up indicate that coil A has more and/or larger voids than coil B.

25.7 TESTING AND SAMPLING INSULATING LIQUIDS

The standard dielectric test for insulating liquids requires the use of an insulated test cup with 1-inch-diameter flat electrodes spaced 0.1 inch apart, an adjustable AC high-potential test set, and a pint sample of the liquid to be tested.[9] A test set with test cup is shown in Figure 25.14. Dielectric test sets are available for manual and automatic voltage adjustment. The test cup should be wiped clean with a clean, dry chamois skin and thoroughly rinsed with an approved safety solvent.

After cleaning, the cup should be filled with a sample of the same cleaning fluid and given a high-potential test. The test voltage should be increased smoothly at a rate approximating 3 kV/s until breakdown occurs. Breakdown is indicated by a continuous discharge across the electrodes or the tripping of the breaker. If the sample withstands a voltage equal to or greater than the established value for the liquid to be

---

[9]American Society for Testing and Materials, *Test for Dielectric Strength of Insulating Oils*, D-877.

FIGURE 25.14
Test set with test cup for testing insulating liquids (courtesy Associated Research Inc.).

tested, the cup is clean enough for test purposes. If it tests less than the established value, the cleaning and testing cycle must be repeated.

Assuming that the cup is in suitable condition, it should be emptied of cleaning fluid, rinsed with a portion of the oil to be tested, and then filled to 20 mm or more above the top of the electrodes. Any air bubbles in the test sample should be allowed to escape from the cup before applying the test voltage. Gently rocking the cup and then allowing it to set for 3 to 5 minutes should get rid of the entrapped air. The test voltage should then be applied and increased smoothly at approximately 3 kV/s until breakdown occurs. Occasional momentary discharges that do not cause a permanent breakdown should be disregarded.

After breakdown occurs, the test cup should be emptied, refilled with a fresh sample of the same oil, and the test repeated. The average of the breakdown voltages of five tests is taken as the dielectric strength of the sample.[10]

To avoid erroneous test results, it is extremely important that the samples of transformer oil be obtained under controlled conditions and properly cared for until tested. Polyethylene or glass containers with glass, cork, or polyethylene stoppers are preferred, and every effort should be made to use them. Rubber or rubber composition stoppers contaminate the liquid and must not be used. The containers should be rinsed

---

[10] Information on other tests of insulating liquids including interfacial tension, power factor, neutralization number, and gas analysis can be obtained from the American Society for Testing and Materials (ASTM).

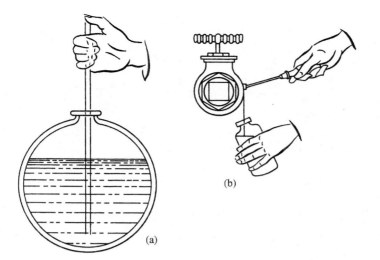

FIGURE 25.15
Methods of sampling insulating liquids: (a) using a thief; (b) from a valve.

with Stoddard solvent or chloroethene, dried, and then washed with strong soap suds, rinsed with distilled water, and baked in an oven for at least 8 hr at 105°C to 110°C. The containers should be corked immediately after drying and remain corked until ready for filling.

Samples of the liquid to be tested can be taken from a test valve, or from the drum by means of a glass or brass "thief," as shown in Figure 25.15. The "thief" should be cleaned in the same manner as prescribed for containers. Samples from outdoor apparatus should be taken on a clear, dry day with adequate protection against windblown dust. To avoid condensation of moisture on the sample, the temperature of the liquid should be no lower than the surrounding air.

Before sampling from a valve, it should be cleaned and then enough liquid allowed to run through it to flush out any moisture or other contaminants trapped in the valve. An 8-hour settling time should be allowed before sampling from a drum. The presence of water can be more easily detected if oil samples are drawn from the bottom of the tank. Each sample should be identified with the serial number of the apparatus from which it was drawn, and whether it was taken from the top or the bottom of the drum or from a valve.

25.8 FILTERING INSULATING LIQUIDS

Insulating oils that test below the minimum acceptable standard should be filtered to remove accumulated sludge and moisture. The oil should be filtered by forcing it through pads of dry filter paper. A filter press for accomplishing this is shown in

FIGURE 25.16
Filter press for cleaning and drying insulating oil (courtesy GE Industrial Systems).

Figure 25.16. The moisture is absorbed by the paper, which also filters out the sludge caused by oxidation of the oil.

Warning

Electrostatic charges can be built up when insulating oils are caused to flow in pipes, hoses, tanks, etc. The static voltages generated may be in excess of 50,000 volts. For safety reasons the tanks, filter press, pipes, and hoses should be grounded during the filtering operation and remain grounded for at least 1 hour after filtering is completed. Furthermore, because electron conduction through dielectric liquids is slow, an arc can occur between the free surface of the liquid and adjacent metal surfaces. Provisions must be made for venting any explosive gases that may accumulate in the receiving tank.

25.9 OTHER INSULATION TESTS

Insulation-resistance tests, dielectric-absorption tests, high-potential tests and power-factor (dissipation-factor) tests are the most common means for determining the general condition of electrical insulation. However, to pinpoint trouble in very large apparatus, many other tests have been devised. Some require very elaborate equipment that is too expensive for most shops. These are slot-discharge tests, surge-comparison tests, corona-probe tests, and interlaminar tests [5]. These tests are highly specialized and are beyond the scope of this book.

SUMMARY OF EQUATIONS FOR PROBLEM SOLVING

$$V_{\text{acceptance proof-test}} = V_{\text{factory proof-test}} \times 0.75 \qquad (25\text{–}1)$$

$$i_{10} = \frac{i_1}{\text{PI}} \qquad (25\text{–}2)$$

$$\text{insulation PF} = \cos\theta = \frac{I_{\text{loss}}}{I_{\text{test}}} \qquad (25\text{–}3)$$

$$\text{insulation DF} = \tan\delta = \frac{I_{\text{loss}}}{I_C} \qquad (25\text{–}4)$$

$$\text{PF} = \frac{\text{DF}}{\sqrt{1 + \text{DF}^2}} \qquad (25\text{–}5)$$

SPECIFIC REFERENCES KEYED TO THE TEXT

[1] Timperley, J. E., and Michalec, J. R., Estimating the Remaining Service Life of Asphalt-Mica Stator Insulation, *IEEE Trans. Energy Conversions,* Vol. 9, No. 4, December 1994.

[2] National Electrical Manufacturers Association, *Motors and Generators,* Sections 3.01, 12.03, 15.48, 20.48, 21.52, 22.52, 23.51.2, 24.48, NEMA Publication No. MG 1–1998.

[3] Institute of Electrical and Electronics Engineers, *Recommended Practice for Insulation Testing of Large AC Rotating Machinery with High Direct Voltage,* ANSI/IEEE STD 95–1991.

[4] Institute of Electrical and Electronics Engineers, *Recommended Practice for Electrical Installations on Shipboard,* IEEE 45–1998.

[5] Institute of Electrical and Electronics Engineers, *Guide for Insulation Maintenance for Rotating Electrical Machinery (5 hp to Less Than 10,000 hp),* ANSI/IEEE STD 432–1998.

[6] Institute of Electrical and Electronics Engineers, *Recommended Practice for Insulation Testing of Large AC Rotating Machinery with High Voltage at Very Low Frequency,* ANSI/IEEE STD 433–1991.

[7] American National Standards Institute, *American National Test Code for Dry Type Distribution and Power Transformers,* ANSI/IEEE C57.12.91–1995.

[8] Institute of Electrical and Electronics Engineers, *Recommended Practice for Testing Insulation Resistance of Rotating Machinery,* IEEE STD 43–2000.

[9] Institute of Electrical and Electronics Engineers, *Guide for Making High Direct-Voltage Tests on Power Cable Systems in the Field,* ANSI/IEEE STD 400–1991.

[10] Nobile, P. A., and LaPlatney, C. A., Field Testing of Cables: Theory and Practice, *IEEE Trans. Industry Applications,* Vol. IA-23, No. 5, September/October 1987.

[11] Institute of Electrical and Electronics Engineers, *Guide for Field Testing Power Apparatus Insulation,* IEEE STD 62–1995.

[12] Institute of Electrical and Electronics Engineers, *Recommended Practice for Measurement of Power Factor Tip-Up of Rotating Machinery Stator Insulation,* IEEE STD 286–1975.

REVIEW QUESTIONS

1. Differentiate between a factory proof-test at the power frequency, an acceptance proof-test at the power frequency, and a maintenance proof-test at the power frequency.
2. What two conditions must be met before making a high-potential test?
3. What is a controlled-DC overvoltage test, and what maintenance function does it serve?
4. Describe briefly the procedure for making a step-voltage test on an induction motor.
5. Explain why an insulation-resistance test and/or a dielectric-absorption test should precede a step-voltage test.
6. Sketch the insulation current-voltage characteristic for induction motors with good insulation and with bad insulation.
7. What precautions must be observed when making a step-voltage test?
8. How is insulation breakdown indicated when making a step-voltage test?
9. Why should the conductors be grounded after a step-voltage test, and how should it be accomplished?
10. What is a grounding stick, and how is it used?
11. Describe the two-voltage insulation-resistance test. What information does it provide?
12. What is the purpose of an AC maintenance proof-test, how is it performed, and how is failure indicated?
13. Under what conditions is an AC proof test justified?
14. What inspections and pretests should be taken before making an AC proof-test?
15. What insulation tests should be made on dry-type transformers?
16. What insulation tests should be made on liquid-filled transformers?
17. Why is high-voltage DC testing of cables preferred over high-voltage AC testing?
18. Explain why a cable should be allowed to cool to ambient temperature before performing a high-voltage test.
19. What precautions should be taken before starting a high-voltage test of cable insulation?
20. Describe the procedure, from start to final discharge, for conducting a high-voltage test on cable insulation.

21. Sketch a set of curves for good insulation and for bad insulation that can be obtained from a high-voltage cable insulation test.
22. What is a dissipation-factor test, and what is its purpose?
23. What is the equation that relates the insulation power factor to the dissipation factor?
24. What is power-factor tip-up, and what is its significance?
25. Differentiate between insulation power factor, and the power factor of a motor or distribution system.
26. Describe the procedure for testing insulating liquids.

PROBLEMS

25–1/2. A 600-V, 500-kVA, 60-Hz, three-phase generator stator is to be given an AC hi-pot maintenance proof-test. What maximum rms test voltage should be used for (a) a power-frequency test; (b) a VLF test?

25–2/2. A 6900-V, 3750-kVA, three-phase, 60-Hz generator stator is to be given an AC hi-pot maintenance proof-test. What maximum rms test voltage should be used for (a) a power-frequency test; (b) a VLF test?

25–3/3. A 500-V, 800-kW, DC shunt generator, resting in an ambient temperature of 30°C, has been cleaned and dried and has passed a dielectric-absorption test. It is to be given a DC maintenance proof-test. Determine the maximum test voltage that should be permitted.

25–4/3. A 240-V, 200-kW, DC shunt generator, resting in an ambient temperature of 30°C, has been cleaned and dried and has passed a dielectric-absorption test. It is to be given a power-frequency maintenance proof-test. Determine the maximum rms test voltage that should be permitted.

25–5/5. A 600-hp, 60-Hz, 575-V, three-phase induction motor stator with Class B insulation is to be given a step-voltage test. No previous test history or manufacturer's data are available. Determine (a) maximum allowable test voltage; (b) absorption-test voltage; (c) voltage step. (d) Prepare a pretest absorption and step-voltage table.

25–6/5. A 3000-hp, 60-Hz, 2300-V, three-phase induction motor stator with Class B insulation is to be given a step-voltage test. No previous test history or manufacturer's data are available. Determine (a) maximum allowable test voltage; (b) absorption-test voltage; (c) voltage step. (d) Prepare a pretest absorption and step-voltage table.

25–7/7. The dissipation factor of a certain transformer, measured on a dissipation factor test set, is 15.6%. Calculate the percent power factor of the insulation.

25–8/7. The dissipation factor of a certain transformer, measured on a dissipation factor test set, is 25%. Calculate the percent power factor of the insulation.

26

Cleaning, Drying, Storing, and Refurbishing Electric Machines and Transformers

26.0 INTRODUCTION

The greatest single cause of electrical failures is the breakdown of insulation. Such failures may be hastened by the absorption of moisture, oil, grease, and dust into the windings and by excessive heat, vibration, overvoltage, and aging.

This chapter includes recommended procedures for cleaning, drying, storing, and refurbishing the insulation of electrical power apparatus.[1]

26.1 WHEN TO CLEAN

The frequency with which insulation is cleaned should be driven by the condition of the equipment (condition-based maintenance) rather than driven by the calendar (time-directed maintenance). Periodic cleaning of the insulation is unnecessary. Not only is it a waste of time and labor, but it may cause the excessive wearing away of insulation. Electrical apparatus should be cleaned when one or more of the following conditions are noted:

1. Visual inspection indicates accumulated dirt on the windings.
2. The operating temperature of the equipment is 10°C to 15°C above its normal operating temperature for the same load and ambient conditions.[2]
3. Insulation resistance or the polarization index indicates the presence of dirt or moisture.

---

[1] The cleaning and lubrication of bearings is covered in Chapter 21.

[2] Reference [1] describes how one company uses a cleanliness ratio obtained by dividing the motor-temperature-rise above the ambient, by the motor-line-current, to determine when cleaning should be performed.

26.2 PRECAUTIONS PRIOR TO CLEANING

All equipment to be cleaned must be disconnected from the power source and the wiring grounded to the frame to eliminate any accumulated static charge. When cleaning large motors, such as the main propulsion motors of electric-drive ships, it is often necessary to climb inside the machine. All pants and shirt pockets should be emptied of all loose articles prior to such entry. If a screw, coin, or other small article were to fall into the machine, it could cause serious damage when starting up.

Do not use protective hand creams containing silicone when cleaning, overhauling, or otherwise working on any part of a DC machine. The presence of silicone vapor, even in minute amounts, will damage the surface film of commutators, causing rapid brush and commutator wear.

26.3 REMOVING LOOSE DUST

Cooling air blowing through electrical machines may contain a mixture of iron dust, copper dust, carbon dust, mica dust, and other atmospheric contaminants. Iron dust is caused by vibration of the steel laminations due to alternating magnetic fields; copper dust, mica dust, and carbon dust are caused by wear of commutators and carbon brushes. The dust, combined with oil and water vapors, is gradually deposited on the windings and in the ventilation slots. The accumulated dirt retards the flow of heat; this causes the buildup of excessive heat, shortening the life of the insulation.[3] The buildup of conducting dust on the insulation may cause tracking, resulting in grounds, burning, short circuits, and a possible flashover.

The preferred method for removing loose, dry dust is with a vacuum cleaner. A nonconducting brush and a thin nonconducting nozzle make excellent attachments for cleaning commutator risers and other hard-to-get-at places. After vacuuming, oil-free, dry compressed air at about 25 psi should be used to loosen and clear out any remaining dust. The use of safety goggles and a dust mask permits close work and encourages a thorough cleaning job. When using compressed air for cleaning, an adequate exhaust must be provided; otherwise it merely rearranges the dust, blowing it off one part of the machine and depositing it on another part or on some other nearby apparatus. Opening both ends of the machine provides an adequate exhaust if compressed air is to be used.

Another very effective means for removing loose dirt is to push a clean, dry lint-free rag or cheesecloth through winding crevices and gently pull it back and forth in a shoe-shine fashion. This should be done along the ledge behind the commutator risers, through the ventilating ducts, through spaces between the end turns, and in the space between the pole iron and the frame. Rags should be changed frequently. A bottle brush with an extension, or a brush used to clean rifle bores, is also very effective. *However, do not use rifle cleaning solvent.*

---

[3] A 10°C increase in temperature, if sustained, will cause the life of the insulation to be cut in half. See Section 23–6, Chapter 23 and Sec.24–3, Ch. 24.

Encrusted dirt that blocks air passages should be removed carefully with a hardwood, plastic, or fiber scraper. A metal scraper will damage the insulation and should not be used.

26.4 CLEANING WITH SOLVENTS[4]

Cleaning Small Motors

Insulation that is coated with an oily film may be cleaned by wiping or rubbing with a piece of cheesecloth or lint-free rag slightly moistened with an approved safety solvent. Since excessive use of solvents will damage the protective coating provided by the varnish, only enough solvent should be used to remove the oily scum, and the insulation then wiped dry with a clean cloth. The windings must not be soaked in solvent. Before using any solvent on electrical insulation, it should be tested on a small spot to determine whether or not it is detrimental to that particular insulation. Solvents used for cleaning electrical apparatus must be OSHA approved.

The immersion and running of electrical machines in solvents should not be done. Although the machine comes out looking shiny and new, the redepositing of conducting dust in unseen crevices may result in shortening the life of the equipment. Immersion of commutators is particularly bad because carbon dust can be carried into inaccessible places. Furthermore, soaking in solvent degrades the insulation, making revarnishing necessary.

Cleaning Large Electric Motors and Generators

Cleaning large electric motors and generators in power generating stations, marine power plants, petrochemical plants and large industrial facilities may be done safely, effectively, and with minimum downtime by using a high-pressure airless washing system.

An airless pressure washing system with vacuum recovery, called the Electrosolve® Cleaning System, is illustrated in Figure 26.1 [1, 2]. The solvent loosens and removes the grime from hard-to-get-at places, such as the ventilating ducts and recesses in rotor and stator windings. A vacuum hose inserted at the bottom of the machine, with the other end connected to a suction pump, sucks up the solvent, along with loosened dirt, carbon dust, and dissolved grease, pumping it into specially designed 55-gallon waste containers for proper disposal. The suction hose is also used to remove loose dust from the windings prior to pressure washing. The cleaning system illustrated in Figure 26.1 has been used on motors as large as 310,000 hp and turbine generators as large as 750 MW.

It is recommended that machines cleaned by airless pressure washing be revarnished before returning to service.

---

[4] Insulations containing silicones, such as Class H insulation, are very sensitive to commercial solvents. Such insulation requires cleaning with water and detergent.

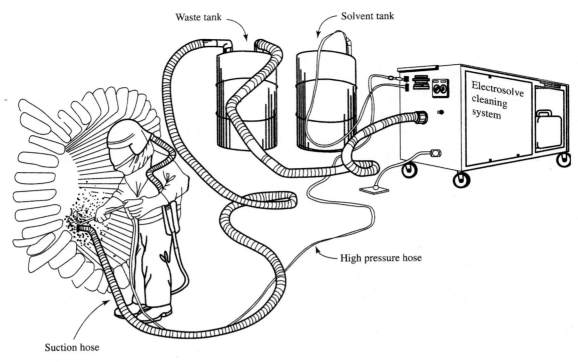

Waste tank

Solvent tank

Electrosolve cleaning system

High pressure hose

Suction hose

FIGURE 26.1
Cleaning with a high-pressure airless washing system (courtesy Electrosolve® Services).

26.5 OTHER CLEANING METHODS

Cleaning with Abrasives

Very hard dirt deposits that are visible but not removable by solvents, or water and detergents, may be cleaned by air blasting with ground corn cobs, peanut shells, or the more abrasive ground walnut shells. Air blasting must be recognized as a drastic measure, because the insulation can be very easily worn away. To prevent damage, the operator must be close enough to the work to see what he or she is doing. It is imperative that the operator wear close-fitting goggles and a dust mask.

Cleaning with Dry Ice

Cleaning by blasting with dry ice is another method for cleaning motors, generators, and switchgear. There is absolutely no residue except for the dirt being removed. A CO_2 miniblast system, illustrated in Figure 26.2, creates a fine crystalline blasting medium from a 120-pound block of dry ice. Nitrogen gas or compressed air is used as a propellant. A specially designed insulated nozzle allows "live" cleaning of pad-mounted switchgears, enabling routine cleaning of switchgear without power shutdown.

FIGURE 26.2
Cleaning by blasting with dry ice (courtesy Alpheus Cleaning Technologies).

26.6 RECONDITIONING SUBMERGED OR FLOODED EQUIPMENT[5]

Electrical equipment that has been submerged or flooded should be disassembled, and the windings washed with fresh hot water. The water should be applied using a high-pressure airless water sprayer. The water temperature should not exceed 90°C (194°F), and the nozzle pressure should not exceed 25 lb/in.^{2}. Cleaning should be done as rapidly as possible, the machine rinsed with fresh warm water, excess moisture wiped off with a clean, dry cloth, and the apparatus baked dry. Note that nonmagnetic rotor retaining rings constructed of manganese-chromium (MnCr) alloys are subject to stress corrosion in the presence of moisture. Such rings must be removed, cleaned, dried, and then tested for pits and cracks using dye penetrants.

---

[5] If possible, the manufacturer should be consulted prior to starting the reconditioning process.

If the windings are covered with mud, or other solid contaminants, a nonconducting detergent should be added to the wash water to help loosen and wash away the dirt.[6]

If the windings were submerged in saltwater, freshwater washing should be continued until salinity tests of the wash water show the insulation to be free of salt; this may take up to 10 hours. Small machines may be soaked for several days in fresh warm water, changing the water frequently. Machines that are dried without flushing out the salt may initially pass an insulation-resistance test, but if shut down for an extended period of time, the salt will reabsorb moisture, and the insulation will fail when put back in service. After a thorough rinsing with freshwater, the residual water should be removed as quickly as possible by wiping and blowing with low-velocity, oil-free, dry compressed air.

Some large DC machines may require removal of the commutator clamping ring to drain any accumulated water; the manufacturer should be consulted before doing this. The machine should then be baked dry with externally applied heat.

However, even with the best treatment and careful tests, a machine that was submerged sometimes fails when put back into service. It is therefore recommended that reconditioned flooded equipment be limited to use in nonvital operations that would not endanger the plant or personnel if the equipment were to fail. After 4 months of continuous service with no failure, and satisfactory insulation-resistance tests, reconditioned equipment can be assumed to be reliable for unrestricted use.

26.7 DRYING MACHINERY WITH EXTERNAL HEAT

The machine should be disassembled, and the component parts dried in a baking oven or by means of an improvised oven of tarpaulin surrounding the machine. A vent at the top of the tarpaulin permits the exit of moisture-laden air. Small disassembled machines are often successfully dried by placing them on top of a boiler.

Drying in a vacuum chamber with external heat is the quickest and most effective method of drying electric apparatus, and can reduce dryout time by 50 to 70 percent. The necessary heat is applied to the outside of the vacuum chamber.

When external heat is used for baking, the temperature of the windings should not exceed 90°C (194°F) as measured by thermometers taped to the coils. The best source of heat for makeshift ovens is an electric heater or radiant-heat lamps; steam heaters or hot-air furnaces also do an adequate job. Drying windings that contain asbestos insulation will take much longer because of the moisture-absorbing properties of asbestos. It may be best to replace those motors.

26.8 DRYING ELECTRICAL MACHINERY WITH INTERNAL HEAT (CIRCULATING CURRENT)

Windings of AC and DC machines that have an insulation resistance greater than 50,000 ohms at room temperature may be baked with internal heat. Current is fed into the windings at low voltage, and the heat developed internally serves to drive off the

---

[6] Examples of detergents used in some repair shops are saltwater soap, and Tech-736 from Cantol Co.

moisture. The required current may be obtained from an external source such as a DC electric welder or generated internally. The current should be increased gradually, over a period of 6 hours, until slightly less than rated current is achieved, and then held at that value until insulation-resistance tests indicate that the drying process is completed. Note that the total drying time may take as long as 20 hours or longer.

A transformer must be under constant surveillance during the drying process, so that emergency action may be taken in the event of fire. A carbon-dioxide fire extinguisher should be available for immediate use.

Too rapid heating of the winding will form steam pockets that may rupture and permanently damage the insulation. Hence, constant supervision and monitoring of winding temperature is required when using circulating current to dry the windings. The internal temperature of the windings must not be allowed to exceed 85°C (185°F) as determined by resistance temperature detectors (RTDs) or by the change in resistance method.[7] When using internal heat, the outside temperature of the winding lags behind the internal temperature. If readings from a thermometer positioned on the outside of the windings are used to monitor the temperature, the temperature as indicated on the thermometer should not be allowed to exceed 77°C (170°F). Mercury thermometers must not be used, because breakage will cause mercury contamination of the windings. Furthermore, if alternating current is used to heat the windings, it will heat the mercury by induced current, resulting in an erroneous indication.

Using a DC Welder

A circuit for using a DC welder to dry the armature windings of three-phase motors and three-phase generators, whether wye or delta, is illustrated in Figure 26.3a. The two welder cables should be shifted every 2 hours; connecting first to T_1 and T_2, then to T_2 and T_3, then to T_3 and T_1, then back to T_1 and T_2, etc. Note that before shifting

---

[7] See Section 1–9, Chapter 1.

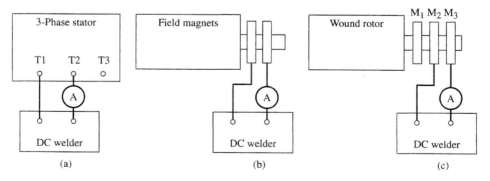

(a)　　　(b)　　　(c)

FIGURE 26.3
Circuits for using a DC welder to dry the windings of motors and generators: (a) three-phase stator windings; (b) rotor windings; (c) wound-rotor windings.

the leads, the current must be reduced to zero, and an insulation-resistance measurement made.

A circuit for using a DC welder to dry the field windings (rotor) of an alternator or synchronous motor is illustrated in Figure 26.3b, and a circuit for drying the rotor of a wound-rotor induction motor using a DC welder is illustrated in Figure 26.3c. Copper bands strapped around the collector rings should be used to conduct the current; if the brushes are used for this purpose, local heating will occur where the brushes contact the rings. This will cause blackening and pitting of the rings and damage to the brushes. Alternating current should not be used for drying the rotor windings, because transformer action will overheat and possibly damage the squirrel-cage or damper windings.

The armature of a DC motor can be dried with circulating current through the brushes if some external means is provided to continuously rotate the armature. If the armature is stationary, localized heating will occur at the brushes. The series field and shunt field are not used, and a DC welder or other adjustable low-voltage source is connected to the armature circuit as illustrated in Figure 26.4. Do not disturb the interpole connections. Mechanical rotation of the armature is started, the welder is energized at its lowest voltage, and the current gradually increased from zero to rated motor current.

Using Applied Low-Voltage AC

Induction motors can also be dried by blocking the rotor and applying approximately 25 percent of its rated AC voltage to the stator.

When drying the stator of a synchronous machine with low-voltage AC, with the rotor blocked so that it does not turn, the field circuit should be short circuited. This will prevent high induced voltage, due to transformer action, from damaging the field-coil insulation. Transformer action may also cause overheating of the damper windings in rotors that are so equipped.

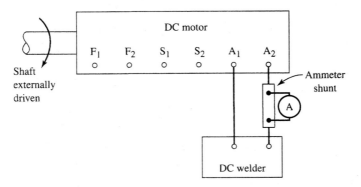

FIGURE 26.4
Using a DC welder to dry a DC armature.

Synchronous propulsion motors for electrically propelled ships may be dried while tied to the dock, by operating the motors at very low speed with weakened field excitation [3].

Using Self-Induced Current

The circuit for drying a synchronous generator with self-induced current is illustrated in Figure 26.5. With the generator stationary, the generator circuit breaker is locked in the open position, and the armature and field cables are disconnected at the machine. The armature is short circuited through a current transformer, using an appropriately sized cable, and the field terminals are connected to a DC welder, or other low-voltage DC source.

Then, with the field switch open, and the DC welder set at its lowest voltage, the prime mover is started and accelerated to below one-half its rated speed. The reduced speed of the machine prevents damage to damp insulation. The field circuit is then closed, and the field current gradually increased until the rated armature current is attained.

DC Cumulative Compound Generators

The normal circuit for a DC cumulative compound generator, rotating in a clockwise direction facing the end opposite the drive, is illustrated in Figure 26.6a. The corresponding drying circuit using self-induced current is illustrated in Figure 26.6b. Note that the series field winding is reversed in the drying circuit, resulting in a differential connection: The series field is in opposition to the shunt field, preventing too rapid a buildup of short-circuit current, which could damage the armature windings.

Procedure

1. With the generator stationary, and the generator panel-board breaker locked in the open position, the armature and field cables are disconnected at the

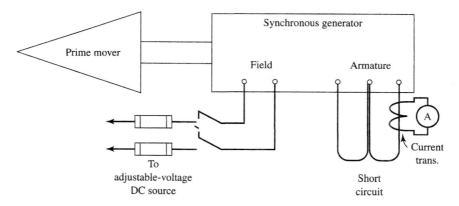

FIGURE 26.5
Drying a synchronous machine with self-induced current.

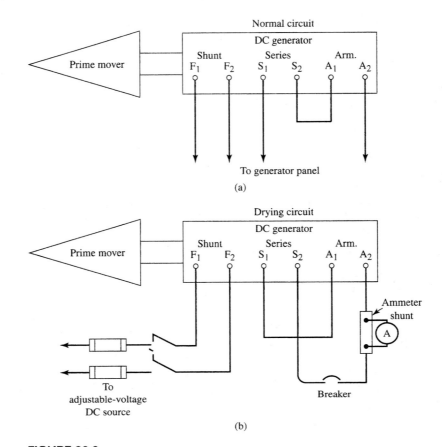

FIGURE 26.6
(a) Normal circuit for a DC cumulative compound generator; (b) corresponding drying circuit using self-induced current. Note that this is a differential connection.

machine. The armature and series field are then connected (with appropriately sized cable) through an ammeter shunt, and a separate adjustable circuit breaker or fuse, as illustrated in Figure 26.6b. The interpole connections must not be disturbed. The circuit breaker protects against a large and rapid buildup of current, which could occur if the series winding is not connected differentially or if the brushes are off neutral. The field terminals are connected to a DC welder or other low-voltage source.

2. With the shunt-field switch open, the circuit breaker open and set at its lowest value,[8] and the DC welder set at its lowest voltage, the prime mover is started and accelerated to less than one-half its rated speed. The reduced speed of the machine prevents centrifugal force from damaging damp or wet insulation.

---

[8] If the circuit breaker is not adjustable, use a low current fuse for the preliminary check.

3. Close the circuit breaker and note if there is a large buildup of armature current due to residual magnetism. If no rapid buildup of current occurs, the differential connection is correct. The shunt-field circuit should then be closed, and the welder voltage gradually increased over a 6-hour period until approximately rated armature current is attained.

Shunt Generators with Compensating Windings

Shunt generators with compensating windings should have the brush rigging shifted a few degrees in the direction of rotation. The resultant slightly drooping voltage characteristic prevents a gradual rise in current (called creepage current) that could attain values high enough to damage the insulation.

26.9 MONITORING THE DRYING PROCESS

The winding temperature should be monitored to make sure that overheating does not occur, and the insulation resistance should be monitored to determine when the drying process is complete.

Insulation-Resistance Measurements

Sixty-second insulation-resistance measurements using a megohmmeter should be recorded every 4 hours during the drying-out process. See voltage guidelines in Table 24.1, Chapter 24. A typical drying curve for a DC motor armature is illustrated in Figure 26.7.

During the first part of the drying operation, the increase in temperature causes a decrease in the indicated values of insulation resistance. Then, with a constant drying temperature, the insulation resistance increases as the moisture is driven out. When the insulation is dried and allowed to cool, the insulation resistance increases to a high value. The plotted values of insulation resistance are not corrected for temperature, because such correction would serve no useful purpose in this case.

The temperature of the winding during the drying period can be determined by the change in resistance method as presented in Section 23–7, Chapter 23.

26.10 CLEANING AND DRYING DRY-TYPE TRANSFORMERS

The windings and air passages of air-cooled transformers (called dry-type transformers) should be cleaned first with a vacuum cleaner, followed by dry, oil-free, compressed air or nitrogen not exceeding 25 psi. All insulating surfaces should be wiped clean with a dry cloth. The use of solvents should be avoided, because they may have a deteriorating effect on the insulation. Strip heaters should be used during periods of prolonged shutdown to avoid condensation of moisture on the windings.

FIGURE 26.7
Typical drying curve for a DC motor armature (courtesy James G. Biddle Co.).

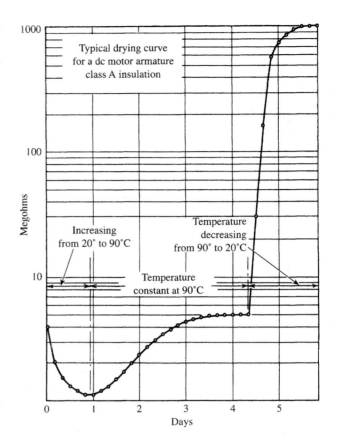

Drying Transformers with External Heat

If insulation-resistance tests indicate that the windings have absorbed moisture or if the windings have been subject to unusually damp conditions, they may be dried by external heat, internal heat, or a combination of both. Drying with external heat is the preferred method and may be accomplished by directing heated air into the bottom ventilating ducts of the transformer housing. The temperature of the air used for drying must not exceed 110°C. If the transformer is small enough, it may be dried in an oven designed for such purposes.

Drying Transformers with Internal Heat

Drying with internal heat is much slower and is safe only if the insulation resistance is 50,000 ohms or more when measured at room temperature of 20°C (68°F). Care must be taken to avoid contact with dangerously high voltages. Drying is accomplished by short-circuiting one winding, preferably the low-voltage winding, and applying reduced voltage to the other winding to obtain approximately 50 to 100 percent of rated current [4]. Some manufacturers recommend drying currents as low as 20 percent of rated current. It is best to consult the manufacturer before attempting internal drying. A transformer must be under constant surveillance during the drying process, so that emergency action may be taken in the event of fire. A carbon-dioxide fire extinguisher should be available for immediate use.

The winding temperature should be maintained between 80°C and 90°C as measured by the change in resistance method, or by spirit thermometers placed in the ducts between the windings. Mercury thermometers must not be used, because induced currents in the mercury will result in an erroneous indication, and breakage will contaminate the transformer.

Circuits for internal drying of transformers are illustrated in Figure 26.8a for single-phase and Figure 26.8b for three-phase units. In both single-phase and three-

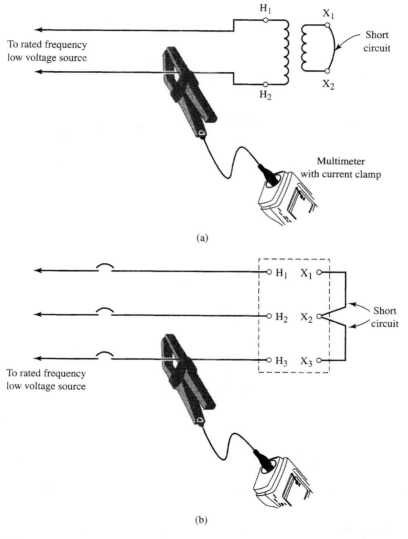

(a)

(b)

FIGURE 26.8

Circuits for drying transformers: (a) single-phase transformer; (b) three-phase transformer.

phase transformers, the low-voltage side is shorted and the high-voltage side connected to a low-voltage source. An adjustable-voltage autotransformer, if available and of sufficient capacity, makes an excellent low-voltage source.

Required Voltage for Internal Drying

The required input voltage to the high side that will result in the desired high-side current (with the low side shorted) can be calculated using the following equation:

$$V_{HS,required} = V_{HS,rated} \times \frac{\%Z_V}{100} \times \frac{\%I_{HSsc}}{100} \qquad (26\text{--}1)$$

where: $V_{HS,rated}$ = rated high-side voltage (V)
  $\%Z_V$ = percent impedance voltage of the transformer
  $\%I_{HSsc}$ = percent high-side short-circuit current desired.

The percent impedance voltage ($\%Z_V$) of the transformer is the percent of rated voltage that will circulate rated current in the transformer with the secondary of the transformer short-circuited. The percent impedance voltage may be obtained from the transformer nameplate or from the manufacturer.

If the percent impedance voltage of the transformer cannot be determined, substituting $\%Z_V = 1.5$, and $\%I_{HSsc} \leq 50$ in Eq. (26–1) should provide an acceptable voltage for the drying current.

EXAMPLE 26.1

A 150-kVA, 60-Hz, 2400—240-volt single-phase dry-type transformer is to be dried with internal heat. The transformer impedance voltage is 1.4%. Determine (a) high-side voltage required to obtain 60% high-side current with the low side shorted; (b) high-side short-circuit current; (c) minimum kVA rating required for the low-voltage source; (d) current in short-circuited secondary; (e) minimum size of cable required to short-circuit the secondary. The circuit is the same as that illustrated in Figure 26.8a.

Solution

a. $V_{HS,required} = V_{HS,rated} \times \dfrac{\%Z_V}{100} \times \dfrac{\%I_{HSsc}}{100}$

$= 2400 \times \dfrac{1.4}{100} \times \dfrac{60}{100} = 20.16 \Rightarrow 20 \text{ V}$

b. $I_{HS,rated} = \dfrac{S_{rated}}{V_{HS,rated}} = \dfrac{150,000}{2400} = 62.5 \text{ A}$

$I_{HSsc} = 62.5 \times 0.60 = 37.5 \text{ A}$

c. $S = V_{HS,required} \times I_{HSsc} = 20.16 \times 37.5 = 756 \text{ VA} \Rightarrow 0.756 \text{ kVA}$

d. From Eq. (10–4),

$$\text{Approximate turns-ratio: } a \approx \frac{V_{HS}}{V_{LS}} = \frac{2400}{240} = 10$$

From Eq. (10–8b),

$$a \approx \frac{I_{LSsc}}{I_{HSsc}} \implies I_{LSsc} \approx I_{HSsc} \times a = 37.5 \times 10 \approx 375 \text{ A}$$

e. From Appendix 2, select 300-MCM cable with Type TW or UF insulation.

EXAMPLE 26.2 A three-phase, 500-kVA, delta-wye, 4160—480Y/277-volt, 60-Hz, dry-type transformer, whose percent impedance voltage is 1.8, is to be dried with internal heat.[9] The circuit is illustrated in Figure 26.9. Determine (a) the high-side voltage required to obtain approximately 40% high-side current with the low side short-circuited; (b) high-side short-circuit phase current; (c) minimum kVA rating required for the low-voltage source; (d) current in the shorting cables; (e) minimum conductor size for the shorting cables.

Solution
The problem will be solved on a single-phase basis. For a delta primary:

$$V_{phase} = V_{line} = 4160 \text{ V}$$

$$S_{phase} = \frac{S_{3phase}}{3} = \frac{500}{3} = 166.67 \text{ kVA}$$

a. $V_{HS,required} = V_{HS,rated} \times \dfrac{\%Z_V}{100} \times \dfrac{\%I_{HSsc}}{100} = 4160 \times \dfrac{1.8}{100} \times \dfrac{40}{100}$

$$= 29.952 \Rightarrow 30 \text{ V}$$

b. The rated high-side phase current is:

$$I_{phase} = \frac{S_{phase}}{V_{phase}} = \frac{166.67 \times 1000}{4160} = 40.06 \text{ A}$$

The high-side short-circuit phase current is

$$I_{HSsc} \approx 0.40 \times 40.06 \approx 16.03 \text{ A}$$

---

[9] See Section 10–8, Chapter 10, for a review of three-phase connections of transformers.

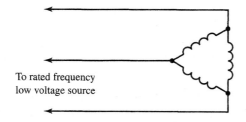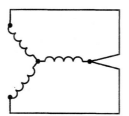

To rated frequency
low voltage source

FIGURE 26.9
Circuit for Example 26.2.

c. $S_{min/phase} = V_{HSreq} \times I_{HSsc} = 29.952 \times 16.03 = 480\,\text{VA}$

$S_{min,3ph} = 3 \times 480 = 1440\,\text{VA} \Rightarrow 1.44\,\text{kVA}$

Thus, a 1.5-kVA, three-phase, 60-Hz source with a line voltage of approximately 30 V is required to heat the transformer.

d. $a = \dfrac{V_{HSphase}}{V_{LSphase}} = \dfrac{4160}{277} = 15.01$

$a = \dfrac{I_{LSsc}}{I_{HSsc}}$

$I_{LSsc} = a \times I_{HSsc} = 15.01 \times 16.03 = 240.56\,\text{A}$

e. From Appendix 2, select AWG-3/0 cable with Type TW or UF insulation.

Drying with External and Internal Heat

A combination of external and internal heat provides the quickest method for drying and also reduces the amount of short-circuit current required.

26.11 CLEANING AND DRYING LIQUID-FILLED TRANSFORMERS [5]

A liquid-filled transformer, illustrated in Figure 26.10, requires more attention than the dry type. The liquid used for the insulating medium is either mineral oil or silicone oil.[10]

These liquids should be checked at least once a year for the presence of moisture and sludge.[11] The accumulation of sludge on transformer coils and in cooling ducts reduces the heat-transfer capability, causing higher operating temperatures.

Liquid-filled transformers should be inspected under the cover at least once every 10 years; the transformer should be de-energized, the lid removed, and enough of the insulating liquid should be drained to expose the top of the core and coils. To prevent the absorption of moisture, do not leave the coils exposed to the air any longer than is necessary to make the inspection.

Minor leaks at welded joints may often be stopped by peening over with a drift and a ball-peen hammer. Leaks in cast iron may be repaired by drilling, tapping, and inserting a pipe plug. Leaks at gaskets may be stopped by pulling up on the bolts or by installing new cork gaskets. Large leaks require the attention of the manufacturer.

---

[10] Insulating liquids, called askerels or PCBs, were used in earlier construction. Due to their hazardous nature they are no longer manufactured.

[11] See Section 25–8, Chapter 25.

FIGURE 26.10
Liquid-filled transformer (courtesy TECO Westinghouse).

Liquid-filled transformers that have been flooded or have otherwise taken in water should be dried by following the manufacturer's instructions. However, if circumstances require that the drying procedure be started immediately, drain the liquid into another container, and follow the procedure outlined for the dry-type transformer. The liquid should be dried with a filter press and tested for dielectric strength before it is pumped back into the transformer tank.

26.12 CARE OF ELECTRICAL APPARATUS DURING EXTENDED PERIODS OF INACTIVITY

The greatest injury done to electrical insulation during periods of idleness or storage results from the absorption of moisture that condenses on the insulation. Energizing a damp winding can break down the insulation, causing a ground or short circuit. Condensation of moisture on electrical apparatus results when the surface temperature of the winding falls below the dew point. Hence, if the temperature within the machine is kept between 5°C and 10°C higher than the ambient, condensation will not take place. Heat can be provided by external or internal space heaters or by low-voltage heating of the windings.

Brushes should be raised off commutator and collector rings to prevent electrolytic action from pitting or wearing flat spots on the metal surfaces. This is especially true in mines and aboard ship where an atmosphere of moist salt air prevails. Shafts should be rotated occasionally to ensure that all bearing surfaces will be covered with protective lubricant.

Space Heaters

Electric lamps may be placed in the housing of open machines to prevent the condensation of moisture. Many generators, motors, and controllers have built-in space heaters for this purpose. A tarpaulin covering may be used to conserve heat and reduce the amount of energy required.

Low-Voltage Maintenance Heating of Windings

If low-voltage heating of windings is used, a voltage of approximately 3 to 7 percent of rated machine voltage should be sufficient to maintain the winding temperature at 3°C to 10°C above the ambient. The low voltage should be applied to the field windings of DC machines and to one stator-phase of AC machines.

Solid-state motor winding heaters small enough to fit in the control box are available for this purpose [6]. The circuit automatically supplies low-voltage heating to the motor windings when the motor-starter contacts are opened (motor off) and automatically removes low-voltage heating when the motor-starter contacts are closed (motor running). The output voltage of the solid-state circuit is factory set to maintain a 3°C to 10°C differential above the ambient temperature, thus protecting against condensation. The heat power supplied to the motor winding is approximately 1 to 3 watts per horsepower.

Warning

It must be assumed that motors equipped with automatic motor-winding heaters will always have voltage present at the motor terminals, even when the motor is stopped. Thus, before servicing such motors, make sure that the circuit breaker ahead of the motor starter is open.

26.13 REVARNISHING ELECTRICAL INSULATION

The revarnishing of electrical insulation should be done only when absolutely necessary. The buildup of many layers of varnish leads to surface crazing and the resultant absorption of dirt and moisture. The windings should be thoroughly cleaned and dried before the application of varnish. Do not mix varnishes from different manufacturers; do not mix different varnishes from the same manufacturer; and do not mix old varnish with new varnish. Use only the recommended thinner.

The best method of revarnishing is to dip the unit in varnish, allow for proper impregnation, drain, and bake in an oven. Armatures too large to fit into a tank may be rolled slowly in a pan of varnish and then baked. Commutators must not be immersed in varnish.

Aboard ship, or in shops where baking ovens are not available or where machines are too large, air-drying varnish may be applied with a sprayer. However, note that spraying can never reach more than 60 to 70 percent of the surfaces, and although it is better than doing nothing, the lack of varnish in some vital areas may cause failure. Before using, the varnish should be strained through several layers of cheesecloth to remove any dirt, lumps, or skin. The air pressure at the spray gun may be anywhere between 20 lb/in.^{2} and 70 lb/in.^{2}, depending on the specific application. An air cleaner should be installed in series with the air hose, between the compressor and the spray gun, to remove any oil or water vapor.

The commutator, slip rings, shaft, and bearings should be protected against the spray with several layers of heavy paper. The operator should wear an approved mask for protection against the noxious vapors. Generally two light coats are recommended, with the second application made when the first coat is no longer tacky. The drying time between coats is generally less than 24 hours and depends on the surrounding temperature. If necessary, the drying may be accelerated by the application of a moderate amount of external heat, provided that the temperature of the apparatus does not exceed 80°C (176°F).

The vapors given off by the varnish are highly explosive. To reduce the risk of fire, the spray gun as well as the apparatus to be varnished should be grounded to a common point before starting. This prevents the buildup of a static charge that may touch off an explosion. A carbon-dioxide type of fire extinguisher should be conveniently located in the spray area.

SUMMARY OF EQUATIONS FOR PROBLEM SOLVING

$$V_{\text{HS,required}} = V_{\text{HS,rated}} \times \frac{\%Z_V}{100} \times \frac{\%I_{\text{HSsc}}}{100} \qquad (26\text{--}1)$$

SPECIFIC REFERENCES KEYED TO THE TEXT

[1] Electrosolve® Services Is Cutting Your Costs to Maintain Electrical Machines, *Journal of the International Ship Electrical and Engineering Service Association*, 1997, pp. 18–20.

[2] Trevor, S. P., Efficient Cleaning Vital for Electrical Equipment, *Power Magazine*, May 1986, pp. 74–77.

[3] *Recommended Practice for Electric Installations on Shipboard*, IEEE STD 45-1998.

[4] *Recommended Practice for Installation, Application, Operation, and Maintenance of Dry-Type General Purpose Distribution and Power Transformers*, ANSI/IEEE C57.94-1982 (R1987).

[5] *Guide for Installation and Maintenance of Oil-Immersed Transformers*, USAS C57.93-1995.

[6] Institute of Electrical and Electronic Engineers, *IEEE Trans. Industry Applications*, Vol IA-11, No. 3, May/June 1975, pp. 287–290.

REVIEW QUESTIONS

1. Is it advisable to clean electrical insulation on an annual basis? Explain.
2. What observable conditions would indicate that the insulation of electrical apparatus requires cleaning?
3. (a) Describe the recommended procedure for cleaning apparatus that is covered with loose, dry dust. (b) How should encrusted dirt be removed?
4. What is the recommended procedure for cleaning insulation that is coated with an oily film?
5. How should insulation containing silicone be cleaned?
6. Why should the cleaning of insulation by air spraying with a solvent be avoided?
7. Explain why the immersion and running of electrical machinery in solvents is not recommended.
8. Is air blasting with abrasives a generally recommended procedure for cleaning insulation? Explain.
9. Outline the procedure to be followed for cleaning and drying submerged or flooded equipment.
10. Describe an improvised method that may be used for drying electrical machinery in your plant or school.
11. What restrictions should be placed on the use of reconditioned flooded equipment?
12. Under what conditions can internal heat be used to dry electrical insulation? How is it applied, and what precautions should be observed?
13. Describe the method used to determine the internal temperature of a winding during the drying period.
14. Why does a thermometer not give a true indication of winding temperature?
15. Describe two methods for revarnishing electrical insulation. What precautions should be observed?
16. What method of revarnishing insulation should be used when a baking oven is not available?
17. Should the insulation of a machine be revarnished every time it is cleaned? Explain.

18. A dry-type transformer that has been idle for 12 months is to be placed into service. The insulation is damp. Describe a method for removing the accumulated moisture using external heat. What is the maximum allowable temperature?

19. If equipment for the external heating of transformers is not available, state the procedure and precautions to be followed when drying with internal heat.

20. Why are spirit thermometers preferred over mercury thermometers when measuring the winding temperature of an energized transformer?

21. What types of insulating liquids are used in liquid-filled transformers?

22. What causes sludging of insulating liquids, and what can be done about it?

23. How can small leaks at the welded joints of a liquid-filled transformer be repaired?

24. What three methods are used to prevent moisture from condensing on the insulation and windings of electrical apparatus during extended periods of inactivity?

25. How many degrees higher than the ambient must the temperature of insulation be in order to prevent condensation of moisture?

26. When a machine is to be left idle for a considerable period, the brushes should be raised off the commutator and collector rings. Why is this necessary?

PROBLEMS

26–1/10. A 100-kVA, 60-Hz, 7620—2400-V, single-phase distribution transformer whose percent impedance voltage is 1.6% is to be dried by short-circuiting the low-voltage side and applying rated high-side current. Sketch the circuit and determine (a) required high-side voltage; (b) high-side short-circuit current; (c) kVA rating required for the low-voltage source; (d) short-circuit current on the low side; (e) minimum size cable required to short-circuit the secondary.

26–2/10. A 75-kVA, 60-Hz, 4160—240-V, single-phase distribution transformer whose percent impedance voltage is 1.5% is to be dried by short-circuiting the low-voltage side and applying 80% rated high-side current. Sketch the circuit and determine (a) required high-side voltage; (b) high-side short-circuit current; (c) kVA rating required for the low-voltage source; (d) current in the shorting cables; (e) minimum size cable required to short-circuit the secondary.

26–3/10. A three-phase, 100-kVA, delta–delta, 2400—240-V, 60-Hz, dry-type transformer with a percent impedance voltage of 1.7% is to be dried by short-circuiting the low-voltage side and applying 30% rated high-side current. Sketch the circuit and determine (a) required high-side voltage; (b) high-side short-circuit current; (c) kVA rating required for the low-voltage source; (d) current in the shorting cables; (e) minimum size cable required to short-circuit the secondary.

26–4/10. A three-phase, 25-kVA, delta–wye, 600—208Y/120-V, 60-Hz, dry-type transformer with a percent impedance voltage of 3.68% is to be dried by short-circuiting the low-voltage side and applying 80% rated high-side current. Sketch the circuit and determine (a) required high-side voltage; (b) high-side short-circuit current; (c) kVA rating required for the low-voltage source; (d) short-circuit current in low side; (e) minimum size cable required to short-circuit the secondary.

27

Operation and Maintenance of Battery Systems for Industrial, Marine, and Utility Operations

27.0 INTRODUCTION

A battery is an electrochemical cell that is used to store chemical energy for conversion later to electric energy. Of the many different types of batteries available, this chapter describes only the principal types in commercial use and presents their characteristics and proper charging methods. The two basic types of cells are primary and secondary. A primary cell, such as those commonly used in flashlights and small radios, is chemically irreversible. Such cells cannot be recharged and must be discarded when no longer usable; their active material is consumed during discharge.

The secondary cell, known as a storage cell, is chemically reversible. Its active materials are not consumed. Such cells can be recharged by passing current through the battery in the reverse direction.

Batteries may be connected in series, parallel, or series–parallel arrangements to provide the necessary voltage and current for a specific application. Battery applications include engine cranking for gasoline and diesel engines; electromotive power for electric cars, industrial trucks, and forklift trucks; ships' emergency power; stationary batteries for uninterruptible power systems used in generating stations and substations; and bulk storage of energy for system overloads (called peaking service). The Southern California Edison Company has the world's largest battery, with more than 8000 lead-acid cells, providing 10 MW of power for 4 hours to handle peak shaving and load leveling [1].

27.1 LEAD-ACID CELL

The lead-acid battery, illustrated in Figure 27.1, has a nominal open-circuit voltage of 2 volts per cell. It has positive plates of lead peroxide and negative plates of sponge lead. The grids that hold the active material may be made of pure lead, lead-antimony

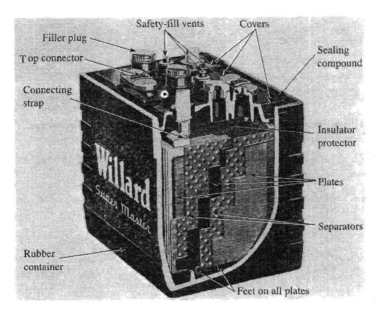

FIGURE 27.1
Automotive battery (courtesy Willard Co.).

alloy, or lead-calcium alloy. Lead-antimony grids maximize performance in high cycling applications and can operate at higher temperatures than lead-calcium grids. Lead-calcium plates have lower watering requirements.

The plates are kept apart by separators made of microporous rubber, fiberglass, perforated plastic or hard rubber, or resin-impregnated cellulose. Since the plates are completely submerged (flooded) in an electrolyte consisting of a solution of sulfuric acid and water, they are called *flooded batteries*.

The container for automotive batteries is made of hard rubber or plastic. The container for a stationary battery is a glass or transparent plastic jar such as that illustrated in Figure 27.2. The container permits visual inspection of the plates and the sedimentation chamber.

The life expectancy of lead-acid batteries used in stationary service varies between 20 and 25 years, depending on plate design, maintenance, and frequency of deep discharging and charging.

During discharge, the lead peroxide in the positive plates and sponge lead in the negative plates are gradually converted to lead sulfate, and the sulfuric acid is converted to water. The charging process extracts the sulfate from the plates, restoring the plates to their original chemical makeup of lead peroxide and sponge lead, and puts the sulfate back in solution as sulfuric acid. This electrochemical reaction is expressed in the following formula:

$$PbO_2 + Pb + 2H_2SO_4 \xrightarrow[\Leftarrow \text{ Charge}]{\text{Discharge} \Rightarrow} 2PbSO_4 + 2H_2O$$

FIGURE 27.2
Stationary battery (courtesy Exide Co.).

where: PbO_2 = lead peroxide (+)
             Pb = lead (−)
             H_2SO_4 = sulfuric acid (electrolyte)
             $PbSO_4$ = lead sulfate
             H_2O = water.

Another type of lead-acid cell, called a *recombinant cell* or *valve-regulated cell,* uses a gelled electrolyte and requires no additional water; the released hydrogen is recombined with oxygen in the cell. However, if sufficient hydrogen pressure builds up, the valve vents it to the atmosphere; this represents a loss of water from the gel that cannot be made up.

Ampere-Hour Capacity

Lead-acid cells are rated in voltage and ampere-hour (A-h) capacity (amperes times hours). The ampere-hour rating is for specified conditions of electrolyte temperature and specific gravity, rate of discharge, discharge time, and low-voltage limit. The low-voltage limit is the voltage below which the cell or battery can no longer supply usable energy to the circuit; that is, motors will not operate, control circuits will not function, etc. The two-cell battery illustrated in Figure 27.3 contains two lead-calcium cells with a total capacity of 3520 A-h; each cell has two positive posts and two negative posts.

FIGURE 27.3
Measuring specific gravity with a hydrometer (courtesy Willard Co.).

27.2 SPECIFIC GRAVITY OF A LEAD-ACID CELL

Specific gravity of the electrolyte is defined as the ratio of the weight of a given volume of electrolyte to the weight of an equal volume of distilled water. The specific gravity of distilled water is 1.000.

Measuring Specific Gravity

The measurement of specific gravity is done with a hydrometer, as illustrated in Figure 27.3. Enough electrolyte should be drawn into the glass barrel to make the hydrometer float. Specific gravity readings should always be taken before the addition of distilled water.[1]

If the level of the electrolyte is low, only distilled water should be added, and the battery should be charged before specific gravity readings are taken. Complete mixing of the distilled water with the electrolyte may takes a day or weeks, depending on the design, and whether it is in a stationary location or in a moving vehicle. Hence, specific gravity readings taken shortly after adding distilled water have no significance.

---

[1] Deionized water, or other approved water, can be used in place of distilled water.

TABLE 27.1
Battery Application and Specific Gravity

Application	Specific Gravity
Industrial trucks	1.275
Automotive service	1.260
Large-engine starting	1.245
Standby and utility service	1.210

Full-Charge Specific Gravity

The design constraints for specific gravity of lead-acid batteries are determined by the particular application. Some of the applications and their approximate specific gravities at full charge, measured at 77°F (25°C), are given in Table 27.1. Batteries with higher specific gravities provide greater ampere-hour capacity in a smaller package, but have a shorter life, and require more frequent watering.

27.3 EFFECT OF TEMPERATURE ON LEAD-ACID BATTERIES

Effect of Temperature on Specific Gravity

The temperature of the electrolyte has considerable effect on its specific gravity. High temperatures cause the volume of the electrolyte to expand, resulting in lower values of specific gravity, whereas low temperatures cause the volume of the electrolyte to contract, resulting in higher values of specific gravity. Thus, if the specific gravity readings are to be of any value in determining the state of charge of a lead-acid battery, corrections to some reference temperature must be made. The temperature of the electrolyte is measured with a thermometer inserted into the vent opening. The thermometer bulb must be completely immersed in the electrolyte for at least 30 seconds and should be read while immersed.

Temperature Correction of Specific Gravity

Specific gravity is always measured to four significant figures, and the three places to the right of the decimal point are called *points*. Thus a specific gravity of 1.285 has 285 points. A convenient rule of thumb is to add 1 point (0.001) to the hydrometer reading for each 3°F above 77°F, and to subtract 1 point from the hydrometer reading for each 3°F below 77°F [2]. Expressing this rule of thumb as an equation,

$$\text{SG}_{77} = \text{SG}_T + \frac{T - 77}{3 \times 1000} \qquad \textbf{(27–1)}$$

where: T = temperature of electrolyte (°F)
SG_T = specific gravity at temperature T
SG_{77} = specific gravity corrected to 77°F.

EXAMPLE 27.1

(a) If the temperature and specific gravity readings of a lead-acid battery are 120°F and 1.240, respectively, what is its specific gravity corrected to 77°F? (b) If the temperature and specific gravity readings of a battery are 30°F and 1.205, respectively, what is its specific gravity corrected to 77°F?

Solution

a. $SG_{77} = SG_T + \dfrac{T - 77}{3 \times 1000} = 1.240 + \dfrac{120 - 77}{3 \times 1000} = 1.254$

b. $SG_{77} = SG_T + \dfrac{T - 77}{3 \times 1000} = 1.205 + \dfrac{30 - 77}{3 \times 1000} = 1.189$

An electronic digital hydrometer, illustrated in Figure 27.4, measures the specific gravity and corrects it to 77°F.

Effect of Temperature on Discharge Capacity

The viscosity of the electrolyte increases with lowering temperatures; at 32°F (0°C) the viscosity is double that at 77°F (25°C), and below 32°F, the viscosity increases more rapidly. Because higher viscosity results in a reduced rate of diffusion of sulfuric acid through the pores of the plates, the discharge capacity of the battery is reduced. This low-temperature effect is very noticeable when cranking an engine in subzero weather; the combined effect of reduced discharge capacity, with the increase in cranking load due to thickened engine oil, makes engine starting more difficult.

When lead-acid batteries are operated in extremely low temperatures, it is imperative that they be kept in a fully charged condition at all times. Low temperatures reduce the useful capacity of a battery, and at 0°F its discharge capacity is reduced to approximately one-half its value at 77°F. Furthermore, the freezing temperature of the electrolyte is affected by the condition of charge.

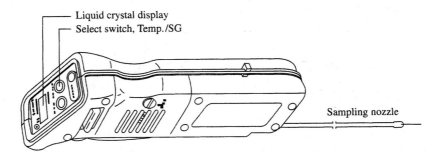

FIGURE 27.4
Electronic digital hydrometer that automatically provides temperature correction (courtesy Storage Battery Systems Inc.).

Freezing of Electrolyte

Freezing of the electrolyte damages the plates and ruptures the case, thus rendering the battery useless. A plot of the approximate temperature at which slush ice forms in the electrolyte vs. specific gravity is provided in Figure 27.5.[2] The measured specific gravity must be corrected to 77°F (25°C).

---

[2]Data points were supplied by Exide.

EXAMPLE 27.2

The specific gravity and temperature of a certain lead-acid cell are 1.179 and 80°F, respectively. Correct the specific gravity to 77°F and then determine the approximate temperature at which slush ice will form in the electrolyte.

Solution
From Eq. (27–1),

$$SG_{77} = SG_T + \frac{T - 77}{3 \times 1000} = 1.179 + \frac{80 - 77}{3 \times 1000} = 1.180$$

From Figure 27.5, slush ice will form at approximately $-10°F$.

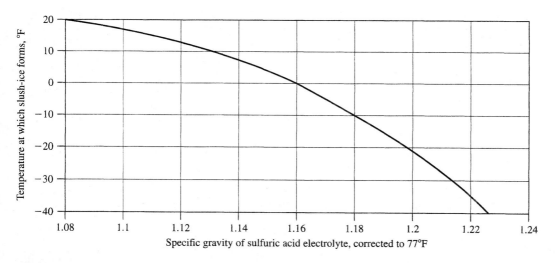

FIGURE 27.5
Approximate temperature at which slush ice forms in the electrolyte vs. specific gravity.

27.4 OPEN-CIRCUIT VOLTAGE VS. SPECIFIC GRAVITY OF A LEAD-ACID CELL

The relationship between open-circuit voltage and specific gravity of a lead-acid cell can be approximated by the following equation [3]:

$$V_{oc} \approx SG + 0.84 \qquad (27-2)$$

where: V_{oc} = open-circuit voltage (V)
SG = specific gravity at battery temperature.

Note Equation (27-2) is valid only if the electrolyte is uniformly diffused throughout the cell.

For greater accuracy when measuring cell voltage, a digital voltmeter should be used.

EXAMPLE 27.3

The open-circuit voltage of a certain lead-acid battery is 2.09 volts. Determine the approximate specific gravity of the electrolyte.

Solution
From Eq. (27-2),

$$SG \approx V_{oc} - 0.84 = 2.09 - 0.84 = 1.250$$

27.5 VOLTAGE BEHAVIOR OF LEAD-ACID CELL DURING DISCHARGE

At the instant a battery starts delivering current, called time-zero, the resistance of the cells cause an internal *IR* drop that results in an immediate reduction in the output voltage of the battery. This is illustrated in Figure 27.6 for a representative 400-A-h lead-acid cell, whose open-circuit voltage is 2.05 V. Furthermore, as illustrated in Figure 27.6, the voltage continues to drop with time as current is drawn.

The continuing drop in voltage is caused by the lowering of specific gravity during discharge. The cell is considered discharged when the voltage has decreased to its minimum usable voltage, called the final voltage, end-of-discharge voltage, or simply the end voltage, for that particular operation. The final voltage for most lead-acid cell applications is 1.75 volts per cell. Note that the IR drop at time-zero is much greater when supplying 102 amperes than when supplying a lesser current.

Discharge Rate

The discharge rate of a storage cell is the number of hours that it can deliver specified amperes continuously, before the voltage drops to its acceptable lower limit (end voltage). The standard discharge rate is the 8-hour rate with a end volt reading of 1.75 V per cell. Thus, a 400-A-h battery can deliver 50 amperes for 8 hours before its voltage drops to 1.75.

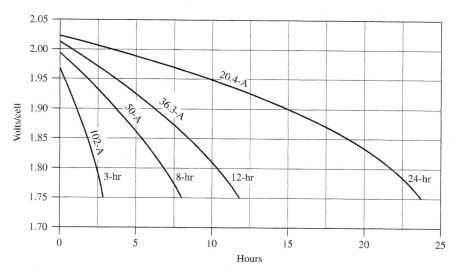

FIGURE 27.6
Voltage behavior of lead-acid cell during discharge.

The higher the discharge rate in amperes, the fewer total ampere-hours a cell can deliver before dropping to 1.75 V, and vice versa. The reason for this behavior is the slow diffusion of the electrolyte; at high discharge rates, the electrolyte in contact with the plates cannot be replenished fast enough. This behavior is illustrated in Figure 27.6, where a discharge rate of 50 amperes can be maintained for 8 hours, for a total of 50 A × 8 h = 400 A-h, whereas 20.4 amperes can be maintained for 24 hours for a total of 20.4A × 24 h = 489.6 A-h, before dropping to 1.75 V per cell.

27.6 STATE OF CHARGE OF LEAD-ACID CELLS AS A FUNCTION OF SPECIFIC GRAVITY

The specific gravity of the electrolyte changes with the state of charge, decreasing when discharging, and increasing when charging. As the battery discharges, the drop in battery ampere-hours is proportional to the drop in specific gravity points, with the specific gravities corrected to 77°F. This is illustrated in Figure 27.7 for a hypothetical 800-A-h automobile battery, whose specific gravity at full charge is 1.260 or 260 points, and when discharged to its acceptable low-voltage limit (end volts), its specific gravity is 1.120 or 120 points.

From the full-charge state to the low-voltage limit, the battery delivers a total of 800 ampere-hours. Note that although energy is still left in the battery when at the low-voltage limit for the specific application, it is not usable because the voltage is too low.

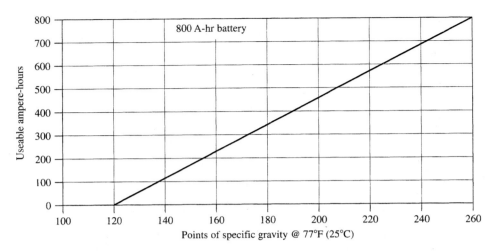

FIGURE 27.7
State of charge of lead-acid cells as a function of specific gravity.

Ampere-Hours/Point

The slope of the line in Figure 27.7 is the ratio of the drop in ampere-hours to the drop in points of specific gravity. This ratio, called *ampere-hours/point*, is used to determine the remaining charge in a battery, after some of its ampere-hours are discharged to a load. Expressed mathematically,

$$(AH/pt) = \left. \frac{AH_{rtd}}{pt_{rtd} - pt_{lv}} \right|_{T=77°F (25°C)} \qquad (27-3)$$

where: AH/pt = ampere-hours per point of specific gravity
AH_{rtd} = nameplate ampere-hr rating (A-h)
pt_{rtd} = specific gravity points at rated charge (77°F)
pt_{lv} = specific gravity points at low-voltage limit (77°F).

The ampere-hours used while supplying current to a load is given as follows:

$$AH_{used} = (AH/pt) \times \Delta pt \qquad (27-4)$$

where: AH_{used} = ampere-hours used
Δpt = change in points due to loss of ampere-hours.

EXAMPLE 27.4   For the battery represented in Figure 27.7, determine (a) ampere-hours per point of specific gravity; (b) amount of charge depleted if the specific gravity drops from its full-charge value to 1.180 after supplying current; (c) remaining charge.

Solution

a. Discharging from 800 ampere-hours to zero usable ampere-hours causes the specific gravity points to change from 260 to 120. Thus,

$$(\text{AH/pt}) = \left. \frac{\text{AH}_{rtd}}{\text{pt}_{rtd} - \text{pt}_{lv}} \right|_{T = 77°F\ (25°C)} = \frac{800}{260 - 120} = 5.71$$

b. $\text{AH}_{used} = (\text{AH/pt}) \times \Delta\text{pt} = 5.71 \times (260 - 180) = 457\ \text{A-h}$

c. AH remaining $= 800 - 457 = 343\ \text{A-h}$

Table 27.2 shows state of charge vs. specific gravity for a representative automobile battery. Note that the state of charge in the table refers to the usable charge available from the battery.

27.7 EFFECT OF ELECTROLYTE LEVEL ON THE SPECIFIC GRAVITY OF LEAD-ACID CELLS

The lowering of electrolyte level in normal usage is caused by evaporation of water. Specific gravity readings taken at low electrolyte levels will be higher than readings taken at the normal level because of the greater concentration of sulfuric acid. Thus the electrolyte level of stationary batteries should be recorded whenever the specific gravity and temperature are measured. Correction factors for low electrolyte levels in stationary batteries are provided by the manufacturer.

After restoration of electrolyte level, with the addition of distilled water, the water should be allowed to diffuse throughout the electrolyte before taking specific gravity readings. Water is so much lighter than the sulfuric acid that it tends to "float" on top of the cell, and can freeze if the temperature is below 32°F. Complete mixing of the electrolyte may take many hours, days, or even weeks, depending on whether the battery is dormant, charging or discharging, or is in a moving vehicle. Changes in temperature during charging or discharging help circulate the electrolyte, and gassing during charging also contributes to the mixing process.

TABLE 27.2
State of Charge vs. Specific Gravity

State of Charge (%)	Specific Gravity at 77°F
100	1.260
75	1.225
50	1.190
25	1.155
0	1.120

27.8 REPLACING SPILLED SULFURIC ACID ELECTROLYTE

If electrolyte is lost because of spilling or bubbling over, it should be replaced with electrolyte of the same specific gravity as that present in the cell. Goggles, rubber gloves, and a rubber apron should be worn when working with electrolyte. Concentrated sulfuric acid (specific gravity 1.835 at 77°F) should be mixed with distilled water in the proportions illustrated in Figure 27.8.[3] *The acid should be poured slowly into the water. Water should never be poured into acid; it will splash violently and may cause serious burns.*

Spilled acid can be neutralized with a solution of baking soda or a dilute ammonia solution. Acid splashed into the eyes should be flushed immediately with large quantities of water, and medical treatment should be administered by a doctor.

The electrolyte should be mixed only in glass, hard rubber, lead, or glazed earthenware containers. After mixing the electrolyte, it should be allowed to cool to room temperature before adding it to the cell.

---

[3] Figure 27.8 is a plot of data supplied by Gould-National Batteries, Inc.

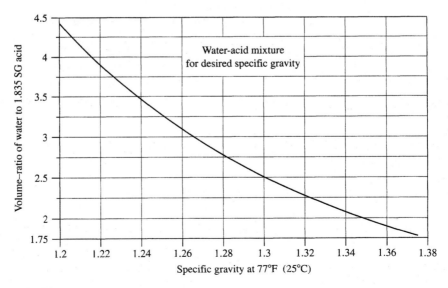

FIGURE 27.8
Curve for determining the water–acid mixture for desired specific gravity.

EXAMPLE 27.5   Determine the volume of water that must be mixed with 0.5 liter of 1.835 specific gravity sulfuric acid in order to have an electrolyte of 1.280 specific gravity.

Solution
From the graph in Figure 27.8, the required volume ratio of water to acid is 2.75. Thus, the required amount of water to be mixed with 0.5 liter of acid is

$$0.5 \times 2.75 = 1.37 \text{ liters of water}$$

Note   To avoid a violent reaction, the acid must be poured into the water, not vice versa.

27.9 CHARGING LEAD-ACID BATTERIES

During the charging process, current is forced into the battery in the opposite direction to that which occurs during discharge. Hence, the positive terminal of the battery must connect to the positive terminal of the charging source. When charging two or more cells or batteries in series, the positive of one must connect to the negative of the other. The rate and method of charge depends on the physical condition of the battery, its state of charge, and its application [4].

Figure 27.9 illustrates some elementary charging circuits. Figure 27.9a is an adjustable-voltage full-wave rectifier; voltage adjustment is obtained through a tap-changing transformer or a slide-wire rheostat. Figure 27.9b is a charging circuit that uses a rheostat and an available DC source.

Local Action and Trickle Charging

Chemical action caused by impurities in the component parts of the plates causes a gradual discharge of the battery even though it is not connected to a load. This behavior, called *local action,* occurs mostly in the negative plates. Local action can cause a lead-acid battery to lose half its charge in 3 or 4 months. Trickle charging uses a very small charging current to offset local action. The trickle charging rate is approximately 0.5 percent of the 10-hour rate. Trickle charging is often used to keep batteries fully charged while in storage, and for emergency lighting.

27.10 PRECAUTIONS WHEN CHARGING LEAD-ACID BATTERIES

When charging lead-acid batteries, the initial charging current should not exceed its ampere-hour rating. For example, a 100-A-h battery should not be charged at a rate greater than 100 amperes. This high rate should taper off as the battery charges. When nearing full charge, the battery absorbs the energy at a slower rate, and the excess energy breaks up the water into its hydrogen and oxygen components; oxygen is released at the positive plate and hydrogen is released at the negative plate. Excessive gassing and excessive temperatures can damage the plates by loosening the active material; the temperature of the electrolyte should not be allowed to exceed 110°F

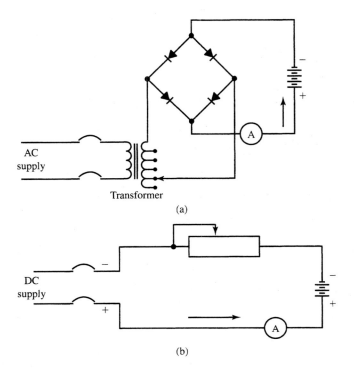

FIGURE 27.9
Some elementary charging circuits: (a) using an adjustable-voltage full-wave rectifier;
(b) using a rheostat and an available DC source.

(43°C). Full charge is attained when the specific gravity of all cells reaches its rated value, and no longer increases over a period of 3 to 4 hours.

Sediment

The many cycles of charge and discharge gradually cause tiny particles of plate material to fall off and settle in a space at the bottom of the container, called the sediment space. The size of the sediment space is designed to be large enough to handle all sediment that may fall during the life of the battery. However, excessive overcharging and discharging at very high rates will cause a more rapid accumulation of sediment. If the accumulation of sediment is high enough to reach the bottom of the plates, the plates will be partially shorted, and the cell will fail to hold a charge.

Mossing

Excess gassing causes shedding of material from the plates. Material shed from the positive plate settles in the sediment chamber at the bottom of the cell. Material shed from the negative plate accumulates in a sponge-like layer of lead at the top of the negative plate. This condition, called *mossing*, can eventually bridge around the separators causing a partial short circuit.

Sulfation

If lead-acid batteries are left in a discharged condition for any great length of time, the lead sulfate may harden and become nonporous, preventing the battery from accepting a full charge. Sulfation will also occur if the battery is consistently undercharged.

27.11 MAINTENANCE OF LEAD-ACID BATTERIES [2]

Although storage batteries require relatively little maintenance, they must not be neglected. Battery covers should be kept clean by damp wiping when necessary. The electrolyte must be maintained at the correct level. Spilled electrolyte should be neutralized with a solution of sodium bicarbonate and water, and then wiped dry with a clean cloth. Dirty or acid-wet tops and sides of batteries will cause grounds that will gradually deplete the battery; they should be wiped with a cloth dampened with sodium bicarbonate solution, and then wiped dry. Flame-arrestor vents, if present, should be cleaned by immersing them several times in distilled water, and then blown out with compressed air. A flame-arrestor vent prevents the propagation of an external flame into the cell, which would explode any hydrogen gas present.

Terminal posts and the contact surfaces of the intercell connectors should be coated with a thin film of anti-oxidant grease approved for battery use. Torquing the bolts to manufacturer's specifications will drive out any surplus grease. A thin film of grease should also be maintained on the body of the terminal posts.

Dirty connections may be cleaned by removing them and brushing with a stiff nonmetallic bristlebrush, being careful not to damage the lead plating. Corroded connectors between cells should be replaced. The handles of all tools used in a battery compartment should be insulated to prevent accidental short circuits.

Batteries that are idle, or are in storage, still require maintenance. They must be kept charged, and distilled water added when required. Failure to maintain the battery in a charged condition can cause irreversible damage to the plates.

27.12 NICKEL-CADMIUM CELLS

The electrolyte in a nickel-cadmium cell (NiCad) is a solution of potassium hydroxide in distilled water at a specific gravity of 1.200 at 60°F. Nickel hydroxide is the active material in the positive plate and cadmium is the active material in the negative plate. The chemical reaction during charge and discharge is represented by the following equation.

$$2NiO(OH) + Cd \xrightarrow[\Leftarrow Charge]{Discharge \Rightarrow} 2NiO + Cd(OH)_2$$

where:
$2NiO(OH)$ = nickel oxyhydroxide (+)
$Cd(OH)2$ = cadmium hydroxide
Cd = cadmium metal (−)
$2NiO$ = nickel oxide.

The electrolyte is potassium hydroxide solution (KOH).

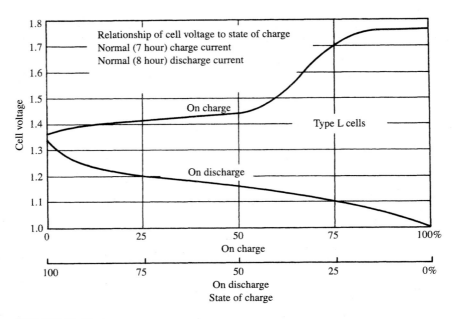

FIGURE 27.10
Relationship of cell voltage to state of charge for a nickel-cadmium battery (courtesy Nickel Cadmium Battery Corp.).

Figure 27.10 illustrates the relationship of cell voltage to state of charge for a nickel-cadmium battery. As illustrated by the curves, if the cell is on charge, and charging at its normal rate, a voltage of 1.7 volts indicates the cell to be 75 percent charged. Similarly, if the cell is supplying a load and discharging at its normal rate, a cell voltage of 1.2 volts will indicate a 75 percent charge. Because the specific gravity of the electrolyte is essentially constant, regardless of charge, the battery condition must be determined with a voltmeter during charge or discharge; a digital voltmeter is preferred.

Charging NiCad Batteries

The charging circuits shown in Figure 27.9 for lead-acid batteries are the same as for NiCad batteries. When charging a nickel-cadmium battery, the temperature of the electrolyte should not be allowed to exceed 115°F (46°C). The battery does not commence to gas until after the first 4.5 hours when charging at the 7-hour rate. No finishing rate is needed. The source of power for charging should have a minimum value of 1.85 volts per cell. Although the specific gravity of the nickel-cadmium battery is unaffected by the state of charge or discharge, it gradually decreases during normal usage. When the gravity falls below 1.170 at 60°F, the electrolyte should be renewed. Operation of the battery below this gravity causes a rapid reduction in its life. A few drops of pure paraffin oil floating on top of the electrolyte in each cell will prevent atmospheric carbon dioxide from contaminating the electrolyte.

Memory

The so-called "memory effect" is an apparent loss of ampere-hour capacity caused by a depressed voltage. During normal charging, cadmium is deposited on the negative plate in the form of tiny crystals. Over time, these tiny crystals combine to form much larger crystals, resulting in less overall surface area for contact with the electrolyte. The net effect is higher cell resistance with lower cell voltage, resulting in lower ampere output to the connected load. Proper cycles of discharge and charge convert the large crystals to tiny crystals, preventing voltage depression and high cell resistance.

The formation of large crystals is encouraged by extended periods of slow charging, by excessive heat, and by highly repetitive cycles of *partial* discharge followed by slow charging. Without complete discharge, the material deep down in the plates never gets worked. Thus, the large crystals never get converted to tiny crystals, and the energy stored in that part of the cell cannot be effectively utilized.

Batteries that suffer from "memory" problems can be reconditioned through specially designed reconditioning equipment, which provides forced discharge followed by full charge. This process converts large crystals to tiny crystals, lowering cell resistance, and reestablishes the proper voltage characteristic [5, 6].

27.13 MIXING ALKALINE ELECTROLYTE FOR NiCad CELLS

Electrolyte lost because of spilling or bubbling over should be replaced with a caustic potash solution of 1.190 to 1.200 specific gravity (60°F). Caustic potash (KOH) is generally supplied as a solid in airtight cans and should be mixed 1 part by weight of KOH to 2 parts by weight of distilled water. The solution should be mixed only in glass, iron, or earthenware containers, using a glass or iron stirring rod. The iron container and stirring rod must not be galvanized or have soldered joints. Great care should be used when handling alkaline electrolyte. The solution is very caustic and dissolves skin as well as other organic matter. It also attacks copper, brass, lead, aluminum, and zinc. Spilled electrolyte should be washed immediately with a large quantity of water and then neutralized with vinegar, citric acid, or a 4 percent solution of boric acid. If alkaline electrolyte gets into eyes, flush continuously with copious amounts of water for 10 or 15 minutes and seek immediate medical attention. Goggles, rubber gloves, and a rubber apron should be worn when changing the electrolyte.

27.14 SAFETY CONSIDERATIONS FOR BOTH LEAD-ACID AND NiCad CELLS

The gases given off during the charging process are an explosive mixture of hydrogen and oxygen. Hence, adequate ventilation must be available and no smoking or carrying of open flames should be allowed in a battery room or compartment. Conspicuously placed NO SMOKING signs must be observed.

Maintaining the electrolyte level at the full mark, by the addition of distilled water, would result in less space inside the battery for the accumulation of explosive gases. Gas inside the battery can be ignited by a spark or flame outside the battery, but near the battery vent. An explosion inside the battery would spray electrolyte over the surrounding areas, seriously injuring personnel, as well as damaging the battery and other nearby apparatus.

Before connecting or disconnecting the terminals of a battery, make sure that all switches are in the OFF position. If one or more switches are in the ON position when making the connection to the battery, a spark may explode any accumulated gas. Rubber mats placed over adjacent terminals will help prevent accidental short circuits.[4] When working in a battery room, maintenance personnel should use non-sparking tools.

27.15 UNINTERRUPTIBLE POWER SUPPLY

Batteries used as part of an uninterruptible power supply (UPS) are operated in the "float mode." That is, the battery, load, and battery charger are always connected in parallel, as illustrated in Figure 27.11a. The charger may be a motor generator set or rectifier.

The minimum practical float voltage per cell for lead-antimony cells is 2.15 V at a specific gravity of 1.215 and is 2.17 V per cell for lead-calcium cells.[5] Prolonged charging at 2.13 V/cell or less will cause sulfation, reducing the life expectancy of the cell. On the other hand, if the float voltage is too high, it will increase the need for repeated watering of the cells and cause increased wear on the plates, shortening cell life. The manufacturer's recommended voltage/cell should be followed when operating in the float mode.[6] Nickel-cadmium cells are usually floated at between 1.4 V and 1.5 V per cell.

During normal operation, as illustrated in Figure 27.11a, the charger supplies the load and feeds current to the battery to make up for losses due to local action and to replace any energy that the battery may have fed to the load. Loads that exceed the charger capacity cause the charger voltage to drop slightly, causing the battery to add its current to the load as illustrated in Figure 27.11b. The battery will be the sole supplier of current if the battery charger fails, as illustrated in Figure 27.11c.

Designing the Uninterruptible Power Supply

Batteries for UPS systems are often connected in series–parallel arrangements to satisfy the voltage and ampere-hour requirements. The number of cells required in a series string is determined from the required system voltage and the desired float volt-

---

[4] For detailed information on short-circuit protection of battery systems, see References [7] and [8].

[5] Because of the two-decimal-place requirement when measuring cell voltage, a digital voltmeter should be used.

[6] See Reference [9] for sizing generator station batteries.

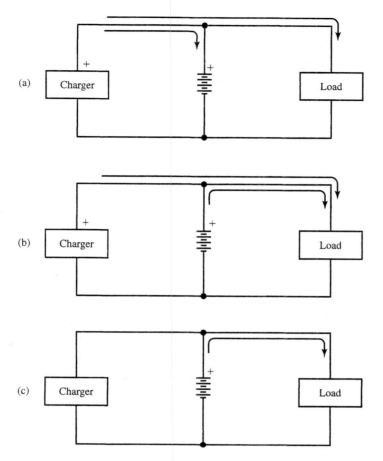

FIGURE 27.11
Uninterruptible power supply: (a) charger supplying power to a load and also charging the battery; (b) both charger and battery supplying power to a load; (c) battery alone supplying power to a load.

age per cell. The number of parallel strings is determined by the ampere-hours required by the system and the ampere-hours per cell. Expressed mathematically,

$$\text{Cells per series string} = \frac{V_{\text{system}}}{V_{\text{cell}}} \qquad (27\text{–}5)$$

$$\text{Number of strings} = \frac{AH_{\text{system}}}{AH_{\text{cell}}} \qquad (27\text{–}6)$$

EXAMPLE 27.6

An uninterruptible power supply is required to supply 1200 A-h at 230 V. The cells available are 400 A-h, lead-acid, and are to be operated at a float voltage of 2.19 V per cell. Determine (a) required number of cells per series string; (b) number of series strings in parallel; (c) total number of cells required.

Solution

a. Cells per series string $= \dfrac{V_{system}}{V_{cell}} = \dfrac{230}{2.19} = 105$ cells

b. Number of strings $= \dfrac{AH_{system}}{AH_{cell}} = \dfrac{1200}{400} = 3$ strings

c. Total cells required $= 105 \times 3 = 315$

The circuit arrangement is shown in Figure 27.12.

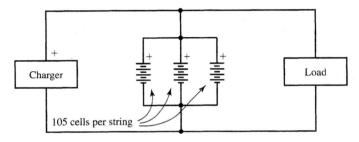

FIGURE 27.12
Circuit arrangement for Example 27.6.

Equalizing Charge

Since no two cells in a battery system are exactly alike, the local action in some cells will be different from that in other cells. In time, this will cause differences in cell voltage and charge. This is particularly true for lead-acid cells. To remedy this condition, the battery should be given an equalizing charge. The equalizing charge is at a higher voltage than the normal float voltage and is generally continued for 35 to 70 hours [2]. The manufacturer's instructions as to voltage, frequency of application, and hours of charge should be followed. Nickel-cadmium batteries have very little local action and thus seldom need an equalizing charge.

Pilot Cell

One cell in a string of series-connected cells is often selected as the typical cell (called the *pilot cell*). The pilot cell's voltage, specific gravity, and temperature are taken as representative of all the cells in that string. However, because of loss of electrolyte when taking specific gravity readings, the cell selected as the pilot cell should be changed quarterly or semiannually.

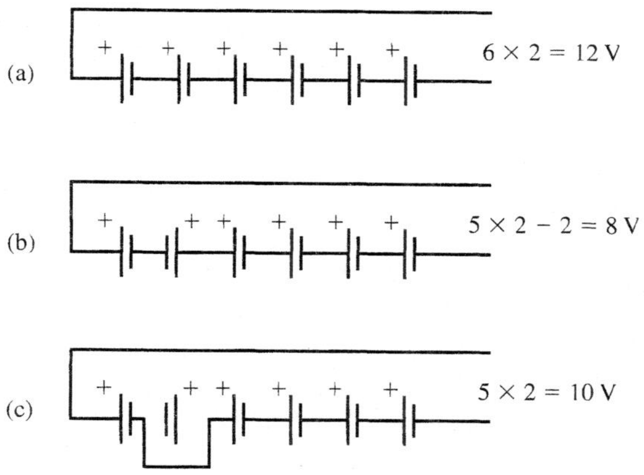

FIGURE 27.13
(a) Normal connections for a string of six 2-V cells; (b) reversed cell in a string; (c) cutting out a reversed cell in a string.

Reversed Cell

If a series-connected string of cells is discharging through a load, and one cell of the string reaches full discharge before the others, the voltage of the discharged cell will drop to zero and then build up in the reverse direction. This condition is caused by faults within the cell itself.

The normal connections for a string of six 2-V lead-acid cells is shown in Figure 27.13a. A reversed cell, shown in Figure 27.13b, will cause a 2-V drop in string voltage resulting in a string voltage of 8 volts.

Prolonged operation of a string containing a reversed cell will cause the temperature of the reversed cell to rise more rapidly than the other cells, and if the string discharge rate is high enough, the temperature of the reversed cell may reach the boiling point.

The reversed cell can be located by conducting polarity checks of each individual cell. Note that cutting out the reversed cell as illustrated in Figure 27.13c will permit operation of the string at 10 V. *Do not bypass the cell by short-circuiting it!*

SUMMARY OF EQUATIONS FOR PROBLEM SOLVING

$$SG_{77} = SG_T + \frac{T - 77}{3 \times 1000} \tag{27–1}$$

$$V_{oc} \approx SG + 0.84 \tag{27–2}$$

$$(AH/pt) = \frac{AH_{rtd}}{pt_{rtd} - pt_{lv}}\bigg|_{T=77°F\ (25°C)} \tag{27–3}$$

$$AH_{used} = (AH/pt) \times \Delta pt \qquad \text{(27–4)}$$

$$\text{Cells per series string} = \frac{V_{system}}{V_{cell}} \qquad \text{(27–5)}$$

$$\text{Number of strings} = \frac{AH_{system}}{AH_{cell}} \qquad \text{(27–6)}$$

SPECIFIC REFERENCES KEYED TO THE TEXT

[1] Walker, Loren H., 10-MW GTO Converter for Battery Peaking Service, *IEEE Trans. Industry Applications,* Vol. 26, No. 1, January/February 1990.

[2] *IEEE Recommended Practice for Maintenance, Testing, and Replacement of Large Lead Storage Batteries for Generating Stations and Substations,* ANSI/IEEE STD 450–1995.

[3] Montalbano, J. F., and Casalaina, R. V., Installation and Maintenance of Lead-Acid Stationary Batteries for Generating Stations, *IEEE Trans. Energy Conversion,* Vol. EC-1, No. 4, December 1986.

[4] *IEEE Recommended Practice for Sizing Large Lead Storage Batteries for Generating Stations and Substations,* IEEE STD 485–1983.

[5] Pensabene, S. F., and Gould, J. W. II, Unwanted Memory Spooks Nickel-Cadmium Cells, *IEEE Spectrum,* September 1976.

[6] Scholefield, C. L., NiCad Seminar, Red Scholefield Associates, July 1996, Gainesville, FL.

[7] Nailen, Richard L., Battery Protection—Where Do We Stand? *IEEE Trans. Industry Applications* Vol. IA-27, No. 4, July/August 1991.

[8] Nelson, John P., Basics and Advances in Battery Systems, *IEEE Trans. Industry Applications,* Vol. 31, No. 2, March/April 1995.

[9] Migliaro, M. W., Sizing Batteries for Generator Stations, *IEEE Trans. Energy Conversion,* Vol. EC-1, No. 4, December 1986.

REVIEW QUESTIONS

1. What is the basic difference between primary and secondary cells?
2. What causes a dry cell to discharge gradually, even though no load is connected to its terminals?
3. What are the principal components of a lead-acid battery?
4. What is specific gravity and how is it measured?
5. What is the relationship between open-circuit voltage and specific gravity?
6. What is meant by points of specific gravity?
7. What effect does temperature have on the specific gravity of the electrolyte and on the discharge capacity of the battery? Explain.
8. What is the relationship between ampere-hour capacity and specific gravity?

9. Explain the drop in voltage that occurs in a cell as it supplies current to a load.
10. What effect does extremely low temperature have on the discharge rate of a lead-acid battery? Explain.
11. What causes the electrolyte level of a battery to lower during normal usage (discount accidental spillage). What effect does this have on specific gravity?
12. Explain why battery voltage drops during discharge.
13. State the correct procedure for mixing acid and water when making battery electrolyte. What precautions should be observed?
14. How much distilled water must be added to concentrated sulfuric acid in order to make a 1-quart mixture having a specific gravity of 1.300?
15. Sketch a circuit for charging a lead-acid battery and indicate the polarities of the battery and charger. What safety precautions should be observed when charging?
16. What causes the accumulation of sediment in a lead-acid battery, and how does it affect the charge?
17. What is *mossing,* what causes it, and how does it harm the battery?
18. What is *sulfation,* what causes it, and how does it harm the battery?
19. How should dirty and corroded terminals of lead-acid batteries be cleaned?
20. What are the principal components of nickel-cadmium batteries?
21. How must an alkaline battery be tested?
22. Why can't a hydrometer be used for testing the condition of an alkaline battery?
23. Sketch a circuit for a charging a NiCad battery, and indicate the polarities of the battery and charger.
24. State the correct procedure for mixing electrolyte for an alkaline battery. What precautions should be observed?
25. State the general safety considerations required during operation and maintenance of lead-acid and NiCad batteries.
26. Sketch an uninterruptible power supply and explain its operation during (a) normal loads; (b) overloads; (c) power failures.

PROBLEMS

27–1/3. The temperature and specific gravity readings of a certain lead-acid cell are 95°F and 1.114, respectively. What is the specific gravity corrected to 77°F?

27–2/3. The temperature and specific gravity of a certain lead-acid cell are 26°F and 1.264, respectively. What is the specific gravity corrected to 77°F?

27–3/3. An automobile stored in a garage for 2 days has an ambient temperature of 65°F. The specific gravity at this temperature is 1.184 . The vehicle will be left overnight in a parking lot, and the weather forecast is for a nighttime temperature of −5°F. Will slush ice form in the battery?

27–4/4. The open-circuit voltage of a certain lead-acid cell is 1.98 V. Determine the specific gravity.

27–5/4. If the specific gravity of a certain lead-acid cell is 1.283, what is its open-circuit voltage?

27–6/6. The specific gravity ranges of a 500-A-h lead-acid cell at rated charge and at its final low-voltage charge are 1.250 and 1.125, respectively. Determine (a) ampere-hours/point; (b) ampere-hours used if the specific gravity is 1.175; (c) remaining charge.

27–7/6. A fully charged 1200-A-h lead-acid cell has a specific gravity of 1.210. The specific gravity at the accepted low voltage is 1.126. Determine (a) ampere-hours/point; (b) ampere-hours used if the specific gravity is 1.140; (c) remaining charge.

27–8/6. An 800-A-h battery with a full-charge specific gravity of 1.280, and an end-voltage specific gravity of 1.123 is discharged at the 10-hour rate for 6 hours. Determine (a) ampere-hours/point; (b) ampere-hours used; (c) specific gravity at the end of the 10-h discharge.

27–9/8. Determine the volume of water that must be mixed with 2 liters of 1.835 SG acid to obtain an electrolyte with specific gravity of 1.250 at 77°F.

27–10/8. Determine the volume of water that must be mixed with 3.2 liters of 1.835 SG acid to obtain an electrolyte with specific gravity of 1.350 at 77°F.

27–11/15. Determine the number of series-connected cells required to obtain 120 V, when operated at a float voltage of 2.14 V/cell.

27–12/15. A 124-V, 1800-A-h uninterruptible power supply is to be designed using 600-A-h cells that are to be floated at 2.34 V/cell. Determine (a) cells per string; (b) number of parallel strings; (c) total number of cells required.

27–13/15. Design a series–parallel arrangement of thirty-six 6-V batteries that will provide an output of 18 V. Sketch the circuit and determine (a) the number of cells per string; (b) number of strings in parallel.

27–14/15. Design a series–parallel arrangement of fifty 12-V batteries that will provide an output voltage of 120 V for an emergency lighting system. Sketch the circuit and determine (a) the number of cells per string; (b) number of strings in parallel.

28

Bonding, Grounding, Earthing, and Ground-Fault Protection of Distribution Systems

28.0 INTRODUCTION

Grounding, bonding, earthing, and ground-fault protection of distribution systems provide protection for operators and for electrical apparatus should a ground fault occur. *A ground fault is the accidental connection to ground of a current-carrying conductor, other than the conductor specified for system grounding.*[1] A ground fault occurring in an inadequately protected system can result in electric shock, arcing, fire, overvoltages, and plant blackouts.

This chapter provides the basics for an understanding of grounding and ground-fault protection used in distribution systems and discusses some of the problems that can occur if the system is not properly grounded.

28.1 EQUIPMENT GROUNDS, BONDING, AND EARTHING

Equipment Grounds

Equipment grounds are the metallic frameworks of switchboards and electrical machinery, metallic switchboxes, metallic cable armor, metallic raceways, metallic cable trays, metallic framework of electronic control systems, etc.

Bonding

For safety purposes, all related equipment grounds must be connected together. This is illustrated in Figure 28.1, where the framework of a motor is connected to the framework of the corresponding motor controller. This is called *bonding,* and the conductors

---

[1]See Section 2.10, Chapter 2, for examples of ground faults.

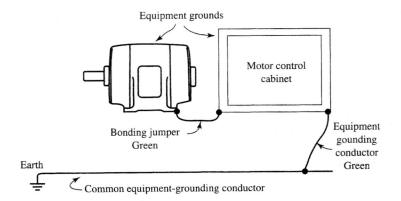

FIGURE 28.1
Bonding related equipment.

used for this purpose are called *bonding jumpers*. The equipment-grounding conductor, illustrated in Figure 28.1, must be connected to a common-equipment-grounding conductor at the service panel, which in turn is connected to earth.

All equipment bonding jumpers and equipment-grounding conductors must be permanently bolted and have sufficient ampacity to safely carry the largest possible ground-fault current that could occur. Sheet-metal screws must not be used for connecting bonding jumpers. Note that bonding and equipment-grounding conductors must not be used to carry load current. Bonding conductors may be bare or insulated. If these conductors are insulated, the color of the insulation should be green or white with a green marker. Green coloring should only be used with bonding and grounding conductors [1].

Proper equipment bonding, and grounding to earth, ensures that the frameworks of all electrical apparatus, driven equipment such as pumps, compressors, etc., water pipes, raceways, and structural steel of buildings will have the same voltage potential. This eliminates hazards such as shock, sparking, or arcing that may otherwise occur from accidental contact between equipment grounds that have different potentials.

Portable and stationary cord-connected appliances often have an extra wire (green bonding conductor) that automatically bonds the framework of the appliance to the equipment-grounding conductor of the system. Bonding occurs when the power cord's three-prong plug is plugged into its corresponding receptacle.

The so-called "double-insulated construction," sometimes used in portable tools, has two layers of insulation that result in lower leakage currents. These tools are useful in areas where a bonding conductor is not available at the receptacle. These cords have a two-prong plug.

Grounding of microprocessor control equipment requires special consideration because of the vulnerability of its electronic power supply to damage from circulating ground currents caused by ground faults and AC welders. To reduce the possibility of damage, the chassis should be connected via a continuous insulated conductor direct to the common equipment-grounding conductor [2].

The common-equipment-grounding conductor for domestic, industrial, and commercial use *must always be connected to earth.* The connection may be made to metallic water pipes, metallic gas pipes, structural steel of buildings, or special earthing electrodes.

Earthing

The primary function of the earth connection is to reduce or eliminate the shock hazard that may occur when an operator makes contact with the earth and with the framework of a machine or other apparatus, at the same time. A properly made earth connection ensures that the equipment-grounding conductor and the earth are at essentially the same potential; a person standing on a concrete floor is effectively connected to earth. The earth connection also serves to limit excessive voltages on circuit conductors that may be caused by lightning, line surges, static discharge, and unintentional contact with higher voltage lines. To provide this protection, the resistance to earth should be kept as low as practical for the particular installation; *the lower the resistance to earth, the safer personnel and equipment will be.*

The resistance to earth includes the resistance of the grounding electrode, the contact resistance between the electrode and the earth, and the resistance of the surrounding earth. The resistance to earth of continuous underground metallic water pipe, metallic gas pipe, or a deep-well metallic casing is generally less than 3 ohms and, if permitted by gas or water authorities, provides excellent earthing. In those areas where underground metallic piping is not available or is not permitted to be used for grounding purposes, other earthing electrodes must be used.

Earthing can be accomplished by using unpainted galvanized pipes, rods, or metal plates as electrodes [3]. The pipes (3/4 in. in diameter) or rods ($\frac{5}{8}$ in. in diameter) should be driven to a depth of at least 8 ft. Each plate electrode should be at least $\frac{1}{4}$ in. thick and have a cross-sectional area of at least 2 ft^2. Where more than one electrode is used, spacing between them should be no less than 6 ft.

The National Electrical Code requires that the resistance to ground of an earthing system shall, where practicable, not exceed 25 ohms. *The 25 ohms is an upper limit,* and much lower values are often required, especially where very large ground-fault currents can occur [1]. Earthing systems of less than 1 ohm are required for large substations and generating plants, whereas earthing systems in the 2- to 5-ohm range are acceptable for industrial and large commercial installations.

Earth resistance is not constant; it is affected by falling water tables, drought, broken earthing electrodes, loose connections, etc. Hence, periodic measurement of earth resistance, and inspection of the connection at the earth electrode, must be made to determine whether the earth resistance is still in the acceptable range and if the connection is still secure. In some cases, chemical treatment of the soil with copper sulfate, magnesium sulfate, or ordinary rock salt may be necessary to lower the earth resistance to the required value.

Detailed procedures for making earth-resistance measurements, along with curves and tables that assist in calculating the specific distances to the auxiliary electrodes, are incorporated in the instruction manuals for the specific test equipment. An in-depth discussion of measuring earth resistivity, ground impedance, and earth surface potentials is given in Reference [4].

> **Warning**
>
> Testing earth connections and making earth resistance measurements can subject personnel to severe shock hazards; a ground fault anywhere on the system could involve a return current to the earth connection under test. When making connections for the test, proceeding with the test, and disconnecting from the test, the operator should wear rubber gloves, if possible stand on an insulated mat, and follow the manufacturer's instructions. Under no circumstances should the test be made while a thunderstorm is in the vicinity or predicted for the vicinity.

28.2 UNGROUNDED DISTRIBUTION SYSTEMS

An ungrounded distribution system has no deliberate connection between the system wiring and the common equipment-grounding conductor. Note however, that the equipment-grounding conductor must be connected to earth, as illustrated in Figure 28.1.

The most significant advantage of an ungrounded system is better service continuity. Accidental contact between one line and ground will not trip a breaker or blow a fuse, and thus will not cause an outage. For this reason the ungrounded system has been used in process industries such as petrochemical, paper, etc., where an accidental shutdown can cause significant financial loss in production and serious damage to production-related equipment. However, note that an ungrounded system is unsuitable for three-phase, four-wire service.

Although an ungrounded system can operate without interruption of service when a ground fault occurs on one phase, the voltage to ground on the unfaulted phases may, under certain conditions, rise to more than six times normal voltage [3]. Such overvoltages increase the shock hazards to personnel and increase the electrical stresses on the unfaulted insulation. High overvoltages have the same adverse effect on insulation as would a prolonged high-potential test made with significantly excessive voltage. The effect of prolonged overvoltages on an operating system could be extensive damage and system outages.

Furthermore, if a first ground fault is not cleared immediately and a second ground fault occurs on another phase, or line of opposite polarity, the resultant short-circuit current may trip one or more circuit breakers. This is illustrated in Figure 28.2 for a three-phase, three-wire ungrounded system. In those instances where the short-circuit current is not high enough to cause tripping, severe damage to machinery, cable, and other connected apparatus may occur by overheating and burning of insulation.

Another safety hazard that can occur in an ungrounded system is that of a motor that "mysteriously" starts, although no START button was pushed, or fails to stop when a STOP button is pushed. The motor may be reacting to a double ground fault. This is illustrated in Figure 28.3, where it is assumed that accumulated dirt, moisture, etc., in a motor control panel caused coil M to be grounded at point 3.

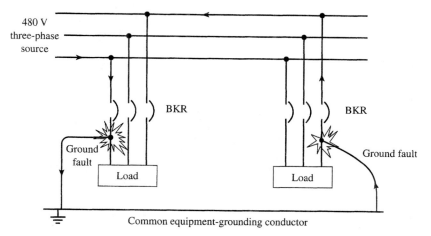

FIGURE 28.2
Effect of two grounds in an ungrounded distribution system.

If a second ground occurs on line L_1, coil M will be energized and the motor will start even though the START button was not pushed. Similarly, if the motor was running and a second accidental ground occurred at L_1, the motor will not stop when the STOP button is pushed. In either case, the results could be disastrous.

28.3 CAUSES OF OVERVOLTAGE IN THREE-PHASE UNGROUNDED SYSTEMS

One or more of the following factors are responsible for overvoltages in an ungrounded three-phase system when a ground fault occurs:

1. A steady-state overvoltage of 73 percent, caused by *distributed capacitance* alone, is always present when a single ground fault occurs.

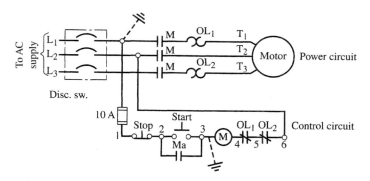

FIGURE 28.3
Effect of two grounds in a motor control circuit in an ungrounded distribution system.

2. A much higher steady-state voltage may be caused by *series resonance* or *partial series resonance,* between the distributed capacitance of the system and the inductance of the faulted circuit.

3. An escalating and very damaging overvoltage may be caused by *repeated intermittent ground faults.*

Effect of Distributed Capacitance

Even though the current-carrying conductors in an ungrounded system are physically isolated from ground by electrical insulation, the system is capacitively coupled to ground by the distributed capacitance between conductors and ground. This is illustrated in Figure 28.4a. Such distributed capacitance between conductor and ground is inherent in motors, generators, transformers, cables, etc.

The symmetrically distributed capacitance between conductor and ground in unfaulted three-phase equipment produces the effect of a wye-connected capacitor bank, as illustrated in Figure 28.4b, and results in balanced line-to-ground voltages equal to the line-to-neutral voltages of the ungrounded system. Operating personnel and electricians must be made aware of the fact that, because of the distributed capacitance, *a normally functioning ungrounded system can produce a severe shock if contact is made between any one of the three lines and ground.*

In the 480-volt, three-phase unfaulted system illustrated in Figure 28.4b, the three lines are capacitively coupled to ground. Thus the three line-to-ground voltages are each equal to the line-to-line voltages divided by 1.73. That is,

$$V_{\text{line-to-ground}} = \frac{V_{\text{line-to-line}}}{1.73} = \frac{480}{1.73} = 277 \text{ V}$$

If a ground fault occurs, such as that illustrated in Figure 28.4c, where line C is accidentally connected to ground, the line-to-ground voltage of the two unfaulted phases will rise to equal the line-to-line voltage. In this example, a single ground fault will cause the steady-state voltage to ground of the unfaulted lines to equal 480 V. This is a 73 percent increase in voltage across the unfaulted insulation that does not include any rise in voltage due to series resonance or arcing. Note that the voltage across the distributed capacitance in Figures 28.4a, 28.4b, and 28.4c is the voltage across the insulation of the unfaulted cable, as well as across the insulation of all connected motors and other apparatus connected to the system.

Effect of Series Resonance

Figure 28.5 illustrates how a ground fault in a coil of a motor, contactor, motor starting reactor, autotransformer, etc., results in a series connection between the distributed capacitance of the unfaulted cable-phases and the inductance of a section of the coil. The resultant resonance rise in steady-state voltage across the insulation may be several times normal voltage, causing severe electrical stresses on the insulation.[2]

---

[2] For more information on series resonance and its effects in electric power systems, see Section 7.2, Chapter 7.

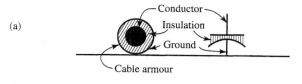

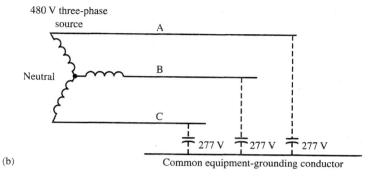

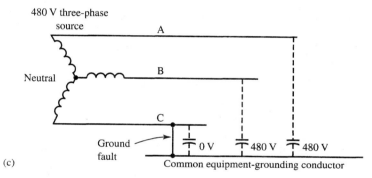

FIGURE 28.4

(a) Armoured cable as a capacitor; (b) distributed capacitance in an ungrounded unfaulted three-phase distribution system; (c) effect of a solid ground-fault on the capacitive voltages in an ungrounded distribution system.

Effect of Intermittent Ground Faults

A vibrating ground or sputtering ground that alternately clears and restrikes can generate very high overvoltages in a normally ungrounded system. A vibrating ground occurs when electrical parts that are in proximity to the framework of a machine, control box, etc., become loose and make vibratory contact to ground. A sputtering ground occurs when the accumulation of dirt, moisture, or other foreign matter results in surface leakage that gradually escalates to carbonized paths (tracking), followed by pinpoint sparking (scintillation), and finally a sputtering arc.

An explanation of voltage buildup in an ungrounded system, caused by intermittent ground faults, is presented with the aid of the equivalent circuit illustrated in

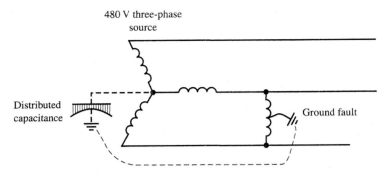

480 V three-phase
source

Distributed
capacitance

Ground fault

FIGURE 28.5
Resonance between distributed capacitance and inductance of electrical apparatus if a
ground fault occurs at the inductive component.

Figure 28.6. The ground to neutral capacitance C_{GN} represents the distributed capacitance of the entire system insulation, and the switch simulates a vibrating ground fault on line A. Under normal (unfaulted) conditions there will be no voltage across capacitance C_{GN}, and for the sake of discussion it will be assumed that the source is a wye-connected, 480-V, three-phase system. Thus,

$$\text{Line-to-neutral volts (rms)} = 480 \div \sqrt{3} = 277 \text{ V}$$
$$\text{Line-to-neutral volts (max)} = 277 \times \sqrt{2} = 391 \text{ V}$$

First Contact to Ground

A contact to ground from line A, simulated by the closing of the switch in Figure 28.6, causes line-to-neutral voltage E_{AN} to be impressed across capacitance C_{GN}, and a capacitive charging current will be present. This is illustrated in Figure 28.7a.

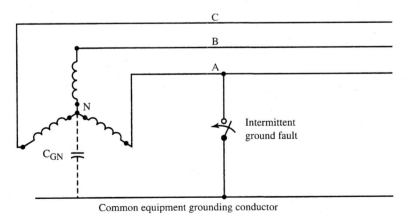

C

B

A

N

C_{GN}

Intermittent
ground fault

Common equipment grounding conductor

FIGURE 28.6
Ungrounded three-phase distribution system with a simulated intermittent ground-fault
on Line A.

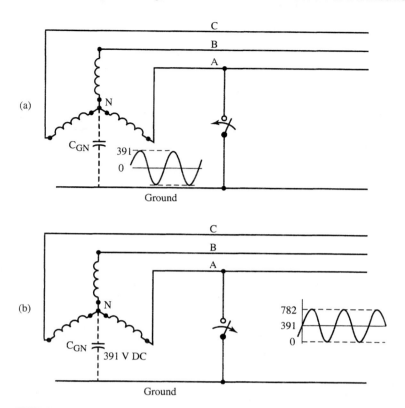

FIGURE 28.7
(a) First contact to ground; (b) first contact to ground breaks.

If the contact to ground breaks, and current ceases at the instant the alternating voltage across the capacitor is at its maximum value, capacitance C_{GN} will be charged to a DC potential of 391 volts. This is illustrated in Figure 28.7b. The 391 volts DC across the capacitor add to the neutral-to-line voltage of phase A, doubling the peak value to 782 volts. This is illustrated on the sine wave in Figure 28.7b, where the voltage across the fault, between line A and ground, cycles between 0 volts and 782 volts. Although not shown, the peak voltage between line B and ground, and between line C and ground, will also be 782 volts.

Second Contact to Ground

If contact to ground reoccurs, a higher voltage will be impressed across capacitance C_{GN}, as illustrated in Figure 28.8a.

If the second contact to ground now breaks at the instant the alternating voltage across the capacitor is at its maximum value, capacitance C_{GN} will be charged at a DC potential of 782 V. The 782 volts DC across the capacitor add to the neutral-to-line voltage of phase A, resulting in a peak value of 1173 V (three times the normal maximum neutral-to-line voltage). This is illustrated on the sine wave in Figure

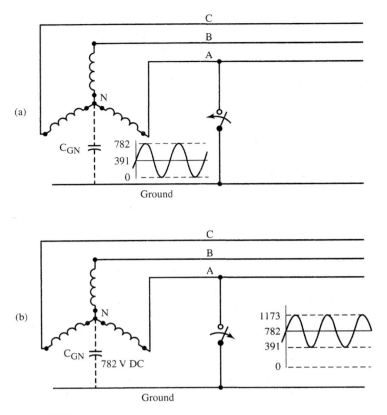

FIGURE 28.8
(a) Second contact to ground; (b) second contact to ground breaks.

28.8b, where the voltage between line A and ground now cycles between 391 volts and 1173 volts.

Rapid restriking of vibrating ground faults or sputtering arc ground faults can quickly escalate the voltage between the three lines and ground to levels that cause insulation breakdowns at other locations in the system. The effect is particularly damaging in confined spaces such as totally enclosed machines, switchgear, etc., where burning insulation and arcing can cause gas pressure to build up. Intermittent ground faults on low-voltage ungrounded neutral systems have been observed to cause overvoltages as high as six times normal voltage [3].

A particularly devastating example of a repetitive restriking ground fault occurred in a West Coast manufacturing plant in the early 1950s. The intermittent arcing ground fault occurred in a motor starter supplied by a 480-volt, three-phase system. The voltage on the system stepped up to more than 1200 volts, and in the 2 hours it took to locate the fault, the insulation on more than 40 motors was damaged! Shortly thereafter, a grounded neutral was installed.

Ungrounded Single-Phase System

An accidental ground in an ungrounded *single-phase* system doubles the steady-state voltage to ground on the unfaulted line. The doubled electrical stresses cause twice the electron leakage through the insulation, hastening its deterioration and thus shortening its life. Figure 28.9a shows the distribution of voltage between the conductors and ground for a 240-volt single-phase supply.

The voltage between each conductor and ground is 120 volts. However, if one conductor is grounded, as indicated by the dotted line in Figure 28.9b, the voltage difference between the unfaulted conductor and ground will rise to 240 volts. The voltage stresses on the insulation of the ungrounded conductor are doubled, and if a weak spot in the ungrounded insulation causes it to blow, a short circuit will result. Note that this doubling of voltage to ground does not include any voltage rise that may occur due to resonance.

28.4 GROUND-FAULT DETECTION IN UNGROUNDED SYSTEMS

Ground-fault detection circuits for ungrounded systems are illustrated in Figure 28.10. In the two-wire or single-phase system (Figure 28.10a), two identical 120-V lamps are connected in series and tied to the two lines. The two lamps should have identical wattage ratings and a voltage rating equal to the line voltage. The junction point of the two lamps is connected to ground through a normally closed spring-return pushbutton. With no ground fault, the voltage across each lamp will be 60 V, and both lamps will burn dimly. A ground fault on line B, illustrated with dotted lines in Figure 28.10a,

FIGURE 28.9

(a) Two conductors of opposite polarity separated from each other and ground by insulation; (b) elementary series circuit formed by insulation resistance (R), metal framework against which the insulation rests, and the conductors.

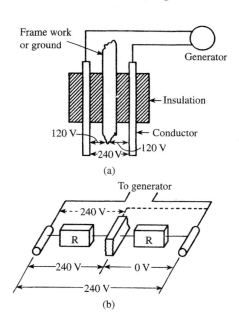

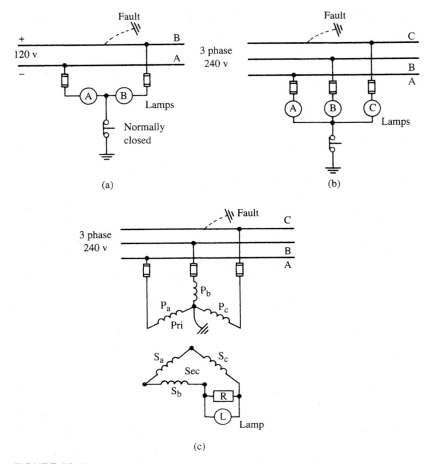

FIGURE 28.10

Ground detection in an ungrounded distribution system: (a) single phase or DC; (b) three-phase low voltage; (c) three-phase high voltage.

short-circuits lamp B, causing it to burn dimly or go out, depending on the severity of the ground, while lamp A burns more brightly. The pushbutton provides a means for comparing the normal and ground-fault indications.

Figure 28.10b shows a ground-detection circuit for a 240-V, three-phase ungrounded system. The three lamps should have identical wattage ratings, and a voltage rating equal to the line voltage (240 V). The junction point of the three lamps is connected to ground through a normally closed spring-return pushbutton. With no ground fault, the voltage across each lamp will be $240 \div \sqrt{3} = 138.7$ volts, and all three lamps will burn dimly. A ground fault on line C, illustrated with dotted lines in Figure 28.10b, short-circuits lamp C, causing it to burn dimly or go out, depending on the severity of the ground, while lamps A and B burn more brightly.

The pushbutton provides a means for comparing the normal and ground-fault indications.

The power rating of the lamps used in lamp-type ground-detecting circuits determines the degree of sensitivity of the ground-detecting system. In general, 25-watt lamps provide the best overall performance. Lower wattage lamps tend to be too sensitive, and higher wattage lamps are not sensitive enough.

Figure 28.10c illustrates a ground-detecting circuit that uses potential transformers in an ungrounded or high-resistance-grounded system. The potential transformers are connected in a wye–delta arrangement. The neutral of the wye is connected to ground, and the delta connection is completed through a paralleled resistor and lamp or a paralleled resistor and alarm relay. With no ground fault on the system, the three secondary voltages are balanced, and no voltage appears across the lamp. However, a ground fault on phase C, for example, will effectively short-circuit primary coil P_c and cause the voltage of secondary coil S_c to be zero. The resultant voltage unbalance in the secondaries will cause the lamp to light, indicating the presence of a ground.

Although the ungrounded systems illustrated in Figure 28.10 are connected to ground via low-wattage lamps or instrument transformers, the grounding effect is insignificant, and they are considered to be ungrounded systems.

28.5 LOCATING GROUND FAULTS IN UNGROUNDED DISTRIBUTION SYSTEMS

Ground faults in an ungrounded distribution system (unless accompanied by a short or an open) are generally located by the process of elimination. The grounded circuit may be determined by pulling switches or tripping breakers on the distribution panel, one at a time, until the ground-detecting device indicates normal. Closing each switch before opening the next keeps the interruption of service to a minimum. The opening of switches feeding vital equipment should be avoided until standby equipment is placed in operation. If this procedure fails, either the ground is in the incoming supply lines or more than one ground is present.

Multiple grounds may be located by opening the switches or breakers, one at a time, and leaving them open, until the ground detector indicates normal. Then, with the grounded circuit left open, the other switches should be closed one after the other until another ground is indicated. The switches to the grounded circuits should be left open, and the procedure continued until every circuit is tested.

Locating the Grounded Conductor

Tracking down the actual location of the grounded conductor may be easily done with a megohmmeter, magneto, or battery and buzzer set. When doing so, the breaker or switch to the grounded circuit should be opened, any fuses removed, and a "DO NOT CLOSE: PERSON WORKING ON LINE" sign hung over the switch or circuit breaker. If possible lock the breaker or switch in the open position.

FIGURE 28.11

Tracking down a ground by the process of elimination.

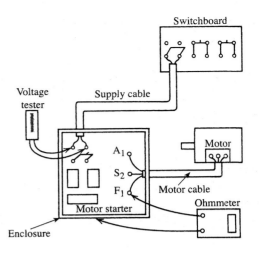

Figure 28.11 illustrates the megohmmeter method for tracking down a ground that may be in a supply cable, motor starter, motor cables, or in the motor itself. The distribution breaker or switch at the switchboard and the disconnect switch or breaker in the motor starter enclosure should be opened and top and bottom terminals of the starter breaker (or switch) tested with an appropriately sized voltmeter to ensure that the incoming power lines are dead. A megohmmeter should then be used to check the cable by testing between the metal framework of the enclosure and the top of the breaker or switch terminals. A zero reading on the megohmmeter indicates a grounded cable.

Assuming the cable tests clear, as indicated by the absence of a zero reading, the test should be made between the starter enclosure and the motor connections at the starter. If a ground is indicated, the motor leads should be disconnected from the starter, and the motor and starter tested individually. Grounds within the motor can be located by means of additional tests; see index for tests on specific apparatus.

A grounded generator can be identified by transferring the load to another machine and tripping the machine in question from the line. If the tripped generator is grounded, removing it from the bus will cause the ground detector to indicate normal.

28.6 SOLIDLY GROUNDED DISTRIBUTION SYSTEMS

A solidly grounded distribution system has one of its current-carrying conductors, called the grounded conductor, deliberately connected to the equipment-grounding conductor. The ground connection is made to either the neutral of a wye-connected system, to one corner of a delta system, or to the center tap of one phase of a delta, as illustrated in Figure 28.12.

Solidly grounded systems are used extensively for commercial and industrial loads. They are also required for residential service. The solidly grounded system provides the highest level of protection from shock hazards, limits the voltage to ground

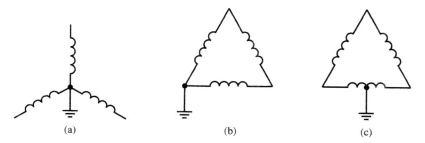

FIGURE 28.12
Solidly grounded systems: (a) at the neutral; (b) at a delta corner; (c) at the center tap of a delta.

during normal operation, lightning, or other surge voltages, and provides protection against arcing when sensitive ground-fault protection is used.

A representative circuit illustrating a three-phase solidly grounded system is illustrated in Figure 28.13. Note that the grounded conductor (colored white or neutral gray) and the equipment-grounding conductor (colored green) are connected to the same point and, thus, are at the same potential. However, *the equipment-grounding conductor must not be used as a normal current-carrying conductor*. The ampacity of the equipment-grounding conductor must be reserved for ground-fault currents.

The grounded conductor is not fused nor is it in series with a circuit breaker.

The principal disadvantage of a solidly grounded distribution system is the severe flash hazard to personnel that will occur if a tool or other conductor accidentally bridges a hot line to the metallic framework of a switchboard, machine frame, water pipe, etc. Ground faults in solidly grounded systems generally cause the opening of breakers or the blowing of fuses as illustrated in Figure 28.14.

Figure 28.15 illustrates a three-wire, single-phase distribution system for lighting and power circuits. The center tap of the transformer secondary is solidly grounded to a water pipe or to a driven ground. The grounded conductor is not fused or in series with a circuit breaker; only the "hot" lines have series-connected fault interrupting equipment. Ground faults in these circuits will cause blown fuses or tripped breakers.

In the case of arcing ground faults, the resultant ground current may not be high enough to operate the overcurrent trip of a breaker, but may be sufficiently high to promote sustained arcing and burning of apparatus. Protection against damage caused by arcing ground faults can be obtained with a ground-sensing relay that operates a tripping coil in the breaker.

28.7 GROUND-FAULT PROTECTION

Ground-fault detection in solidly grounded, three-phase, wye-connected systems can be accomplished by grounding the neutral through a window-type current-transformer (CT) as illustrated in Figure 28.16. The CT secondary is connected to a relay or ammeter and relay. An ammeter will indicate the presence and severity of the ground fault,

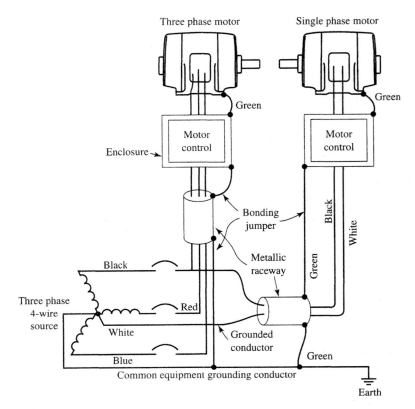

FIGURE 28.13
Representative circuit illustrating a solidly-grounded system feeding single-phase and three-phase circuits.

and a relay will actuate the tripping mechanism of the breaker. Window-type CTs designed for this purpose are called *current sensors* or *zero-sequence CTs,* and the ground fault currents are called *zero-sequence currents*.

Another method for ground-fault tripping uses a CT to enclose all three lines and the neutral (if present), as illustrated in Figure 28.17. With no ground fault present, all current flows out and returns through the window of the CT. Thus, the net magnetic flux in the core of the CT will be zero, and no current will be induced in the secondary.

If a ground fault occurs between line L_3 and the motor frame, as illustrated in Figure 28.17, some of the current from line L_3 arcs to the motor frame, returning through the equipment-grounding conductor. Since the ground-fault-current component from line L_3 does not flow back through the other lines, the net magnetic flux in the CT is no longer zero, and a current proportional to the magnitude of the ground-fault current will be induced in the CT secondary. The secondary current actuates a relay that trips the breaker. Window CTs that are sensitive enough to detect a few milliamperes of ground-fault current are available. Although ground-fault relays using

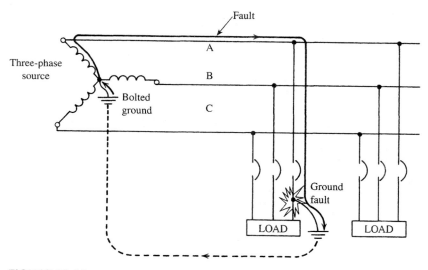

FIGURE 28.14
Ground fault on a normally grounded system.

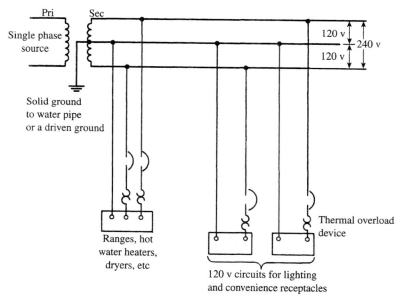

FIGURE 28.15
Three-wire, single-phase distribution system for lighting and power circuits.

FIGURE 28.16
Current transformer and ammeter used for
ground detection.

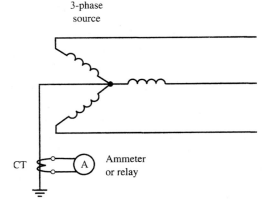

CTs can respond rapidly to clear ground faults once they occur, they cannot prevent
ground faults from occurring. Ground-fault relays protect equipment from further
damage only after a ground fault occurs.

28.8 LOCATING HIGH-RESISTANCE GROUND FAULTS
IN SOLIDLY GROUNDED DISTRIBUTION SYSTEMS

High-resistance ground faults may be located by means of a clamp-on ammeter. When
clamped around all active conductors feeding a circuit, the ammeter will indicate the
phasor sum of all the enclosed currents. The equipment-grounding conductor (green)
is not an active conductor and must not be included in the clamp. If no ground fault
exists, the meter will read zero. If a ground fault is present, the phasor sum of the currents enclosed by the clamp will not be zero, and the ammeter will indicate the presence of a fault current.

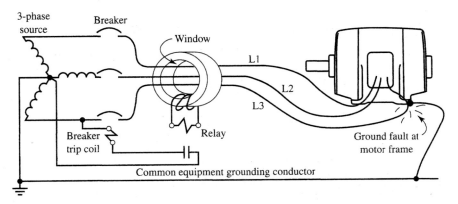

FIGURE 28.17
Circuit for ground-fault tripping.

FIGURE 28.18
Procedure for locating a ground fault in a solidly grounded system.

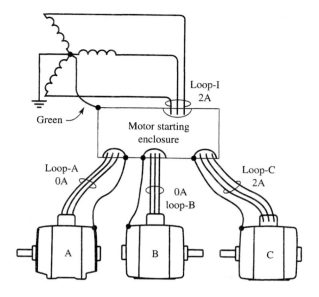

A representative system showing the procedure for locating a ground fault in a solidly grounded system (with a ground fault in motor C) is illustrated in Figure 28.18. The four loops surrounding the active conductors in each circuit show the conductors that must be enclosed by the clamp-on ammeters.

Tests to locate the ground fault should begin at the power source, and proceed step by step until the fault is found. Referring to Figure 28.18, the clamp-on ammeter at loop I indicates 2 amperes in the loop enclosing the incoming lines, indicating a ground-fault in the distribution panel or in one of the downstream circuits.

Testing of the downstream circuits shows zero leakage in loop A, zero leakage in loop B, and 2-ampere leakage current in loop C. The ground fault is in motor C or the lines leading to it.

28.9 OTHER METHODS OF SYSTEM GROUNDING

Other methods of system grounding, such as resistance grounding, reactor grounding, resonant grounding, and grounding through a grounding transformer are presented and discussed in Reference [2].

SPECIFIC REFERENCES KEYED TO THE TEXT

[1] *National Electrical Code 1985,* National Fire Protection Association.
[2] Crane, R., Programmable Controllers and Distributive-Control Installation Practices, *IEEE Industry Applications Society Newsletter,* March/April 1986, pp. 8–11.

[3] *Recommended Practice for Grounding of Industrial and Commercial Power Systems,* IEEE STD 142–1991.

[4] *Guide for Measuring Earth Resistivity, Ground Impedance, and Earth Surface Potentials of a Ground System,* IEEE STD 81–1983.

REVIEW QUESTIONS

1. Explain why it is necessary for the framework of all electrical apparatus to be grounded to a common point.
2. Differentiate between equipment grounds, bonding, common equipment-grounding conductors, and earth.
3. How are low-resistance connections to earth made?
4. What factors affect earth resistance, and what is the upper allowable limit for earth resistance?
5. What is an ungrounded distribution system, and what are its advantages and disadvantages?
6. Why is it important that a single ground fault in an ungrounded distribution system be located and cleared as soon as possible?
7. Explain how distributed capacitance in an ungrounded system causes overvoltage when a ground fault occurs.
8. Explain how series resonance in an ungrounded system can cause overvoltage when a ground fault occurs in a motor, reactor, autotransformer, etc.
9. Explain how a vibrating ground in an ungrounded system can cause overvoltage when a ground fault occurs.
10. Explain how a ground fault in an ungrounded single-phase system doubles the voltage across the insulation of the ungrounded line.
11. (a) Sketch a two-lamp ground detection circuit for a 240-volt ungrounded single-phase system. (b) Specify the recommended voltage and wattage rating of the lamps. (c) What is the voltage across each lamp if no fault is present? (d) What is the voltage across each lamp if a solid ground fault occurs on one line?
12. (a) Sketch a three-lamp ground detection circuit for a 480-volt ungrounded three-phase system. (b) Specify the recommended voltage and wattage rating of the lamps. (c) What is the voltage across each lamp if no fault is present? (d) What is the voltage across each lamp if a single solid ground fault occurs on one line?
13. What effect does the power rating of the lamps used in ground detector systems have on the sensitivity of the detector?
14. (a) Sketch a ground detection circuit that uses potential transformers for a 2300-volt ungrounded three-phase system. (b) Explain the behavior of the circuit without a ground fault, and with a ground fault.
15. Using an appropriate sketch, describe the correct procedure for locating a ground fault in an ungrounded distribution system.
16. What is a grounded distribution system, and what are its advantages and disadvantages ?
17. Sketch a circuit showing a solidly grounded distribution system. What are its advantages?

29

Protection Against Sustained Overloads and Short Circuits

29.0 INTRODUCTION

Excessive current can destroy insulation, melt conductors, cause explosions and fires, kill and maim operating personnel, and in general raise havoc with the immediate surroundings. Furthermore, the man-hours and production time lost as a result of this type of failure can never be replaced. Protection against damaging overcurrent caused by overloads and short circuits can be provided by correctly applied and properly maintained fuses, circuit breakers, and relays.

The protective devices discussed in this chapter are limited to fuses and molded-case circuit breakers used in low-voltage AC distribution systems (no more than 600 volts).

29.1 FUSES

A fuse is an overcurrent protective device containing a circuit-opening fusible link directly heated by the passage of current through it. Excessive current melts the fusible link opening the circuit. Fuses are selected on the basis of operating voltage, rated current, and interrupting rating.

When operating in an ambient temperature of no more than 55°C (131°F), fuses should be able to carry rated current indefinitely [1].

Figure 29.1 illustrates two types of cartridge fuses used in low-voltage distribution systems. The single-element fuse shown in Figure 29.1a is used for short-circuit protection in lighting and general power applications. The fuse shown in Figure 29.1b is called a dual-element fuse. It consists of normal fuse links, similar to those shown in Figure 29.1a, in series with a time-delay element.

The path of current through the dual-element fuse is through two links and a spring-loaded soldered joint (time-delay element). The links provide short-circuit protection. The

623

FIGURE 29.1
(a) Single-element fuse; (b) dual-element fuse; (c) fuse blowing at short circuit; (d) fuse blowing with overload (courtesy Cooper Bussmann Mfg. Co.).

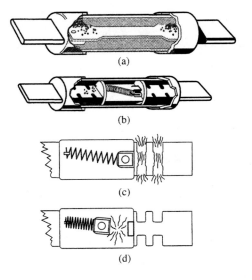

spring-loaded soldered joint provides protection against motor overloads while allowing a high current of short duration for motor starting. The time delay is designed to permit a transient starting current of 200 to 300 percent of the rated motor current and yet provide protection against a sustained overload equal to 125 percent of the rated current. When a short circuit occurs, the high current causes the fuse links to blow, as shown in Figure 29.1c. However, with a sustained overload, the fuse links remain intact, but the heat generated by the overload current softens the low-melting-point solder that connects the link to a center piece of copper. The spring pulls the connector away and the circuit is opened; this is shown in Figure 29.1d.

Interrupting Rating

The interrupting rating of a fuse is the highest direct current or rms alternating current that it can successfully interrupt. If the short-circuit current exceeds the interrupting rating, the fuse may explode violently, causing an electrical fire and severe damage to the equipment it was supposed to protect. Fuses of the correct type and size provide good dependability and protection. They have no latches, coils, bimetallic strips, or other moving devices common to circuit breakers and hence require little maintenance.

Paralleled Fuses

Under no circumstances should fuses be paralleled without specific approval from the manufacturer. This practice is dangerous and should be avoided. The total inductive energy stored in the circuit at the instant of interruption may be increased due to the overall lowered resistance of the paralleled fuses. It is also possible that the current will not be evenly divided between the fuses due to impedance variations. The fuse carrying the higher share of the current may explode.

Fuse-Blowing without Apparent Cause

Common causes for blown fuses, other than short circuits and sustained overloads, are poor contact due to loose fuse clips, poor contact within the fuse, improper location of the fuse in extremely hot surroundings, and excessive vibration. Fuses that blow because of poor contact can generally be detected by discolorations or charring at the defective region. When replacing blown fuses, be sure that the fuse clips are straight, tight, and in good contact with the fuse. If the spring clips have lost their grip, they should be replaced. Clip clamps shown in Figure 29.2 can be used as a temporary fix until an overhaul period permits replacement of the fuse holder.

When fuses of the correct size and type are found to blow frequently, look for trouble within the circuit. Do not overfuse. A fuse larger than necessary endangers the apparatus it is supposed to protect.

Fuse Time-Current Characteristic

A representative time-current characteristic that illustrates the average melting time for a 15-ampere low-voltage cartridge fuse, operating under overload conditions in a 77°F (25°C) ambient temperature, is illustrated in Figure 29.3. Time-current curves for specific fuses are available from the manufacturer.

29.2 TESTING LOW-VOLTAGE FUSES

Fuse testing should be done with the circuit dead. If the circuit is dead, the fuses can be removed and tested with an ohmmeter, megohmmeter, magneto, or battery and buzzer test set.

However, if the testing must be done on an energized circuit, a voltmeter or mechanical voltage tester of the correct voltage rating should be used. Use of light bulbs should be avoided, because there is always the danger of breakage from striking or overvoltage. The mechanical voltage tester, shown in Figure 29.4, has the additional advantage of differentiating between alternating and direct current; when applied across an AC circuit, a slight vibration may be felt. Some mechanical voltage testers are equipped with a small neon light to determine polarity on DC systems. Do not use your fingers to test fuses.

FIGURE 29.2
Clip clamps for securing fuses to weak fuse clips (courtesy Ideal Industries Inc.).

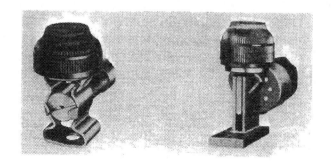

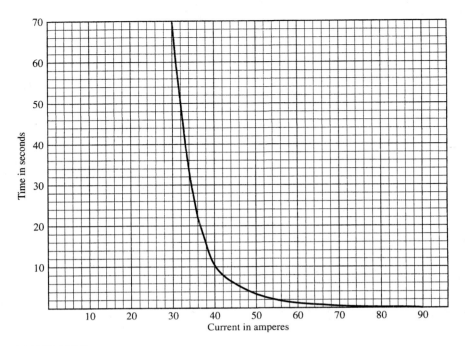

FIGURE 29.3

Representative time-current characteristic for a 15-ampere low-voltage cartridge fuse.

To check for blown fuses in a single-phase or DC system, the switch should be closed, and the voltage tester applied across the tops of both fuses, as shown in Figure 29.4a, and then across the bottoms of both fuses. If voltage is indicated across the tops but not across the bottoms, or vice versa, one or both of the fuses are blown. To locate the faulty fuse, the leads of the tester should be placed across the "hot" ends of both fuses; then one test point should be moved down the fuse to the other end, as in Figure 29.4b. If the tester indicates voltage, the fuse is good.

This test can also be used for checking the three line fuses of a three-phase system, provided that the load is disconnected. If the load consists of a three-phase motor, the starter must be in the "stop" position; a lightly loaded three-phase motor will continue to operate with one fuse blown and will generate a voltage for the blown fuse, thereby preventing detection by a voltage tester. A continuity tester, such as a battery and buzzer test set, megohmmeter, or ohmmeter, provides a foolproof method for testing fuses in a three-phase system; the circuit must be *de-energized* and the tester applied across each fuse, or the fuses must be removed and then tested. When fuses are removed or replaced, the circuit should be de-energized and an insulated fuse puller used.

29.3 CURRENT-LIMITING FUSES

A current-limiting fuse (CLF) differs from those having only a high interrupting rating in that the melting and arcing time is so brief that the actual let-through current is much less than the maximum available short-circuit current. Whereas the ordinary

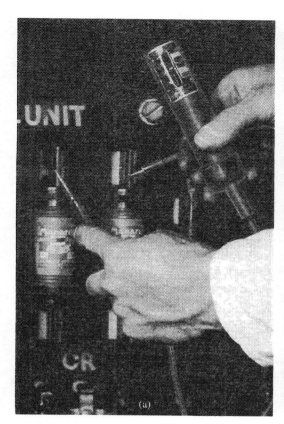 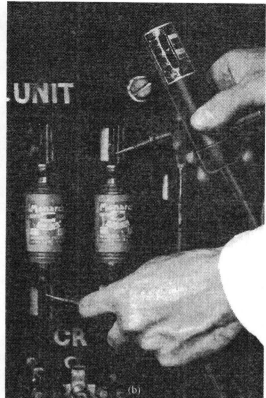

FIGURE 29.4
Method of testing low-voltage fuses: (a) testing across fuse tops; (b) testing diagonally.

high-interrupting-rating fuses let through several half-cycles of short-circuit current before circuit interruption occurs, the current-limiting fuse clears the fault in a small fraction of a cycle.

A current-limiting fuse, shown in Figure 29.5, has many silver links connected in parallel between heavy copper end blocks. The end blocks act as a heat sink to absorb, radiate, and conduct away the heat generated in the links. The links are embedded in pure quartz sand inside a heavy synthetic-resin-glass cloth-laminated tube that spaces the end blocks and seals them against the emission of gas or flames.

Although the silver links have a high normal current-conducting ability, excessive current, such as that caused by a short circuit, almost instantaneously raises the temperature of the links to the melting point. The resultant arc, formed in the surrounding quartz sand, transforms the sand into glass and the arc is extinguished. The silver-melting process on a high-magnitude fault is so rapid that the fault current does not have time to reach its maximum possible peak value but is "limited" to a relatively low value, as shown in Figure 29.6.

FIGURE 29.5
Current-limiting fuse with outer case and sand filler removed, showing the silver links and the heavy copper end blocks (courtesy Chase-Shawmut Co.).

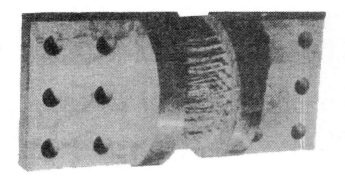

The peak of the triangle represents the peak let-through current, also called the cutoff current. The triangular-shaped section illustrates how a current-limiting fuse prevents the current from reaching destructive values. The available short-circuit current is the short-circuit current that would occur if a non-current-limiting fuse were used.

29.4 MOLDED-CASE CIRCUIT BREAKER

A molded-case circuit breaker, shown in Figure 29.7a, is a nonadjustable air circuit-breaker assembled as an integral unit in a supporting and enclosing housing of insulating materials. It is used for circuit protection in low-voltage distribution systems. Breakers of this type are able to provide overload protection, as well as short-

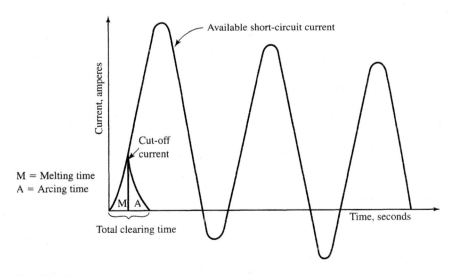

FIGURE 29.6
Comparison of available short-circuit current and actual short-circuit current passed by a current-limiting fuse.

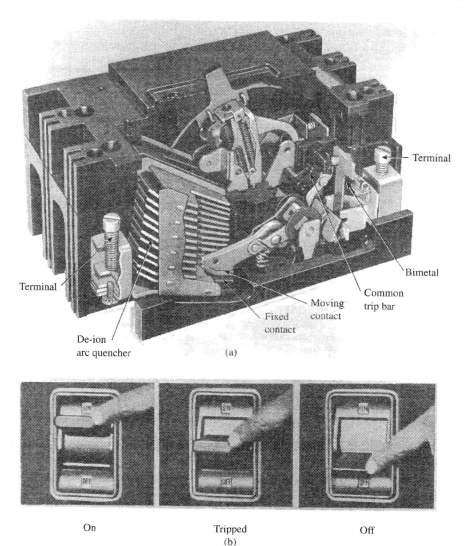

Terminal

Bimetal

Common
trip bar

Moving
contact

Fixed
contact

Terminal

De-ion
arc quencher

(a)

On

Tripped
(b)

Off

FIGURE 29.7
Three-pole molded-case circuit breaker: (a) cutaway view; (b) position of handle indicates status of breaker 'ON', 'TRIPPED,' or 'OFF' (courtesy TECO Westinghouse).

circuit protection, for conductors, motors, control equipment, lighting circuits, and heating circuits. The different position of the breaker handle, illustrated in Figure 29.7b, indicates the status of the breaker: 'ON', 'TRIPPED', or 'OFF'.

The arc chute containing a de-ion arc quencher is illustrated in Figure 29.7a. It consists of parallel steel plates partially surrounding the contacts and enclosed by a fiber wrapper or ceramic supports. As the contacts open, the current arcing between

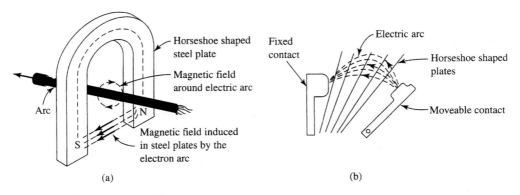

FIGURE 29.8
(a) Interaction of magnetic fields in the arc quencher; (b) movement of arc through the arc quencher.

the stationary contact and the moving contact induces a magnetic field in the horseshoe-shaped steel plates. The interaction of this field with the magnetic field surrounding the arc causes the arc to move upward into the horseshoe-shaped plates, where it is split into a series of smaller arcs, cooled, and extinguished.[1]

The interaction of the magnetic fields in the arc quencher and the upward movement of the arc through the horseshoe-shaped plates are shown in Figures 29.8a and 29.8b, respectively.

Molded-case breakers are classified by frame size, ampere rating, and interrupting rating. The frame size, expressed in amperes, is the largest ampere rating available in the group. The ampere rating is the value of current that the breaker will carry continuously in an ambient temperature of 40°C without either tripping or exceeding the permissible temperature rise. The interrupting rating of a breaker is the maximum rms current that it can interrupt without damage to the breaker. The very high magnetic forces produced by short circuits can damage the breaker if the current exceeds its interrupting rating.

29.5 TRIP UNITS FOR MOLDED-CASE BREAKERS

Trip units for molded-case breakers may be thermal, magnetic, thermal-magnetic, or electronic (solid state). The trip unit activates the tripping mechanism of the breaker, causing the breaker to open.

Thermal Trip

A thermal trip unit provides protection against sustained overloads. The thermal trip consists of a bimetal element connected in series with the breaker contacts as shown in Figure 29.9a.

---

[1] For more detailed explanations of the interactions of magnetic fields and the resulting magnetic forces, see Section 3.5, Chapter-3.

FIGURE 29.9

(a,b) Thermal trip; (c,d) magnetic trip; (e,f) thermal-magnetic trip (courtesy TECO Westinghouse).

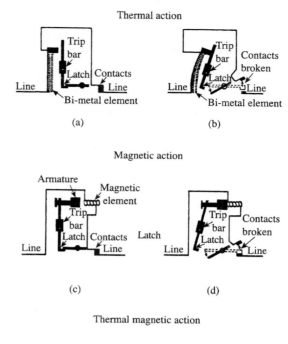

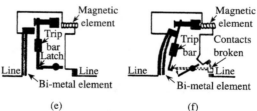

The bimetal element consists of two bonded strips of metal having different rates of thermal expansion. Excessive current passing through the bimetal element will cause it to deflect, as shown in Figure 29.9b. This trips the latch that opens the contacts. The left strip of the bimetal has a greater rate of linear expansion than the right strip, causing the element to deflect to the right when heated. Thermal-acting trips have an inverse-time characteristic. That is, they provide a long time delay before tripping when a light overload occurs and a faster response for heavy overloads. The thermal trip element may be fixed or adjustable, depending on the type of breaker and the frame size.

Magnetic Trip

Protection against short circuits is accomplished by an electromagnet, as shown in Figure 29.9c. The line current passes through the magnet coil, causing a magnetic pull on a movable iron slug called the armature. During short-circuit conditions the magnet becomes strong enough to attract the armature, as shown in Figure 29.9d; the armature trips the latch and opens the breaker contacts. This action is "instantaneous" in that there is no deliberately built-in time delay. The magnetic trip element may be fixed or adjustable, depending on the type of breaker and the frame size. Adjustable magnetic trips have adjustment knobs on the front of the breaker that change the size of the air gap between the armature and the electromagnet, thereby changing the current required for tripping.

Thermal-Magnetic Trip

Thermal-magnetic breakers provide instant tripping action on short circuits but allow momentary overloads, such as high starting currents in motors and initial surge currents in lighting circuits. Figure 29.9e illustrates the arrangement for a combination thermal-magnetic breaker, and Figure 29.9f illustrates combined thermal and magnetic action.

Ambient Compensation

For those applications where a thermal-acting breaker must be located in a region of unusually high, low, or fluctuating temperature, an ambient-compensating trip, illustrated in Figure 29.10a, must be used.

Compensation is obtained by using a bimetal tripping bar to counteract the effect of ambient temperature changes on the load-sensitive bimetal element, as shown in Figure 29.10a. Figure 29.10b illustrates a condition of overload in a high ambient environment. Note how the compensating bimetal changes its position to prevent tripping. Figure 29.10c illustrates the effect of an overload in a normal ambient environment.

Solid-State Tripping Circuits

Molded-case breakers with solid-state tripping circuits use preprogrammed microprocessors in place of thermal and magnetic trips. Solid-state sensing circuits compare the actual circuit current with data preprogrammed into the breaker's microprocessor.

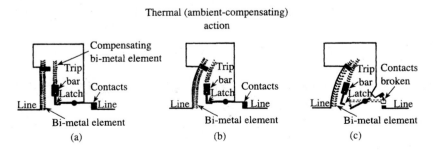

Thermal (ambient-compensating) action

FIGURE 29.10
(a) Thermal trip with ambient compensation; (b) overload in a high ambient environment; (c) overload in a normal environment (courtesy TECO Westinghouse).

Then, if an overcurrent or short circuit occurs, a low-energy electrical signal from the microprocessor energizes a coil that trips the circuit breaker. Depending on the design, the microprocessor may provide fixed or adjustable long-time current pickup with fixed or adjustable long-time delay; fixed or adjustable short-time current pickup with fixed or adjustable short-time delay, and fixed or adjustable instantaneous current pickup. A removable rating plug electronically sets the continuous current rating for the breaker.

The usual function of the long-time-delay element is to provide overload protection for conductors and connected apparatus. The function of the short-time-delay element is to provide a short delay on a fault current to give selectivity to other breakers, that is, to give other breakers closer to the fault an opportunity to trip. The instantaneous trip provides protection under high short-circuit currents. Two settings each are required to completely define the long-delay characteristic and the short-delay characteristic, namely, a pickup current setting and a time-delay setting. The instantaneous characteristic is defined by the pickup setting alone. No intentional time delay is provided for the instantaneous trip. To provide adequate protection, the instantaneous pickup should be set as low as possible, without causing nuisance tripping; if the instantaneous setting is higher than the available short-circuit current, the instantaneous trip will never operate.

Shunt Trips

In addition to overcurrent tripping devices, circuit breakers are often equipped with low-voltage trips and shunt trips. The shunt trip may be actuated by auxiliary devices, such as a power-directional (reverse power) relay, a ground relay, a phase-balance relay, or a remote pushbutton. For example, circuit breakers for AC generators usually have shunt trips that are actuated by a power-directional relay to trip the generator from the bus in the event that it is motorized. Similarly, breakers for DC generators in parallel use reverse-current trips to open the respective generator breaker if its current is reversed.

29.6 INTERRUPTING RATING OF A BREAKER

The interrupting rating of a breaker is the *rms value of current at rated rms voltage that the breaker can safely interrupt without damage to itself.*

To prevent damage or complete destruction of a circuit breaker when a short circuit occurs, the breaker must be able to withstand the high mechanical forces caused by the short-circuit current. The magnitude and direction of these forces on the current-carrying parts of the breaker are the same as those produced in adjacent cables of opposite polarity.[2] These forces, proportional to the square of the current, cause a considerable strain on the breaker components. If the short-circuit current is greater than the interrupting rating of the breaker, the tremendous mechanical forces that are

---

[2] See Section 3.3, Chapter-3.

produced on the adjacent breaker poles can literally tear it apart or, even worse, cause a violent explosion and fire.

Reclosing a breaker after a fault occurs could be dangerous if the interrupting rating of the breaker is inadequate. The operator may be seriously injured or killed by the resulting explosion if the breaker was damaged in the initial trip. The fault must be located and repaired before reclosing the breaker.

Figure 29.11 illustrates the damage to a switchboard in an industrial plant caused by current in excess of the interrupting rating of a molded-case circuit breaker. The switchboard consisted of sixteen 600-ampere circuit breakers fed from a 5000-kVA, 600-volt transformer bank. The only fault protection for the transformer bank and switchboard was a circuit breaker at the utility substation on the 33-kV line.

A one-line diagram for the distribution system is shown in Figure 29.11b. The available secondary-fault current was calculated to be 80,000 A, whereas the interruption rating of the distribution circuit breakers was only 25,000 A. When the molded-case breaker opened, it could not interrupt the high short-circuit current, and the breaker exploded. Short-circuit damage and the burned-off studs on the line side of the circuit breakers are shown in Figure 29.11a. The 1/4-inch by 3-inch copper bus was

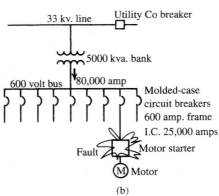

FIGURE 29.11
(a) Switchboard damaged by electromechanical forces caused by short-circuit current;
(b) one line diagram of the distribution system (courtesy Mutual Boiler and Machinery Insurance Co.).

badly distorted from the mechanical forces developed by the fault current. Ten of the 16 circuit breakers were burned beyond repair and had to be replaced. The plant was out of service for 2 days to make temporary repairs.

In many low-voltage distribution systems, possible short-circuit currents as high as 100,000 amperes are common. Because fault currents of this intensity exceed the interrupting rating of the largest molded-case breakers, some form of backup protection is required. To avoid the use of larger and more expensive breakers capable of withstanding the large thermal and magnetic stresses, special high-interrupting-rating current-limiting fuses (CLFs) are used.

A molded-case breaker with a built-in current limiter is shown in Figure 29.12a. The thermal and magnet elements are the same as previously illustrated for the standard breakers. When a very high short-circuit current occurs, the silver links of the current limiter melt, opening the circuit; simultaneously, the magnetic trip functions to open the breaker contacts and thus aids in clearing the short circuit. A wire holding a plunger against the pressure of a spring melts when the silver links melt, causing the plunger to extend outward, holding the trip bar in the unlatched or open position. This makes it impossible to close the breaker until the blown limiter is replaced. Interlocks prevent relatching the breaker if a limiter is omitted, and will also open the circuit breaker contacts if an attempt is made to remove the limiter housing assembly with the breaker in the 'ON' position. Figures 29.12b and 29.12c illustrate a good limiter and a blown limiter, respectively.

29.7 TRIPPING CURVES

Figure 29.13 illustrates a representative average tripping curve for a 300-A molded-case breaker of the type shown in Figure 29.12. The curve is a plot of tripping time versus overload current. The upper left-hand segment of the curve represents the thermal tripping (average time delay) of the breaker as a result of a sustained overload condition. As indicated by the curve, the breaker will not trip at currents of 300 A or less. For currents ranging from 325 A to 3000 A, the time delay for tripping will vary inversely with the current. Thus, with a current of 325 A, it will take over 2 hours to trip, whereas with a current of 2500 A the average tripping time will be approximately 12 seconds.

The abrupt change in the curve to a vertical segment for currents in excess of 3000 A represents "instantaneous" magnetic tripping of the breaker due to short-circuit currents. The broken line segment in the lower right-hand part of the characteristic curve represents operation of the current limiter. The point at which the current-limiter curve crosses the lower portion of the magnetic trip curve is called the *crossover point,* and the magnitude of the short-circuit current at this point is called the *crossover current.* At values of current less than the crossover current, the magnetic trip interrupts the fault without operation of the current limiter. At values of current greater than the crossover current, the current limiter clears the fault.

As indicated by the characteristic curves, the protective actions of the inverse time-delay thermal trip, the "instantaneous" magnetic trip, and the current limiter are coordinated so that overcurrents and low-magnitude faults are cleared by thermal

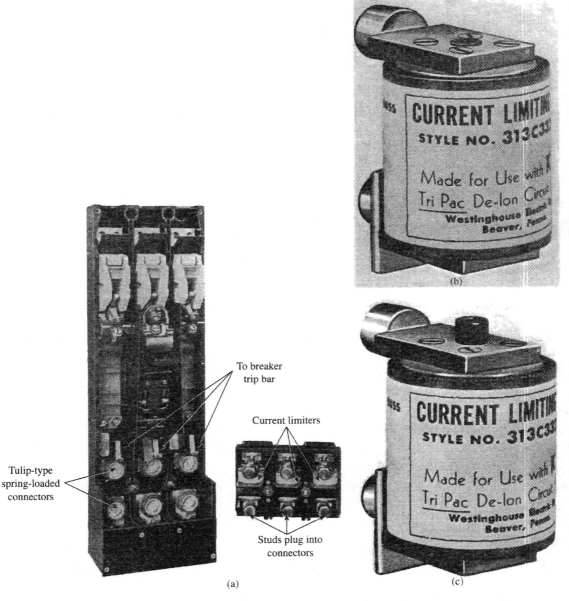

FIGURE 29.12
(a) Molded-case circuit breaker with a built-in current limiter; (b) good limiter; (c) blown limiter (courtesy TECO Westinghouse).

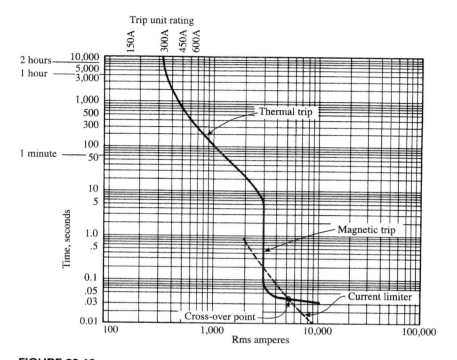

FIGURE 29.13

Representative average tripping curve for a 300-A molded-case breaker with thermal, magnetic, and current-limiting action (courtesy TECO Westinghouse).

action; "normal" short circuits are cleared by magnetic action, and abnormal short circuits are cleared by the current limiter.

Band Curves

Band curves illustrated in Figure 29.14 are used to indicate the calibration limits of a thermal-magnetic trip breaker. For a given overload current at 25°C, the breaker should clear the fault in the designed clearing time within the band bounded by the maximum and minimum characteristics. The thermal component of the breaker establishes the long time-delay region; the band curves extend from 100 percent rated current to 1000 percent rated current. Combined thermal and magnetic action occurs between 1000 percent and 2000 percent rated current; this is the transition region. For currents in excess of 2000 percent rated current, tripping is essentially "instantaneous" by magnetic means. The table in the inset in Figure 29.14 gives the average unlatching times for different values of current when tripping magnetically, with a maximum interrupting time of 0.01 seconds.

For example, a 15-ampere breaker whose characteristic is shown in Figure 29.14 should trip in not less than 16 seconds and not more than 55 seconds on a 200 percent

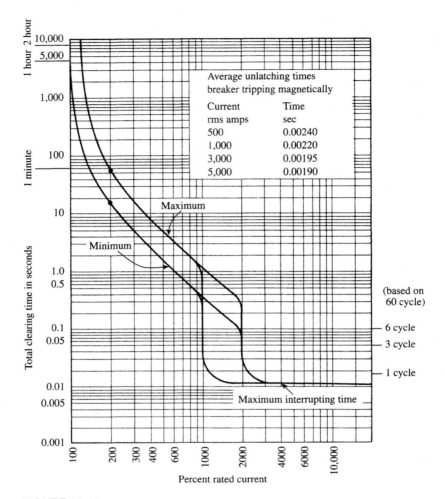

FIGURE 29.14
Band curves used to indicate the calibration limits of a 15-A ambient compensated thermal-magnetic breaker (courtesy TECO Westinghouse).

overload (30 A). Band curves are very useful in maintenance work when checking the calibration of overload trips.

29.8 MOTOR BRANCH-CIRCUIT PROTECTION

Figure 29.15 illustrates the typical motor branch circuit. It consists of a motor, motor controller, branch-circuit conductors, and a power supply.

The *maximum* current rating of motor branch-circuit protective devices depends on the full-load current rating of the motor, and whether the choice is a non-time-delay

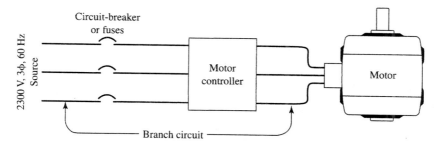

FIGURE 29.15
Motor branch circuit.

fuse, a dual-element fuse, an instantaneous trip breaker, or an inverse-time-delay breaker. A table of maximum ratings or settings of branch-circuit protective devices for different types of motors is given in Appendix 5.

EXAMPLE 29.1

Determine the maximum current rating for a dual-element time-delay fuse that could be used as the motor branch-circuit protective device for a 60-hp, 460-V, three-phase induction motor started at full voltage.

Solution
From Appendix 9, the full-load current of the motor is 77 A. From Appendix 5, the maximum rating for a dual-element time-delay fuse protecting the motor is 175% of rated current. Thus, for these conditions, the maximum rating of a dual-element fuse is

$$77 \times \left[\frac{175}{100} \right] = 134.8 \text{ A}$$

29.9 COORDINATION OF OVERCURRENT PROTECTIVE DEVICES

The complete protection of an electrical system against overcurrents and short circuits requires the coordination of all breakers, fuses, and overcurrent relays so that only the protective device nearest the fault will trip.

This is accomplished through a selective tripping arrangement whereby the delay-time and current-pickup settings of the breakers are adjusted to avoid overlapping of the time-current characteristics. In addition, each breaker and fuse in the system must be fully rated; that is, they must have interrupting ratings equal to or greater than the available short-circuit current in series with it. To ensure adequate protection, the selection and coordination of protective device settings must be done by design engineers knowledgeable in the field.

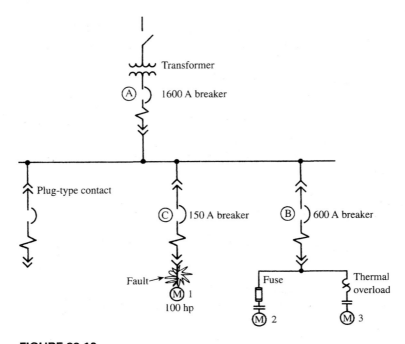

FIGURE 29.16
One-line diagram for a typical selective tripping system (courtesy GE Industrial Systems).

Figure 29.16 is a one-line diagram for a typical system involving circuit breakers and fuses in a fully protected selective tripping arrangement. Breakers A and B have long-time and short-time trips, and breaker C has a long-time and instantaneous trip.

The corresponding coordination curves are shown in Figure 29.17. If a fault of approximately 5000 amperes (50 × 100) occurs between breaker C and motor 1, only breaker C will open. This is indicated in Figure 29.17a, which illustrates breaker C clearing the fault in 0.03 second or less. Thus, continuity of service will be maintained for the rest of the system. However, if for some reason breaker C fails to trip, breaker A will open somewhere between approximately 48 seconds and 140 seconds.

If a fault occurs on the branch circuit for motor 2, only fuse D in that branch circuit will open. This is indicated by the time-current characteristic of the fuse in Figure 29.17b.

29.10 MOTOR CONTRIBUTIONS TO SHORT-CIRCUIT CURRENT

The magnitude of the short-circuit current, in the case of the simple system shown in Figure 29.18a, depends on the overall impedance of the circuit as measured from the place of short circuit to the source and includes the impedance of the source itself. However, if the system includes induction motors that are running at the time

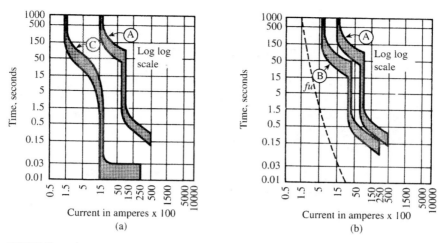

FIGURE 29.17
Coordination curves for the selective tripping system of Figure 29.16 (courtesy GE Industrial Systems).

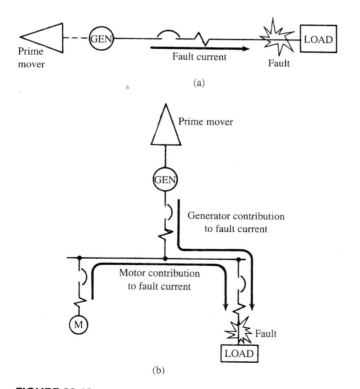

FIGURE 29.18
(a) Fault current to a short circuit; (b) generator and motor contributions to a fault.

a short occurs, as shown in Figure 29.18b, they will contribute to the short-circuit current.

The current contributed by induction and synchronous motors to the fault at the instant the short circuit occurs is approximately equal to 3.6 times and 4.8 times their respective full-load currents. Considering the many motors and the complexity of the distribution networks in many plants, the determination of the available short-circuit current in different parts of the system is not an easy task and is beyond the scope of this book.[3]

29.11 MAINTENANCE AND TESTING OF OVERCURRENT RELAYS AND BREAKERS

Unless checked periodically, there is no assurance that a properly installed and correctly rated overload relay or circuit breaker will continue to provide the same degree of protection that was afforded at the time of installation. There are cases on record where the bimetal elements in thermal overload relays became insensitive to heat from metal fatigue and failed to operate under sustained overloads of 300 percent of rated heater current. The maintenance technician should check for corrosion, accumulation of dirt and other foreign matter that block free movement of the tripping element, loose or missing parts, and improper substitution of heater element or trip coil. All of these indicate signs of trouble that must be corrected immediately.

Overcurrent relays and breakers can be tested by passing a specified overcurrent through the heater or overload coil and observing the time required for it to trip. The actual tripping time should then be compared with the manufacturer's current-time characteristic. It is recommended that breakers and overload relays be tested for proper tripping at 150, 300, and 600 percent of rated current.

To test an overcurrent relay, it should be disconnected from the power line and connected to a low-voltage high-current power source, as shown in Figure 29.19a.

A variable low-voltage autotransformer can be used to provide the adjustable high current for the test, and a stopwatch should be used to determine the tripping time. A preliminary adjustment of the current should be made by connecting the two heavy test leads 1 and 2 together, closing the autotransformer breaker, and adjusting the autotransformer until the clamp-on ammeter indicates a current slightly higher than the desired test current (the heater, when inserted, will cause the test current to drop slightly). Then open the autotransformer breaker and connect leads 1 and 2 to overload heater OL1. To start the test, simultaneously close the autotransformer breaker, start the stopwatch, and quickly readjust the current if required. Stop the watch when the contacts open. Repeat the test for OL2. If the tripping time is not within the manufacturer's specified limits, as indicated by the time-current curve, the heater may be defective or incorrectly sized, or the latching device may be damaged.

---

[3] See Reference [2] for an introduction to short-circuit analysis.

FIGURE 29.19
(a) Testing a thermal overcurrent trip;
(b) testing a circuit breaker.

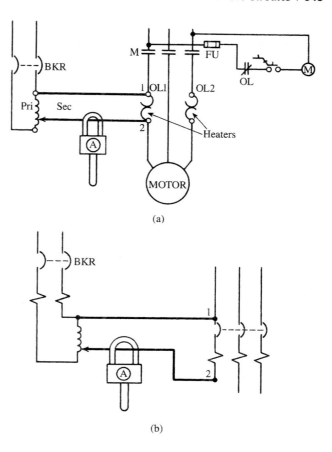

(a)

(b)

Magnetic overload trips and circuit breakers can be tested in a similar manner. A simple testing circuit for circuit breakers is shown in Figure 29.19b. Each pole should be tested separately, even though the common tripping bar should cause all breaker poles to open when any one overcurrent trip is actuated.

Portable self-contained breaker-testing units complete with a high-current power source, ammeter, and timer are available and are very useful in maintenance programs.

In those applications where circuit breakers are kept in either the closed or open position for extended periods of time, they should be "exercised" periodically by operating the tripping mechanisms several times in succession. The tripping should be done electrically using the shunt trip, or manually if the breaker does not have a shunt trip. However, before exercising the breaker, be sure that all power lines leading to the breaker are de-energized.

The main contacts of circuit breakers should be inspected for wear and misalignment. Contacts roughened by service should not be filed smooth unless large projections of metal or beads of metal, formed by arcing, prevent good contact. If filing is necessary, it should be done with a smooth mill file, and should closely follow the original shape of the contact. Alignment can be checked by obtaining a print of the contacts; placing carbon paper and a thin sheet of tissue paper between the contacts and then closing and opening the breaker will leave an imprint on the tissue. If the impression indicates that less than three-fourths of the normal area of the contacts is touching, an adjustment is needed. The manufacturer's manual should be consulted for information on adjustment.

SPECIFIC REFERENCES KEYED TO THE TEXT

[1] *Low Voltage Cartridge Fuses*, NEMA FU1-1986.
[2] *Recommended Practice for Protection of Industrial and Commercial Power Systems*, IEEE STD 242–1991.

REVIEW QUESTIONS

1. What are some of the injurious effects that can be caused by excessive current?
2. How does a fuse provide protection against sustained overcurrents? How is a fuse rated?
3. What is meant by the *interrupting rating* of a fuse? What can happen if the available short-circuit current is greater than the IC rating of the "protecting" fuse?
4. Describe the characteristics of the following: an NEC fuse, a dual element fuse, and a current-limiting fuse. How are they constructed?
5. What are the common causes for blown fuses other than short circuits?
6. Describe a method that can be used to check for blown fuses without energizing the circuit.
7. Describe a method that utilizes the circuit voltage when testing for blown fuses in a low-voltage circuit (that is, the fuses are tested while the power is on).
8. What are molded-case breakers? How are they rated, and where are they used?
9. How does a molded-case breaker provide protection against (a) sustained overloads and (b) short circuits?
10. Explain the operation of the de-ion arc quencher.
11. Define the interrupting rating of a breaker.
12. What can happen if a circuit breaker is called on to clear a short-circuit current that is considerably in excess of its interrupting rating?
13. How does the current limiter, used with some molded-case breakers, provide a current-limiting action?
14. What are circuit-breaker band curves? Of what use are they in maintenance testing?
15. What are the respective functions of long-time delay, short-time delay, and instantaneous elements in an air circuit breaker?
16. What are the advantages of selective tripping?

17. What factors determine the magnitude of the short-circuit current that can occur at some point in a system?
18. Describe a test procedure that can be used to check the calibration of a circuit breaker.
19. What should a maintenance technician check for when inspecting (a) overcurrent relays and (b) circuit breakers?
20. What simple test can be used to determine the area of contact between the movable and fixed contacts of a breaker?

PROBLEMS

29–1/8. Determine the maximum current rating of an inverse-time breaker that could be used as the motor branch-circuit protective device for a 100-hp, 575-V, design E, three-phase induction motor.

29–2/8. Determine the maximum current rating of a non-time-delay fuse that could be used as the motor branch-circuit protective device for a 240-V, 75-hp, DC motor.

29–3/8. Determine the maximum current rating of an instantaneous type breaker that could be used as the motor branch-circuit protective device for a 240-V, three-phase, design C induction motor. The current at full load is 75 A.

29–4/8. Determine the maximum current rating for a dual-element time-delay fuse that could be used as the motor branch-circuit protective device for a 40-hp, 460-V, three-phase, design B induction motor.

30

Maintenance of Switchgear and Other Miscellaneous Electrical Apparatus

30.0 INTRODUCTION

Miscellaneous electrical apparatus (MEA) is the insurance term used to describe a collection of electrical apparatus, that includes equipment such as circuit breakers, rectifiers, main disconnect cabinets, load centers, distribution panels, instrument transformers, wire, cables, and bus structures. The MEA list does not include rotating machinery, power and distribution transformers, furnace and rectifier transformers, and induction feeder regulators.

Inadequate maintenance of switchboards and other miscellaneous electrical apparatus can result in short circuits, explosions, fire, bent and distorted bus bars, damaged instruments, loss of productivity, and injury or death to operating personnel.

Discounting lightning-induced fires, the primary causes of fire in electrical systems are overheating of conductors and connections, and arcing faults between conductors of opposite polarity or between conductors and ground. Most electrical fires are preventable if operators, maintenance personnel, and electricians adhere to the guidelines and wiring codes of the National Electrical Code (NEC) [1, 2].

Warning

Before working on a switchboard, make sure that all incoming power to the board is off. If possible, all breakers and switches that supply power to the board should be locked in the OFF position. Signs tied to the breakers and switches should state clearly that these devices must not be closed without approval from the head of the maintenance department.

30.1 GENERAL MAINTENANCE OF SWITCHGEAR

When possible, switchgear should be scheduled out of service on a 1- or 2-year basis for a thorough inspection, for cleaning of all bus and cable connections, and for the testing and calibration of critical components, such as circuit breakers, switches, relays, and instruments. The frequency of inspection should be dictated by the severity of plant operation and the degree of atmospheric contamination.

Insulation-resistance tests should be made from line to ground and from line to line. Voltmeters and other line-to-line and line-to-ground indicators should be in the OFF position, and all circuit breakers must be open, before making such tests. The voltage rating of the megohmmeter should not exceed 1500 V, unless otherwise specified by the manufacturer. Since there are no specific minimum insulation-resistance values for switchboards, the condition of the insulation should be determined by comparison with previous readings. However, a reading of less than 1 $M\Omega$ should be considered unsafe. The readings should always be taken at shutdown while the equipment is still warm and hence above the dewpoint, and with the same instrument. The insulation resistance should be recorded along with the ambient temperature and humidity.

Clean All Switchgear

Accumulated dirt absorbs moisture and provides a conducting path across the insulation. This in turn causes carbonized paths, called *tracking,* that may eventually cause a flashover. A flashover can cause sustained arcing that can migrate throughout the switchboard. Use a vacuum cleaner to clean loose dust from all creepage surfaces of insulating materials and check for evidence of carbonized paths. Compressed air should be avoided since it merely redeposits the dirt in other areas, with the possibility of making matters worse. Embedded dirt can be cleaned with lintless cloths and an approved safety solvent.

Switchgear operating in very hot atmospheres with high sulfur concentrations should be checked for the growth of metallic whiskers. Figure 30.1 shows a filament type of growth resembling steel wool on the silver-plated spring-loaded finger contacts of a circuit breaker. Continued growth of the whiskers would eventually result in a connection between the contacts of opposite polarity, causing a flashover.

Switches that have silver-to-silver contacts should be resilvered if the plating has worn off. Silver-to-silver switches have lower contact resistance and therefore operate much cooler than their copper-to-copper counterparts. Silver plating compounds for resilvering contacts in place are relatively safe and easy to use.

Check Ventilation

Ventilation ducts, air scoops, etc., that are used to obtain outside air for cooling switchgear should be checked for entry of water that may occur during rain- or snowstorms. This is of particular concern in shipboard installations where air scoops with a 90° elbow ride in a large ball bearing and have a wind-activated rudder to trim the opening away from the wind. If the bearing freezes from lack of grease or accumulated dirt and paint, and the scoop is frozen in the forward position, large amounts of

FIGURE 30.1
Fine filament growth of silver, resembling steel wool, on the silver-plated spring-loaded finger contacts of a circuit breaker (courtesty Mutual Boiler and Machinery Insurance Co.).

rain, snow, and ocean spray will land on bus bars, circuit breakers, instruments, and control equipment, with disastrous results.

30.2 OVERHEATING OF CONTACTS AND CONNECTORS

Connections and connectors that are loose, dirty, corroded, or improperly mated will result in high resistance at the points or surfaces of contact. If the connections involve power apparatus, excess heat energy generated in the high-resistance connection may cause burning of the insulation near the connection, or melting of the conductor at the junction with severe arcing and burning of nearby conductors and enclosures.

Heat generated by loose connections at the terminals of a molded-case breaker can cause the thermal overload to trip, even though the circuit is not overloaded. Similarly, loose connections at fuse clips may cause the fuse to blow at less than the fuse rating.

Care of Sliding Contacts

Unless specified otherwise by the manufacturer, knife-blade switches and rheostat buttons may be lubricated with a light coating of clean light grease such as Vaseline®. Be sure to wipe off any grease that may be deposited on the insulation between the rheostat buttons.

Knife switches that carry high current will operate at a lower temperature if lubricated with a light grease containing pure silver powder, such as Conducto-Lube.[1] The conducting lubricant prevents balling up and freezing of the switch when in heavy use. Furthermore, by eliminating hot spots, it prevents the clips of knife switches from overheating and losing tension.

Glowing Electrical Connections

In laboratory experiments with loose connections in residential-type branch circuits using aluminum wire or copper wire connected with *steel binding screws,* the connections were observed to overheat to the point of glowing [3]. Glowing occurred when the temperature at the loose connection reached 1022°F to 1112°F. Since glowing at loose connections was demonstrated to occur at rated and below-rated current (even as low as 0.3 A), it would not cause fuses to blow or circuit breakers to trip. However, glowing electrical connections were shown to generate enough heat to raise the temperature of outlet boxes to 450°F, to char insulation and wood paneling, to deform plastic wall plates, and to ignite combustible material that came in contact with the glowing connection. Based on the above-mentioned experiments it is conceivable that glowing electrical connections could be the cause of some unexplained fires. It is interesting to note that repeated attempts to develop glows with loose connections using aluminum or copper wire and *brass binding screws* were unsuccessful.

Connections in Residential Circuits

Binding screws used on outlets and switches in residential circuits are deliberately colored brass, silver, or green to indicate the respective hot, neutral, and bonding connections. Thus, it is difficult to determine whether the binding screws are brass or steel without using a magnet. However, wiring devices listed by the Underwriters Laboratories as CO/ALR have large *brass* binding screws, a large surface area for the bonding plate, and may be used with copper or aluminum wire. Any problems with aluminum wire connections are caused almost entirely by improperly made branch-circuit connections to branch fuses, branch circuit breakers, wall receptacles, and wall switches.

Load testing of branch circuits can be accomplished at receptacles, using instruments such as the handheld digital wiring analyzer shown in Figure 30.2. A microprocessor measures the voltage under no load, and again under a 15-ampere (or 20-ampere) self-contained instrument load, and then calculates and displays the percent voltage drop in the circuit. Excess voltage drop indicates damaged conductors, poor connections, poor splices, overlong circuits, or undersized wires.[2] Any of these circuit deficiencies can cause overheating, with the possibility of fire. The three top LEDs in Figure 30.2 indicate wiring conditions: normal, no ground, polarity reversed, no hot line, no neutral line, or hot and ground lines reversed.

---

[1] Conducto-Lube is a trademark of the Cool-Amp Conducto-Lube Company.

[2] The National Electric Code (NEC), Article 210–19, recommends that the maximum total voltage drop, on both feeders and branch circuits to the farthest outlet, not exceed 5 percent to obtain reasonable efficiency of operation.

FIGURE 30.2
Handheld digital wiring analyzer (courtesy
Industrial Commercial Electronics Inc.).

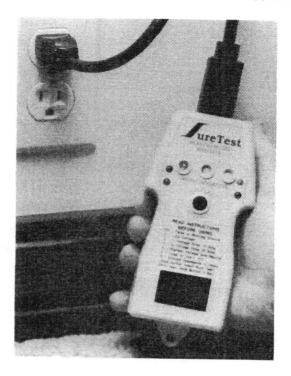

A display of "FG" on power up indicates a suspected false ground. A false ground is defined as a short between the ground and neutral wires very close to the outlet.

Warning

A false ground is a safety hazard.[3]

The pushbutton on the circuit tester shown in Figure 30.2 is used to switch the instrument to different test modes: line volts; percent voltage drop at 15 A; percent voltage drop at 20 A; neutral-to-ground voltage; estimated load on line in amperes; ground impedance in ohms; ground-fault circuit-interrupter test, seconds to trip.

30.3 OVERCURRENTS AND FAULT CURRENTS IN SWITCHGEAR

Assuming the conductors and associated insulation are correctly sized and installed in accordance with the National Electrical Code for the specific application, the generation of excessive heat within the conductors can only be caused by excessive currents

---

[3] See Chapter 28 for more information on equipment grounding.

due to overloads or short circuits, or by an adverse change in the environmental conditions. If excessive currents are caused by overloads they are called *overcurrents,* if caused by the breakdown of insulation between conductors of opposite polarity or between a conductor and ground they are called *fault currents.* Fault currents are the more dangerous of the two and, if not quickly interrupted, may cause the temperature of the conductor to reach its melting point: 1083°C (1981°F) for copper, and 660°C (1219°F) for aluminum.

Overcurrents

If due to severe overcurrent, the heat energy generated in an insulated conductor causes its temperature to rise significantly above its rated operating temperature,[4] the insulation may char, melt, smoke, or burn, depending on its chemical makeup.

If the combined resistance of cables and connected apparatus limits the overcurrent to a value just below the trip setting of the circuit breaker or fuse, and the breaker or fuse is oversized, the cable will overheat along its entire length with the possible ignition of nearby combustible materials, burning of cable insulation, "popping" of the cable insulation due to built up gas pressure, and/or melting of the conductors with severe arcing and burning of the conductor and nearby enclosures.

Arcing Faults

Arcing faults caused by line-to-line or line-to-ground shorts in switchgear are not easily extinguished. Such arcs, once started, travel in a direction away from the power source. As the arc migrates downstream, it burns everything in its path. In some faults, additional arcs called secondary arcs emanate from the initial fault area, and follow the first arc downstream.[5]

Arcing faults can be caused by the accumulation of dirt and moisture, rodents, loose or swinging wires, accidents, and voltage surges. Thus good housekeeping practice can be very helpful in reducing the probability of arcing faults.

Bolted Faults

Metal-to-metal faults, often called *bolted faults,* occur when bare conductors of opposite polarity make solid contact with each other. Most bolted faults are caused by cable wiring errors such as a bolted-in reverse-phase connection or a bolted-in dead short. Hence, very thorough checks must be made after connecting or reconnecting cables and before applying power. The current in a bolted fault is substantially greater than that in an arcing fault. The extremely high electromechanical forces produced by a bolted fault can rip switchgear apart.[6] The very high current can melt conductors, con-

---

[4] The rated operating temperature of different types of insulating materials is given in Appendix 1.

[5] The theory behind arc migration is illustrated and discussed in Section 3.6, Chapter 3.

[6] The theory behind destructive magnetic forces is illustrated and explained in Section 3.3, Chapter 3.

verting the bolted fault into an arcing fault, which, in turn, may cause a total burndown of the distribution switchgear.

30.4 ALUMINUM CABLE

Special care must be taken when making connections and joints with aluminum cable. The aluminum should be wire brushed, and then immediately coated with electric joint compound (inhibitor paste) to prevent corrosion. Wire brushing removes the oxide film that is quick to form when aluminum is exposed to air. Although the oxide film is very thin and almost invisible, it does cause a relatively high contact resistance that can result in overheating, arcing, and burning at the cable connection or joint. If the application of joint compound is delayed, a second wire brushing will be required.

Aluminum cable connections require pressure-type connectors using compression or flat bar connectors. Flat bar connections such as lug-to-bus or bus-to-bus connections can be made with high-strength aluminum bolts, aluminum nuts, and wide aluminum washers as shown in Figure 30.3a. With other than aluminum bolts, flat steel washers and Belleville conical spring washers must be used as shown in Figure 30.3b; *the nut should be tightened until the Belleville washer is flat.* However, in all cases manufacturer's instructions and the National Electrical Code must be followed [1, 4].

Retightening of aluminum connections should be done only when they are obviously loose or are overheating. Excessive and repeated tightenings of properly made connections could eventually damage the conductor, connector, or both. If the correct bolt and correct Belleville washer are used, and the Belleville washer is flat, the connection is tightened properly.

Despite the need for special care, when properly installed and maintained, aluminum connections should provide many years of reliable service.

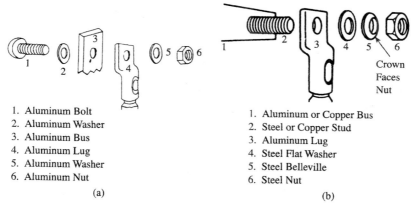

1. Aluminum Bolt
2. Aluminum Washer
3. Aluminum Bus
4. Aluminum Lug
5. Aluminum Washer
6. Aluminum Nut

(a)

1. Aluminum or Copper Bus
2. Steel or Copper Stud
3. Aluminum Lug
4. Steel Flat Washer
5. Steel Belleville
6. Steel Nut

(b)

FIGURE 30.3

Aluminum cable connections: (a) when all components are made of aluminum; (b) with other than aluminum bolts (courtesy The Aluminum Association).

30.5 VISUAL INSPECTION FOR LOOSE CONNECTIONS

Insurance company statistics [5] indicate that the most frequent primary cause of failure in MEA is loose connections. Loose or poorly made connections could result from improper factory assembly, improper field assembly, or improper connectors. However, even though properly assembled, electrical connections can work loose by vibration, or by thermal expansion and contraction during normal operation. A bad contact coupled with heavy current will cause hot spots and may cause the connection to melt. The results could be disastrous.

Thus, when cleaning switchgear, a careful inspection should be made of all bus and cable connections for evidence of overheating. A special effort should be made to inspect wires that are subject to heat from resistors. Discolorations indicate poor contact and the generation of heat. Replace all hookup wire that has burned, discolored, or cracked insulation. Where possible, small-diameter hookup wire connected to meters, switches, and controls may be given a slight tug to check if the connection is tight. Loose bus connections are best tightened with a torque wrench to match manufacturer's specifications.

Do not automatically retighten connections unless there is evidence of overheating.

30.6 INFRARED INSPECTION FOR OVERHEATED CONNECTIONS

Infrared inspection is a nondestructive method for locating hot spots in electrical distribution and power connections at circuit breakers, fuse clips, fuses, bus structures, switchgear, cable feeders, motors, generators, etc. It can thus provide an early warning of potential failure [6]. Loose connections in distribution and power apparatus radiate a greater amount of infrared energy than do good solid connections. The two basic types of infrared testing are infrared spot testing and infrared imaging.

Infrared Spot Testing

This test uses an infrared thermometer to determine the temperature of a connection without physically touching it. When aimed at a connection, the infrared thermometer displays the temperature of that spot.

Infrared Imaging

Infrared imaging is useful when large areas with many connections are to be checked. All objects radiate infrared energy in proportion to their temperature. An infrared imager converts the heat radiation of the object into a visual picture, which can be photographed, viewed on a TV-type screen, recorded on a video disk or on videotape. Portable thermal imaging equipment, such as that shown in Figure 30.4, can be purchased or rented, or an inspection service can be contracted to perform the desired tests.

FIGURE 30.4
Portable thermal imaging equipment (courtesy Land Infrared Co.).

The thermographic image is shown in different shades of gray or multicolors representing different temperature levels; all areas with the same shade of gray, or same color, have the same temperature. An isotherm shows all areas of the image that have the same temperature (same shade of gray) and indicates the actual temperature difference from a known temperature reference. The isotherm level may be set at some known reference temperature or the ambient temperature. The controls are then varied throughout the temperature range until it reaches the region (shade or color) of the unknown temperature; the temperature rise is then read.

A significant advantage of a thermographic scan is that it can be made while the equipment is in operation, permitting quick identification of heating due to overload, loose connections, or inadequate ventilation. Beginning hot spots can be repaired at the next scheduled shutdown; serious overheating may require immediate action such as reducing load or a brief shutdown for repair.

30.7 MAINTENANCE OF HIGH-VOLTAGE INSULATORS AND BUSHINGS

A high-voltage insulator is a highly glazed ceramic support used to separate high-voltage lines from each other and from ground. High-voltage bushings are hollow ceramic tubes used to insulate conductors that pass through a metal tank or enclosure.

FIGURE 30.5
High-voltage bushing with cable for a transformer (courtesy GE Industrial Systems).

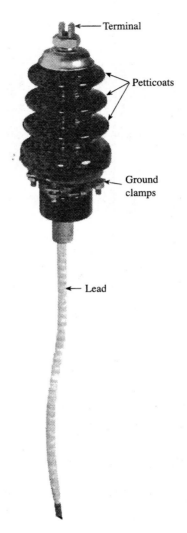

Terminal

Petticoats

Ground clamps

Lead

Figure 30.5 illustrates a high-voltage bushing for a transformer. The petticoat construction increases the creepage distance (the shortest distance measured on the surface of the insulator, between parts of different polarities).

Deposits of dirt on high-voltage insulators may cause a flashover and should be removed by periodic cleaning. This is particularly important in areas where there is a prevalence of salts, conducting dusts, or other contaminants in the surrounding air. However, before attempting to test or clean a bushing, it must be de-energized and then grounded to discharge any stored energy.

Cleaning of bushings and insulators can be accomplished by rubbing with a cloth moistened with water or an ammonia solution. Special porcelain cleaning compounds are available from the manufacturer for stubborn cases that do not respond to simple cleaning methods. The insulator must be thoroughly rinsed with clean fresh water before returning to service. Chipped or abraded porcelain can be repaired. The sharp edges should be honed smooth, and the defective areas painted with an insulating varnish to provide a glossy finish. Fine hairlike cracks in the surface of an insulator must be repaired as soon as detected; the accumulation of dirt and moisture in the cracks can cause leakage current that may result in a flashover.

The insulation resistance of a bushing can be tested with a 2500-volt megohmmeter that has a 20,000-megohm range. Bushing resistance should be high, and any value below 20,000 megohms indicates the need for reconditioning. Low megohm readings indicate moisture in the bushing or a deposit of dirt on the porcelain surface. To make the test, the bushing or standoff insulator should be disconnected from the power circuit, the terminal grounded to discharge any stored energy, and then the megohmmeter leads connected between the terminals and ground.

30.8 MAINTENANCE AND TESTING OF SHUNT CAPACITORS

Shunt capacitors are capacitors used in AC distribution systems for power-factor improvement. They are called shunt capacitors because they are connected in parallel (shunt) with the distribution system.

An example of a shunt capacitor used in AC distribution systems for power-factor improvement is shown in Figure 30.6a. It is composed of series groups of parallel-connected elements immersed in an insulating liquid. The internal bleeder resistor is used to discharge the capacitor to 50 volts or less in 1 to 5 minutes after disconnecting it from the line. Figure 30.6b illustrates the typical internal connections of a power capacitor.

Shunt capacitors (rated in kvars) such as those shown in Figure 30.6, are not easily tested with a megohmmeter; the high capacitive effect requires considerable charging before any appreciable indication is made. Large capacitors must be tested by checking their current against the rated value and/or power-factor test values. A maintenance inspection of power capacitors used for power-factor improvement should always include a check of the fuses, ventilation, voltage, and ambient temperature; the life of a capacitor is shortened by overheating, overvoltage, and physical damage. Ceramic bushings and other insulated surfaces should be kept clean.

Warning

Large capacitors can store a considerable amount of energy for a long time. Hence, before cleaning, touching, or disconnecting a large capacitor, open the switch or circuit breaker and allow the internal bleeder resistor to discharge the capacitor. As a safety precaution, after 5 minutes of bleeder discharge, continue the discharge through a heavy-duty external 50,000-ohm resistor for another 5 minutes. Discharging should be done between terminals and between terminals and case. Do not depend entirely on internal bleeder resistors; they may become open circuited.

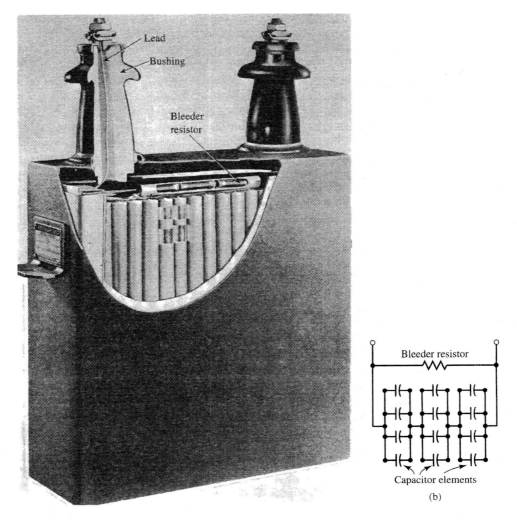

FIGURE 30.6
(a) Capacitor for power-factor improvement (courtesy TECO Westinghouse); (b) internal connections of capacitor elements.

Although small capacitors may be discharged by short-circuiting their terminals with a copper strap or heavy cable, large capacitors may be damaged by this practice; the very large mechanical forces caused by the high outrush currents may destroy them. Under no circumstances should a bank of large capacitors be discharged through short-circuiting. The sudden release of a large amount of energy could vaporize or explode the shorting device, causing injury to personnel as well as damage to the capacitors.

Capacitors that are leaking insulating liquid should be patched by soldering a patch over the hole, using a rosin-flux solder. After patching, leaky capacitors should be returned to the manufacturer for repair or replacement.

Shunt capacitors are rated in rms terminal voltage, frequency, and reactive power. In accordance with NEMA standards, "*capacitors shall give not less than the rated reactive power at rated terminal voltage and frequency, and not more than 115% of this value, measured at 25°C uniform case and internal temperature*" [7].

Capacitors used for power-factor improvement can be tested at rated voltage for internal shorts or opens by using the circuit shown in Figure 30.7a. To prevent a high-transient surge current from damaging the ammeter, it should be hooked on the conductor *after* the breaker or switch is closed.[7]

---

[7] Because of low values of current, the test should be done with a digital ammeter.

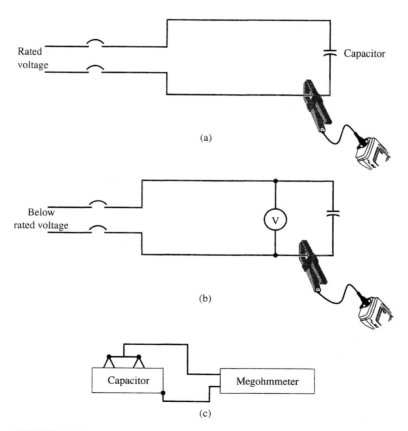

(a)

(b)

(c)

FIGURE 30.7

Testing power capacitors for (a) shorts and opens at rated voltage; (b) shorts and opens at below-rated voltage; (c) insulation-resistance test between capacitor plates and case.

The rated current for a good capacitor can be determined from its nameplate data by substituting into Eq. (30–1)

$$I_{rated} = \frac{Q_{rated} \times 1000}{V_{rated}} \qquad (30–1)$$

where: $\quad I_{rated}$ = rated current (A)
$\qquad\quad V_{rated}$ = rated voltage (V)
$\qquad\quad Q_{rated}$ = rated reactive power (kVA or kvar).

Ammeter readings greater than 115 percent of rated current, when tested at rated voltage and rated frequency, indicate a short in one or more elements; ammeter readings less than the rated current indicate an open in one or more elements [7].

If rated voltage is not available, the capacitor can be tested at other than rated voltage by including a voltmeter with a clamp-on ammeter, as shown in Figure 30.7b. The test current, measured at rated frequency but at other than rated voltage, is adjusted to rated voltage by using Eq. (30–2):

$$I_{test(adj)} = I_{test} \times \frac{V_{rated}}{V_{test}} \qquad (30–2)$$

where: $\quad V_{test}$ = test voltage (V)
$\qquad\quad I_{test}$ = test current (A)
$\qquad\quad I_{test(adj)}$ = test current adjusted for rated voltage (A).

EXAMPLE 30.1 A 480-volt, 300-kVA, 60-hertz capacitor, tested at rated voltage and frequency, draws a current of 760 amperes. Is the capacitor in serviceable condition?

Solution
The rated current, as determined from nameplate data, is

$$I_{rated} = \frac{Q_{rated} \times 1000}{V_{rated}} = \frac{300 \times 1000}{480} = 625 \text{ A}$$

Allowable maximum current is $625 \times 1.15 = 718.75$ A. The test current exceeds the allowable 115 percent limit, which means that one or more elements are shorted, and the capacitor must be replaced.

EXAMPLE 30.2 A 480-volt, 200-kVA, 60-hertz capacitor, tested at 350 volts and 60 hertz, draws a current of 255 amperes. Is the capacitor in serviceable condition?

Solution

$$I_{rated} = \frac{Q_{rated} \times 1000}{V_{rated}} = \frac{200 \times 1000}{480} = 416.67 \text{ A}$$

$$I_{test(adj)} = I_{test} \times \frac{V_{rated}}{V_{test}} = 255 \times \frac{480}{350} = 349.71 \text{ A}$$

The test current adjusted to rated voltage is less than the rated current, which means that one or more of the elements is open, and the capacitor should be replaced.

Ground-Insulation Test

To determine the insulation resistance between the conductors and the metal case, connect one terminal of a megohmmeter to both line terminals of the capacitor and the other terminal of the megohmmeter to the case, as shown in Figure 30.7c. The resistance measured between terminals and case should be not less than 1000 MΩ.

30.9 EXTINGUISHING ELECTRICAL FIRES [8]

The preferred procedure for fighting electrical fires (Class C fires) is to de-energize the circuit and then apply an approved extinguishing agent such as CO_2, Halon, or dry chemical. The carbon-dioxide (CO_2) extinguisher offers the advantage of extinguishing the fire, cooling the apparatus, leaving no residue, and having no adverse effect on the insulation and metal parts, and may be used on live circuits. Hence, it is the preferred extinguishing agent for most electrical fires. However, when applied in confined spaces, such as the engine room of a ship, the CO_2 should be purged with fresh air before people are allowed to enter.

Halogenated extinguishing agents such as Halon-1301 (bromotrifluoromethane) and Halon-1211 (bromochlorodifluoromethane) are odorless, colorless gases that can be used to fight electrical fires in machinery spaces. However, only Halon-1301 is recommended for fires involving electronic computers and control rooms.

Dry chemical extinguishers are satisfactory and may be used on live circuits. However, the chemical leaves a residue and does not cool the apparatus as effectively as does the CO_2 extinguisher.

Water sprinklers and steam smothering apparatus may be used only after the circuit is de-energized. Such systems have been built into large machines and are very effective. The serious drawback to their extended usage is the time required to clean, dry, and test the apparatus before it can be placed back in service.

However, regardless of the type of extinguishing agent used, the apparatus should be thoroughly inspected and tested after the fire is out. The cause of the fire should be determined, and corrective action should be taken before the apparatus is put back in service. If no damage is apparent, and an insulation-resistance test indicates normal operation, the apparatus may be started at reduced load and carefully observed as load is increased.

SUMMARY OF EQUATIONS FOR PROBLEM SOLVING

$$I_{\text{rated}} = \frac{Q_{\text{rated}} \times 1000}{V_{\text{rated}}} \tag{30-1}$$

$$I_{\text{test(adj)}} = I_{\text{test}} \times \frac{V_{\text{rated}}}{V_{\text{test}}} \tag{30-2}$$

SPECIFIC REFERENCES KEYED TO THE TEXT

[1] *National Electrical Code 1996,* National Fire Protection Association.
[2] An Analysis of MEA Failures, *The Locomotive,* Vol. 64, No. 8, The Hartford Steam Boiler Inspection And Insurance Co., 1985.
[3] Neese, W. J., and Beausoliel, R. W., *Exploratory Study of Glowing Electrical Connections,* National Bureau of Standards PB-273577, October 1977.
[4] Aluminum Alloy Building Wire Installation and Design Guide, Aluminum Association Inc., 1997.
[5] Should You Be Doing Electrical Maintenance?, *The Locomotive,* Vol. 72, No. 2, The Hartford Steam Boiler, Inspection, and Insurance Co., Spring 1998.
[6] Bruce III, C. W., Infrared Detection of Hot-Spots in Energized Electrical Equipment, *IEEE Trans. Industry Applications,* Vol. IA-15, No. 3, May/June 1979.
[7] *Shunt Capacitors,* NEMA CP-1, National Electrical Manufacturers Association, 1988.
[8] *Guide for Substation Fire Protection,* ANSI/IEEE STD 979–1994.

QUESTIONS

1. Discounting lightning-induced fires, what is the primary cause of fire in electrical systems?
2. What precautions should be observed before working on a switchboard?
3. Outline an inspection program for checking switchgear. Include a list of the apparatus to be inspected and the types of tests to be made.
4. How should switchgear be cleaned?
5. Why should molded-case breakers that have not been operated for an extended period be exercised, and what precautions should be observed?
6. What are the benefits of lubricating sliding contacts on high-current switchgear?
7. What can cause glowing electrical connections in residential-type branch circuits?
[8] Differentiate between overcurrents and fault currents.
9. What damage can severe and prolonged overcurrent have on conductor insulation and on nearby installations?
10. (a) What is an arcing fault? (b) What sustains the arc? (c) Why is the arc fault so very destructive in switchgear? (d) What are some of the causes of arcing faults?
11. What is a bolted fault, and what effect can it have on switchgear?

12. (a) What special care must be taken when making connections and joints with aluminum cable? (b) What is a Belleville washer, what is its purpose, and how is it used?

13. State two reasons why properly assembled electrical connections can work loose.

14. What is an infrared inspection, how is it made, and how does it fit into a preventive maintenance program?

15. What is the correct procedure for discharging power capacitors, and what precautions should be observed?

16. How are power capacitors tested?

17. Describe the different methods for extinguishing fires of electrical origin.

PROBLEMS

30–1/8. A 240-volt, 50-kVA, 60-hertz capacitor, tested at rated voltage and frequency, draws a current of 150 amperes. What is the condition of the capacitor?

30–2/8. A 480-volt, 60-kvar, 60-hertz capacitor, tested at 480 volts and 60 hertz, draws a current of 135 amperes. What is the condition of the capacitor?

30–3/8. A 2400-volt, 200-kvar, 60-hertz capacitor, tested at 2400 volts and 60 hertz, draws a current of 65 amperes. What is the condition of the capacitor?

30–4/8. A 480-volt, 35-kvar, 60-hertz capacitor, tested at 125 volts and 60 hertz, draws a current of 26 amperes. What is the condition of the capacitor?

30–5/8. A 216-volt, 25-kvar, 60-hertz capacitor, tested at 120 volts and 60 hertz, draws a current of 83.3 amperes. What is the condition of the capacitor?

31

Cost–Benefit Relationship of Preventive Maintenance

31.0 INTRODUCTION

Preventive maintenance is the orderly routine of inspecting, testing, cleaning, drying, refinishing, adjusting, and lubricating electrical apparatus. To be effective, a good preventive maintenance program must provide for planned shutdowns during periods of inactivity or reduced plant operations for the purpose of cleaning, adjusting, or overhaul. This ensures continuity of operation and reduces the danger of breakdown at peak loads. Through such a program, problems can be detected in their early stages, and corrective action can be taken before extensive damage is done.

The objective of preventive maintenance is to reduce the number of avoidable breakdowns. This is particularly important in processing plants, assembly lines, power plants, aircraft, ships, etc., where the failure of a relatively minor component can disrupt the entire operation or cause a catastrophic disaster. Add to this the probability of lawsuits that may arise from injury to personnel and failure to keep commitments because of an avoidable breakdown, and the cost of not having a preventive maintenance program becomes staggering.

Additionally, the safety of electrical systems is also the concern of the Occupational Safety and Health Act, which mandates that electrical systems must be maintained in a safe condition and that the associated maintenance procedures themselves must be safe. The law provides rigid standards as well as heavy fines and even imprisonment for noncompliance. A good preventive maintenance program can be expensive; it requires properly trained personnel, adequate shop facilities, special tools specific to the equipment, technical manuals, record-keeping capabilities, and the stocking of consumables (e.g., rags, lubricants, cleaning fluids, personal protective equipment). However, the cost of downtime resulting from avoidable outages may amount to 10 or more times the actual cost of a maintenance program.

Figure 31.1 illustrates the damage that can be done when preventive maintenance is neglected. The burned core and coils were caused by a short circuit in the stationary

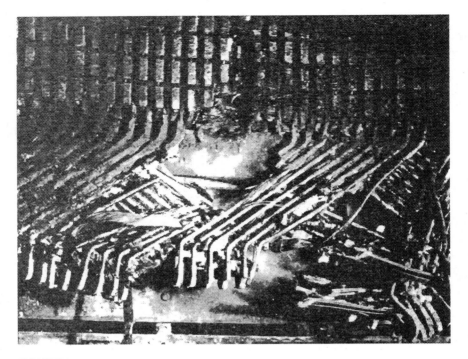

FIGURE 31.1
Burned coils and core caused by a short circuit in the stationary armature of a waterwheel generator. The generator had been in service for 6 years without cleaning (courtesy GE Industrial Systems).

armature of an open-type waterwheel generator. The generator had been operating in a dirty environment for 6 years without cleaning.

In a study made by an electric utility, using 10,000 failures of electric utility control and monitoring equipment as a sample, it was determined that 25 percent of the failures could have been avoided by preventive maintenance, and 65 percent could have been determined prior to failure by diagnostic testing [1].

31.1 ECONOMIC FACTORS

The cost of preventive maintenance is directly related to the frequency of testing, inspecting, lubricating, and adjusting. A greater frequency of maintenance results in increased plant reliability, but must be paid for in increased maintenance costs; a lesser frequency of maintenance results in lower maintenance costs but results in increased and more costly failures. This is shown graphically in Figure 31.2, where the cost of failure and the cost of preventive maintenance are plotted against frequency of maintenance for a critical machine in a hypothetical continuous process plant.

The total-cost curve is the cost of maintenance plus the cost of failure. As indicated by the total-cost curve, each apparatus has a frequency of maintenance at which the total cost is a minimum. The function of management is to seek this minimum value [1].

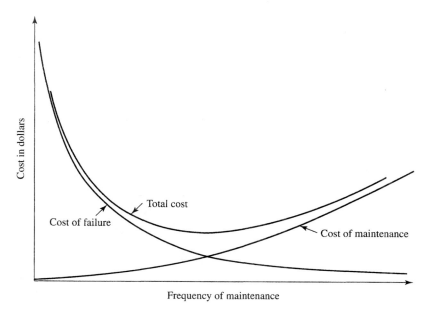

FIGURE 31.2
Plot of dollar cost vs. frequency of maintenance.

Planned maintenance programs should be designed to maximize the benefit per dollar outlay; paying too much for maintenance is wasteful, but paying too little can be catastrophic. To avoid production loss, especially in continuous process systems, management must be reasonably certain that the frequency and methods for inspection, test, and adjustment of critical machinery and controls are based on a carefully planned, financially sound program.[1]

The Low-Bid Trap

Think total cost when considering purchasing on the basis of the lowest bid. The cheapest product is not necessarily the most economical. Factors that must enter the decision process are service life, annual operating cost, annual maintenance cost, availability of replacement parts, and the manufacturer's long-term reputation and support. As postulated by Ruskin (1819–1900), "It is unwise to pay too much, but it is worse to pay too little. When you pay too much, you lose money—that is all. When you pay little, you sometimes lose everything, because the thing you bought was incapable of doing the thing it was bought to do. The common law of doing business prohibits paying a little and getting a lot—it can't be done. If you deal with the lowest bidder it is well to add something for the risk you run. And if you do that you may have enough to pay for something better."

---

[1] See Chapter 32 for more details on a preventive maintenance program.

31.2 FAILURE RATE

No matter how much care is devoted to the design, construction, and installation of electrical apparatus, its life expectancy will follow a failure-rate curve having the general shape shown in Figure 31.3. The failure-rate curve, also called a mortality curve or bathtub curve, is a plot of failure-rate λ (lambda) vs. service life, where:

$$\lambda = \frac{\text{Total number of failures}}{\text{Total operating hours}} \qquad (31\text{--}1)$$

The left section of the curve in Figure 31.3 represents the early-failure rate, called infant mortality. The infant mortality may be due to manufacturing defects, improper installation, wrong or defective wiring, misapplication, or poor workmanship during a recent equipment overhaul. As these failures are corrected, the failure rate decreases and remains fairly constant throughout the useful life of the apparatus.

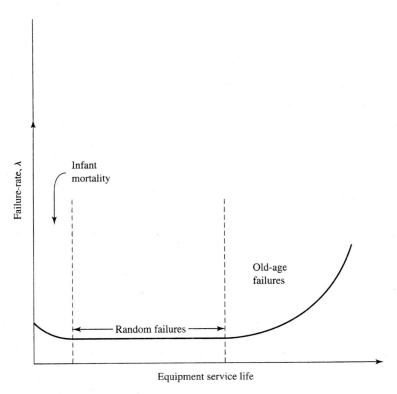

FIGURE 31.3
Failure rate vs. service life.

The center section, called the constant-failure rate, is due to chance failures, or random failures, that occur suddenly without warning and cannot be predicted by testing, monitoring, or inspection. Examples of random failures are connections working loose, wires breaking, springs breaking, bolts breaking or loosening, lightning surges, operator errors, mechanical damage from impact, shorted capacitors, water leaks, and oil leaks. Although random failures cannot be predicted, the number of such failures can be significantly reduced through the use of:

- Effective training programs for all operators and maintenance personnel,
- more reliable equipment and components,
- lightning surge protection, splash guards, etc., and
- effective inspection and adjustment methodologies.

The right section of the curve is the wear-out failure rate, and represents old age failures. The wear-out failures start occurring after the "useful life" of the apparatus has been exceeded. Wear-out failures are a result of a gradual wearing and aging process that starts as soon as the apparatus is placed in operation. Wear-out failures in moving parts are caused by friction, vibration, stress, heat, dirt, and moisture. Wear-out failures in insulation are caused by thermal, mechanical, and electrical aging of insulating materials. In most cases wear can be monitored by inspection, measurement, and diagnostic testing, and corrective action can be taken before failure occurs.

Because wear-out failures generally occur gradually, the useful life of electrical apparatus can be extended through the implementation of a good preventive maintenance program. Catastrophic failures can be minimized, if not eliminated, through the proper lubrication of wearing parts, replacement of worn components, continuous alignment checks of motor to driven unit, or the renewal or reconditioning of electrical insulation.

Weak components that can wear out or deteriorate too rapidly should be redesigned or replaced by heavy-duty components. In addition, the operating environment and material composition should always be considered when selecting a replacement part. Salty, moist, or caustic atmospheres can wreak havoc on electrical apparatus not suitable for operating in those conditions.

31.3 SPECIFYING WEAR LIMITS AND OPERATING LIMITS

Wear Limits

Parts that wear, such as carbon brushes, bearings, contactors, insulation, and springs, should be assigned a specific wear limit that signals when replacement, adjustment, or reconditioning is necessary. For example, when a carbon brush wears down to less than some assigned minimum allowable dimension, replace it; when a contact wipe becomes less than some assigned minimum, replace it; when insulation resistance to ground becomes less than some assigned minimum, recondition it; when the bottom air gap between rotor and stator of a sleeve-bearing motor falls below a specified value, replace the bearings; etc.

Operating Limits

Electrical apparatus is designed to operate within specified limits of voltage, frequency, ambient temperature, and load. Undervoltage, overvoltage, voltage unbalance, overcurrent, under- or overfrequency, excessive ambient temperatures, excessive load, and excessive vibration all take their toll in affecting equipment performance and shortening its useful life.

These vital signs are excellent indicators of the "health" of an apparatus. Baseline readings of vital signs should be taken for all critical equipment when first put into service; these signs should be monitored periodically and compared with the baseline data for indications of potential problems. A little extra load when operating continuously at peak power could cause an outage, a small percentage voltage unbalance can cause overheating, an increase in vibration can cause mechanical damage, etc.

The initial wear and operating limits for electrical apparatus can be obtained by consulting with the manufacturer and by referencing the appropriate regulatory standards. Later, these limits can be refined as plant personnel acquire experience in the monitoring, evaluation, and repair of their specific machinery. Depending on the service reliability level of the electrical apparatus and its role within the plant, plant managers have been known to adopt even tougher wear and operating standards than those recommended by the manufacturer or prescribed by regulatory bodies.

31.4 ASSIGNING SERVICE RELIABILITY LEVELS

As a first step in preparing an electrical preventive maintenance program, all motors, generators, power and distribution transformers, induction voltage regulators, and miscellaneous electrical apparatus (MEA)[2] should be assigned a service reliability level based on their role in the manufacturing or operating process. To provide the same level of maintenance for vital and nonvital equipment is a waste of money.

The following categories of service reliability (SR) will be helpful in developing an economically sound preventive maintenance program:

SR-0: *Equipment whose loss would not affect safety or production.* This category includes room air conditioners, electric fans, space heaters, drinking water coolers, etc. Such equipment should be given corrective maintenance; repair or replace when breakdown occurs.

SR-1: *Equipment whose loss would have only limited effect on production, and would not affect safety or cause an extended outage.* This category should be provided with a relatively low-priority preventive maintenance program.

---

[2]MEA is the insurance term used to describe all nonrotating electrical apparatus except power and distribution transformers and induction feeder regulators. It includes equipment such as switchboards, bus structures, switches, circuit breakers, enclosures, mercury arc rectifiers, constant-current transformers, instrument transformers, wires, cables and feeders.

SR-2: *Equipment whose loss would cause an extended outage, loss of product, serious damage to other equipment, and/or injury to personnel.* Such equipment should be provided with a very comprehensive, high-priority preventive maintenance program.

The frequency of inspection, test, cleaning, and lubrication for all apparatus whose assigned service reliability level is SR-1 or SR-2 must be determined before scheduling can be done. This information can be obtained from equipment instruction manuals, from the manufacturer, and from experience. Assigning service reliability levels is a very important phase of the planning process. When requesting information, the manufacturer should be informed of the duty cycle, type of load (continuous, impact, vibrating, etc.), and the environment in which the apparatus must operate (clean, dusty, oily, dry, moist, hot, etc.).

31.5 RECONDITION, REPAIR, REBUILD, OR RENEW

The primary duties of a preventive maintenance department are prevention, reconditioning, and repairing. Rebuilding and renewing are not a part of preventive maintenance, are often contracted out, and should not be charged against the preventive maintenance budget. Note the following definitions:

To *recondition* is to restore to a good or satisfactory condition. This may require thorough cleaning, drying, revarnishing, tightening, lubricating, adjusting, making minor repairs or replacements (small contacts, brushes, springs, etc.), and final testing.

To *repair* is to restore to a sound condition after decay or damage.

To *rebuild* (overhaul) is to replace all worn parts with new or repaired components.

To *renew* is to restore to a new condition by replacing all items that are subject to wear or deterioration with new components.

31.6 SPARE PARTS

An adequate and carefully selected assortment of spare parts is essential to a good maintenance program and will provide insurance against prolonged shutdowns. Having the right spare parts on hand can become an absolute necessity, especially if the apparatus is old, foreign made, or has a manufacturer with a history of backlog orders or a poor parts delivery program.

Overstocking and understocking should be avoided. Overstocking results in excessive inventory costs as well as losses due to obsolescence when equipment is replaced. Understocking places production in jeopardy if an outage occurs. Very often replacement parts for one machine can be used for many other units in the same plant. In such instances it is unwise to stock spares for each machine alone, but to follow the manufacturer's recommendations concerning the minimum stock for a group of machines. In some cases stocking a spare motor may be a trivial expense compared to production losses if the motor has to be rewound. Recommended renewal parts required for vital machinery are generally available from the manufacturer.

Group Replacement

Think total cost when replacing worn parts. It is generally far more cost effective, from the standpoint of man-hours consumed, to replace an entire set of contacts in a control panel even though only one or two exceed their wear limits. Likewise, if one brush in a four-brush motor is sparking because of wear, it is more cost effective to replace all four at the same time. If failure was caused by a weak component, then all like components should be modified or redesigned as soon as possible.

31.7 STAFFING AND TRAINING

Maintenance and operating personnel should be selected on the basis of qualifications for the job. Assignments should not be made on a hit-or-miss basis to those who happen to be idle at the moment. Much equipment has been ruined by overzealous but inexperienced personnel. The maintenance department must not be a catch-all for personnel who could not adjust elsewhere. The complexity of control systems in automatic and semiautomatic plants requires well-educated and skilled personnel. Very often, because of the newness and complexity of the maintenance requirements, special training must be provided by the company or at manufacturers' schools.

Defects will occur and equipment will break down. However, a dedicated and knowledgeable technician can, by taking appropriate action, extend the life of components subject to wear and replace them before they reach the end of their useful life.

Maintenance Training

The most effective maintenance technicians are those trained to expect defects everywhere, who constantly look, listen, smell, feel, measure, and test for them. The maintenance force must be kept abreast of the latest techniques and developments in the field. This may be done by visiting plant maintenance shows, subscribing to appropriate periodicals, and attending in-house classes, panel discussions, and lectures. A file of equipment manufacturer's bulletins and instructions should be available to all maintenance personnel. Only modern testing equipment, good tools, and the latest methods should be used.

In those cases where special skills that are not available in-house are required or expensive diagnostic test equipment is necessary, outside help must be used.

Operator Training

Operators must be trained in what to do during an apparent fault or emergency condition—and also what *not* to do. Case histories from insurance company files illustrate the need for more comprehensive operator training. Operator training must include what *not* to do if equipment does not start when the start button is pushed, circuit breaker or switch is closed, etc. Part of the training should include real-life cases of failures, and how such failures could have been avoided. The following are real-life examples of how lack of training caused minor faults to escalate into major destruction.

A circuit breaker feeding a large transformer tripped out. The operator tried several times to reclose it, but each time it tripped out. The operator called his supervisor, who in turn called the plant electricians; all were unsuccessful in keeping the breaker closed. After 19 recorded reclosures of the breaker, the transformer was totally destroyed.

In another incident, the circuit breaker tripped when an operator pushed the START button for a 100-hp synchronous motor. The operator reset the breaker, and again pushed the START button with the same results. The operator did not try to determine the cause of tripping; he kept resetting the breaker and pushing START until a bright flash of light occurred, and smoke poured from the machine. All stator coils had to be replaced.

Plant personnel must be made aware of the fact that a circuit breaker trips because of overcurrent or a fault, and that *reclosing into a fault can be disastrous*. The cause of tripping must be determined, and corrective action taken, before the breaker is reclosed.

These avoidable outages are generally classified as operator errors. Operator errors include actions taken by the operator that were based on carelessness, poor judgment, and unfamiliarity with proper operating procedures including what not to do when equipment does not start. A properly trained operator will be better equipped to take appropriate emergency action to prevent further damage to apparatus and associated equipment should a failure occur.

31.8 SAFETY: A MANAGEMENT PRIORITY

Safety is everyone's responsibility, and it should be a management priority. Safe operation of machines and safety to personnel should be paramount in all considerations when planning a preventive maintenance program.

All electrical wiring should conform to the recommendations of the National Board of Fire Underwriters as presented in the National Electrical Code. The ergonomics of plant layout and design play a large role in the relationship between man and machine. Even the best maintenance procedures cannot be put into practice unless personnel can safely access or work around the machinery.

Machinery and other equipment should be installed with a view toward ease of accessibility and maintenance. Piping systems containing water, steam, or other liquid mediums should not have any valves, fittings, or joints located over electrical apparatus unless such equipment is classified as drip-proof or splash-proof. Similarly, electrical equipment should not be placed into operation in areas where explosive atmospheres congregate (e.g., granaries, refineries) unless such equipment is classified as explosion-proof.

Only approved safety solvents should be used for cleaning insulation, bearings, and other parts.

Switches and circuit breakers should always be closed completely and with a swift firm action. This will enable the quick operation of a fuse or breaker if a fault exists. Partly closing the switch or tickling the contacts "to see if everything is all right" is dangerous. The high resistance of a partly closed switch may cause severe burning and arcing at the contacts and a possible explosion.

Adequate overload and short-circuit protection are absolute necessities for safe operation. Obsolete switchgear can cripple an industrial plant by failing to clear a short circuit. The interrupting capacity of all breakers and fuses must be sufficient to clear the fault rapidly and without damage to itself.

The incoming electrical service and the plant distribution system should be restudied whenever sizable additions are made to the electrical load. The ampacity of the feeder cables should be checked to determine whether or not it will carry the additional load without overheating, and a new coordination study of all protective devices in the system should be made to ensure a high degree of reliability and protection.

Disconnect switches and breakers should be open and padlocked before servicing any apparatus, and a DO NOT CLOSE, PERSON WORKING ON LINE sign hung on the switch and/or breaker. Each maintenance electrician should have his own individual padlock and individual key to ensure that someone else does not inadvertently energize the circuit. If more than one technician is working on the system, multiple padlocks with individual keys should be used. The incoming lines to the apparatus must be tested with a voltmeter or other voltage tester of the correct range for proof that the circuit is dead, and all capacitors connected to the apparatus must be discharged before any repair work or adjusting is done.

Should an emergency arise and it becomes necessary to work on live circuits, the following steps should be taken:

- Provide only qualified personnel who are aware of the dangers involved.
- Personnel working on live circuits should wear safety goggles, rubber soled shoes (without nails), insulating gloves, and a long-sleeve shirt.
- Insulate personnel from ground by means of an insulating mat.
- Provide ample lighting.
- All tools should be insulated so that only a small amount of the "business end" of the tool is exposed.
- Do not wear loose clothing, metal buckles, key chains, watches, or rings.
- Provide a CO_2 or other approved fire extinguisher.
- Have a person stationed at the circuit breaker or switch.
- Personnel should not work alone on or near energized equipment. If a person is injured, the other person nearby could take appropriate life-saving action.
- Have personnel standing by that are trained in cardiopulmonary resuscitation.

SPECIFIC REFERENCE KEYED TO THE TEXT

[1] Sheliga, Douglas J., Calculation of Optimum Preventive Maintenance Intervals for Electrical Equipment, *IEEE Trans. Industry Applications*, Vol. IA-17, No. 5, September/October 1981.

REVIEW QUESTIONS

1. What is preventive maintenance?
2. Plot a set of representative curves that show the cost of failure and the cost of preventive maintenance versus frequency of maintenance. Explain the significance of the curves.
3. (a) Define failure rate. (b) Sketch and explain a representative failure-rate curve.
4. Discuss the advantages of specifying wear limits. Give three practical examples of wear limits.
5. Why should electrical apparatus be assigned operating limits? Give two examples of electrical operating limits.
6. What does the acronym *MEA* mean?
7. Describe three basic service reliability levels, and state the degree of preventive maintenance required for each.
8. Who is responsible for safety in a preventive maintenance environment?
9. What should be done before any work is done on electrical apparatus?
10. What are the precautions one should take when performing emergency work on energized circuits?

32

Scheduling, Conducting, and Evaluating a Preventive Maintenance Program

32.0 INTRODUCTION

Scheduling, conducting, and evaluating a preventive maintenance program for factories, ships, shore-side power plants, etc., is a complex and costly process. Such a program must take into consideration the required reliability level of each specific apparatus, the required reliability of the overall system, and the overall cost of downtime. To be effective, scheduling should include the monitoring of vital signs (current, voltage, temperature, vibration, etc.), diagnostic testing, record taking and interpreting, and a thorough investigation of failures. Furthermore, scheduling and maintenance procedures should be reviewed periodically to ensure that safe, efficient, and reliable operation is provided at minimum cost.

This chapter provides an introduction to the significant factors involved in scheduling, conducting, and evaluating a preventive maintenance program.

32.1 THE SCHEDULING STRATEGY

When initiating a preventive maintenance program, don't try to include everything at once. Start with equipment that would cause the greatest economic loss if breakdown occurred, and include only those pieces of equipment that are absolutely necessary to the operation of the plant. The economic loss should take into account factors such as the cost of downtime, overtime pay, possible loss of product, and loss of market.

If not already available, a system one-line diagram showing all circuit breakers, disconnects, motors, generators, transformers, etc., must be prepared. The diagram should include numbers corresponding to the identifying numbers assigned to and marked on the actual apparatus. Don't overplan; preventive maintenance should not be a make-work program. Don't try to protect against every possible fault. Consider the most likely problems. Determine from the maintenance technicians and in-house records the items that failed in the past, what caused the failure, how frequently it occurred, how often it should be checked, and what should be done to reduce the frequency of failure. Remember, if the annual cost of preventive maintenance for a particular item exceeds the cost of repair and downtime for that item, preventive maintenance for that item is not economically justifiable; replace it when it fails.

Decisions on scheduling the frequency of maintenance for a particular apparatus, in a particular environment, set of operating conditions, and required service reliability, require comparison of data obtained directly from the manufacturer, from maintenance supervisors, from existing records of defects for that apparatus, and from outside consultants.

As previously mentioned, and shown in Figure 31.2 of Chapter 31, too high a frequency of maintenance or too low a frequency of maintenance is a waste of money. The frequency of maintenance should be adjusted up or down as experience dictates; increase the maintenance frequency if frequent breakdowns occur, and vice versa. Scheduling critical equipment on the basis of actual running time (using running-time meters) instead of calender time will ensure a higher frequency of inspection for greater usage, and a reduced frequency of inspection for lower usage. The net result would be fewer breakdowns and lower maintenance costs. New equipment should be monitored frequently until the break-in period is over, and it is working well. Then gradually increase the time between inspections. Old equipment that is near the end of its useful life should have more frequent inspections and will require more adjustments to offset wear.

Advance plans should be made for the swift repair of major equipment breakdowns; the expected downtime, required tools, spare parts, and manpower needed, and the preventive maintenance testing that could be done on other equipment idled by the major breakdown should be determined. When making advance plans for such contingencies, a reliable service shop should be contacted to determine to what extent their personnel and equipment could expedite repairs. Outside service shops have a wide variety of facilities and services to cover almost all types, sizes, and ages of electrical equipment. They can also be of service in checking the calibration of instruments, relays, and tripping mechanisms of breakers, as well as checking insulation of machines, transformers, cables, etc.

Equipment in production lines or continuous processes should be inspected and lubricated while operating, and their vital signs monitored. Maintenance personnel should look, listen, smell, and feel for abnormalities in operation, and such vital signs as voltage, current, temperature, and vibration should be measured. All other maintenance procedures such as testing, adjusting, and cleaning must be done during a scheduled shutdown. Some companies find it convenient to close down completely for a "plant vacation" to provide time for carefully planned maintenance.

Maintenance supervisors should plot the vital-sign data of critical machines to determine trends: vibration displacement, bearing temperature, machine temperature, ambient temperature, current, insulation resistance, etc., should be plotted when measured. Even though the vital signs do not exceed the specified operating limits, the trend may provide an early warning of impending trouble. By detecting and correcting troubles before they become serious, the cost of maintenance and downtime can be reduced.

As a final note, apparatus should not be disassembled for maintenance inspection more frequently than specified by the manufacturer. Every disassembly and assembly increases the probability of damage.

32.2 THE PREVENTIVE MAINTENANCE ROUTINE

The preventive maintenance routine should include inspecting, testing, cleaning, drying, monitoring, adjusting, exercising components, and making minor repairs or corrective modifications for the purpose of minimizing failures. Other maintenance responsibilities include troubleshooting faults and investigating the cause of failures.

Physical Inspection

Look, feel, smell, and listen; the apparatus may be "saying something" about a defect in progress. Maintenance inspectors should look for and report dirt and other contaminants such as oil films on windings, commutators, slip rings, etc.; clogged ventilating ducts, water leaks, oil leaks, and grease leaks; burn marks or discoloration of insulation and paint; smoke, sparking, arcing, or glowing; adverse atmospheric influences such as excessive or inadequate humidity, higher than normal ambient temperatures, and abrasive or other contaminants in the air.

Correct and report loose bolts and loose connections; tightening a loose connection, when spotted, can save many hours of troubleshooting for an intermittent fault. In the case of open machines, look for evidence of coil movement, loose wedges and ties, corona discharge, and checked, cracked, or brittle insulation. The same inspection should be made on cable where it enters a machine.

Feel the frames of rotating equipment and the frames of control equipment for any unusual vibration (protect hands from hot equipment). Inspect for hot or noisy bearings, broken or loose fans and belts, etc. A lot of burned out windings are a result of bearing failures, misalignment, bent shafts, etc. Report all unusual odors in the general vicinity of electrical apparatus.

Listen for unfamiliar sounds that could indicate trouble. Report rubbing and clicking sounds, unusual magnetic noises in rotating machines, and chattering, buzzing, or unusual humming from relays or control enclosures.

Monitoring Vital Signs

Performing maintenance work only when actually needed, not by a fixed schedule, minimizes labor costs and reduces downtime. This is called condition monitoring, on-line monitoring, or predictive monitoring. A maintenance program that provides for routine monitoring of vital signs such as voltage, current, temperature, and vibration can often detect developing faults before they can cause an outage. Preventive maintenance monitoring equipment that will detect the most frequent causes of failure are voltmeters, clamp-on ammeters, frequency meters, megohmmeters, vibration meters, and thermographic survey instruments.

On-line monitoring, combined with a computer-based expert system is in use in a number of industries. Although it requires an investment in sensors, data logging equipment, computer software, and additional training, the benefits can outweigh the investment costs. For examples of applications that use condition monitoring see References [2–6].

Diagnostic Testing

When physical inspection or monitoring of vital signs indicates trouble, diagnostic tests may have to be made to identify the specific areas of difficulty. Diagnostic tests on insulation and troubleshooting techniques for specific apparatus are detailed in the appropriate chapters.

32.3 RECORD TAKING AND INTERPRETATION

An essential element of a scheduled preventive maintenance program is the keeping of adequate records. Records provide a visual guide for monitoring, directing, and evaluating the cost effectiveness of maintenance activities, as well as providing proof of compliance with OSHA requirements. A minimum of four records are required: a schedule control chart, an equipment card, a history card, and a work card.

The *schedule control chart* should provide a comprehensive view of the entire year's schedule at a glance, and thus facilitate work assignments.

The *equipment card* should include complete nameplate data of a specific apparatus, location in plant, purchase price, installation cost, date of installation, manufacturer's part numbers for each replaceable part, connection diagram number, and telephone number of manufacturer. The equipment card should also indicate the number of starts per 24-h period and whether the load is fairly constant or cyclic. Surging loads such as that caused by reciprocating compressors cause cyclic movement of the coils that tend to loosen wedges, ties, etc., and can cause chafing of insulation; motors driving this type of load require more frequent inspections, more frequent monitoring of vital signs, and more frequent diagnostic tests than do motors with nonpulsating loads such as fans and blowers. The specified wear limits and operating limits for the apparatus should be entered on the equipment card.

The *history card* should include the date and findings of each inspection, the date of each periodic test, the test data, any adjustments or repairs, hours used, cost of materials, and the name of the person in charge of the specific inspection, test, or repair.

The *work card* is tailored to specific apparatus and applications and should be issued each day to maintenance technicians; a separate work card should be issued for each apparatus and should specify the required maintenance activity. Only enough work cards that can be handled effectively in an 8-h period should be issued; allowance should be made for "coffee breaks" and minor interruptions.

32.4 INTERMITTENT FAULTS

Intermittent faults are faults that appear and disappear in a random manner. Such faults are generally caused by vibration or thermal expansion and contraction. A typical example of an intermittent fault is the random blinking of an automobile headlight or taillight caused by loose connections that make and break by vibration when the car is in motion.

Determining the cause of an intermittent fault, after it "clears itself," can often be a very time-consuming and labor-intensive chore. However, if failure of equipment can result in injury to personnel, or where an unpredicted shutdown can cause significant financial loss, the fault (intermittent or not) *must be located and repaired.*

A recommended procedure to follow when troubleshooting intermittent faults in complex control systems is to list what does and what does not function when the fault occurs; this can often narrow the target area. Examine the circuit diagram to determine what components are in the target area, and then review any pertinent descriptive literature that states the function and behavior of each component. Concentrate first on those components that appear to be the most likely cause of the specific intermittent operation. Look for loose connections, broken wires, corrosion, accumulation of foreign matter, hairline cracks in resistors, shorted capacitors, open or shorted operating coils, weak or broken springs in contactors or relays, etc. A light tug on a connection will determine whether or not it is loose. An ohmmeter connected across a resistor while applying a back and forth twisting motion will easily indicate a hairline crack.

The following is an example of how inadequate maintenance and an intermittent fault combined to cause a generator explosion and a cable fire that resulted in the total loss of a ship. The brief analysis of the accident is based on a review of depositions and testimony.

The ship propulsion system consisted of a 6000-hp, 80-pole, 2300-V synchronous propulsion motor driven by a 5400-kW, two-pole, turbine-driven synchronous generator. The propulsion motor was directly connected to the generator through a set of manually operated reversing contactors. There was no circuit breaker or overload trip between the motor and the generator because the motor load could not exceed the turbine rating. The motor and generator were excited by the same exciter generator. Protection against short circuits and grounds were provided by phase-balance relays and ground-fault relays, respectively. A switching arrangement permitted the field of the exciter generator to be controlled automatically through either one of two automatic voltage regulators or through manual control using a rheostat.

While operating at sea, a failure in the automatic voltage regulating system caused loss of exciter voltage. This cut the excitation to the propulsion motor and generator. The resultant loss of load caused the turbine generator to accelerate. A sluggish speed governor did not respond quickly enough to slow the turbine down; it was finally stopped by the emergency overspeed trip. The propulsion system could not be restarted using either one of the two automatic voltage regulators, and the ship had to resume its voyage on manual control (rheostat).

When the ship arrived in port, the excitation was switched from manual to automatic and it worked! The trouble was identified as an intermittent fault somewhere in a control circuit common to both regulators. The intermittent fault was not located, and the sluggish governor was not repaired. Thus, the ship sailed again with a sluggish governor and an intermittent fault "lurking" in a circuit common to both automatic voltage regulators.

While under way in smooth seas, the intermittent fault struck again, killing the excitation to the propulsion generator and motor. However, this time neither the sluggish governor nor the emergency overspeed trip was able to control the turbine.

It continued its acceleration to a speed that caused the generator to explode and the ship was set on fire. A lesson to be learned from this experience is that intermittent faults that can jeopardize safety or result in significant loss must be located and corrected.

32.5 INVESTIGATING FAILURES

Investigation to determine the cause of failure should be commenced as early as possible, and all equipment that is directly or peripherally involved with the failed apparatus should be examined for its possible role in the failure. Every effort must be made, consistent with safety, to avoid destroying or altering evidence. In the case of failures that caused explosions, fire, loss of life, or injury to personnel, it is essential that, as soon as possible, a professional photographer be hired to take a complete set of photographs of the damaged apparatus from all possible angles. The position of the camera should be noted for each photograph, for example, "looking down on the generator brush rigging from the port side of the engine room."

Try to determine the sequence of events that led to the failure. Don't make assumptions. Review the respective history cards, and question witnesses, operators, and maintenance personnel separately. Determine if there was any unusual behavior, noise, smoke, vibration, etc., prior to the failure. Was there a power failure or lightning stroke shortly before the failure? Were any modifications made to the system? Were operating instructions changed? Was a critical component near the end of its useful life?

Since there is a lesson to be learned from every outage and every accident, the results of the investigation should be discussed with all maintenance and operating personnel, so that a recurrent failure can be prevented [1].

32.6 SPECIAL CONSIDERATIONS FOR SHIPBOARD ELECTRIC POWER APPARATUS

Marine electrical power apparatus must operate in a hostile environment (vibration, salt air, and oily vapors) and is often far from shore facilities when breakdown occurs. Breakdowns will occur, and often it is a small component such as a loose wire or bad contactor that can escalate into arcing, fire, or explosion. Thus, it is imperative that an adequate preventive maintenance program be planned by management, in consultation with the chief engineer and port engineer. Emphasis must be placed on electric power apparatus that is essential to the safety of the ship and its cargo; safety of personnel is paramount.

The shipboard preventive maintenance program should include a reasonable schedule of routine inspections, routine tests, and adjustments for all vital apparatus; all work must be done by qualified personnel. Maintenance of nonvital apparatus should be done on a time-available basis. The specific inspections, tests, spare parts requirements, etc., and scheduling frequency should be based on manufacturer's recommendations, modified to take into consideration the hostile environment and the lack of shore-side repair facilities when at sea.

32.7 EVALUATING PREVENTIVE MAINTENANCE PROGRAMS

Preventive maintenance programs should be evaluated annually to determine which maintenance tasks do not contribute to preventing outages, and where additional maintenance is required. If an item such as a bearing, contact, or brush has to be replaced frequently, the cause must be determined. It may be an indication of a more serious problem.

Analysis of recorded data from inspections, monitoring, and diagnostic testing should be used to determine if and when reconditioning, repairing, rebuilding, or renewing of apparatus is required. When it is determined that a motor requires renewing, be sure to compare the cost of a new motor with the cost of a rewind; in the case of motors 25 hp and below, the cost of a new energy-saving motor may be less than the cost of an energy-efficient rewind.

SPECIFIC REFERENCES KEYED TO THE TEXT

[1] *Manual on the Investigation of Fires of Electrical Origin*, NFPA-907M, National Fire Protection Association, 1983.

[2] Homce, Gerald T., and Thalimer, John R., Reducing Unscheduled Plant Maintenance Delays—Field Test of a New Method to Predict Electric Motor Failure, *IEEE Trans. Industry Applications*, Vol. 32, No. 3, May/June 1996.

[3] Schoen, R. R., Habetler, Kamran, F., and Bartheld, R. G., Motor Bearing Damage Detection Using Stator Current Monitoring, *IEEE Trans. Industry Applications*, Vol. 31, No. 6, November/December 1995.

[4] Schoen, R. R., Lin, B. K., Habetler, T. G., Schalag, J. H., and Farag, S., An Unsupervised, On-Line System for Induction Motor Fault Detection Using Stator Current Monitoring, *IEEE Trans. Industry Applications*, Vol. 31, No. 6, November/December 1995.

[5] Riley, C. M., Lin, B.K., Habetler, T. G., and Schoen, R. R. A Method for Sensorless On-Line Vibration Monitoring of Induction Machines, *IEEE Trans. Industry Applications*, Vol. 34, No. 6, November/December 1998.

[6] Cash, M. A., Habetler, T. G. and Kliman, G. B., Insulation Failure Prediction in AC Machines Using Line-Neutral Voltages, *IEEE Trans. Industry Applications*, Vol. 34, No. 6, November/December 1998.

A.1

Table 310–16 of *NEC*

ALLOWABLE AMPACITIES OF INSULATED CONDUCTORS RATED 0 THROUGH 2000 VOLTS, 60°C THROUGH 90°C (140°F THROUGH 194°F) NOT MORE THAN THREE CURRENT-CARRYING CONDUCTORS IN RACEWAY, CABLE, OR EARTH (DIRECTLY BURIED), BASED ON AMBIENT TEMPERATURE OF 30°C (86°F)

Size AWG or kcmil	Temperature Rating of Conductor					
	60°C (140°F)	75°C (167°F)	90°C (194°F)	60°C (140°F)	75°C (167°F)	90°C (194°F)
	Types TW, UF	Types RHW, THHW, THW, THWN, XHHW, USE, ZW	Types TBS, SA, SIS, FEP, FEPB, MI, RHH, RHW-2, THHN, THHW, THW-2, THWN-2, USE-2, XHH, XHHW, XHHW-2, ZW-2	Types TW, UF	Types RHW, THHW, THW, THWN, XHHW, USE	Types TBS, SA, SIS, THHN, THHW, THW-2, THWN-2, RHH, RHW-2, USE-2, XHH, XHHW, XHHW-2, ZW-2
	Copper			Aluminum or Copper-Clad Aluminum		
AWG 18	—	—	14	—	—	—
16	—	—	18	—	—	—
14*	20	20	25	—	—	—
12*	25	25	30	20	20	25
10*	30	35	40	25	30	35
8	40	50	55	30	40	45
6	55	65	75	40	50	60
4	70	85	95	55	65	75

A.1 Continued

	Temperature Rating of Conductor					
	60°C (140°F)	75°C (167°F)	90°C (194°F)	60°C (140°F)	75°C (167°F)	90°C (194°F)
Size AWG or kcmil	Types TW, UF	Types RHW, THHW, THW, THWN, XHHW, USE, ZW	Types TBS, SA, SIS, FEP, FEPB, MI, RHH, RHW-2, THHN, THHW, THW-2, THWN-2, USE-2, XHH, XHHW, XHHW-2, ZW-2	Types TW, UF	Types RHW, THHW, THW, THWN, XHHW, USE	Types TBS, SA, SIS, THHN, THHW, THW-2, THWN-2, RHH, RHW-2, USE-2, XHH, XHHW, XHHW-2, ZW-2
	Copper			Aluminum or Copper-Clad Aluminum		
3	85	100	110	65	75	85
2	95	115	130	75	90	100
1	110	130	150	85	100	115
1/0	125	150	170	100	120	135
2/0	145	175	195	115	135	150
3/0	165	200	225	130	155	175
4/0	195	230	260	150	180	205
kcmil (MCM) 250	215	255	290	170	205	230
300	240	285	320	190	230	255
350	260	310	350	210	250	280
400	280	335	380	225	270	305
500	320	380	430	260	310	350
600	355	420	475	285	340	385
700	385	460	520	310	375	420
750	400	475	535	320	385	435
800	410	490	555	330	395	450
900	435	520	585	355	425	480
1000	455	545	615	375	445	500
1250	495	590	665	405	485	545
1500	520	625	705	435	520	585
1750	545	650	735	455	545	615
2000	560	665	750	470	560	630

Reprinted with permission from NFPA 70-2002, the *National Electrical Code®*, Copyright © 2002, National Fire Protection Association, Quincy, MA 02269. This reprinted material is not the complete and official position of the National Fire Protection Association, on the referenced subject, which is represented only by the standard in its entirety. *National Electrical Code®* and *NEC®* are registered trademarks of the National Fire Protection Association, Inc., Quincy, MA 02269.

Note 1: Ambient correction factors for ambients other than 30°C (86°F) are in Appendix 3.

Note 2: Descriptions and applications of insulations for the designated insulation code letters are in Section 310–13 of the *National Electrical Code*.

*For these conductors, overcurrent protection by fuses or circuit breakers shall not exceed: For copper conductors: 15A for AWG-14, 20A for AWG-12, and 30A for AWG-10. For aluminum or copperclad aluminum conductors: 15A for AWG-12 and 25A for AWG-10.

A.2

Table 310–17 of *NEC*

ALLOWABLE AMPACITIES OF SINGLE-INSULATED CONDUCTORS RATED 0 THROUGH 2000 VOLTS IN FREE AIR, BASED ON AMBIENT AIR TEMPERATURE OF 30°C (86°F)

		Temperature Rating of Conductor					
		60°C (140°F)	75°C (167°F)	90°C (194°F)	60°C (140°F)	75°C (167°F)	90°C (194°F)
Size AWG or kcmil		Types TW, UF	Types RHW, THHW, THW, THWN, XHHW, ZW	Types TBS, SA, SIS, FEP, FEPB, MI, RHH, RHW-2, THHN, THHW, THW-2, THWN-2, USE-2, XHH, XHHW, XHHW-2, ZW-2	Types TW, UF	Types RHW, THHW, THW, THWN, XHHW	Types TBS, SA, SIS, THHN, THHW, THW-2, THWN-2, RHH, RHW-2, USE-2, XHH, XHHW, XHHW-2, ZW-2
		Copper			Aluminum or Copper-Clad Aluminum		
AWG	18	—	—	18	—	—	—
	16	—	—	24	—	—	—
	14*	25	30	35	—	—	—
	12*	30	35	40	25	30	35
	10*	40	50	55	35	40	40
	8	60	70	80	45	55	60
	6	80	95	105	60	75	80
	4	105	125	140	80	100	110
	3	120	145	165	95	115	130
	2	140	170	190	110	135	150
	1	165	195	220	130	155	175

A.2 Continued

		Temperature Rating of Conductor				
	60°C (140°F)	75°C (167°F)	90°C (194°F)	60°C (140°F)	75°C (167°F)	90°C (194°F)
Size AWG or kcmil	Types TW, UF	Types RHW, THHW, THW, THWN, XHHW, ZW	Types TBS, SA, SIS, FEP, FEPB, MI, RHH, RHW-2, THHN, THHW, THW-2, THWN-2, USE-2, XHH, XHHW, XHHW-2, ZW-2	Types TW, UF	Types RHW, THHW, THW, THWN, XHHW	Types TBS, SA, SIS, THHN, THHW, THW-2, THWN-2, RHH, RHW-2, USE-2, XHH, XHHW, XHHW-2, ZW-2
		Copper		Aluminum or Copper-Clad Aluminum		
1/0	195	230	260	150	180	205
2/0	225	265	300	175	210	235
3/0	260	310	350	200	240	275
4/0	300	360	405	235	280	315
kcmil (MCM) 250	340	405	455	265	315	355
300	375	445	505	290	350	395
350	420	505	570	330	395	445
400	455	545	615	355	425	480
500	515	620	700	405	485	545
600	575	690	780	455	540	615
700	630	755	855	500	595	675
750	655	785	885	515	620	700
800	680	815	920	535	645	725
900	730	870	985	580	700	785
1000	780	935	1055	625	750	845
1250	890	1065	1200	710	855	960
1500	980	1175	1325	795	950	1075
1750	1070	1280	1445	875	1050	1185
2000	1155	1385	1560	960	1150	1335

Reprinted with permission from NFPA 70-2002, the *National Electrical Code®*, Copyright© 2002, National Fire Protection Association, Quincy, MA 02269. This reprinted material is not the complete and official position of the National Fire Protection Association, on the referenced subject, which is represented only by the standard in its entirety.

Note 1: Ambient correction factors for ambients other than 30°C (86°F) are in Appendix 3.

Note 2: Descriptions and applications of insulations for the designated insulation code letters are in Section 310–13 of the *National Electrical Code*.

*For these conductors, overcurrent protection by fuses or circuit breakers shall not exceed: For copper conductors: 15A for AWG-14, 20A for AWG-12, and 30A for AWG-10. For aluminum or copperclad aluminum conductors: 15A for AWG-12 and 25A for AWG-10.

A.3

Temperature Correction Factors for Appendix 1 and Appendix 2

FOR AMBIENT TEMPERATURES OTHER THAN 30°C (86°F)

Ambient Temp. (°C)	60°C (140°F)	75°C (167°F)	90°C (194°F)	Ambient Temp. (°F)
21–25	1.08	1.05	1.04	70–77
26–30	1.00	1.00	1.00	78–86
31–35	0.91	0.94	0.96	87–95
36–40	0.82	0.88	0.91	96–104
41–45	0.71	0.82	0.87	105–113
46–50	0.58	0.75	0.82	114–122
51–55	0.41	0.67	0.76	123–131
56–60	—	0.58	0.71	132–140
61–70	—	0.33	0.58	141–158
71–80	—	—	0.41	159–176

Reprinted with permission from NFPA 70-2002, the *National Electrical Code®*, Copyright © 2002, National Fire Protection Association, Quincy, MA 02269. This reprinted material is not the complete and official position of the National Fire Protection Association, on the referenced subject, which is represented only by the standard in its entirety.

Note: Multiply the allowable ampacities shown in Appendix 1 and Appendix 2 by the appropriate correction factor shown in this table.

A.4

USAS and RETMA Standard Color Code for Small Tubular Resistors

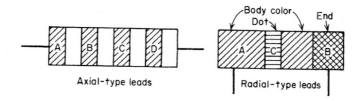

Axial-type leads

Radial-type leads

Color	Band A, 1st Figure	Band B, 2nd Figure	Band C, Remaining Figures	Band D, Tolerance
Black	0	0		
Brown	1	1	0	
Red	2	2	00	
Orange	3	3	000	
Yellow	4	4	0,000	
Green	5	5	00,000	
Blue	6	6	000,000	
Violet	7	7	0,000,000	
Gray	8	8	00,000,000	
White	9	9	000,000,000	
Gold	..	..	....	± 5%
Silver	..	..	....	± 10%
No color	..	..	....	± 20%

EXAMPLE. If the color bands of a resistor are as follows: band A yellow, band B green, band C orange, and band D gold, the value of the resistor is 45,000 ohms ± 5 percent.

A.5

Table 430–52 of *NEC*

MAXIMUM RATING OR SETTING OF MOTOR BRANCH-CIRCUIT SHORT CIRCUIT AND GROUND-FAULT PROTECTIVE DEVICES

Type of Motor	Percentage of Full-Load Current			
	Nontime Delay Fuse[a]	Dual Element (Time-Delay) Fuse[a]	Instantaneous Trip Breaker	Inverse Time Breaker[b]
Single-phase motors	300	175	800	250
AC polyphase motors other than wound rotor				
Squirrel cage—other than Design E or Design B energy efficient	300	175	800	250
Design E or Design B energy efficient	300	175	1100	250
Synchronous[c]	300	175	800	250
Wound rotor	150	150	800	150
Direct current (constant voltage)	150	150	250	150

Reprinted with permission from NFPA 70-2002, the *National Electrical Code*®, Copyright © 2002. National Fire Protection Association, Quincy, MA 02269. This reprinted material is not the complete and official position of the National Fire Protection Association, on the referenced subject, which is represented only by the standard in its entirety.

Note: For certain exceptions to the values specified, see 430.54, of *NEC*.

[a] The values in the Nontime Delay Fuse column apply to Time-Delay Class CC fuses.

[b] The values given in the last column also cover the ratings of nonadjustable inverse time types of circuit breakers that may be modified as in 430.52(C) of *NEC*, Exception No. 1 and No. 2.

[c] Synchronous motors of the low-torque, low-speed type (usually 450 rpm or lower), such as are used to drive reciprocating compressors, pumps, and so forth, that start unloaded, do not require a fuse rating or circuit-breaker setting in excess of 200 percent of full-load current.

A.6

Full-Load Current in Amperes, Direct-Current Motors

The following values of full-load currents are average DC quantities and are for motors running at base speed.

hp	Armature Voltage Rating*					
	90 V	120 V	180 V	240 V	500 V	550 V
¼	4.0	3.1	2.0	1.6		
⅓	5.2	4.1	2.6	2.0		
½	6.8	5.4	3.4	2.7		
¾	9.6	7.6	4.8	3.8		
1	12.2	9.5	6.1	4.7		
1½		13.2	8.3	6.6		
2		17	10.8	8.5		
3		25	16	12.2		
5		40	27	20		
7½		58		29	13.6	12.2
10		76		38	18	16
15				55	27	24
20				72	34	31
25				89	43	38
30				106	51	46
40				140	67	61
50				173	83	75
60				206	99	90
75				255	123	111
100				341	164	148
125				425	205	185
150				506	246	222
200				675	330	294

* These are average DC values.

A.7

Table 430–148 of *NEC*

FULL-LOAD CURRENTS IN AMPERES, SINGLE-PHASE ALTERNATING-CURRENT MOTORS

The following values of full-load currents are for motors running at usual speeds and motors with normal torque characteristics. Motors built for especially low speeds or high torques may have higher full-load currents, and multispeed motors will have full-load current varying with speed, in which case the nameplate current ratings shall be used.

The voltages listed are rated motor voltages. The currents listed shall be permitted for system voltage ranges of 110 to 120 and 220 to 240 volts.

Horsepower	115 Volts	200 Volts	208 Volts	230 Volts
1/6	4.4	2.5	2.4	2.2
1/4	5.8	3.3	3.2	2.9
1/3	7.2	4.1	4.0	3.6
1/2	9.8	5.6	5.4	4.9
3/4	13.8	7.9	7.6	6.9
1	16	9.2	8.8	8.0
1 1/2	20	11.5	11.0	10
2	24	13.8	13.2	12
3	34	19.6	18.7	17
5	56	32.2	30.8	28
7 1/2	80	46.0	44.0	40
10	100	57.5	55.0	50

Reprinted with permission from NFPA 70, *National Electrical Code*, Copyright © 2002, National Fire Protection Association, Quincy, MA 02269. This reprinted material is not the complete and official position of the NFPA on the referenced subject, which is represented only by the standard in its entirety.

A.8

Table 430–148 of *NEC*

FULL-LOAD CURRENT, TWO-PHASE ALTERNATING-CURRENT MOTORS (FOUR-WIRE)

The following values of full-load current are for motors running at speeds usual for belted motors and motors with normal torque characteristics. Motors built for especially low speeds or high torques may require more running current, and multispeed motors will have full-load current varying with speed, in which case the nameplate current rating shall be used. Current in the common conductor of a two-phase, three-wire system will be 1.41 times the value given.

The voltages listed are rated motor voltages. The currents listed shall be permitted for system voltage ranges of 110 to 120, 220 to 240, 440 to 480, and 550 to 600 V.

hp	Induction-Type Squirrel-Cage and Wound-Rotor (A)				
	115 V	230 V	460 V	575 V	2300 V
½	4	2	1	0.8	
¾	4.8	2.4	1.2	1.0	
1	6.4	3.2	1.6	1.3	
1½	9	4.5	2.3	1.8	
2	11.8	5.9	3	2.4	
3		8.3	4.2	3.3	
5		13.2	6.6	5.3	
7½		19	9	8	
10		24	12	10	
15		36	18	14	
20		47	23	19	
25		59	29	24	
30		69	35	28	
40		90	45	36	

A.8 Continued

hp	Induction-Type Squirrel-Cage and Wound-Rotor (A)				
	115 V	230 V	460 V	575 V	2300 V
50		113	56	45	
60		133	67	53	14
75		166	83	66	18
100		218	109	87	23
125		270	135	108	28
150		312	156	125	32
200		416	208	167	43

A.9

Table 430–150 of *NEC*

FULL-LOAD CURRENT, THREE-PHASE ALTERNATING-CURRENT MOTORS

The following values of full-load currents are typical for motors running at speeds usual for belted motors and motors with normal torque characteristics.

Motors built for low speeds (1200 rpm or less) or high torques may require more running current, and multispeed motors will have full-load current varying with speed. In these cases, the nameplate current rating shall be used.

The voltages listed are rated motor voltages. The currents listed shall be permitted for system voltage ranges of 110 to 120, 220 to 240, 440 to 480, and 550 to 600 volts.

Horsepower	Induction-Type Squirrel Cage and Wound Rotor (Amperes)							Synchronous-Type Unity Power Factor[a] (Amperes)			
	115 Volts	200 Volts	208 Volts	230 Volts	460 Volts	575 Volts	2300 Volts	230 Volts	460 Volts	575 Volts	2300 Volts
½	4.4	2.5	2.4	2.2	1.1	0.9	—	—	—	—	—
¾	6.4	3.7	3.5	3.2	1.6	1.3	—	—	—	—	—
1	8.4	4.8	4.6	4.2	2.1	1.7	—	—	—	—	—
1½	12.0	6.9	6.6	6.0	3.0	2.4	—	—	—	—	—
2	13.6	7.8	7.5	6.8	3.4	2.7	—	—	—	—	—
3	—	11.0	10.6	9.6	4.8	3.9	—	—	—	—	—
5	—	17.5	16.7	15.2	7.6	6.1	—	—	—	—	—
7½	—	25.3	24.2	22	11	9	—	—	—	—	—
10	—	32.2	30.8	28	14	11	—	—	—	—	—
15	—	48.3	46.2	42	21	17	—	—	—	—	—
20	—	62.1	59.4	54	27	22	—	—	—	—	—
25	—	78.2	74.8	68	34	27	—	53	26	21	—
30	—	92	88	80	40	32	—	63	32	26	—
40	—	120	114	104	52	41	—	83	41	33	—

A.9   Continued

Horsepower	Induction-Type Squirrel Cage and Wound Rotor (Amperes)							Synchronous-Type Unity Power Factor[a] (Amperes)			
	115 Volts	200 Volts	208 Volts	230 Volts	460 Volts	575 Volts	2300 Volts	230 Volts	460 Volts	575 Volts	2300 Volts
50	—	150	143	130	65	52	—	104	52	42	—
60	—	177	169	154	77	62	16	123	61	49	12
75	—	221	211	192	96	77	20	155	78	62	15
100	—	285	273	248	124	99	26	202	101	81	20
125	—	359	343	312	156	125	31	253	126	101	25
150	—	414	396	360	180	144	37	302	151	121	30
200	—	552	528	480	240	192	49	400	201	161	40
250	—	—	—	—	302	242	60	—	—	—	—
300	—	—	—	—	361	289	72	—	—	—	—
350	—	—	—	—	414	336	83	—	—	—	—
400	—	—	—	—	477	382	95	—	—	—	—
450	—	—	—	—	515	412	103	—	—	—	—
500	—	—	—	—	590	472	118	—	—	—	—

Reprinted with permission from NFPA 70, *National Electrical Code*, Copyright © 2002, National Fire Protection Association, Quincy, MA 02269. This reprinted material is not the complete and official position of the NFPA on the referenced subject, which is represented only by the standard in its entirety.

[a] For 90 and 80 percent power factor the preceding figures shall be multipled by 1.1 and 1.25, respectively.

A.10

Multiplying Factors for Converting DC Resistance to 60-Hz AC Resistance

	Multiplying Factor			
	For Nonmetallic Sheathed Cables in Air or Nonmetallic Conduit		For Metallic Sheathed Cables or All Cables in Metallic Raceways	
Size	Copper	Aluminum	Copper	Aluminum
Up to 3 AWG	1.000	1.000	1.00	1.00
2	1.000	1.000	1.01	1.00
1	1.000	1.000	1.01	1.00
0	1.001	1.000	1.02	1.00
00	1.001	1.001	1.03	1.00
000	1.002	1.001	1.04	1.01
0000	1.004	1.002	1.05	1.01
250 MCM	1.005	1.002	1.06	1.02
300 MCM	1.006	1.003	1.07	1.02
350 MCM	1.009	1.004	1.08	1.03
400 MCM	1.011	1.005	1.10	1.04
500 MCM	1.018	1.007	1.13	1.06
600 MCM	1.025	1.010	1.16	1.08
700 MCM	1.034	1.013	1.19	1.11
750 MCM	1.039	1.015	1.21	1.12
800 MCM	1.044	1.017	1.22	1.14
1000 MCM	1.067	1.026	1.30	1.19
1250 MCM	1.102	1.040	1.41	1.27
1500 MCM	1.142	1.058	1.53	1.36
1750 MCM	1.185	1.079	1.67	1.46
2000 MCM	1.233	1.100	1.82	1.56

Answers to Problems

Chapter 1

1. (a) 20 μS, (b) 0.00215 S, (c) 55.556 S, (d) 0.01923 μS, (e) 2500 S
2. (a) 243.9 Ω, (b) 0.3077 Ω, (c) 5.538 mΩ, (d) 1 mΩ, (e) 2.5 Ω
3. 0.21995 mΩ
4. 19.923 mΩ
5. (a) 159.155 kcmil, (b) 13.033 mΩ
6. (a) 5092.96 kcmil, (b) 0.6109 mΩ
7. (a) 16 Mcmil, (b) 1.0625 mΩ
8. 0.14641 mΩ
9. (a) 15.557 $\mu\Omega$, (b) 19.835 $\mu\Omega$
10. 74.2 Ω
11. 27.947 Ω
12. 89.92°C
13. 169.64 Ω
14. 269.6 Ω
15. 142.10 Ω
16. 72.22°C
17. 82.4°F
18. 20°C
19. 27.93°C
20. (a) −230°C, (b) 228.17 Ω
21. 0.01496 Ω
22. 0.08698 Ω
23. 208.8 A
24. 39 A
25. 88 A

Chapter 2

1. 58.6 V
2. 96.3 V
3. 26.25 μA
4. 39 kΩ
5. 51 Ω
6. 300 Ω
7. (a) 5.4 mA, (b) 0.108 V, 1.08 V, 10.8 V, 108 V
8. (a) 2 A, (b) 60 V
9. (a) 8.53, (b) 116.96 V, (c) 1.536 V
10. (a) 220 Ω, (b) 0.573 A, (c) 30.9 V
11. (a) 4 A, (b) 2 A
12. (a) 3.31A, (b) 0.828 A
13. 2.0 Ω
14. 0.464 Ω
15. (a) 44.44 Ω, (b) 4.95 Ω, (c) 1000 Ω, (d) 9.99 mΩ
16. (a) 0.6 Ω, (b) 66.67 A
17. 4.3 A
18. (a) 99.97 mΩ, (b) 30 A
19. (a) 4 A, 6 A, 2 A, (b) 12 A
20. (a) 4 A, (b) 30 Ω
21. 19.1 Ω
22. (a) 391.67 Ω, (b) 612.8 mA
23. (a) 245 Ω, (b) 2.45 A
24. (a) 7 A, (b) 12 kW
25. (a) 0.5 A, 0.833 A, 1.25 A, (b) 2.583 A, (c) 1.333 A
26. 5.56 mΩ
27. 393.7 W
28. (a) 0.492 Ω, (b) 1771.2 W, (c) 29.52 V, (d) 210.48 V
29. (a) 35.33 A, (b) 20.95 V, (c) 219 V, (d) 740.31 W (e) 404.6 A
30. (a) 200 V, (b) 16 W, 24 W, 40 W, 400 W
31. (a) 24 Ω, (b) 10 A, (c) 24 Ω
32. (a) 20 A, (b) 5 Ω, (c) 16 kWh
33. (a) 500 W, (b) 4.167 A, (c) 28.8 Ω
34. 60 A for 5 s
35. 80 A for 0.5 s
36. (a) 104 Ω, (b) 2.31 A, (c) 553.8 W, (d) 60 V, (e) 4.43 kWh
37. (a) 3.6 A, (b) 21.6 V, (c) 77.76 W, (d) 1555.2 Wh
38. (a) 1.2 A, (b) 7.2 V (c) 8.64 W, (d) 172.8 Wh
39. 2.08 A
40. (a) 20.325 A, (b) 0.1018 Ω, (c) 248.1 V, (d) 2436.8 A

Chapter 3

1. conductors will swing parallel to each other
2. sketch

3. CCW
4. (a) sketch, (b) 352.8 lb
5. 720 lb

Chapter 4

1. (a) 14.25 A, (b) 1881.2 A-t, (c) 1383.26 A-t/Wb
2. (a) 0.4 Wb, (b) 2 Ω
3. 0.08 Wb
4. 2.352 H
5. 9.375 H
6. 250 A-t/Wb
7. (a) 3.375 A, (b) 5625 A-t/Wb, (c) 810 W
8. 917.3 V
9. (a) sketch, (b) 2.84 A, (c) 41.18 A/s
10. 118 H
11. 0.773 H
12. 20 H
13. 0.96 H
14. (a) 1.104 A, (b) 38.4 J
15. (a) 20 A, (b) 12 Ω
16. (a) 10.4 A, (b) 2274.5 J, (c) 2393.7 W
17. (a) 10.256 A, (b) 284 J, (c) 2461.5 W
18. (a) 0.2 s, (b) 1 s, (c) 240 A
19. (a) 3 A, (b) 2 s, (c) 1.896 A
20. (a) 0.625 s, (b) 0.5 s, (c) 6 A, (d) 13.41 Ω
21. (a) 0 A, (b) 1 A, (c) 0.632 A, (d) 1.264 V, (e) 20 Ω
22. (a) 10 s, (b) 2 s, (c) 2.4 A, (d) 1.5168 A, (e) 4.416 V

Chapter 5

1. 0.266 C
2. 917.65 V
3. 101.08 J
4. 357.88 J
5. 1.726 μF
6. 220 μF
7. 4.323 μF
8. 316 μF
9. (a) 10 s, (b) 50 s
10. (a) 126.4 V, 1.84 A (b) 2.396 J, (c) 6 J
11. (a) 2.5 MΩ, (b) 18 μJ
12. (a) 15 kΩ, (b) 200 J, (c) 328 V (d) 27.08 J

Chapter 6

1. (a) 459.62 A, (b) 0.01667 s
2. (a) 339.41 V, (b) 0.04 s

3. (a) 188.5 Ω, (b) 1.273 A, (c) 1.8 A
4. (a) 11.31 Ω, (b) 9.55 A, (c) 13.5 A, (d) 2.74 J
5. (a) 12 A, (b) 16.97 A, (c) 53 mH
6. (a) 2.653 Ω, (b) 45.23 A
7. (a) 15.91 Ω, (b) 13.07 A, (c) ∞, (d) 0 A
8. (a) 5.305 Ω, (b) 45.24 A, (c) 339.4 V
9. (a) 12.81 Ω, (b) 9.37 A, (c) 0.0212 H
10. (a) 6.283 Ω, (b) 3.537 Ω, (c) 10.37 Ω, (d) 43.39 A, (e) 15.36°
11. (a) 1.768 Ω, (b) 376.99 Ω, (c) 375.75 Ω, (d) 0.5535A, (e) V_L = 208.68 V, V_C = 0.9789 V, V_R = 11.07 V, (f) 86.95°
12. (a) 2.236 Ω, (b) 107.3 A, (c) 322 V, (d) 0 A, (e) 240 V
13. (a) I_R = 12 A, I_C = 15.83 A, I_L = 10.61 A, (b) 13.09 A, (c) 9.169 Ω, (d) 23.52°
14. (a) 2.4 Ω, (b) 100 A
15. (a) 3.529 Ω, (b) 113.3 A, (c) 28.07°
16. (a) I_R = 60 A, I_L = 120 A, I_C = 48 A, (b) 93.72 A, (c) 2.56 Ω, (d) L = 5.3 mH, C = 530.5 μF
17. (a) I_R = 24 A, I_L = 15.91 A, I_C = 20.36 A, (b) 24.4 A, (c) 4.916 Ω
18. (a) 145.91 A, (b) 1.645 Ω, (c) 0 A
19. (a) 40 A, (b) 24 A, (c) 30 A, (d) 40.47 A

Chapter 7

1. (a) 419.9 Hz, (b) 44 A, (c) 7545.9 V, (d) 7545.9 V
2. (a) 634 Hz, (b) 19.23 A, (c) V_L = V_C =114.92 V, V_R = 120 V
3. (a) I_R = 12 A, I_L = 24.48 A, I_C = 7.917 A, (b) 20.46 A, (c) 5.866 Ω, (d) 12 A, (e) 105.52 Hz
4. (a) I_R = 100 A, I_L = 1.48 A, I_C = 135.7 A, (b) 167.39 A, (c) 2.867 Ω, (d) 100 A, (e) 6.267 Hz, (f) ∞
5. 54.39%
6. 25.94%
7. (a) 44.8 Ω, (b) 13.39 A, (c) V_L = 12.52 V, V_C = 612.5 V, V_R = 8.73 V, (d) 628.74 A, (e) V_L = V_C = 4114.8 V, V_R = 410 V

Chapter 8

1. 16.012 kVA
2. 17.86 kvar
3. (a) 447.72, (b) 312.51 kvar, (c) 34.91°
4. (a) 3001.15 kVA, (b) 78.6%, (c) 38.15°
5. (a) 125 kVA, (b) 75 kvar
6. (a) 14 kW, (b) 14.28 kvar, (c) 45.57°
7. (a) 195 kW, (b) 156.14 kvar, (c) 249.81 kVA, (d) 78.06%
8. (a) 34 kW, (b) 32 kvar, (c) 46.69 kVA, (d) 72.82%
9. (a) 119.24 W, (b) 32.51 var, (c) 123.59 kVA, (d) 96.48% leading
10. 26.12 kW
11. (a) 81.09 kW, (b) 94.29 kVA, (c) 48.11 kvar
12. (a) 17.55 kW, 15.01 kvar, (b) 24.57 kW, 17.21 kvar, (c) 42.127 kW, (d) 32.2 kvar, (e) 53.03 kVA, (f) 79.43%

13. (a) 2602.3 W, (b) 77.45%, (c) 2125.4 kvar
14. (a) $I_R = 40\,A$, $I_L = 20\,A$, $I_C = 12\,A$, (b) 2.94 Ω, (c) 4800 W, (d) 960 var, (e) 4895 var, (f) 40.79 A
15. (a) 12.55 kW, (b) 11.32 kvar, (c) 16.9 kVA, (d) 74.25%, (e) 38.41 A, (f) 6.275 kWh
16. 13.05 kvar
17. (a) 20.54 kvar, (b) 309 μF, (c) 460 V
18. (a) 1.672 kVA, (b) 1.316 kW, (c) 78.73%, (d) 56.5 μF, (e) 5.98 A, (f) 4.68 A

Chapter 9

1. (a) 26.56 A, (b) 26.56 A, (c) 265.58 V
2. (a) 8.32 A, (b) 14.41 A, (c) 208 V
3. (a) 450.3 V, (b) 260 V
4. (a) 16.15 Ω, (b) 13.58 A, (c) 13.58 A
5. 500 kVA
6. (a) 41.18 kVA, (b) 87.42%
7. (a) 12.47 kVA, (b) 10.725 kW, (c) 6.36 kvar
8. (a) 18.445 kW, (b) 21.104 kVA, (c) 10.255 kvar, (d) 26.49 A
9. (a) 31.2 kW, (b) 8.4 kvar, (c) 32.31 kVA, (d) 96.56%
10. (a) 16.2 kW, (b) 10.28 kvar, (c) 19.12 kVA, (d) 84.42%
11. (a) 19.92 kVA
12. (a) 19.92 kVA, (b) 14.94 kW, (c) 13.17 kvar
13. (a) 27.24 kVA, (b) 11.92 kvar, (c) 89.92%
14. (a) 65 kW, (b) 75.99 kvar, (c) 128.3 A, (d) 44.5 kvar, (e) 466.46 μF, (f) 92.66 A
15. (a) 293.5 kvar, (b) 16.22 μF, (c) 80.53 A
16. (a) 65 kW, (b) 75.99 kvar, (c) 125.5 A, (d) 44.51 kvar, (e) 558 μF

Chapter 10

1. (a) 110 V, (b) 0.5 V/t
2. 40 t
3. 22.11 A
4. 102.27 A
5. (a) 30 V, (b) 3 A, (c) 0.15 A
6. (a) 100 V, (b) 1.667 A, (c) 0.4167 A
7. (a) 20.7 A, (b) 62.1 A, (c) 1200 t
8. 184 V, 149.5 V
9. (a) 38.4 A, (b) 30.72 A
10. 83.33%
11. (a) 90 V, (b) 16 A, (c) 7.2 kW
12. (a) 2.95, (b) 58.7 A
13. (a) 479.8 V, (b) 41.64 A, (c) 24 A, (d) 361 A, (e) 361 A
14. (a) 4, (b) 162.5 A, (c) 129.47 kVA
15. (a) 87 kVA, (b) 42.47 kVA
16. (a) 5, (b) 5.016, (c) 0.3135%
17. (a) 6.933, (b) 6.78, (c) −2.22%
18. (a) 5.75, (b) 5.769, (c) 0.334%

Chapter 11

1. 1800 rpm
2. (a) 2 poles, (b) 3499.2 rpm, (c) 100.8 rpm
3. (a) 1800 rpm, (b) 75 rpm, (c) 0.0417
4. 147.53 lb-ft
5. 221.29 lb-ft
6. 16.01 hp
7. $73.5 \text{ A} \leq I < 82.7 \text{ A}$
8. $734.8 \text{ A} \leq I < 826.6 \text{ A}$
9. $2657 \text{ A} \leq I < 2952 \text{ A}$
10. (a) 2 poles, 8 poles, (b) no
11. (a) 4 poles, 8 poles, (b) yes
12. (a) 6 poles and 12 poles, (b) yes

Chapter 13

1. (a) 1163.4 lb-ft, (b) 840.57 lb-ft, (c) -27.75%
2. (a) 562.7 lb-ft, (b) 455.8 lb-ft, (c) -19%
3. yes
4. 355 V
5. 2876 A
6. 1028 A
7. 346.3 A
8. 60.5 A
9. 725.4 A
10. (a) 20.6 A, (b) 23.5%
11. (a) 42 A, (b) 15.2%
12. (a) 4.66%, (b) 43.5%, (c) 90°C, (d) 129°C
13. (a) 3.56%, (b) 25.3%, (c) 105°C, (d) 131°C
14. (a) 6.32%, (b) 79.97%, (c) 115°C, (d) 207°C
15. (a) 433.3 V, (b) 3.76%, (c) 0.84, (d) 21 hp
16. (a) 450.7 V, (b) 4.73%, (c) 0.78, (d) 15.6 hp
17. (a) 2273 V, (b) 3.23%, (c) 0.875, (d) 175 hp
18. (a) 83.3 hp, (b) 383.3 V
19. (a) 41.7 hp, (b) 200 V
20. (a) 90 hp, (b) 264 V

Chapter 14

1. (a) 2025 rpm, (b) 67.5 Hz
2. 1200 rpm
3. 600 rpm
4. 58.33 Hz
5. 50 Hz
6. 12 poles
7. 800 V, 120 Hz
8. 184.75 A

9. 50 A
10. (a) 450.3 V, (b) 260 V
11. (a) 265.58 V, (b) 460 V
12. 61.2 Hz
13. 61.5 Hz

Chapter 15
1. yes
2. yes
3. no

Chapter 16
1. 174 μF
2. 288 μF

Chapter 17
1. 835.6 V
2. (a) 255.6 V, (b) 172.5 V
3. 1894.7 rpm
4. 12.5%
5. (a) 261.98 V, (b) 9.158%
6. (a) 800 A, (b) 257.28 V, (c) 2.91%
7. 618 V
8. 244.8 V

Chapter 18
1. (a) 585.2 A, (b) 1.1 A
2. 0.773 Ω
3. 217.96 V
4. (a) 30.18 Ω, (b) 219.6 V, (c) 0.513 Ω
5. (a) 3.87 A, (b) 140 A, (c) 229.6 A, (d) 32.174 kW, (e) 1084 W, (f) 929 W
6. (a) 3.92 A, (b) 335 A, (c) 234.6 A, (d) 78.617 kW, (e) 1.3 kW, (f) 941.2 W
7. (a) 667.6 lb-ft, (b) 123 kW, (c) 90.98%
8. (a) 228.3 lb-ft, (b) 41.76 kW, (c) 89.3%
9. 6.25%
10. 1176.4 rpm

Chapter 23
1. 96.3°C
2. 104.6°C
3. 0.2815 Ω
4. 55.1°C
5. (a) 0.02 Ω, (b) 0.06 Ω
6. (a) 0.436 Ω, (b) 0.872 Ω

Chapter 24
1. 240 MΩ
2. 84.85 MΩ
3. satisfactory condition
4. 100 MΩ
5. 3.3 MΩ
6. (a) 32 MΩ, (b) 5.16 MΩ, (c) satisfactory
7. (a) 640 MΩ, (b) 100 MΩ, (c) good
8. good
9. defective
10. (a) 1.08, (b) no, minimum is 2.0
11. (a) 3.0, (b) yes, minimum is 2.0
12. (a) 2.4, (b) yes, minimum is 2.0

Chapter 25
1. (a) 900 V, (b) 1467 V
2. (a) 10350 V, (b) 16780 V
3. 1275 V
4. 360 V
5. (a) 1275 V, (b) 425 V, (c) 38.25 V
6. (a) 5865 V, (b) 1955 V, (c) 175.95 V
7. 15.41%
8. 24.25%

Chapter 26
1. (a) 121.9 V, (b) 13.12 A, (c) 1.6 kVA, (d) 41.67 A, (e) AWG-10-RHW
2. (a) 49.9 V, (b) 14.4 A, (c) 0.720 kVA, (d) 250 A, (e) 340 kcmil-TW
3. (a) 12.24 V, (b) 4.17 A, (c) 0.153 kVA, (d) 72.2 A, (e) AWG-6-TW
4. (a) 17.7 V, (b) 11.11 A, (c) 588.8 VA, (d) 55.5 A, (e) AWG-8-TW

Chapter 27
1. 1.12
2. 1.247
3. NO
4. 1.14
5. 2.123 V
6. (a) 4, (b) 300 A-h, (c) 200 A-h
7. (a) 14.29, (b) 1000 A-h, (c) 200 A-h
8. (a) 5.096, (b) 480 A-h, (c) 1.1858
9. 6.5 l
10. 6.4 l
11. 56 cells
12. (a) 53 cells/string, (b) 3 strings, (c) 159 cells

13. (a) 3 cells/string, (b) 12 strings
14. (a) 10 cells/string, (b) 5 strings

Chapter 29
1. 247.5 A
2. 382.5 A
3. 600 A
4. 91 A

Chapter 30
1. open
2. satisfactory
3. open
4. shorted
5. shorted

Index